Millimeter-Wave Circuits for 5G and Radar

Discover the concepts, architectures, components, tools, and techniques needed to design millimeter-wave circuits for current and emerging wireless system applications. Focusing on applications in 5G, connectivity, radar, and more, leading experts in radio-frequency integrated circuit (RFIC) design provide a comprehensive treatment of cutting-edge physical-layer technologies for RF transceivers – specifically RF, analog, mixed-signal, and digital circuits and architectures. The full design chain is covered, from system design requirements to building blocks, transceivers, and process technology. Gain insight into the key novelties of 5G through authoritative chapters on massive MIMO and phased arrays, and learn about the very latest technology developments, such as FinFET logic process technology for RF and millimeter-wave applications. This is an excellent reference and essential reading for high-frequency circuit designers in both academia and industry.

Gernot Hueber is Chief Scientist RF at Silicon Austria Labs leading the RF Research Unit. Previously, he held various positions in industry and academia in the field of wireless communications. He has published extensively on RFICs, and has over 60 patents in the field.

Ali M. Niknejad is a Professor in the Department of Electrical Engineering and Computer Sciences at the University of California, Berkeley, and Faculty Director of the Berkeley Wireless Research Center. He is the author of *Electromagnetics for High-Speed Analog and Digital Communication Circuits* (Cambridge University Press, 2007) and a Fellow of the Institute of Electrical and Electronics Engineers (IEEE).

The Cambridge RF and Microwave Engineering Series

Peter Aaen, Jaime Plá and John Wood, *Modeling and Characterization of RF and Microwave Power FETs*
Dominique Schreurs, Máirtín O'Droma, Anthony A. Goacher and Michael Gadringer (Eds), *RF Amplifier Behavioral Modeling*
Fan Yang and Yahya Rahmat-Samii, *Electromagnetic Band Gap Structures in Antenna Engineering*
Enrico Rubiola, *Phase Noise and Frequency Stability in Oscillators*
Earl McCune, *Practical Digital Wireless Signals*
Stepan Lucyszyn. *Advanced RF MEMS*
Patrick Roblin, *Nonlinear RF Circuits and the Large-Signal Network Analyzer*
Matthias Rudolph, Christian Fager and David E. Root (Eds), *Nonlinear Transistor Model Parameter Extraction Techniques*
John L. B. Walker (Ed.), *Handbook of RF and Microwave Solid-State Power Amplifiers*
Anh-Vu H. Pham, Morgan J. Chen and Kunia Aihara, *LCP for Microwave Packages and Modules*
Sorin Voinigescu, *High-Frequency Integrated Circuits*
Richard Collier, *Transmission Lines*
Valeria Teppati, Andrea Ferrero and Mohamed Sayed (Eds), *Modern RF and Microwave Measurement Techniques*
Nuno Borges Carvalho and Dominique Schreurs, *Microwave and Wireless Measurement Techniques*
David E. Root, Jason Horn, Jan Verspecht and Mihai Marcu, *X-Parameters*
Earl McCune, *Dynamic Power Supply Transmitters*
Hossein Hashemi and Sanjay Raman (Eds), *Silicon mm-Wave Power Amplifiers and Transmitters*
T. Mitch Wallis and Pavel Kabos, *Measurement Techniques for Radio Frequency Nanoelectronics*
Giovanni Ghione and Marco Pirola, *Microwave Electronics*
Isar Mostafanezhad, Olga Boric-Lubecke and Jenshan Lin (Eds), *Medical and Biological Microwave Sensors*
Richard Carter, *Microwave and RF Vacuum Electronic Power Sources*
José Carlos Pedro, David Root, Jianjun Xu and Luis Cotimos Nunes, *Nonlinear Circuit Simulation and Modeling*
Gernot Hueber and Ali M. Niknejad, *Millimeter-Wave Circuits for 5G and Radar*

Millimeter-Wave Circuits for 5G and Radar

Edited by

GERNOT HUEBER
Silicon Austria Labs, Austria

ALI M. NIKNEJAD
University of California, Berkeley

CAMBRIDGE
UNIVERSITY PRESS

University Printing House, Cambridge CB2 8BS, United Kingdom

One Liberty Plaza, 20th Floor, New York, NY 10006, USA

477 Williamstown Road, Port Melbourne, VIC 3207, Australia

314-321, 3rd Floor, Plot 3, Splendor Forum, Jasola District Centre, New Delhi – 110025, India

79 Anson Road, #06–04/06, Singapore 079906

Cambridge University Press is part of the University of Cambridge.

It furthers the University's mission by disseminating knowledge in the pursuit of
education, learning, and research at the highest international levels of excellence.

www.cambridge.org
Information on this title: www.cambridge.org/9781108492782
DOI: 10.1017/9781108686396

First published 2019
Reprinted 2019

Printed in the United Kingdom by TJ International Ltd. Padstow Cornwall

A catalogue record for this publication is available from the British Library.

Library of Congress Cataloging-in-Publication Data
Names: Hueber, Gernot, 1972- | Niknejad, Ali M., 1972-
Title: Millimeter-wave circuits for 5G and radar / edited by Gernot Hueber, Silicon Austria Labs, Austria,
Ali M. Niknejad, University of California, Berkeley.
Description: Cambridge, United Kingdom ; New York, NY, USA : Cambridge University Press, 2019. |
Series: The Cambridge RF and microwave engineering series | Includes bibliographical references and
indexes.
Identifiers: LCCN 2018060438 | ISBN 9781108492782 (hardback)
Subjects: LCSH: Millimeter wave devices. | Wireless communication systems–Equipment and supplies. |
Radar circuits.
Classification: LCC TK7876.5 .M45 2019 | DDC 621.3848–dc23
LC record available at https://lccn.loc.gov/2018060438

ISBN 978-1-108-49278-2 Hardback

"The 5G cellular standard has been under development for a number of years and is projected to support speeds far greater than those of earlier generations. This ambitious goal relies on innovations at all levels of abstraction, and is well served by the publication of *Millimeter-Wave Circuits for 5G and Radar*. Written by nearly 30 experts in the field, this book provides a great overview of the state of the art and will benefit those who wish to advance their knowledge of millimeter-wave circuits."

Behzad Razavi,
UCLA

"A book of landmark importance for practitioners of 5G radio frequency (RF) system and RF circuit design covering millimeter-wave and frequency division duplex (FDD) transceivers. It covers the essential topics of transceiver system design, beam forming, and circuit design for advanced 5G and radar systems."

Kamal Sahota,
Qualcomm

"This book is an excellent read with topics ranging from architecture to layouts, and the concepts are illustrated by test results from manufactured circuits in advance technology nodes. Leading industry and academic researchers give a comprehensive overview of system aspects as well as in-depth state-of-the-art circuit design solutions. The scope is also very timely, as integration of advanced radio and radar transceivers is the key enabling technology for 5G communication and automotive application hardware."

Sven Mattisson,
Ericsson

"This book is on the 5G system and radar, which are both part of our future indeed. A wide range of experts have been brought together to discuss the design of high-frequency circuitry for 5G and radar. Both the system level and the transistor level are addressed in great detail. It includes 5G system considerations and transceivers, digital phase-locked loops (PLLs), voltage-controlled oscillators (VCOs), power amplifiers, etc. Actually, in systems of such complexity, few circuits of importance can be missing. Linearity and noise considerations are omnipresent as well. The last chapter provides the trade-off between planar complementary metal-oxide semiconductors (CMOS) and Fin Field-effect transistors (FinFETs). This is a choice that each designer faces sooner or later. As a result, this book is a must for whoever wants to play a design role in the coming 5G or radar systems."

Willy Sansen,
KU Leuven

Contents

Contributors *page* xiv
Foreword xvii

1 Introduction 1
Gernot Hueber and Ali M. Niknejad

1.1 5G 1
1.1.1 What Is 5G? 1
1.1.2 A Brief History of the Gs 2
1.1.3 Do We Need 5G? 4
1.2 Radar 10
1.3 A Circuit Designer's Perspective 13

2 5G Transceivers from Requirements to System Models and Analysis 18
Aarno Pärssinen

2.1 RF Requirements Inspired by 5G System Targets 18
2.2 Radio Spectrum and Standardization 20
2.3 System Scalability 22
2.4 Communications System Model for RF System Analysis 24
2.5 System-Level RF Performance Model 29
2.5.1 Transmitter 29
2.5.2 Receiver 31
2.5.3 Antenna Array 33
2.5.4 Transceiver Architectures for RF and Hybrid Beamforming 36
2.6 Radio Propagation and Link Budget 38
2.7 Multiuser Multibeam Analysis Example 43
2.8 Conclusion 51

3 MU-MIMO and Massive MIMO for 5G Radios 55
Antonio Puglielli, Greg LaCaille, Elad Alon, Borivoje Nikolić, and Ali M. Niknejad

3.1 Spatial Processing: Untapped Potential 55
3.2 MIMO Technology Overview 56

3.2.1 Spatial Multiplexing with Antenna Arrays 58
3.2.2 MIMO: Exploiting Multipath Propagation 60
3.2.3 Channel Rank 62
3.2.4 Multiuser MIMO (MU-MIMO) 63
3.3 Conventional MIMO Processing 64
3.3.1 Channel Estimation 64
3.3.2 Linear Beamforming 65
3.3.3 ML and Near-ML Receivers 66
3.3.4 Successive Interference Cancellation 67
3.3.5 MU Downlink: Dirty Paper Coding 68
3.3.6 Massive MIMO: High-Order MU-MIMO 68
3.4 System Architecture for Large Arrays 69
3.4.1 State of the Art 70
3.4.2 A Scalable Beamforming-Aware Array Architecture 71
3.5 Impairments in Large Arrays 76
3.5.1 Synchronization 76
3.5.2 Reciprocity and Channel Estimation 77
3.5.3 Transmit Linearity 78
3.6 Conclusion 78

4 **RF and Millimeter-Wave Full-Duplex Wireless for 5G and Beyond** 84
Harish Krishnaswamy and Tolga Dinc

4.1 Overview of Full-Duplex 84
4.2 Millimeter-Wave Full-Duplex Applications 86
4.2.1 Millimeter-Wave Backhaul 86
4.2.2 Milimeter-Wave FD Relaying 86
4.2.3 Millimeter-Wave Vehicular Radar 87
4.2.4 5G Small-Cell Base Stations 88
4.2.5 Virtual Reality (VR)/Augmented Reality (AR) Headsets 88
4.3 Full-Duplex Challenge and System Considerations 88
4.4 Self-Interference Suppression Techniques 92
4.4.1 Antenna Suppression 92
4.4.2 Shared Antenna Interface 94
4.4.3 Integrated Low-RF FD Radios 98
4.4.4 Integrated Millimeter-Wave FD Radios 103
4.5 Conclusion 106

5 **Flexible Integrated Architectures for Frequency Division Duplex Communication** 112
Lucas A. Calderin, Sameet Ramakrishnan, Elad Alon, Borivoje Nikolić, and Ali M. Niknejad

5.1 Introduction 112
5.2 Approaches to Self-Interference Cancellation 113

5.2.1 Hybrids 113
5.2.2 Active Cancellation 114
5.3 System Concept and Architecture 115
5.3.1 Conceptual Overview 115
5.4 System Implementation Considerations 117
5.4.1 DAC Power Consumption 117
5.4.2 System Thermal Noise 118
5.5 System Degradation 123
5.6 Transmitter 124
5.7 Cancellation DAC Design 126
5.7.1 DAC Linearity 126
5.7.2 DAC Thermal Noise Cancellation 127
5.8 Quantization Noise Measurements 130
5.8.1 Channel Memory 130
5.8.2 PA Dynamic Nonlinearity 132
5.9 Measurement Results 135
5.10 Conclusion 143

6 **Scalable RF and Millimeter-Wave Multibeam Approaches** 146
Arun Natarajan

6.1 Large-Scale Phased and MIMO Arrays 146
6.2 Reconfigurable Spatial Filtering 147
6.2.1 MIMO Spatial Filtering at RF 148
6.3 N-Path Spatiospectral Filtering 149
6.4 Scalable mm-Wave Packaging 152

7 **Millimeter-Wave Radar SoC Integration in CMOS** 162
Piet Wambacq, Davide Guermandi, André Bourdoux, and Jan Craninckx

7.1 Introduction 162
7.2 Frequency-Modulated Continuous-Wave Radar 164
7.3 Phase-Modulated Continuous-Wave Radar 166
7.4 Comparison between FMCW and PMCW 168
7.4.1 Sensitivity to Phase Noise and Flicker Noise 168
7.4.2 TX Orthogonality for MIMO Radar 168
7.4.3 Interference Robustness 168
7.4.4 IF Bandwidth and ADC 168
7.4.5 TX-to-RX Spillover 169
7.4.6 Waveform Generation and Linearity 169
7.4.7 Other Aspects 169
7.5 Link Budget for a PMCW Radar 169
7.5.1 Link Budget for Single-Antenna TX and RX and MIMO Systems 170
7.5.2 LO Phase Noise 172

7.6 MIMO Techniques for PMCW Radars 172
7.6.1 TX Orthogonality by Sequence Engineering: Different TX Use Different Sequences 173
7.6.2 TX Orthogonality with Outer Code: All TX Use the Same Sequence 173
7.6.3 Comparison of the Two Approaches and Implementation 173
7.7 Analog and Millimeter-Wave Circuits 174
7.7.1 Frequency Generation 175
7.8 Experimental Results 181
7.8.1 Module and Antenna Design 181
7.8.2 Circuit-Level Measurements 182
7.8.3 Radar System Measurements 186
7.8.4 Conclusions and State-of-the-Art Comparison 189

8 CMOS Transceiver Design for Ultra-High-Speed Millimeter-Wave Wireless Communications 193
Kenichi Okada and Rui Wu

8.1 Introduction 193
8.2 60 GHz CMOS Transceiver Architecture 194
8.2.1 Challenges and Design Considerations 194
8.2.2 Direct-Conversion Transceiver Architecture 201
8.3 Circuit Implementation of Key Building Blocks 202
8.3.1 Local Synthesizer 202
8.3.2 Transmitter 205
8.3.3 Receiver 211
8.3.4 Calibration Techniques 217
8.4 Measurement Results of Transceiver Chips 224
8.5 Conclusion 236

9 Phased Arrays for 5G Millimeter-Wave Communications 243
Bodhisatwa Sadhu and Leonard Rexberg

9.1 The Role of mm-Wave in 5G Communications 243
9.2 Introduction to Beamforming 244
9.2.1 Beamforming as a Fourier Transform 244
9.2.2 Beam Shaping and Beam Steering 249
9.2.3 2D Antenna Array 250
9.3 Desired Features of Millimeter-Wave Phased Arrays 252
9.3.1 Accurate Beam Control 252
9.3.2 Architecture Scalability 253
9.3.3 Dual-Polarized Operation 254
9.3.4 Small Solution Footprint 255
9.3.5 Orthogonal Phase and Gain Control 257
9.4 Exemplary Si-Based Millimeter-Wave Phased Array 258

9.4.1 Circuit Details 258
9.4.2 Measurement Results 263
9.5 Conclusion 270

10 Millimeter-Wave Frequency Synthesis Based on Frequency Multiplications 273
Payam Heydari

10.1 Introduction and Motivation 273
10.2 Design of a Silicon-Based Ka-Band PLL 279
10.3 Design of a W-Band ILFT 280
10.3.1 Harmonic Generation of HBT 280
10.3.2 Circuit Design of the mm-Wave ILFT 281
10.3.3 Measurement Results 286
10.4 Design of a W-Band Silicon-Based HBFT 287
10.4.1 Circuit Design of the mm-Wave HBFT 287
10.4.2 Measurement Results 289
10.5 Design of a Transformer-Based CMOS ILFT 291
10.5.1 Harmonic Generation of an MOS Transistor 291
10.5.2 Millimeter-Wave T-ILFT Structure 293
10.5.3 Measurement Results 296
10.6 Comparisons and Discussions 300
10.7 Conclusions 302

11 Digitally Intensive PLL and Clock Generation 305
Wanghua Wu and R. Bogdan Staszewski

11.1 Introduction to Digitally Intensive PLL 305
11.2 Multirate DPLL-Based Frequency Modulator Architecture 309
11.3 High-Resolution mm-Wave DCOs 312
11.3.1 Distributed Switched Metal Capacitor Bank for mm-Wave DCOs 312
11.3.2 Transformer-Coupled Fine-Tuning Bank 315
11.3.3 A 60 GHz DCO Design Example 317
11.4 Time-to-Digital Converter 319
11.5 Digital Calibration Techniques for High RF Performance 324
11.5.1 DCO Gain Calibration and Linearization 324
11.5.2 Mismatch Calibration of the Fine-Tuning Bank 326
11.5.3 Synchronization in a Multirate System 327
11.5.4 Experimental Results 328
11.6 Built-In Self-Test and Built-In Self-Characterization for DPLL 331
11.6.1 Critical Signals in DPLL for BIST and BISC 333
11.6.2 Snapshotting Internal Signals for Debugging 336
11.6.3 DCO Tuning Step Analyzer 337
11.7 Another Approach: DTC-Assisted DPLL Architecture 337

12 Practical VCO Design 347
Mohyee Mikhemar

12.1 LO Design 347
12.1.1 LO Architectures 348
12.1.2 Impact of LO Architecture on VCO Requirements 349
12.2 Fundamentals of VCO Design 350
12.2.1 Improving Noise Factor by Avoiding Triode Operation 354
12.3 VCO Frequency Scaling 356
12.4 Design Procedure 358
12.5 Practical Considerations in VCO Design 359
12.5.1 Tail Tuning and Bypass Capacitance 360
12.5.2 Kickback from the First VCO Buffer 362
12.5.3 Resilience to Pulling 364
12.5.4 AM/PM Conversion in Small Tuning-Range VCO Designs 364
12.5.5 Bias Circuit Design for VCO 365
12.6 Conclusion 366

13 CMOS Power Amplifier Design for 5G Mobile Applications 369
Yang Zhang and Patrick Reynaert

13.1 Introduction 369
13.2 5G RF Front-End Requirement 369
13.2.1 Quantify the Signal Quality 370
13.2.2 Signal Influenced by PA Nonlinearities 373
13.3 Power Amplifier Basics 374
13.3.1 Transistor Optimization for PAs 375
13.3.2 Passive Device in CMOS 375
13.4 Impedance Transformation and Power Combining 379
13.4.1 PA Nonlinearity 381
13.4.2 Linearity Enhancement Technology 382
13.5 Design Example of a 40 nm CMOS PA 384
13.5.1 Power Transistor with Source Degeneration Inductor 384
13.5.2 Design 385
13.5.3 Measurement Results 389
13.6 Conclusion 395

14 FinFET Process Technology for RF and Millimeter-Wave Applications 400
Hyung-Jin Lee and Bernhard Sell

14.1 Overview of FinFET Technology 400
14.2 Unique Properties of FinFET Technology for RF/mm-Wave Design Consideration 404
14.2.1 Transistor Scaling and Performance 404

14.2.2 Nonlinear Gate Resistance by 3D Structure 406
14.2.3 Fin Self-Heating 408
14.3 Assessment of FinFET Technology for RF/mm-Wave 410
14.3.1 Parasitics and RF Performance 411
14.3.2 Noise Performance 412
14.3.3 Gain and Noise Matching at the mm-Wave Frequency 414
14.4 Design Methodology for RF/mm-Wave Performance Optimization with FinFET 416
14.4.1 Wireless Design Consideration in Cascade Chain 416
14.4.2 Optimizing *NF* with G_{max} for LNA within Self-Heat Limit 417
14.4.3 Gain per Power Efficiency 420
14.4.4 Linearity for Gain and Power Efficiency 423
14.4.5 Neutralization for mm-Wave Applications 425
14.5 Design Example for an mm-Wave Amplifier with the Proposed Design Methodology 427
14.6 Conclusion 429

Author Index 432
Subject Index 433

Contributors

Elad Alon
University of California at Berkeley, US

André Bourdoux
Imec, Belgium

Lucas Albert Calderin
Apple, US

Jan Craninckx
Imec, Belgium

Tolga Dinc
Kilby Labs, Texas Instruments, US

Davide Guermandi
Imec, Belgium

Payam Heydari
University of California, Irvine, US

Gernot Hueber
Silicon Austria Labs, Linz, Austria

Harish Krishnaswamy
Columbia University, New York, US

Greg LaCaille
University of California at Berkeley, US

Hyung-Jin Lee
Intel Corporation, Portland, US

Thomas Lee
Stanford University, US

Mohyee Mikhemar
Broadcom Limited, US

Arun Natarajan
Oregon State University, US

Ali M. Niknejad
University of California at Berkeley, US

Borivoje Nikolić
University of California at Berkeley, US

Kenichi Okada
Tokyo Institute of Technology, Japan

Aarno Pärssinen
Oulo University, Finland

Antonio Puglielli
Zendar, US

Sameet Ramakrishnan
Amazon, US

Leonard Rexberg
Ericsson, Sweden

Patrick Reynaert
KU Leuven, Belgium

Bodhisatwa Sadhu
IBM T. J. Watson Research Center, US

Bernhard Sell
Intel Corporation, Portland, US

R. Bogdan Staszewski
University College Dublin, Ireland

Piet Wambacq
Imec, Belgium and Vrije Universiteit Brussel, Belgium

Rui Wu
Tokyo Institute of Technology, Japan

Wanghua Wu
Samsung Semiconductor, US

Yang Zhang
KU Leuven, Belgium

Foreword

Around New Year's Day of 2002, the number of wireless subscribers exceeded the number of wireline subscribers for the first time in history, marking our becoming a wireless species. The number of wireless subscriptions continued to grow until by 2017 it had actually exceeded the population of the earth. By any measure, wireless has been one of the most spectacularly successful technologies in history. Consumers purchase five million cellular phones daily and use them to exchange one million information-free text messages every second. The average person now has instant access to the cumulative extracorporeal intellectual treasure generated by our ancestors over millennia. The cell phone allows us to order goods and services from all over the planet, at any time and from any place. The most potent sovereigns of prior centuries could not even dream of such capabilities for themselves, let alone for all of humanity. And yet we want more.

Given that virtually everyone with the ability to operate a cell phone now has one, it is natural to ponder what form *more* could possibly take. Predictions ("especially about the future") are notoriously unreliable and often absurdly humorous in retrospect, but perhaps in the history of wireless we may find useful clues about likely futures. After all, even when history doesn't repeat, it sure seems to rhyme, to paraphrase Mark Twain.

Wireless has evolved through three distinct ages, each characterized by its network topology. Station-to-station spark telegraphy found its niche in maritime communications, where it had no competition. The role of Marconi equipment (and an operator who lost his life) in saving over 700 passengers of RMS *Titanic* in 1912 testified dramatically to the transformation that wireless could bring. A transoceanic voyage would no longer be the equivalent of disappearing from the planet for the duration. Wireless telegraph equipment was installed so rapidly that thousands of ships and stations had been connected by the advent of World War I in this first age of wireless.

As revolutionary as was that first age, the station-to-station topology forced by the use of Morse code and complex equipment limited deployment to less than about 10,000 nodes - the sum total of the number of ships and stations. The next age of wireless happened almost by accident, driven by engineers who sought to convey the voice wirelessly. The unprecedented appropriation of *broadcasting* ("the spreading of seeds") from the lexicon of farmers speaks to the fierce velocity with which the technology took hold. The asymmetrical topology of broadcasting supported a large number of passive listeners with inexpensive and simple receivers, enabling programs to be heard by millions of people. This second age of wireless quickly transcended the kiloscale limits of spark-wireless to achieve megascale reach.

The Second World War introduced millions of soldiers to people-to-people wireless technology with walkie-talkies and other military communications devices. After returning to civilian life, they wondered why they couldn't have the same convenience that they had enjoyed during the war. Service providers such as AT&T in the United States responded to the demand by experimenting half-heartedly and often skeptically with people-to-people wireless communications for 30 years. Several nations finally began wide-area deployment of analog cellular systems throughout the 1970s and 1980s. Digital systems eventually replaced those early analog systems, and by 2013 voice traffic constituted only a minority of the bits conveyed; we had become a digital wireless species.

The evolution of wireless saw a three-order-of-magnitude jump in scale at each transition, from spark's station-to-station kiloscale to today's people-to-people gigascale connectivity. Thus having covered all possible permutations of *stations* and *people*, one might argue that wireless has reached the end of history. While that is certainly a possibility, an optimistic view is that it is an improbability. It is nonetheless sobering to observe that sustaining the evolution of wireless along historical trajectories would require an increase in connectivity to the terascale.

The much-hyped *Internet of Things* (IoT) has the potential to drive such an increase in connectivity, even though the name sometimes evokes images of one's refrigerator conspiring with the toaster against the blender. Since, for the moment at least, only humans have credit cards, any new age of wireless must present something of compelling value to people. The grandiose-sounding *Internet of Everything* (IoE) is sometimes used to distinguish this human-centered view from the more impersonal-sounding IoT.

The 5G wireless networks currently under development aren't betting solely on the IoE's success to justify deployment, but they are wisely accommodating the possibility. Whatever the future of wireless, tomorrow's networks will have to support vast increases in aggregate data rate, achieved in part by exploitation of the huge, untapped millimeter-wave spectrum and by the use of cells of ever-smaller radii. Recognition that millimeter-wave signals have considerable utility beyond communication has led to expectations that applications such as radar and other sensing, and perhaps even wireless power delivery to IoE devices, will be served as natural and inevitable consequences of 5G deployments. The smartphones and other conversants of the fourth age of wireless will thus surely possess advanced capabilities that will make today's devices appear primitive in comparison. This book provides a comprehensive guide to solving the challenging problems that stand between today and that glorious future.

Thomas H. Lee

1 Introduction

Gernot Hueber and Ali M. Niknejad

1.1 5G

A lot of the focus of this book is on 5G, so you may be wondering, what exactly is 5G? And, perhaps more importantly, how does it impact me as a circuit designer? Hopefully we can answer the first question in this chapter, and leave the rest of the book to address the second one.

1.1.1 What Is 5G?

The term "5G" has been around for a while as it is really a marketing term. People were talking about 5G even before anyone knew what 5G was going to be about. Even today, if you ask five different people, "What is 5G?" you may get more than five answers! Well, the name is naturally 5G because it is the "Fifth-Generation" mobile network standard. Ultimately, 5G will be defined by standardization bodies such as 3GPP (3rd Generation Partnership Project), and even then the concept of 5G will evolve. The reason that it's so difficult to pin down a clear definition for 5G is that it's going to be a worldwide network standard for the next decade, and there's a long wishlist of new technology elements that people want to see in 5G, and then there's the reality of building and deploying a new network and keeping costs and power consumption at a reasonable level. So 5G is a compromise between our dreams for the next-generation radio versus the reality of what is technologically feasible and economically viable.

5G technology is positioned to address all of the shortcomings of 4G technology. In particular, people envision "everything in the cloud," which can offer a desktop-like experience on the go, immersive experiences (lifelike media everywhere), ubiquitous connectivity (intelligent web of connected things), and telepresence (real-time remote control of machines) [1]. To address these new application scenarios from a mobile device, the following "rainbow of requirements" shown in Figure 1.1 have been defined: (1) peak data rates up to 10 Gbps, (2) cell edge data rate approaching 1 Gbps, (3) cell spectral efficiency close to 10 bps/Hz, (4) Mobility up to 500 km/h, (5) cost efficiency that is 10 to 100 times lower than 4G, (6) a latency of 1 ms, and finally, and perhaps most importantly, (7) over 1 M simultaneous connections per km^2 [1,2].

Before we dive into the details, it's useful to have a very brief history lesson.

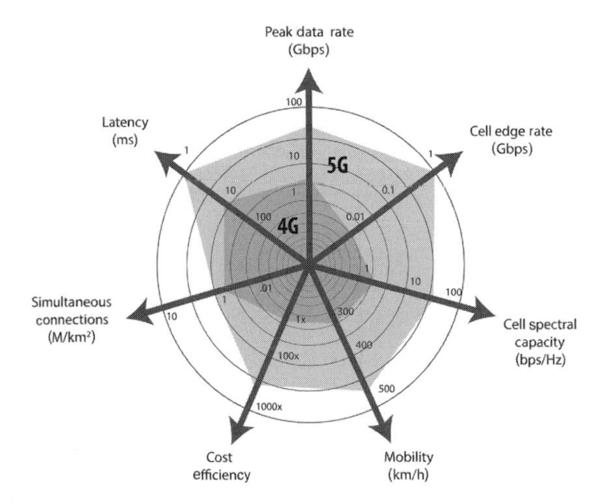

Figure 1.1 The 5G rainbow of requirements, adapted from [2].

Figure 1.2 Evolution of humankind alongside wireless communication technology [3].

1.1.2 A Brief History of the Gs

Some of us are old enough to remember the days of brick-sized phones and analog mobile communication, the so-called 1G era and the Advanced Mobile Phone System (AMPS), first deployed in 1979 (see Figure 1.2). The system was analog and operated originally in the 850 MHz frequency band. The channel bandwidths were only 60 kHz and it was intended for voice communication. One important distinction to note is that 1G systems were circuit switched, so once a call was activated, the spectrum was allocated to a user, even if both sides of the link were silent.

In the early 1990s, the 2G generation took over and offered digital communications for the first time, including the ability to use Time Division Multiple Access (TDMA). In most parts of the world, 2G and the term GSM (Global System for Mobile communications) were synonymous, which used 200 kHz per channel, and Gaussian Minimum Shift Keying (GMSK) modulation (constant envelope) for power amplifier (PA) efficiency. But in addition to the GSM standard, a second-generation AMPS standard called Digital AMPS (D-AMPS), also referred to as TDMA, was in operation (IS-54 and IS-136). At the same time, Qualcomm was actively selling a new radio access technology known as Code Division Multiple Access (CDMA), and these radios were standardized as IS-95.

There were some 2.5G systems that used packet switching, known as General Packet Radio Service (GPSR), as opposed to circuit switching, which allowed the system to offer more efficient spectral access. This meant that more time slots could be allocated on demand, and the latency and data rate depended on the number of users connected to a base station. By today's standards, 2.5G systems were dog slow, topping in at 50 Kbps. At first no one was really using mobile for data, and this didn't seem to be an issue. But the increasing popularity of mobile communication drove the need for more bandwidth and more speed. This is where the 2.75G standard evolved and offered EDGE (Enhanced Data Rates for GSM Evolution), offering theoretical speeds of 1 Mbps, by using 8-PSK encoding (three bits per symbol).

Interestingly, the first iPhone was released in 2007, 16 years after the introduction of 2G, and it was still a 2G device. For those of us lucky enough to have owned a first-generation iPhone, the experience was both amazing and also tortuous because of the slow network speeds due to 2G limitations and also due to the fact that in dense urban environments, everyone was all of a sudden trying to access the network for Internet connectivity at the same time. These early smartphones, especially the iPhone, were heavy users of data, and they really showed the world that the 2G network was not good enough. Other devices at the time were already using 3G technology, which came in many shapes and sizes.

In the late 1990s and early 2000s, the 3G networks started to operate and offered improved data rates by increasing the bandwidth of channels (up to 5 MHz) and adopting spread spectrum techniques and higher-order constellations (16- and 64-QAM) and multiple-input and multiple output (MIMO) techniques. The Universal Mobile Telecommunications Service (UMTS) radios were introduced as hybrid 2G/3G UMTS/GSM radios. Sometimes these systems were referred to as W-CDMA systems, due to the use of a wideband code division multiple access technique. Data rates increased to 384 Kbps in the original systems, and evolutions pushed the data rates higher to Mbps regions with High Speed Packet Access (HSPA) and HSPA+ offering up to 168 Mbps in downlink and 22 Mbps in the uplink. The adoption of multiple bands meant more complex front-end circuitry, wider bandwidths, and therefore more linearity to handle more complex modulation schemes. In parallel, the CDMA2000 standard (IS-2000) offered peak data rates of 14.7 Mbps using 1.23 MHz of channel bandwidth. Unfortunately, the CDMA2000 and UMTS/HSPA radios were standardized by different committees and were not interoperable, making phones not only region-specific but also carrier-specific.

Today we are living and fully immersed in the 4G world of LTE, or Long Term Evolution, the "winner" technology that is ubiquitous worldwide. One of the requirements for 4G was to offer over 100 Mbps of peak data rate for highly mobile access and approximately 1 Gbps for low mobility access. The Samsung Galaxy Indulge was the world's first LTE smartphone starting on February 10, 2011 [1]. To move toward these lofty goals in power transfer and low latency, LTE networks were all Internet Protocol (IP) packet switching, employed very dynamic network architectures for optimum sharing of network resources, offered scalable bandwidths from 1.4 MHz up to 20 MHz, and distributed these resources on demand using Orthogonal Frequency Division Multiple Access (OFDMA) [4]. The spread spectrum techniques widely used in 3G systems were abandoned in favor of OFDMA, or the division of a wide bandwidth into smaller bands, modulation of the subcarriers at a much lower rate, and the use of a cyclic prefix in the guard band, thereby circumventing frequency-dependent fading and intersymbol interference. Using many subcarriers also allows the base station to optimize resource allocation by allocating spectrum resources in both time and frequency slots. More efficient turbo codes and MIMO techniques also improved the link quality.

One well-known pitfall with OFDMA is that the composite multicarrier signal has a very high peak-to-average power ratio (PAPR), which spells disaster for power amplifiers, requiring high back-off and linearization. These issues are well known to the power amplifier community as WiFi networks adopted OFDM (Orthogonal Frequency Division Multiplexing) as early as 1999 and the introduction of 802.11a. To circumvent this high PAPR, and single-carrier FDMA (SC-FDMA) for the uplink reduces PAPR. This slightly complicates the transmitter and requires frequency domain equalization in the receiver.

While most 4G systems converged on LTE, providing compatibility in theory, in practice the number of LTE bands exploded covering from 450 to 3600 MHz and both frequency division duplex (FDD) and time duplex (TDD) access. This meant that designing a "worldwide" LTE phone would be a formidable task due to the number of different front-end components required to cover disparate frequency bands and access modes (FDD versus TDD). LTE-Advanced (LTE-A) is an extension of LTE with new features including up to 8×8 MIMO and 128 quadrature amplitude modulation (QAM) in the downlink and carrier aggregation of contiguous and noncontiguous spectrum allocations, allowing up to 100 MHz of aggregated bandwidth. This means a device with LTE-A has a theoretical peak download data rate of 3 Gbps [5]. While this rate is impressive, in practice most users never reach these peak data rates.[1]

As evident in this brief history, each generation of mobile standards has embraced the latest advancements in communication theory and technology, in particular advances in coding, multicarrier modulation and wider bandwidths, and MIMO techniques to enable ever-increasing data rates and more efficient and dynamic networks. Each generation lasted about a decade, and it is a small miracle today that we can all enjoy watching our favorite cat videos from virtually anywhere on the planet.

1.1.3 Do We Need 5G?

So why do we need 5G? LTE and WiFi are amazing technologies that have served us well. Will the investment in a new network pay off? First, let's consider the new generation of users of wireless technology. A typical 12 year old today was born with

[1] The coeditor of this book has obsessively tested his phone all over the Bay Area and topped out at 162 Mbps downlink and 43.5 Mbps uplink.

a smartphone or tablet in her vicinity for most of her life. She may have never even experienced Internet blackout as a whole generation of parents replaced the TV with the smartphone/tablet as the de facto caregiver. The TV was limited in mobility whereas a smartphone can be carried anywhere and offer not only videos, but countless games and other forms of entertainment that only this new generation can understand.[2] This generation has a different relationship with bandwidth because they constantly stream video. Students prefer watching lectures online, especially because they can slow down and speed up the lecture and look up things while watching. To give a simple but illustrative example, the coeditor of this book was telling his daughter about paper and how it's actually a fibrous material that looks like a thin layer of spaghetti under an electron microscope. Before finishing his sentence, his daughter was watching such videos on YouTube. What surprised the coeditor was that she went directly to YouTube rather than to an Internet search engine or to Wikipedia.

Now let's try to imagine what a kid will do in 10 years when trying to understand something new, such as an internal combustion engine works. Hopefully this will be an ancient relic that arouses her curiosity since electric propulsion will completely displace such engines. She'll slip on her virtual reality or augmented reality goggles, or perhaps use a holographic projector to show the engine. She'll be able to rotate the engine, look at the different parts, and then with a simple gesture, she'll be able to blow out the engine into thousands of parts. She can then put back the engine and just look at a few components, say inside the engine block, and play with the pistons and see how they move up and down and generate a rotational motion through the crank shaft. She'll be able to learn a tremendous amount in a short period of time. Clearly, this data will have to be downloaded from the Internet and played back in real time. Maybe she's repairing a classic automobile and needs to see the 3D images again while she's in the garage. Remember that a single base station will need to serve hundreds or thousands of curious kids, all at the same time.

In certain situations, the demands on the network will explode. Imagine a classroom full of thousands of students learning anatomy. The professor will have a virtual cadaver in front of him and he'll be making incisions and demonstrations of different parts fit together. Every student will have his or her own virtual cadaver as well. In fact, there's no need to use an inanimate body, because a virtual body that is alive and moving is much more interesting, for example to understand how muscles and connective tissue work together to enable different motions. In this scenario, we have thousands of simultaneous three-dimensional (3D) high-definition (HD) connections, all in the same geographic location. This is clearly beyond the capabilities of both WiFi and 4G networks today.

At the Berkeley Wireless Research Center (BWRC), we looked at these issues and considered a blank slate to imagine what should the next generation of wireless look like. In December of 2013, we codified our vision with the xG network, shown in Figure 1.3. Our vision is for a new network that utilizes a massive number of antennas in access points to allow a high degree of spatial multiplexing to many different disparate devices,

[2] Such as watching others play video games or watching someone playing with slime.

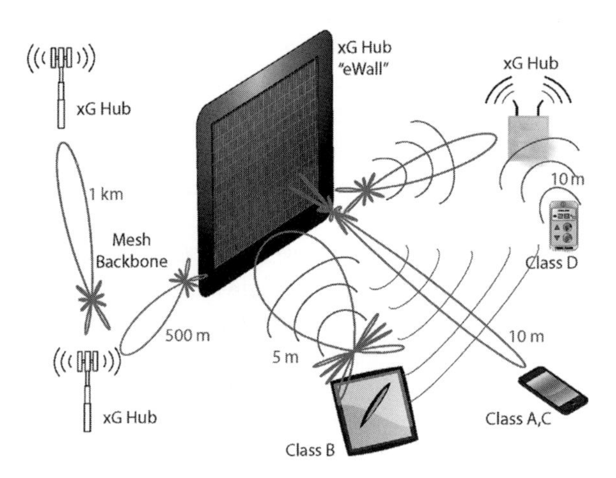

Figure 1.3 The BWRC "xG" vision for the next-generation wireless communication system (December 2013).

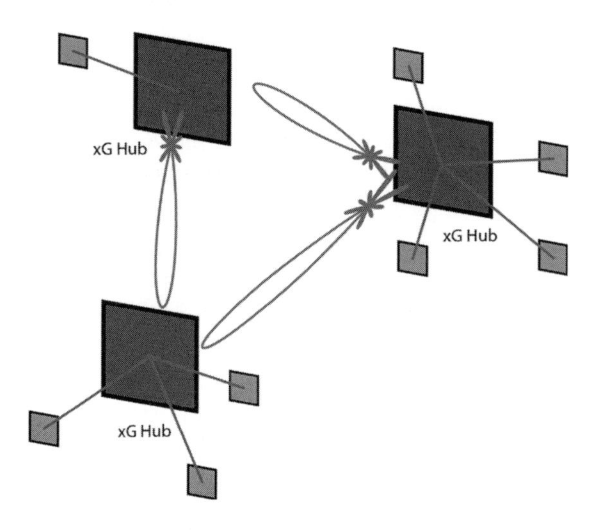

Figure 1.4 Wireless backhaul using phased arrays and mesh networking can reduce the cost of deployment of a 5G system by obviating the need for fiber connectivity.

from cell phones and tablets to Internet of Things (IoT) devices. Multiple RF and mm-wave frequency bands are used in a complementary fashion to form both sharp and broad beams. Also, most importantly, these access points self-backhaul by forming a hierarchical wireless mesh network (Figure 1.4), avoiding the need to use cables or fiber to form the backhaul network. In such a way, the network can grow organically to serve the demands for wireless traffic. The access point can wake up, identify other nodes in the network, and begin routing traffic on demand, with links going up or down in a dynamic fashion, much like the original vision for the Internet and the need for packet switching. In parallel, people started dreaming of 5G and what it should encompass. Many people came to the realization that to serve these visions, we need to utilize higher-frequency bands to realize higher spectral efficiencies and to circumvent interference, and the idea that 5G would also operate in the mm-wave bands was born.[3]

5G Wishlist

Given this xG vision, which is more or less the same as what people were thinking for 5G, let's enumerate our wishlist a bit more carefully. Are today's networks fast enough both in terms of speed of transmission and latency per user? While a lot of progress has been made on speed, even exceeding 1 Gbps, these improvements are mostly for marketing and don't bear out in practice. But nevertheless, being able to get a mobile wireless connection over 10 Mbps is quite impressive and certainly sufficient for many applications such as video. The problem is that many times we cannot get sufficient coverage and we are all too familiar with video streams coming to a screeching halt at just the right moment. The other issue with today's networks is the latency is typically tens to hundreds of milliseconds long, and sometimes even longer. The latency is also unpredictable, making it difficult to design a closed-loop control system. For this reason, many applications that could benefit from wireless technology have not embraced wireless. Examples include industrial control, semiautonomous driving, multiuser gaming, and virtual reality and augmented reality devices driven from the cloud.

While speed is definitely a great marketing specification, another revolution is under way, the proliferation of low-cost devices with wireless connectivity. This is the well-known and much anticipated IoT revolution, which requires very small footprint and low-power wireless connectivity, and in most cases the speed is not an issue. More important than speed is the power consumption. Today people are adding Bluetooth, Bluetooth Low Energy (BTLE), Zigbee, WiFi, or other radios for wireless connectivity. These radios are short range and cannot actually connect to the Internet without a nearby access point (such as a WiFi router connected to the Internet). Why not just put LTE radios in such devices? The problem with LTE is cost and power consumption, and a lack of a clear business model. For example, many smart watches today have an LTE radio inside but suffer from poor connectivity and require frequent recharging, and each device requires registration with the carrier (and a not-so-insignificant fee per month). Clearly this does not lend itself well to IoT, where we imagine thousands of devices operating on small coin cell batteries.

This brings us to another point. WiFi technology has advanced tremendously in the past 20 years, and for a long time there was a clear boundary between mobile carrier connectivity and wireless connectivity with WiFi and Bluetooth. But today the boundary is blurring, and in many cases these technologies compete. In a crowded café, dozens of users are streaming video from the Internet and one may find that LTE outperforms WiFi. LTE technology operates in licensed spectrum and interference is managed much better than in WiFi unlicensed spectrum, where the access point may only have control over a subset of devices operating in the same band. In many ways, both WiFi and

[3] Samsung, "Pioneer in 5G Standards, Part 1: Finding the 'Land of Opportunity' in 5G Millimeter-Wave." http://bit.ly/2GBDoiA.

mobile standards have converged, for example the use of OFDM to manage equalization in a wideband channel, power control for interference mitigation, similar modulation and coding schemes, and MIMO. LTE even now operates in unlicensed bands, and carriers are encouraging users to use the WiFi infrastructure to relieve traffic demands on the operator. So why should we have two standards if they are so similar? While it's unlikely that WiFi and LTE will ever merge into one standard (politics alone will prevent this from happening), we could wish for more interplay and compatibility between the radios. Too often we are frustrated by our wireless devices not connecting to the Internet only to find that the WiFi has taken over without truly connecting to the Internet. Many users have to actually manually shut off their WiFi on a daily basis to prevent their phones from connecting to a weak network. The situation has worsened because traditional broadband carriers are trying to compete with the wireless carriers by deploying citywide outdoor WiFi networks.

All of these problems arise because today's mobile networks are simply not up to the task of serving the exponentially growing needs of our modern devices. The spectral capacity of today's wireless networks are in fact operating close to Shannon capacity limits, and MIMO techniques are not as effective in outdoor channels (see Chapter 3). In dense urban environments, this is especially problematic because of high population densities (about 7,282 people in a square kilometer in San Francisco). If 10% of the population is actively watching videos at a given time in a given square kilometer (25 Mbps per HD stream), then the base station has to have a capacity of over 18 Gbps. To serve that much data with a 100 MHz swath of spectrum translates into a spectral efficiency of 180 bits/Hz, which is impossible without enormous signal-to-noise ratio (SNR) in a single channel scenario (not using MIMO). Base stations could be deployed over increasingly smaller areas to solve this problem, but then we are plagued with interference and cost barriers. On the other hand, massive MIMO demonstrations have already showed nearly 100 bits/Hz of spectral capacity in a multiuser MIMO scenario, which is a technique that can improve the aggregate capacity of a system rather than the per-user capacity, and this is an exciting technology on our wishlist for 5G. The other approach is to just go to higher carrier frequencies where wider bandwidths make it easier to serve high data rates. Higher frequencies have propagation issues but offer the ability to use beam-forming to reduce interference as well.

Finally, let's consider the enormous cost to deploy a new network, especially a network with an order of magnitude more base stations to serve dense urban environments. Such an investment should pay off in less than a decade to allow the operators to be profitable. This means that the cost of base stations has to go down, especially in terms of rents on property, backhaul access, and electricity costs. Since mm-wave radios are shorter range, one can anticipate a 10 $\times$ densification of base stations, which must be accommodated by a 10 $\times$ reduction in building a base station. For this to happen, wireless backhaul is a must, as many locations cannot be served by fiber without tearing up the streets and installing new access. Also, wireless backhaul using a phased array, rather than a dish, is clearly advantageous to reduce the setup cost for a new base station. A phased array can dynamically find other nodes and point the beam appropriately, whereas a fixed point-to-point link requires precision antenna alignment. Even a massive

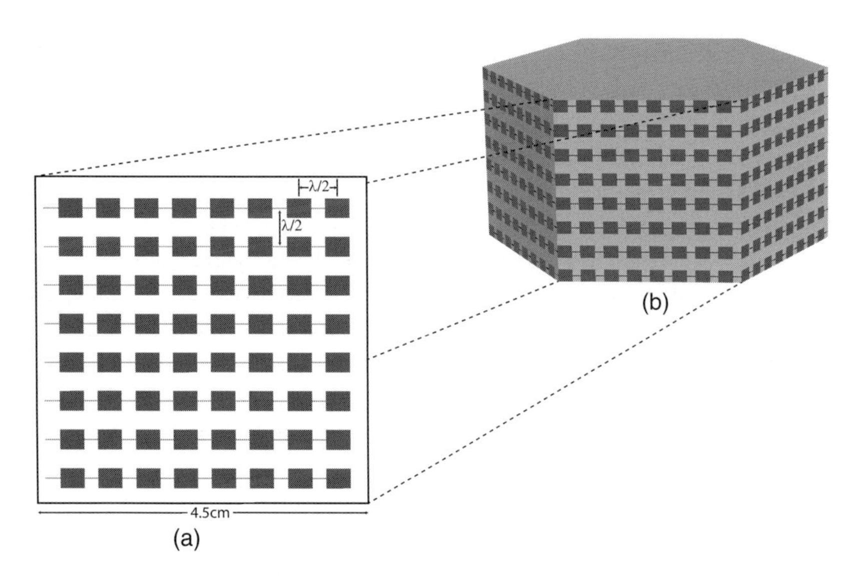

Figure 1.5 An array of 64 patch antennas only occupies an area of 4.5 cm by 4.5 cm as shown in (a). In (b), an array of such panels forms a four-sector base station that can serve thousands of users simultaneously using multiple spatial beams.

array of antennas in mm-wave bands does not occupy much area. Take a linear array consisting of 64 elements, or antenna subarrays, each with 8 elements, as shown in Figure 1.5. Even at 28 GHz, one of the lowest frequency mm-wave bands, the size of the array is $8 \cdot \lambda/2 \times 8 \cdot \lambda/2$ or about 4.5 cm × 4.5 cm. A base station may consist of a half a dozen of such panels, which means the entire base station could fit in a cube with an edge.

The Cloud

Today we have an enormous amount of data moving from edge devices (say your mobile phone) all the way up the cloud, a room full of servers running the applications. These data have to move back and forth, and it means a lot of data transport over hundreds of kilometers and also a lot of latency. For example, using the web service cloudping.info, the measured ping speed from a mobile phone to the Amazon Web Services is around 50 ms, whereas the ping speed to the carrier is only 25 ms. Clearly any applications such as gaming or augmented reality require millisecond delays, both for health concerns (to avoid making people dizzy) and to make the experience more real. If we could run applications much closer to edge devices, we could greatly improve the latency. This is exactly what people are proposing in industry, putting the servers in base stations, or rather moving base stations into server racks. To keep costs low and allow base station densification, the base station is split into a remote radio head and then backend processing is moved offsite into a server rack. This architecture has other benefits, such as making the network more software defined and flexible. Traffic to/from remote radio heads can be managed on the fly, serving demand (a stadium during the game) when and where it's needed.

Recently there's a lot of buzz around the concept of full duplex communication. Full duplex means a radio can transmit and receive at the same time, in the same bandwidth. Traditionally this was achieved with a circulator or isolator, or a nonreciprocal element. A circulator is a three-port device that allows both the PA and low-noise amplifier (LNA) to be connected to the antenna without any interference (or only a small amount of the transmitter signal leaks into the receiver). A four-port hybrid can do the same thing, at the cost of insertion loss. A practical circulator has loss too, but there's no fundamental limit to how low this loss can be. On the other hand, more importantly, circulators are bulky and narrowband, and cannot be integrated into a chip due to the need for nonlinear magnetics. Recently the coauthors of this book have demonstrated new CMOS-compatible architectures for circulators, and these are described in detail in Chapter 4. Another active cancellation approach, which is applicable to novel full-duplex systems and also traditional frequency division duplex (FDD) systems is presented in Chapter 5. FDD is common today and allows simultaneous transmit and receive in two nearby bands by the application of a sharp filter, or duplexer, to provide isolation. These filters are also band-specific and difficult to integrated into CMOS. This chapter will take a different route and use active impedance synthesis to cancel the transmit signal in the receive band.

While much of the buzz around 5G is in the new mm-wave bands, such as 28 GHz and 39 GHz, the 60 GHz band will likely play an equally important role as an unlicensed spectrum, much as WiFi today plays a complementary (and sometimes competing) role to LTE. The amount of bandwidth available in the 60 GHz band is enormous, and we are witnessing multiband radios that can pump tens to hundreds of gigabytes per second through this spectrum. This capability will enhance local area networks and provide backhaul mesh networking to 5G systems. In Chapter 8, we describe the latest chipset, which can push the limits of CMOS in the 60 GHz band to demonstrate record data transfer speeds.

1.2 Radar

Advanced Driver Assistance Systems (ADAS) are all systems to support the driver for safety and enhance driving convenience. There is a strong and important focus on safety, as many if not most accidents are a result of human behavior or error. Consequently, the ultimate goal of ADAS is to avoid any kind of accidents or collisions, by facilitating automated systems ranging from obstacle detection (e.g., vehicle, parking, pedestrian, etc.) to traffic sign detection and driver monitoring (e.g., drowsiness) or communication (car-to-car, car-to-infrastructure; see Figure 1.6). A key technology of ADAS is the detection of any kind of obstacles by specialized radar systems.

The use cases for automotive radar are diverse and include the following scenarios that demand specific requirements on the detection device:

Adaptive cruise control (ACC) is applicable in normal driving conditions to adapt the drive speed to the cars ahead as well to detect obstacles in the far distance to avoid

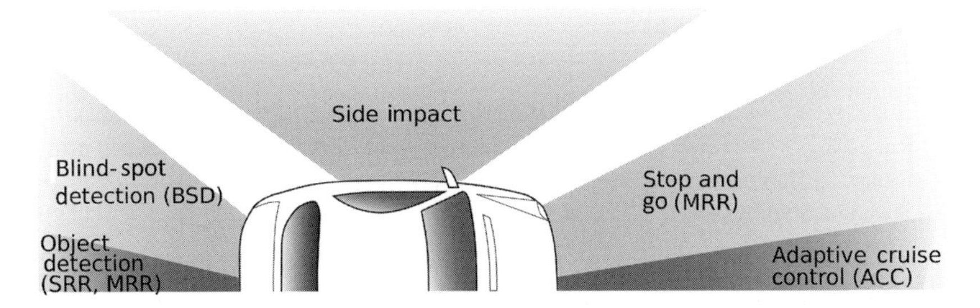

Figure 1.6 Future ADAS systems make use of multiple dedicated sensors for dedicated use-case scenarios and as part of a *sensor fusion*.

any accidents. This is the use case with the most demanding requirement in SNR and range (up to ∼200 m). ACC is addressed by long-range radar (LRR) 77 GHz radar systems at the front of a vehicle.

Blind spot detection (BSD) is a feature to detect other vehicles located to the driver's side and rear with the purpose to warn the driver of vehicles hardly visible and, in turn, to avoid potential collisions. With a range requirement of up to ∼20 m, both, 24 GHz and 77 GHz are applicable.

Short-range radar (SRR) is covered by 24 GHz systems located around the car (front, rear, side, or all four corners) and are used, e.g., for BSD, stop-and-go, or parking assistant applications, all of them at lower driving speed and lower total range (< 20 m)

Medium-range radar (MRR) is used for BSD as well as for stop-and-go scenarios, with limited range (up to 40 m). Sensors are, depending on the specific use case, mounted around the car (four corners) for 24 GHz systems. However, 77 GHz will be used for medium range as well.

Long-range radar (LRR) is the radar technology (77 GHz) that is applied for ACC. The sensors are mounted in the front of the vehicle to allow detections of other vehicles or obstacles ahead.

As of today's developments and available products, there are dedicated solutions that are highly optimized for the specific scenarios. Looking into the next generations, following a holistic approach, it is expected to see higher complexity by combining individual techniques into a *sensor fusion*. Consequently, specific dedicated solutions that are highly optimized for a specific purpose are combined (e.g., LRR, LiDAR, cameras) such that an extensive coverage of all use cases can be achieved. In turn, the challenge for sensor fusion is the real-time data aggregation of multiple sensors into a single holistic ADAS system.

Though currently 24 GHz [7–9] car radar solutions have been widely deployed, in current new designs 77 GHz is used [10–12]. However, it shall be noted, that regulation bodies are considering to allow for additional frequency bands beyond the 100 GHz, e.g., 134/136–141 GHz, and the European Technical Standards Institute (ETSI) is reviewing ultrawide band (UWB) radio determination applications within the frequency range between 120 to 260 GHz [13].

Table 1.1 Requirements for automotive radar systems.

Parameter	Value (target)	Value (max)	Unit
Range R	>100	250	m
Range resolution ΔR	0.25	100	m
Speed resolution Δv	5		km/h
Antenna gain G_{ant}	>10		dBi
System sensitivity RX_{sens}	<70		dBm
Tx effective isolated radiated power (EIRP) (77 GHz/79 GHz)	25	50/33	dBm
Bandwidth BW (77 GHz/79 GHz)	0.8/2	1/4	GHz

Table 1.2 Requirements for automotive long range radar receivers [14,15].

Parameter	Value	Unit
Stopband attenuation	>65	dB
Spurious-free dynamic range (SFDR)[a]	< -70	dBc
RX chain gain (RX,in to BB)	70	dB
Overall noise figure	~13	dB
Effective number of bits (ENOB) (at the BB-interface)	>12	bit
Minimum input signal $P_{RX,min}$	−100	dBm
SNR_{min}	16	dB

[a] Including intermodulation products (P_{in} at −20 dBm).

System-level specifications are shown in Tables 1.1 and 1.2 with a focus on range and speed resolution, and some key specifications for receivers and transmitters are given in Tables 1.2 and 1.3, respectively.

From a silicon technology perspective, there are two main trends. First technology widely used is SiGe BiCMOS. Advantages of SiGe BiCMOS include the developments of novel technologies with high f_{max} for currently available $f_{max} = 500$ GHz to higher $f_{max} = 700$ GHz and beyond is ongoing [16–20].

Alternatively, and in scope of monolithic integration with signal processing and digital control, is to use CMOS technology [21–23], which is a key advantage of CMOS. The development and predictions of CMOS available gate-length and maximum frequency is shown in Figure 1.7.

Another key advantage of CMOS technology is its ability to use the digital logic in a monolithic integration. Hence, the RF transceiver can be perfectly integrated with the baseband and application processing units, which allows highly efficient codesign of the RF, baseband (BB), and application processor (AP) and their interaction while peripheral parts (e.g., power management unit (PMU)) can be shared. This approach benefits from the increase of digital logic and static random-access memory (SRAM) transistor density increasing from $9{,}725\,\mathrm{Mt/cm^2}$ (CMOS 15 nm half-pitch, 2018), $15{,}437\,\mathrm{Mt/cm^2}$ (CMOS 11.9 nm half-pitch, 2020), to $24{,}505\,\mathrm{Mt/cm^2}$ (CMOS 9.5 nm

Table 1.3 Requirements for automotive long range radar transmitters based on [14,15].

Parameter	Value	Unit
Phase noise PN	−94	dBc/Hz at 1MHz
Frequency ramps T_{ramp}	20	µs
Frequency range Δf	2	GHz
Antenna gain G_{ant}	20	dBi
Minimum output power $P_{TX,min}$	10	dBm

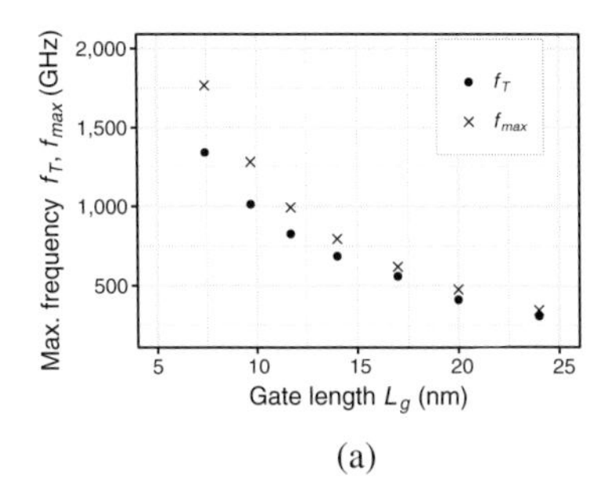

(a)

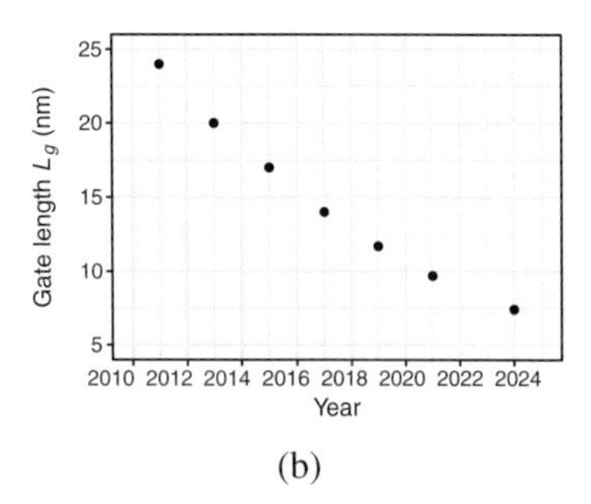

(b)

Figure 1.7 CMOS scaling as per the International Technology Roadmap for Semiconductors (IRS) [25] with (a) the expected f_T and f_{max} and (b) the projected RF-CMOS nodes.

half-pitch, 2022), considering the microprocessor unit (MPU)/application-specific integrated circuit (ASIC) MPU/ASIC technology roadmap data [24].

1.3 A Circuit Designer's Perspective

So we have a long wishlist for 5G, radar, and communication devices, and we are seeing many of our wishes coming to fruition in the standards bodies. The focus of this book is of course more on the circuits rather than the application. But we cannot design circuits without understanding the big picture first. This is why we start the book with an introduction to the system aspects of 5G (Chapter 2) as well for radar (Chapter 7). We also discuss the technology behind beamforming, MIMO, and massive MIMO in Chapter 3, phased arrays (Chapter 6), and full-duplex operation (Chapters 4 and 5). The rest of the book delves into circuit design details, including important topics related to frequency synthesis, power generation, and phased arrays. To achieve higher data rates, the industry has embraced more complicated modulation schemes such as 16-QAM, 64-QAM, and even 1024-QAM in 802.11ax. This requires extremely linear transmitters and receivers, and error vector magnitude (EVM) degradation will come from every source, especially the voltage-controlled oscillator (VCO)/phase-locked loop (PLL). High-frequency mm-wave VCOs are usually designed in stages, including a frequency multiplication stage and a more traditional PLL. But due to the high multiplication ratio from the crystal oscillator (XTAL), e.g., from 10 MHz to 28 GHz means the phase noise increases by nearly 70 dB, requiring careful optimization of the synthesized signal. The VCO will likely have lower noise than the multiplied reference, so the VCO design, the PLL loop bandwidth, and carrier recovery strategy require cooptimization. Many chapters of this book are dedicated to the understanding of these issues, in particular VCO design in Chapter 12, frequency synthesizer design in Chapter 10, and the all-digital PLL approach covered in Chapter 11.

As alluded to earlier, mm-wave systems need beamforming to improve the SNR and to overcome the high path loss, but this is actually a great benefit of these systems, as they constrain energy to a beam rather than spreading the energy out over a wide area. This means more efficient transmitters and less interference, allowing higher spatial multiplexing and reuse of spectrum in adjacent sectors. Beamforming and MIMO technology are described in detail in Chapters 3, 6, and 9, with an emphasis on the circuit, package, and antenna side provided in Chapter 9.

Another well-known barrier to moving to higher frequencies is the poor efficiency of power amplifiers. In Figure 1.8, we show the PA efficiency and output power from silicon technology (SiGe and CMOS), which has some obvious and alarming trends [26]. The output power drops significantly and so does the efficiency. One saving grace is that directional communication improves the efficiency by focusing it in a narrow beam, rather than spreading it omnidirectionally or over a wide sector, and the sharper the beam, the more efficient the link. But nevertheless, to overcome the high path loss at mm-wave frequencies requires higher power, and single-digit efficiency numbers are very bad. Keep in mind that the actual efficiency is the average back-off efficiency, which is at least 3 to 6 dB back-off from the 1 dB compression point, depending on the modulation scheme and the number of users being served. Class B amplifiers have a linear back-off efficiency with output voltage amplitude, whereas Class A amplifiers have an efficiency that degrades with output power. So a 6 dB power back-off means the efficiency is 4 $\times$ lower, going from, say, 20% to 5%. There are two approaches to solve the PA transmitter problem. One approach is to use the best technology at hand, say, INP/GaAs or newer GaN devices that have much higher output power capability and better efficiency. PAs with a power density exceeding 10 W/mm have been demonstrated at 40 GHz in GaN with 33% efficiency [27]. The other approach is to use a larger array so that the power per element can fall down to levels served by CMOS. We expect that both approaches will find their place in practice, and we cover CMOS PA design at high frequencies in Chapter 13.

Finally, without new technology nodes to address these high frequencies, 5G would be nothing more than hot air. Technology advancements pushed CMOS and SiGe from an RF technology to one of the finest mm-wave technologies, allowing several dozen front-ends to be integrated onto a single chip. In Chapter 14 we will discuss the latest generation of Fin Field-effect transistor (FinFET) CMOS devices and their RF performance. Throughout the book, we also cover technology aspects of circuit design, both passives and actives, and how they impact the design of building blocks. Thanks to these

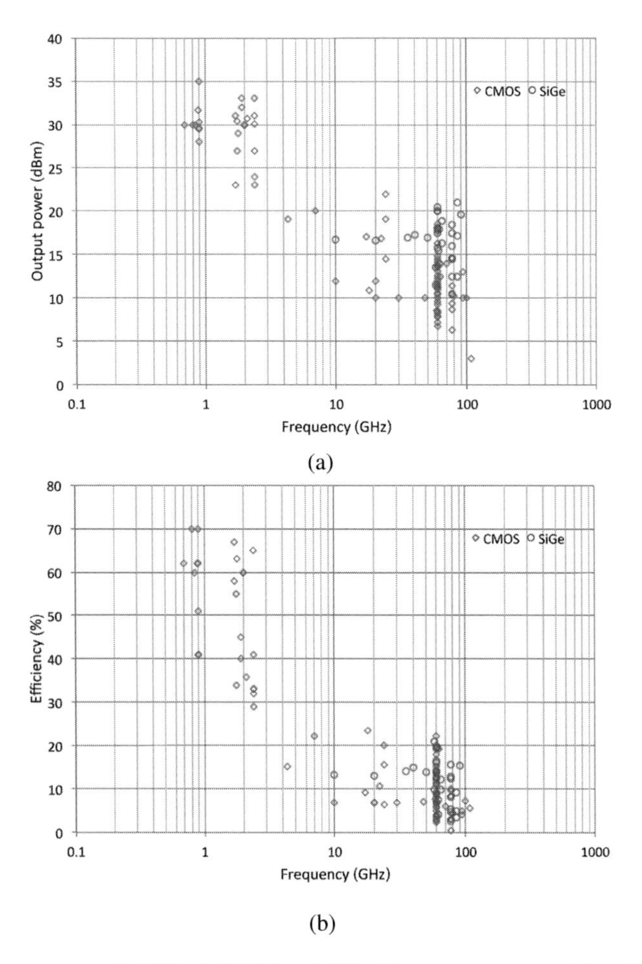

Figure 1.8 Published CMOS and SiGe (a) output power and (b) efficiency versus frequency [6].

technology advancements, large arrays are no longer only in the purview of the military, but will become a mass technology. As an RF and system engineer, you are responsible for making this dream a reality, so please pull up your sleeves and get ready to get your hands greasy.

References

[1] C. J. Zhang, "Realizing Massive MIMO in LTE-Advanced and 5G," The Brooklyn 5G Summit, 2015. [Online]. Available: https://ieeetv.ieee.org/ieeetv-specials/brooklyn-5g-2015-realizing-massive-mimo-in-lte-advanced-and-5g

[2] IMT Vision, "Framework and overall objectives of the future development of IMT for 2020 and beyond," in *18th Meeting of Working Party 5D*, Ho Chi Minh City, Viet Nam, February 2014.

[3] *Using LaTeX for Your Thesis*. [Online]. Available: www.youtube.com/watch?v=r9JZgbZ1pVk

[4] *4G*, Wikipedia. [Online]. Available: https://en.wikipedia.org/wiki/4G

[5] *LTE Advanced (LTE-A)*, Wikipedia. [Online]. Available: https://en.wikipedia.org/wiki/LTE_Advanced

[6] A. M. Niknejad, S. Thyagarajan, E. Alon, Y. Wang, and C. Hull, "A circuit designer's guide to 5G mm-wave," in *Custom Integrated Circuits Conference (CICC), 2015 IEEE*, IEEE, 2015, pp. 1–8.

[7] *Short Range Devices; Transport and Traffic Telematics (TTT); Ultra-Wideband Radar Equipment Operating in the 24,25 GHz to 26,65 GHz Range; Harmonised Standard Covering the Essential Requirements of Article 3.2 of Directive 2014/53/EU*, ETSI Std. ETSI EN 302 288, Rev. 2.1.1, May 2017. [Online]. Available: www.etsi.org/deliver/etsi_en/302200_302299/302288/02.01.01_60/en_302288v020101p.pdf

[8] *Operation within the Bands 902-928 MHz, 2400-2483.5 MHz, 5725-5875 MHZ, and 24.0-24.25 GHz*, FCC Std. 47 CPR 15.249.

[9] *Operation of Wideband Vehicular Radar Systems within the Band 23.12-29.0 GHz*, FCC Std. 47 CPR 15.252.

[10] *Short Range Devices; Transport and Traffic Telematics (TTT); Short Range Radar Equipment Operating in the 77 GHz to 81 GHz Band; Harmonised Standard Covering the Essential Requirements of Article 3.2 of Directive 2014/53/EU*, ETSI Std. ETSI EN 302 264, Rev. 2.1.1, February 2017. [Online]. Available: www.etsi.org/deliver/etsi_en/302200_302299/302264/02.01.01_30/en_302264v020101v.pdf

[11] *Operation of Level Probing Radars within the Bands 5.925–7.250 GHz, 24.05–29.00 GHz, and 75–85 GHz.*, FCC Std. 47 CPR 15.256.

[12] K. Ramasubramanian, K. Ramaiah, and A. Aginskiy, "Moving from legacy 24 GHz to state-of-the-art 77 GHz radar," Texas Instruments, Whitepaper, 2017. [Online]. Available: www.ti.com/lit/wp/spry312/spry312.pdf

[13] T. Weber, "Frequency regulation – global update for automotive radars," in *Conference on New Features for Automotive Radars*, Stuttgart, Germany, February 2018. [Online]. Available: https://cept.org/files/10899/Presentation%20Frequency%20regulation%20update%20Automotive%20Radars%2020-22%20February%202018%20Stuttgart.pptx

[14] R. Stuhlberger, "RFIC concepts for future integrated automotive radar sensors," in *Workshop on "RFIC Design for Automotive Radar" at 2018 IEEE Radio Frequency Integrated Circuits Symposium*, IEEE, June 2017.

[15] J. Lee, "Circuits and systems of millimeter-wave automotive radars," in *Workshop on "RFIC Design for Automotive Radar" at 2018 IEEE Radio Frequency Integrated Circuits Symposium*, IEEE June 2017.

[16] "DOTSEVEN: towards 0.7 terahertz silicon germanium heterojunction bipolar technology," EU FP7 Funding Project. [Online]. Available: www.dotseven.eu/

[17] "TowARDs Advanced Bicmos NanoTechnology platforms for rf and thz applicatiOns," EU ECSEL Funding Project. [Online]. Available: https://cordis.europa.eu/project/rcn/210525_en.html

[18] "RTN7735PL 77GHz radar transmitter," Datasheet, Infineon Technologies.

[19] "RASIC RXS8160PL: radar 77/79 GHz RF millimeter wave IC," Datasheet, Infineon Technologies.

[20] "MR3003 high-performance 77 GHz radar transceiver," Datasheet, NXP Semiconductors.

[21] D. Guermandi, Q. Shi, A. Dewilde, V. Derudder, U. Ahmad, A. Spagnolo, I. Ocket, A. Bourdoux, P. Wambacq, J. Craninckx, and W. V. Thillo, "A 79-GHz 2 × 2 MIMO PMCW

radar SoC in 28-nm CMOS," *IEEE Journal of Solid-State Circuits*, vol. 52, no. 10, pp. 2613–2626, October 2017.

[22] "TEF810X fully-integrated 77 GHz radar transceiver," Datasheet, NXP Semiconductors.

[23] B. P. Ginsburg, S. M. Ramaswamy, V. Rentala, E. Seok, S. Sankaran, and B. Haroun, "A 160 GHz pulsed radar transceiver in 65 nm CMOS," *IEEE Journal of Solid-State Circuits*, vol. 49, no. 4, pp. 984–995, April 2014.

[24] *2012 ITRS Tables ORTC*, ITRS Std., 2012. [Online]. Available: www.itrs2.net/2012-itrs.html

[25] *2012 ITRS Tables RFAMS*, ITRS Std., 2012. [Online]. Available: www.itrs2.net/2012-itrs.html

[26] A. M. Niknejad, D. Chowdhury, and J. Chen, "Design of CMOS power amplifiers," *IEEE Transactions on Microwave Theory and Techniques*, vol. 60, no. 6, pp. 1784–1796, 2012.

[27] U. K. Mishra, L. Shen, T. E. Kazior, and Y.-F. Wu, "GaN-based RF power devices and amplifiers," *Proceedings of the IEEE*, vol. 96, no. 2, pp. 287–305, 2008.

2 5G Transceivers from Requirements to System Models and Analysis

Aarno Pärssinen

Implementation constraints coming from Radio Frequency Integrated Circuit (RFIC) technologies and other RF components set boundary conditions for new radio systems that try to stretch data rates, power consumption, cost and range to new extremes. Some of these techniques may be absolutely needed for optimal solutions, but in many cases achieving sufficient performance in the extreme conditions may challenge research community for a long time. Requirements for RFIC solutions are evaluated using selected examples for some of the key design aspects in 5G.

2.1 RF Requirements Inspired by 5G System Targets

Motivation for the research and development of fifth-generation wireless communications systems is coming from several targets that specifically include higher data rates, more reliable connectivity, and faster response times over the network. This, of course, should be done with better system efficiency and scalability to different use-case requirements from extremely high data rates to fast real-time control applications. Framework for the future targets is globally consolidated in International Telecommunications Union (ITU) vision "IMT for 2020 and beyond" [1]. The targets of the vision are summarized in Table 2.1. Those will not directly give specific requirements for RF design but guidelines that can be translated to more detailed goals. In this chapter, challenges and more practical RF design constraints will be considered that address the goals of making forthcoming cellular radio systems 10 or even a hundred times better than current state-of-the-art based on Long Term Evolution Advanced (LTE-A) [2]. 5G systems that are addressing ITU targets will be deployed both at conventional frequency bands below 6 GHz, at sub-mm-wave (<30 GHz) and mm-wave (here 30–100 GHz) bands. Frequency range above 20 GHz is of interest in this book.

Targets as given in Table 2.1 are of great importance for many very different applications. Therefore, we should not assume that all of these stringent targets will be met in one specific use case at the same time. Instead, there is at a system level intention to address the scalability issue such that very different applications should be able to use the same network and many times also exactly the same infrastructure with appropriately scaled requirements for different classes of devices operating in the network. Some extremes of applications include interactive and personalized high-definition live streaming requesting very broadband and low latency, and at the other end various

Table 2.1 IMT-2020 system capability targets against IMT-A.

System parameter	IMT-A	IMT-2020	Impact to RF design
Peak data rate (Gbps)	1	20	Bandwidth, carrier frequency, EVM
User experienced data rate (Mbps)	10	1000	Range, noise, spectrum sharing, EVM
Spectrum efficiency	1×	2×–5×	EVM, linearity, noise, spatial/spectral filtering
Mobility (km/h)	350	500	Timing
Latency (ms)	10	1	Digital signal processing, protocols
Connection density (x/km^2)	10^5	10^6	Interference
Network energy efficiency	1x	100x	From system to RF circuit details
Area traffic capacity ($Mbps/m^2$)	0.1	10	Interference

sensor or monitoring activities that occasionally send a small amount of data but call for very long battery lifetime and sometimes also very good reliability. In addition, there are more generic goals regarding efficiency of spectrum and energy usage that need to be understood holistically when different applications are utilizing the wireless network and specifically depending on the type of device operating in the network. In principle, many RF requirements and specifications can be extracted from the top-level goals in a relatively straightforward manner. However, the feasibility to achieve the performance goals with reasonable power consumption and cost is a problem that is far from trivial and leads sometimes to controversy between different system-level targets. This chapter will highlight some key trade-offs taking into account many physical constraints where RF processing sets stringent boundary conditions to performance especially when cost and power consumption are carefully considered. On the other hand, some requirements are not very relevant for RF performance behavior or the impact is limited only to time accurate control.

We can understand RF circuitry from a data processing perspective as a bit pipe with very small processing delay. For example, low latency is one of the key parameters for real-time operation in various applications from reliable, remote control of factory operations to gaming utilizing virtual reality. The challenge to achieve those goals is more related to higher layers of radio processing, networking, and coding of the application. RF delay coming from filter group delay and also RF-related digital processing in the L1 bit pipe are practically negligible in this equation. Also mobility applies mostly to digital processing related to the Doppler effect but also to beam search and tracking. Of course, RF timing accuracy and beam control accuracy in phased arrays can't be fully omitted in case of mobility, but in practice the delays originating from actual RF processing are small and can be considered as normal control functions that should have appropriate time synchronization anyway. Focus in this chapter will be mostly on peak data rate and link range as well as capacity of a single radio connection. Those have been key drivers through several generations of broadband wireless access both in local area and in cellular networks. However, other parameters, including connection density, area traffic capacity, and total spectrum efficiency (beyond single connection), are also heavily related to radio interference and linearity requirements. Determining specifications for those is a more complex procedure requiring very comprehensive simulations at the network level that are rarely done for a wide range of various frequency channels. For example, due to that traditionally intermodulation specifications have been based on simpler worst-case scenarios where known worst-case interferers are placed at a certain distance from the receiver. In mm-wave communications, highly directive links will make the scenario very different compared to conventional assumption of omnidirectional antennas.

2.2 Radio Spectrum and Standardization

Discussion of the availability of radio spectrum at bands beyond 6 GHz for mobile 5G use was initiated as the capacity at lower bands has been intensively utilized at unlicensed, license-exempt, and licensed bands. The spectrum at millimeter wave for personal communications was first-time utilized at 60 GHz ISM band with 802.11ad standard for broadband communications up to 7 Gbps data rate [3]. Due to absorption peak around 60 GHz, the band is not very suitable for long range but can serve well local-area indoor use cases. Further development toward higher data rates is currently ongoing in 802.11ay for license-exempt bands above 45 GHz [4].

For licensed use, World Radiocommunication Conference 2015 (WRC-15) started to consider new frequency allocations for IMT in worldwide scale at a frequency range between 24.25 and 86 GHz [5]. Several frequency bands were named for further allocation studies within the range. The studies are to be completed before 2019 conference (WRC-19). However, there has been significant pressure to speed up 5G development and therefore find frequency allocations at a faster pace. Focus is on frequency regions that can be allocated at least nationally for the purpose, and in many cases, but not always, they are aligned with WRC-15 study bands. In standardization, 3GPP has been already taken opportunity for new bands into account, and the first version of standard for 5G New Radio (NR) physical (PHY) layer has been approved and made public in TS38 series [6] in December 2017. In addition to several bands below 6GHz that are targeted for 5G operations, three bands in the range between 24 and 40 GHz have been defined for sub-mm-wave and mm-wave operations using time-division duplexing (TDD). 3GPP has agreed to adopt the same waveform based on Orthogonal Frequency Division Multiplexing (OFDM) and scalable numerology in both uplink and downlink up to at least 52.6 GHz [7].

The first 60 GHz wireless routers have been in the market for a while based on 802.11ad technology. However, along 5G technology much more spectrum at mm-wave regime will be taken to use for wireless applications. In the first phase, available bands up to 40 GHz are of the greatest interest in cellular networks for connections with user equipment while higher bands will be considered later. For wireless backhaul and fixed links, several bands, for example 71 through 76 and 81 through 86 GHz, are available already at mm-wave. The wide range of spectral opportunities means that in RF design not only programmability at a single band is of interest. Scalability over a large frequency range needs to be understood when 5G system and the implementation impacts will be considered. Although many techniques that are developed for radios operating below 6 GHz could be considered, their adaptation to mm-wave range requires different approach. For example, use of integrated resonators may become more common to achieve decent gain at mm-wave. But design trade-offs differ due to frequency-dependent Q-values of lumped elements and better opportunity to use transmission lines when frequency increases. If we only look at physical and technology constraints over a couple of octaves from 24 GHz up to 86 GHz, we can observe very different trade-offs in overall form factor, transistor performance including power delivery capability and noise performance of each technology of interest, and link range that could be achieved per antenna element. In addition, very wideband signals from several hundreds of MHz up to 1 GHz or even more will also lead to significant digital processing payload and demand for powerful and potentially power-hungry A/D- and D/A converters. Despite well-known issues due to large envelope content of OFDM signals [8], the benefits at system level in terms of scalability and spectral efficiency seem to overcome RF- and PHY-related challenges in 5G waveforms. This means that additional dynamic range due to envelope content for all RF processing elements needs to be taken into account similarly as in LTE, including impacts both to crest factor and to SNR or error vector magnitude (EVM) requirements of higher-order modulations. These aspects will not cause performance trade-offs only in the transmitters due to PA linearity but also in receivers or generally in different blocks in transceivers as discussed in [9] and [10]. Therefore, also a standardization proposal to recognize performance degradation as a function of frequency in receiver noise figure due to physical constraints of semiconductor circuits has been made in 3GPP [11].

A new aspect in standardization at sub-mm-wave and mm-wave communications is related to the directivity of the communications. Therefore, whenever power limits of the transmitter needs to be defined, effective isotropic radiated power (EIRP) instead of total radiated power (TRP) will be used as a measure [6]. In phased arrays, the other new aspect is measurement method. Conducted measurements become difficult if not impossible for a very large number of antennas. Therefore, testing needs to be done in most cases using over-the-air (OTA) techniques. That results in new challenges in prototyping and when defining the specifications. Depending on the scenario, tests can be done to the direction of interest, or especially in interference tests as total radiated interference power integrated over the whole sphere. Therefore, testing method is of importance when studying specification documents. The other challenge is very flexible numerology in 5G NR parametrization. That applies to subcarrier spacing, type of the device, etc. Therefore, it is not possible to show in a simple table specification requirements in a manner that would give a thorough view of the scalability that 5G systems may provide. In the first version of the RF specification from December 2018, channel bandwidths up to 400 MHz were determined, but one can expect those to be extended in the future. For sub-mm-wave and mm-wave operating bands, the following three bands are defined for NR: 26.5 through 29.5 GHz, 24.25 through 27.5 GHz, and 37.0 through

40.0 GHz. All of them are TDD bands. Also, an option for carrier aggregation (CA) at these bands is already being discussed. This means that many new bands and details will be seen in the future when 5G NR is evolving to a mature standard. However, the first version gives already a solid view of the approach and many key parameters. In a later section, there are examples of the 5G specifications as well as a case study that is done for a wider bandwidth than the first version of 5G NR will support. They will give good understanding of some of the key parameters and are valid for the analysis taking into account the scalability of the system now and in the future.

2.3 System Scalability

One major challenge of wireless systems we have today is the complex set of various parameters from signal waveform to signal processing nonidealities in any RF processing. As system scalability is one of the key targets in 5G communications, it is evident that properties and controls must be engineered such that they provide a broad set of options for system optimization. They need to be feasible and hopefully also sensible, taking physical limitations into account. Some earlier examples indicate many opportunities for major rethinking rescued for the future. Bluetooth (BT) could be extended to much longer battery lifetime in the Bluetooth Low Energy (BLE) extension with the original target being one-year battery lifetime using a single button cell battery [12]. That was achieved finally with reasonably small changes to RF performance requirements but even more so with significantly improved time domain control, optimizing both device discovery and active operation modes with RF performance and power consumption in mind. For example, the so-called advertisement mode, where communication was initiated by the lower-power device like a sensor node rather than by listening regularly, given that the master has a larger battery, appeared to be the more optimal solution for battery lifetime. Even more protocol and RF-level flexibility has been needed in the fourth-generation long-term evolution (LTE) systems in standardization in 3GPP [13]. The original specification required in practice a two-channel receiver in downlink for diversity and MIMO and scalability of channel bandwidths from 1.4 to 20 MHz. Later the communications speed was extended, increasing both bandwidth and the possible number of MIMO channels. To provide means of finding capacity flexibly from different parts of the radio spectrum, CA techniques are included to the specification. Different component carriers (CC) can be located at adjacent frequency channels (contiguous CA), other intraband channels (noncontiguous CA), or different bands (interband CA). This was later extended to more than two component carriers in the specification. Although from an RF implementation and power consumption perspective some of these features are far from being optimal, backward compatibility and practical aspects in global and regional spectrum management have led to scalability needs beyond RF optimality. This evolution is continuing in 5G specifications, especially at bands below 6 GHz but also giving directions at sub-mm-wave (or sometimes called as centimeter-wave [cm-wave]) and mm-wave regions.

In machine-to-machine (M2M), or what 3GPP is calling machine type communications (MTC), scalability to extremely long battery lifetime (up to 10 years as a target) and long range has called for new features in standards included as LTE-M (LTE for Machines) and LTE-NB (LTE narrowband) [14] versions. Opportunity for extremely long battery lifetimes is not realized in time domain scaling only. In addition, lower bandwidths down to 200 kHz, removal of the requirement of the second receive path, and operation in half-duplex mode in FDD bands are elements needed to meet the set targets of battery lifetime and cost. Hence, three different elements, i.e., enabling lower sampling speeds, minimizing RF hardware, and removing the strongest blocker (the device's own TX signal) facilitate a low-power RF design. Still one can not expect scalability down to microwatt range systems like those needed, for example, in implantable devices [15]. That is due to inevitable need for decent output power and a low-noise figure to achieve long link range, and tolerance to out-of-channel and out-of-band interferers as expected is in cellular networks.

The targets of future 5G inherently embed all that scalability and preferably extends it even more. But in many cases, like IoT for extremely long battery life, we need to assume strong inheritance from LTE legacy and sensibly limit certain applications to frequencies at the lower range of the spectrum, that is, at and below 2 GHz. That is feasible even for small form factors, as recent antenna designs have demonstrated feasibility of small form factors for wearable devices even at bands below 1 GHz [16]. Due to evident challenges achieving very low-power consumption at bands operating at tens of GHz, we will limit the discussion here mostly to systems targeting for very high throughput (in 3GPP called enhanced mobile broad band, eMBB). Therefore, is it assumed that scaling to lower data rates happens by mostly sharing a channel in time domain, scaling the bandwidth or the number of OFDM subcarriers (or actually resource blocks having several subcarriers each), or efficiently coding the symbols. Only time domain techniques can result in significant power consumption reductions in RF processing. Although some recent results indicate promising results in mm-wave receivers [17] it is quite far-fetched to assume that mm-wave bands could be used efficiently in applications requiring extremely long battery lifetimes.

Scalability in cm-wave and mm-wave wireless systems will require quite different approaches on top of existing techniques. Because range depends directly on the antenna aperture that is relative to physical size, scaling is practically a function of the carrier frequency and leads to a need for antenna arrays as antenna size per element scales down with the frequency. This leads to use of directive antennas and in many cases to phased arrays that provide new constraints and opportunities to scale the performance of the system beyond the typical assumption of an almost uniform radiation pattern of antennas at 6 GHz and below at least in mobile devices. Of course, directive antennas have already been used in base stations and other fixed infrastructure devices such as backhaul links and point-to-point. However, mm-wave communications demand even more extensive use of dynamic scalability, adding a new dimension of complexity to the radio system design, especially when mobility is taken into account. As scaling will have a strong impact on RF implementation, the following sections of the chapter are strongly motivated to analyze the impact of mm-wave scalability in detail, omitting many other aspects of scalability in 5G radio systems.

2.4 Communications System Model for RF System Analysis

Before studying the scalability aspects in detail, it is necessary to evaluate and address the accuracy of the used models and their impact on system performance. Furthermore, as the nature of the RF is nonlinear, the models have to depend on the input and output power levels. The latter is of specific importance as many simulators in the analysis of communications system include only the noise of the receiver and maximum output power of the transmitter. This approach is for many scenarios too simplistic for the RF behavior limiting the validity only to certain extremes in link distances and easily omits inevitable constraints of physical implementation at other power levels. On the other hand, communication systems typically try to model waveforms in an accuracy that goes well beyond the tolerances of RF components and thus leads to optimization that is not necessarily needed at the system level, as impact is smaller than inaccuracies coming from a too simple RF model. To bridge this gap, a highly abstracted model is needed that can provide necessary parameters for scalability but omit the fine-tuning of all the possible knobs in the same environment.

The system model will include baseband, RF, antenna, and channel propagation models as shown in Figure 2.1 [18]. Each of them are based on signal-to-noise ratio (SNR) and in the case of RF also on absolute power levels. Those can be split to smaller parts and refined one at a time if detailed analysis of specific insights will be needed. Abstraction level of different models can vary internally, but in the interfaces root mean square (rms) power levels will be utilized without phase relation. That simplifies the analysis and keeps the top level simple. In most cases, also internal models are just scalar numbers, but for example, definition of MIMO rank requires inevitably not only scalar amplitude information but also phase and delay parameters as a complex number in channel models if we go beyond ideal dual-polarized antennas in the line-of-sight (LOS) condition. Thus, we can for example search for different propagation paths in a selected environment in a separate analysis. And once all necessary paths for MIMO processing are found, it is possible to calculate RF transceiver specifications for each path separately and evaluate at which conditions those could be received with sufficient SNR.

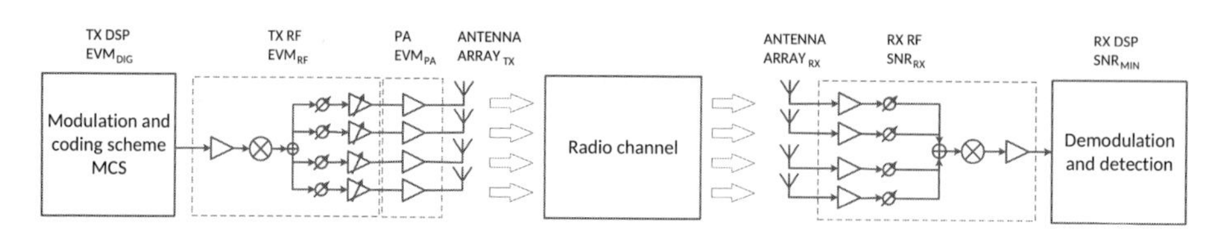

Figure 2.1 Communications system model. RF beamformer is a conceptual representation without data converters. Beamforming can be made either in analog or digital domain in the analysis. (©2017 IEEE. Reprinted, with permission, from *IEEE Transactions on Antennas and Propagation*.)

Table 2.2 Examples of possible modulation and coding schemes with minimum link-level SNR.

	SNR_{min} (dB)				
Coding rate	BPSK	QPSK	16-QAM	64-QAM	256-QAM
1/3	2.2	5.2	12.7	19.2	25.2
1/2	4.0	7.0	14.5	21.0	27.0
2/3	5.2	8.2	15.7	22.2	28.3
3/4	5.8	8.8	16.3	22.8	28.8
5/6	6.4	9.2	16.7	23.2	29.2
7/8	6.6	9.4	16.9	23.4	29.4
1	7.0	10.0	17.5	24.0	30.0

Although in mm-wave communications the number of radio paths is likely quite small in practice due to the propagation environment, one must not automatically assume that each orthogonal MIMO path will be constructed only from one directed beam at a time. Therefore, eigenmode analysis using, for example, singular value decomposition (SVD) is a necessary step in system analysis. The analysis typically assumes linearity, although nonlinearity in circuits may also have an impact on the final result.

The most significant simplification is that all digital baseband processing is abstracted to a single SNR value for each modulation and coding combination. If we assume that system can tolerate a bit-to-error rate (BER) of 10^{-3}, one can construct results as in Table 2.2. It is a total minimum SNR for the whole link, including both receiver and transmitter nonidealities as discussed later. In many cases, it is assumed that one of them is the limiting factor. However, in high SNR scenarios, that is not valid in general and assumption of the split between receiver and transmitter must be made. Packet-based wireless systems are also often specifying such a requirement either as packet error rate (PER) or throughput instead of BER. In those cases, an equivalent SNR can be defined for each specified signal scenario with separate system simulations using an appropriate communications system model for improved accuracy. However, for system analysis and first-order link range estimation, simple models using an average white Gaussian noise (AWGN) channel and repetitive coding are a sufficient starting point. At lower frequencies that have a rich scattering environment, a fixed fading margin on the top of a line-of-sight estimate is typically adopted. However, for phased arrays and directive beam patterns at mm-wave, the approach is more problematic as it won't take spatial filtering of the antenna arrays in a generic manner into account. However, it is not very likely that two separate propagation paths received by the same antenna array would have roughly the same amplitude and opposite phase causing steep multipath fading for the signal. Therefore, it is more appropriate to analyze individual propagation paths or a cluster of propagation paths between transmit and receive nodes and replace statistical propagation models with map-based approaches as in [19].

Maximum achievable data rate (R_{max}) for any coding scheme for an OFDM modulated data can be calculated using the following formula:

$$R_{max} = r_{code} N_{SC} f_{SCS} \log_2(M) \frac{t_{OFDM}}{t_{CP} + t_{OFDM}} \tag{2.1}$$

where r_{code} is the coding rate, N_{SC} is the number of subcarriers in an OFDM symbol, f_{SCS} is the subcarrier spacing, M is the modulation order (i.e., number of constellation points in the symbol), t_{OFDM} is the OFDM symbol length (inverse of subcarrier spacing), and t_{CP} is the length of the cyclic prefix (CP). As cyclic prefix does not contain payload data, the last term is calculating the ratio of the actual payload period to the total time of the OFDM symbol with CP. If we take 5G NR numerology into use for a 400 MHz channel BW, we can calculate maximum theoretical data rates for different modulations using the widest bandwidth defined so far for the 5G NR system. We use here 120 kHz subcarrier spacing. In that case, maximum number of resource blocks (RB) is 264 for a 400 MHz channel [6]. Each RB contains 12 subcarriers leading to a total of 3,168 subcarriers occupying bandwidth of 380.16 MHz. The length of the cyclic prefix is 0.583 μs. When using (2.1), we can calculate maximum data rates for different coding rates and modulations for the cases defined in Table 2.2. Results are shown in Table 2.3.

One can see immediately that a 10 Gbps maximum rate is a very challenging target for 5G even at 400 MHz bandwidth. Rank 4 MIMO with uncoded 64-QAM ($4 \times$ 12.13 Gbps) will be still below 9 Gbps. It is expected that in the future 3GPP must specify wider bandwidths, like in the 802.11 family, if the speed target is to be reached in practice. In Table 2.3, also 256-QAM modulation is shown, although the SNR requirement, especially for uncoded traffic, will be very difficult to achieve at mm-wave communications over the link, taking both transmitter and receiver nonidealities into account. It is obvious from Table 2.3 that the other important data rate target being 0.1 to 1 Gbps at cell edge is much easier to achieve from a waveform perspective. However, in addition network optimization has to consider range (i.e., path loss) also in non-line-of-sight conditions, interference, and overall system efficiency.

In the following sections, a slightly different parameter set is being used in the analysis with 900 MHz effective bandwidth. That is originated from earlier working assumptions for a 5G system and used, for example, in [18]. The two parameter sets are given for comparison in Table 2.4. They are very close in practice and therefore the analysis results given in this chapter can be applied to understand potential 5G signal bandwidths almost up to 1 GHz. Moreover, the analysis throughout this chapter includes

Table 2.3 Maximum achievable data rates for a 400 MHz OFDM channel based on 5G NR numerology and for 900 MHz uncoded data used in the examples.

Channel BW (MHz)	Coding rate	Maximum achievable data rate, R_{max} (Gbps)				
		BPSK	QPSK	16-QAM	64-QAM	256-QAM
400	1/3	0.12	0.24	0.47	0.71	0.95
	1/2	0.18	0.36	0.71	1.07	1.42
	2/3	0.24	0.47	0.95	1.48	1.90
	3/4	0.27	0.53	1.07	1.60	2.13
	5/6	0.30	0.59	1.19	1.78	2.37
	7/8	0.31	0.62	1.25	1.87	2.49
	1	0.36	0.71	1.42	2.13	2.85
900	1	0.76	1.53	3.06	4.59	6.11

Table 2.4 Comparison of Pre-5G and 5G NR parameter sets.

Parameter	Ref	5G NR	Unit
Subcarrier spacing	120	120	kHz
OFDM symbol length	8.33	8.33	μs
CP length	0.5	0.586	μs
Number of occupied subcarriers	7,500	3,168	
Number of nonoccupied subcarriers	692	924	
FFT size	8,192	4,092	
Channel bandwidth	900	380.16	MHz
Protocol efficicency	90	100	%

a protocol overhead of 10% for control and retransmissions in the data rate estimates for the 900 MHz channel. That overhead is included in data rate analysis without further analysis of protocol behavior in time domain.

The numbers given in Table 2.2 are SNR requirements for the whole link, and in practice this requirement needs to be shared between transmitter and receiver in any practical radio link. In standardization scenarios, this is solved such that transmitter EVM specifications are defined for some selected modulation and coding schemes (MCS) and then tested for transmitter only in test benches that have significantly better performance in the test equipment than in an actual receiver. Then performance is solely dominated by the transmitter. In case of the receiver testing, the transmitter MCS scheme or receiver input power level is set such that the transmitter will not have significant impact on the measured result. This is very practical approach that allows separate specifications for both ends of the link.

But if the complete data rate vs. range performance needs to be modeled in practical network conditions, the performance at both ends should be modeled at the same time. In practice, link adaption selecting the most appropriate MCS for each radio condition in a cellular network is a very complex procedure due to the high number of independent parameters. If we simplify assumptions and neglect most of the parameters including impact of co-channel and out-of-channel interference, a quite straightforward functional model for link behavior can be created. This can be utilized to estimate link range especially in high data throughput, high SNR cases, when both transmit and receive behavior must be addressed simultaneously. Split for a simple link adaptation model is done as shown in Figure 2.2. In the given model, LTE requirements for a 5G transmitter were taken from binary phase shift keying (BPKS) to 64-QAM modulation, because no requirements were set for mm-wave transmitters at the time of the analysis. 256-QAM performance for the higher data rates is defined such that somewhat balanced values can be defined for transmitter and receiver. EVM values are shown in Table 2.5, and they can be translated to SNR simply using

$$SNR_{TX} = 20\log_2(M)\left(\frac{1}{EVM}\right) \tag{2.2}$$

where EVM is given here as absolute value. Although EVM is fundamentally a vector, we assume here that it can be utilized as scalar (i.e., rms value) in the system analysis.

Table 2.5 Transmitter EVM requirements for the transmitter in the analysis.

Modulation	EVM (%)
BPSK	17.5
QPSK	17.5
16-QAM	10.0
64-QAM	5.0
256-QAM	2.4

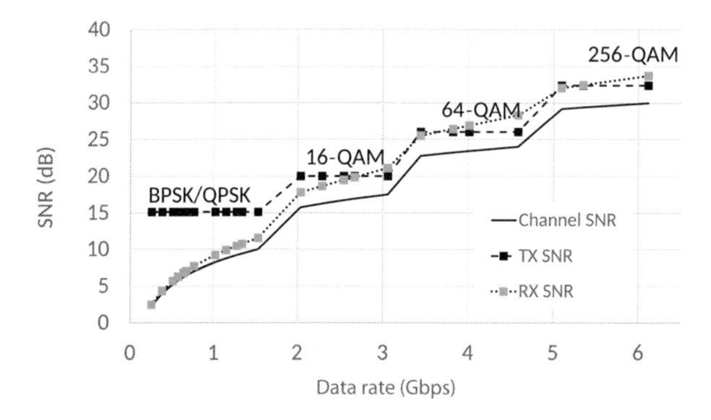

Figure 2.2 SNR requirements for the link and possible split between transmitter and receiver as a function of maximum achievable data rate for one stream transmission in the case of 900 MHz effective channel bandwidth. (©2017 IEEE. Reprinted, with permission, from *IEEE Transactions on Antennas and Propagation*.)

In a properly designed transmitter, EVM is mostly dominated by the PA nonlinearity taking into account the necessary crest factor for back-off and modulation. Those are then determining the maximum output power of a transmitter that can deliver certain power for each modulation. That level is independent of the coding. Therefore, a model where EVM and thus output power is fixed for each modulation makes sense. Of course, the effect of phase noise, etc., needs to be taken also into account, as explained later. Then the receiver SNR and coding scheme will determine the maximum data rate and range for each modulation and coding combination. For example, minimum link SNR for each MCS given in Table 2.2 leads to the specific data rate in Figure 2.2 for 900 MHz channel bandwidth. At the same time, we can calculate the SNR requirement for the receiver based on the SNR requirements of the link and transmitter according to

$$SNR_{link} = \frac{1}{\frac{1}{SNR_{TX}(P_{TX})} + \frac{1}{SNR_{RX}(P_{RX})}} \tag{2.3}$$

where transmitter and receiver SNR values (in absolute scale) are dependent on the absolute output and input power levels, respectively. Minimum receiver SNR values in Figure 2.2 are calculated based on that. We can see that at low data rates, transmitter specification contributes very little to link SNR, but as the data rates increase, SNR requirements will become difficult to achieve at both ends of the link. Therefore, proper balance between transmitter and receiver nonidealities is a necessity when extreme throughput performance is searched. As SNR is a highly implementation-dependent quantity, the performance balance depends also on the specific application. In cellular systems, we can expect somewhat better performance from the base station compared to mobile as it has a fixed power source and thus a less stringent power budget. However, the most dominant difference is typically higher output power in the downlink transmission that will allow also higher data throughput for the same distance in downlink compared to uplink when similar spectral resources (e.g., MIMO rank) are available. Especially in TDD systems, proper symmetry is essential at least for pilots determining the radio channel properties due to reciprocal channel assumption.

2.5 System-Level RF Performance Model

As given in (2.3), SNR is dependent on the absolute power level in the radio interface. Relation is fundamentally complex as it includes internal partitioning of the receiver and transmitter and all physical nonidealities from noise to nonlinearity and quantization in analog and digital signal processing. It should have also gain control functionality embedded to the model for the widest achievable dynamic range both in the transmitter and in the receiver. Fortunately, in properly designed transceivers, the actual SNR behavior as a function of the power follows closely to a rather limited set of constraints that are different at low, medium, and high signal levels. We will discuss some of these aspects before link range for any radio path can be determined, taking into account power-dependent SNR, physically constrained antenna properties, path loss, and of course bandwidth of the transmitted channel.

Physical implementation constraints of different RF blocks can be summed to an SNR-based approach as a function of absolute power levels based on rms values. The model neglects phase response or other very fast time domain effects similarly as an rms-based cascade rule for calculating a third-order intercept point of a receiver or a transmitter. Also in this case, it is a reasonable assumption. When many uncorrelated or weakly correlated parameters are summed up we don't have to take the worst-case assumption of correlated phases into account. Model abstraction can be made, depending on the needs, in very different ways, but here the focus is on the simplicity and capability of bridging the gaps to other parts of the system. That enables codesign of RF processing as part of a complete wireless system. The SNR model naturally embeds also all distortion elements.

2.5.1 Transmitter

In general, both the transmitter and the receiver models are split to three regions: (1) noise limited performance, (2) nonidealities linearly dependent on the signal level, and

(3) the region where nonlinear effects dominate. However, in the case of the transmitter, the noise-limited region may be practically nonexistent if the power control requirement is reasonable. Region (2) includes, for example, in-channel phase noise, in-phase/quadrature (I/Q) mismatch, and quantization effects for fixed amplitude signal processing in the digital domain. In the case of the transmitter, the nonlinear region requires a complex modeling approach, especially in the design phase when amplitude modulation (AM)-phase modulation (PM) and AM-AM distortion and in many cases also memory effects, need to be considered in detail. However, once that is done we can plot EVM for the modulated signal as a function of the output power. The system-level model depends of course on the accuracy and validity of the behavioral model in specific cases, but at the same time it provides a straightforward approach to use different abstractions from behavioral model to experimental circuit in the same manner. With this approach, we can simplify the transmitter model to two regions and transition between them. At low signal levels, EVM is dominated by the quantization of the digital signal and other linearly signal-dependent nonidealities. This gives the best possible EVM that a transmitter can achieve. At the transition region, nonlinear effects will pick up, and at high signal levels nonlinear distortion dominates the performance. The system model in EVM can be given as

$$EVM_{TX} = \sqrt{EVM_{DIG}^2 + EVM_{RF}^2 + EVM_{PA}^2(P_{TX})} \tag{2.4}$$

where EVM_{DIG} and EVM_{RF} represent digital and analog nonidealities, respectively. $EVM_{PA}(P_{TX})$ includes all output-level dependent nonlinearities. In this model, the PA driver and other nonlinear effects of the transmitter are embedded in the PA for simplicity. As the PA term should dominate the properly partitioned transmitter, instead of the driver stage, this is in many cases a sufficient approximation. Of course, a more detailed model can include both nonlinearities separately, but in that case also the absolute gain of the PA must be determined separately, which is immediately even more dependent on specific implementation. The two first terms in (2.4) are assumed to be fixed for any signal level because in many cases digital-to-analog (D/A) converter output level is fixed and phase noise and possibly I/Q imbalance dominate the rest of the RF chain in the transmitter. In the case of digital power control, this is not fully valid. But if the EVM for the signal level is determined to be correct for the most complex modulation (largest SNR requirement), then error for other signal types is likely small in all practical circumstances. An example of a transmitter model is shown both as SNR and EVM in Figure 2.3. SNR illustrates better system-level impact, as EVM given in percentage is the typical way of categorizing PA performance. In the example, digital EVM is 1%, RF nonidealities are assumed to contribute 0.75 %, and PA has power-dependent behavior that is based on a commercial device data sheet [20] with an estimated front-end implementation margin of 3 dB for losses, etc., subtracted from the output power. The example describes a high-power device in the sub-mm-wave range. However, similar behavior is generically applicable to any power class where PA is the only dominant source at high-power levels while other nonidealities dominate in back-off conditions.

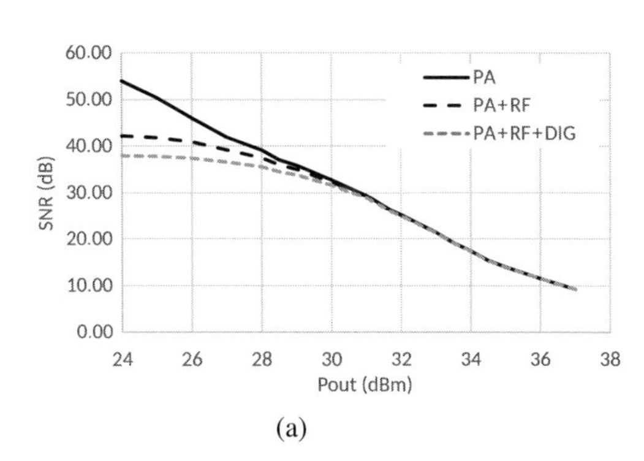

(a)

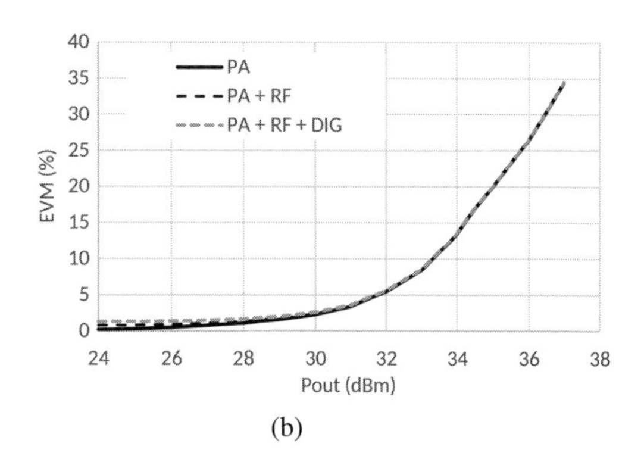

(b)

Figure 2.3 Transmitter nonidealities as a function of the PA output power (a) in SNR and (b) in EVM. The former shows also impact of different nonidealities in the signal path from PA to digital. (©2017 IEEE. Reprinted, with permission, from *IEEE Transactions on Antennas and Propagation*.)

2.5.2 Receiver

Radio receivers are to certain extent more complex to model because out-of-channel interference depends on external sources that can't be controlled comprehensively by the link or even the whole communications system on its own when several operators are occupying adjacent channels within the same band. Adjacent channel behavior can be analyzed in advanced network-level simulators, but taking all possible harmonics and nonlinear combinations into account makes practical analysis very complex. Therefore, requirements for nonlinearity in receivers are typically based on a limited set of *worst-case* test scenarios in standards. Also in link-level analysis, parameters defining selectivity can be initially omitted and considered separately as special cases. Of course, this may not be fully true in densely populated networks even if the links are directive by nature. But we assume here that the contribution to the system performance is in most out-of-channel scenarios negligible and the focus on link-level performance is only at the carrier frequency of interest. This also leads to a rather straightforward receiver model. In many radio link models, receiver performance is typically modeled only with thermal noise as

$$SNR_{RX,n} = P_{RX}(\text{dBm}) - \left(10\log_{10}(kTB) + NF(\text{dB})\right), \tag{2.5}$$

where P_{RX} is the received power in dBm, k is the Boltzmann's constant, T is the temperature (typically 290 K), B is the system/modulation bandwidth, and NF is the cascaded noise figure (NF) of the whole receiver. This is a valid assumption at low signal levels, but as the signal level gets higher, SNR peak dominated by other nonidealities will be reached and therefore a noise-only model can give too optimistic results, especially when high SNR, i.e., high throughput, is modeled. The other nonidealities include once again phase noise, I/Q mismatch, etc. As those are signal-level-dependent quantities, their combined effect will be seen as a maximum level of SNR that can't be exceeded.

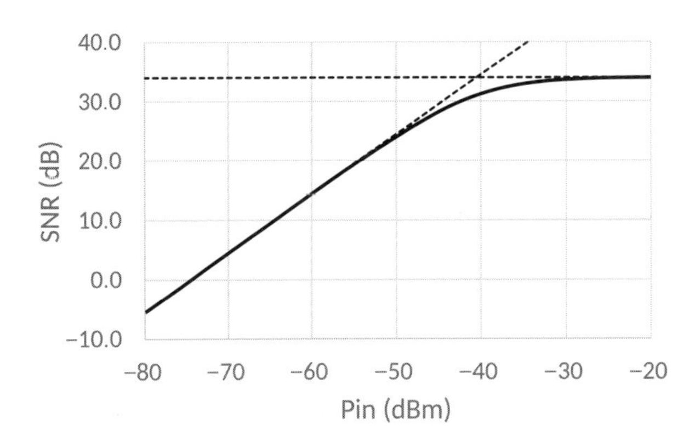

Figure 2.4 Receiver SNR model.

If we assume that receiver compression will not become an issue and automatic gain control (AGC) gain steps have almost negligible effect on SNR, the receiver model can be simply expressed as

$$SNR_{RX} = \frac{1}{\frac{1}{SNR_{RX,n}(P_{RX})} + \frac{1}{SNR_{peak}}}, \tag{2.6}$$

where SNR_{peak} is the maximum achievable SNR. SNR_{RX} includes also analog-to-digital converter (ADC) quantization noise that is typically almost negligible with proper gain partitioning. An example of a receiver model is given in Figure 2.4 for 900 MHz effective channel bandwidth, 10 dB noise figure, and peak SNR of 34 dB.

The impact of the RF modeling approach has been studied by comparing maximum link range for different data rates in [21]. Parameters in the analysis are slightly different from those in the other case of this chapter, but similar conclusions can be drawn. Three different models were compared. Model 1 in Figure 2.5 describes the case where the transmitter has been fixed and with the highest possible output power that can be used to achieve the maximum SNR for the highest order of modulation. In the receiver, only the noise figure limits the performance. Model 2 sets the peak SNR for the receiver as in Figure 2.4. We can observe that at very high data rates, the link range reduces drastically as expected because receiver peak performance contributes significantly to link SNR and thus directly to link margin impacting the range. Model 3 has scalable EVM as a function of transmitter output power according to Figure 2.3. In that case, we see a much larger link range at low data rates because we can push more power through from the same PA as the EVM requirement is relaxed. This example shows that with overly simple RF transceiver models, system-level modeling can easily lead either to too optimistic or too pessimistic conclusions. However, the models can be improved with a very limited set of key RF parameters that dominate the performance at various power regions. It also shows the impact of link adaptation to the system performance. Link adaptation is a cross-layer optimization process with a large set of parameters and highly abstracted view to RF performance. Therefore, results as in Figure 2.5 are not directly applicable to a generic case. However, this analysis confirms the value of

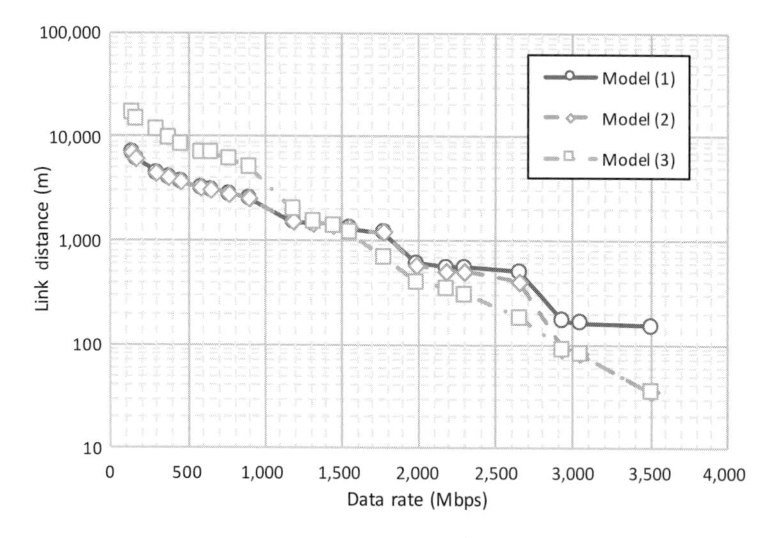

Figure 2.5 Comparing different RF modeling approaches for link distances as a function of data rate [21]. (©2016 IEEE. Reprinted, with permission, from *Proceedings of the European Microwave Conference [EuMC].*)

modulation and coding scheme–dependent power control if other parameters such as cochannel interference are not dominant for the channel capacity.

Before going to link range analysis, we can already make some observations of the challenges that 5G will face when 10 Gbps is a target. First, SNR targets for both transmitter and receiver will become very challenging if 256-QAM modulation is adopted as in Figure 2.2. The transmitter and receiver models as shown are made such that 256-QAM would be feasible. However, achieving that is definitely not straightforward at the mm-wave range. Especially, frequency synthesizers targeted for mm-wave transceivers have suffered from relatively low SNR performance rarely exceeding 30 dB. Recent work has reported EVM of almost –35 dB for a 28 GHz system for a phased lock loop targeted for sliding intermediate frequency (IF) architecture [22]. This indicates that two PLLs at both ends of the link will dominate the performance and all other nonidealities should have a much smaller contribution. This is not an easy target, especially in mobile devices. Therefore, 256-QAM might be possible in the future, but a minimum SNR of 64-QAM with some coding gain is likely a more feasible target for the highest achievable data rates when the first commercial 5G systems are launched. This would mean more than five orthogonal MIMO channels at 400 MHz bandwidth for 5G NR to achieve 10 Gbps. The target is definitely not easy and predicts a need to scale up the bandwidth in the future.

2.5.3 Antenna Array

In mm-wave systems, antenna pattern generation, and thus also system-level modeling differ significantly from the lower frequencies in cellular systems. At low GHz range, sectored antennas are de facto in a base station but mobile side assumptions, at least in simple models, rely on omnidirectional radiation patterns. At higher frequencies, decent link range requires improved antenna gain to compensate for the higher propagation loss per antenna element due to reduced aperture of the radiating element. Also implementing omnidirectional antennas in mobile platforms is a practical problem. Hence, directive antennas at both ends of the link should be considered. That results in additional complexity in system analysis, or at least more detailed assumptions should be made on how the system actually works in practice. The first aspect is related to beam tracking and alignment. Beam scanning is a challenge on its own in this context, and we assume that it has been done successfully for all available beams using pilot signals with low SNR requirements. This section is based on the assumption that we know the direction of all beams precisely. Of course, some implementation margin is required as there might be some errors in direction estimates as well as how precisely the beams are aligned. For example, if we use the typical 3 dB aperture width assumption for the beam precision, the worst-case scenario leads to 6 dB loss in link budget. That is large and likely not acceptable for the system. Therefore, 1 dB (or even smaller) loss per beam is a more appropriate requirement leading to 2 dB total implementation margin in the link budget. The most practical way to manage beam alignment error is using an implementation margin in simple system models. In addition, the impact on the precision requirement of the phased array and beam direction estimate is heavily dependent on the size of the array. Beam pattern of a linear array is

$$E(\Theta) = \sum_{i=1}^{n} a_i e^{-j\left(i2\pi\frac{D}{\lambda}\sin\Theta + \Phi_i\right)} \tag{2.7}$$

where a_i is the amplitude excitation of the ith antenna element, D is the distance between antenna elements, λ is the wavelength of the signal, Θ is the direction of the input or output signal, and Φ_i is the phase steering angle of the ith antenna element as shown in Figure 2.6. The precision requirement increases when we move from simple beamformers toward massive MIMO and so-called pencil beams. That can be visualized when comparing 4 and 16-element linear arrays having typical $\lambda/2$ distance between elements in Figure 2.7. However, as the number of elements is very large in massive MIMO systems, the uncorrelated errors between signal paths will average out. Hence, the beam precision requirement does not directly impact component precision but actually relaxes that.

Spatial filtering is an essential mechanism to separate data signals in dense networks. Therefore, an even larger number of elements in horizontal (or elevation) dimension might be needed in the future, leading to very narrow pencil beams, especially at very high frequencies. This leads to tight requirements both for precision in beam tracking and in synthesis of optimized beam patterns for multiple signals based on beam search results. From a hardware perspective, these two are to a certain extent independent processes as beam patterns for tracking and actual communications are likely performed as separate steps, for example in pilot-based tracking. Techniques to combine beam tracking and synthesis in analog domains are also being proposed [23]. However, those may also become very complex in large, multibeam scenarios.

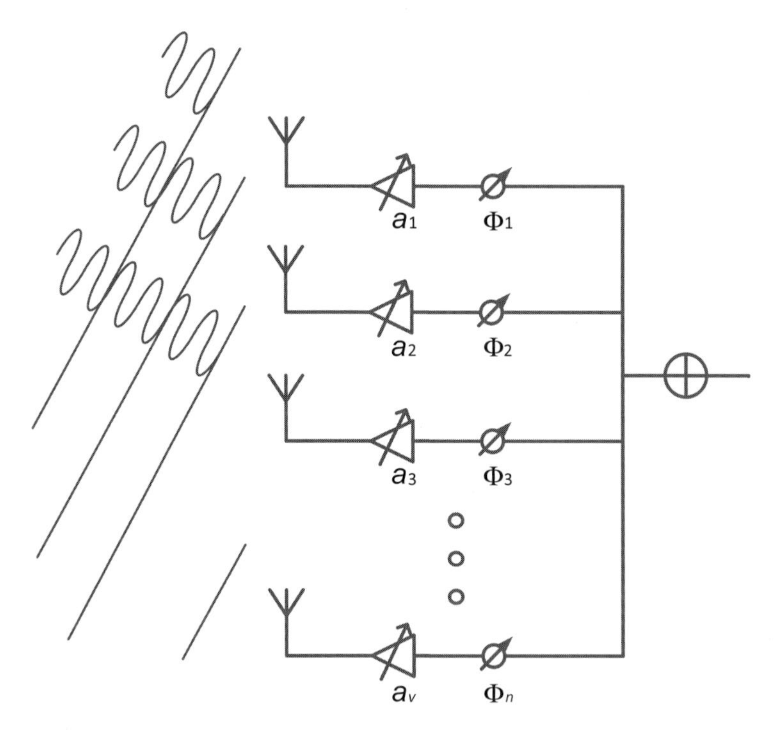

Figure 2.6 Principle of a phased array.

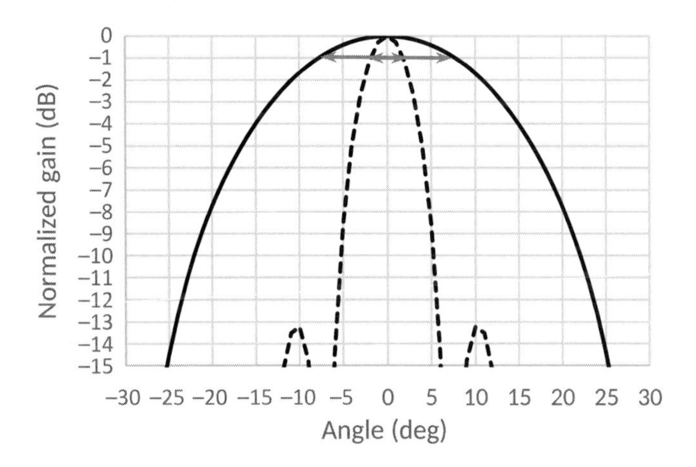

Figure 2.7 Difference in beamwidth between 4 (dashed line) and 16- (solid line) element linear array. Array gain is normalized to 0 dB in both cases.

Properties of antenna arrays and beam synthesis can be modeled in most cases based on the well-known theoretical background discussed, for example, in [24]. RF literature is mostly focusing on uniform antenna weighting, i.e., the same average output power in all antennas when evaluating the array performance. However, amplitude tapering techniques can provide significantly better spatial filtering to reduce interference outside the main lobe. Tapering causes a trade-off between EIRP and spatial filtering because in the case of amplitude control, power amplifiers in the array are driven in different back-off conditions. That can be utilized to assist linearization in analog arrays as in [25] to reduce the penalty of additional back-off and thus poorer efficiency. In addition, recent work on spatial notch filters [26] for interference mitigation will provide new aspects to beam synthesis in the analog domain that can be also modeled as part of system analysis if necessary. Interference management and system optimization in networks with directive links is a complex topic for the system analysis. Despite interference management's importance in dense 5G networks, the discussion in this chapter is limited to the link-level performance analysis instead of thorough interference analysis to keep some simplicity in the system model.

If we now neglect beam pattern and look at how individual beams can be modeled, the most compact model can include only antenna gain in the direction of interest for link budget analysis. Once directions of the various beams, i.e., peak values of received energy, are known, we can assume that each signal (or beam) is having its own and independent antenna array that can be directed the most optimal way. That is, of course, a highly optimistic view but helps to understand the best possible link condition in each case. Impact of antennas can be described in a linear system by multiplying the array gain with the gain of individual element in the array as given in logarithmic scale

$$G_{array}(\mathrm{dB}) = 10\log_{10}(N_{ANT}) + G_{element}(\mathrm{dB}), \tag{2.8}$$

where N_{ANT} is the total number of antennas in the array and $G_{element}$ the gain of each antenna element. The model is valid as long as the coupling between antenna elements is reasonably low and has little impact on the total performance. For more accurate modeling, the complete antenna array can be simulated using 3D design tools that are widely adopted in the antenna community as in [18]. Or, as with any level of modeling, in this approach also the antenna array model can be replaced with experimental data. This approach can be combined in the system model with the receiver model in a straightforward manner if we assume that all or at least most of the noise contribution is coming from the elements that are in front of the combining node in a phased array. As this is not likely the case, a bit more complex approach when calculating the impact of RF combining to array noise could be considered. Readers are referred to [27] for details.

2.5.4 Transceiver Architectures for RF and Hybrid Beamforming

In 5G systems, the assumption is that at least the base station can support multiple users and data streams simultaneously. And as the target is to support a 10 Gbps data rate for individual users, multiple MIMO streams would be required also in a mobile device. Mapping the number of streams to the data rate was discussed earlier in this chapter. There are several different architectural approaches to achieve the targets. The simplest way in line-of-sight conditions is to use dual-polarized antennas that will double the needed hardware but with proper design can support two independent streams without complex digital processing. In more versatile radio channel conditions, the calculation of orthogonal signals requires more sophisticated signal processing. We will discuss that briefly as related to the radio channel model in the following section. But if we look at the issue here from a receiver or transmitter architecture perspective, we can see major differences in implementation and how it can limit MIMO processing in practice.

If we would like to have full flexibility to weight all antennas independently for each MIMO stream in an optimal manner, we need a solution where all signal paths, i.e., data streams, will have individual amplitude and phase control words when connected to each antenna. This leads to a classical fully digital MIMO processing where all data streams are connected to all antennas without any boundary conditions. In fully digital architecture, all antennas have complete receive and transmit paths from antenna to digital conversion and front-end processing before MIMO paths are combined or separated. Hence, fully digital antenna processing will have all degrees of freedom (i.e., full flexibility) to optimize system capacity using, for example, maximum rate combining (MRC) or zero forcing (ZF) schemes [28]. Benefits of this approach are evident, but power consumption when processing wideband digital signals becomes easily a major problem in the case of tens or hundreds of antennas. If we assume that the number of supported data streams is much smaller than the number of antennas, this will become a system optimization issue. We have basically two independent problems to be solved here. The one is the number of independent data streams that needs to be supported, which can also be large in base stations. The other is solely related to link range. The number of data streams can be still much smaller compared to the number of antennas that are required to achieve sufficient array gain to compensate for link losses. That is a practical reason why other approaches for beamforming are of interest. The range aspect will be discussed in the next section in detail.

However, a simple, fully RF or analog beamformer can only spatially separate (i.e., filter) signals from different directions as given in Figure 2.8a. If we assume that spatial filtering in that case results in orthogonal data streams, each beamformer will operate as an independent MIMO channel. Although this model is easy to understand from an RF signal processing perspective, it has problems as well. Each beam would have totally independent, parallelized RF hardware that can be inefficient, as the beamformer in Figure 2.8a needs to be multiplied by the number of data streams. And in that case, the sidelobes will cause significant degradation of SNR in certain directions when independent beams at different directions start to interfere with each other. However, this approach is utilized here for simplicity when individual link budgets for different beams are calculated one at a time. This is definitely too optimistic but shows the performance border that can't be exceeded if we have certain maximum power per transmitter path, noise figure in the receiver, and a limited number of antennas in use.

As in a common scenario for 5G, the number of MIMO channels or data streams is considered to be smaller than the number of antennas. In that case, a subarray-based hybrid beamforming approach as in Figure 2.8b is considered as a decent compromise between analog beamforming and fully digital antenna processing with each antenna having a separate signal path to digital. The limitation of this approach is that each of the four subarrays shown in Figure 2.8b should be steered exactly to the same direction if maximum array gain is targeted. In other cases, some degradation for each beam is expected. And if all the subbeams are directed similarly, we have one degree of freedom less to precode orthogonal data streams. From an implementation perspective, this approach is straightforward, and subarray-based hybrid beamforming resembles a sectored antenna approach with the opportunity to steer sectors dynamically based on the traffic.

From a signal processing perspective, an RF architecture that connects all precoded MIMO streams to all antennas with independent phase and amplitude weights at the antenna node would provide an ideal solution that will not limit the degrees of freedom to optimize the transmission. This is of course with the assumption that RF performance is fully ideal as well. The RF cross-connected hybrid beamforming as in Figure 2.8c has been recently demonstrated with a limited number of data paths and antennas in a receiver [29]. Large-scale implementation of fully connected RF architecture will have quite obvious challenges in the case of massive MIMO. Connecting all antennas to all mm-wave transceivers will become very complex, and compensating losses at high frequencies when connecting over large arrays requires a substantial amount of battery power. Therefore, signal processing of hybrid beamforming for partially connected arrays becomes attractive as in [30].

From an RF performance perspective, each digitally precoded signal stream that is converted to an analog domain using a D/A-converter (and then in the receiver back to digital) will further experience additional RF array gain in all architectures described in Figure 2.8. Therefore, as individual signals they experience the RF phased array and model described above is valid. As digital precoding may split user data streams to multiple beams and each RF subarray may transmit or receive several MIMO streams at the same time, the analysis on the dynamic range requirement becomes rather complex in practice. For example, assume that we have two different streams passing the same RF subarray and the same PAs with nonequal power levels. Then nonlinearity of the PA will be dominated by the stronger component deteriorating also the EVM of the smaller one [31]. This must be taken into account when modeling multistream, multibeam transmissions, especially in the systems that are interference limited from a capacity perspective. However, simple superposition of different signals and independent analysis as is done in most cases in this chapter is a much simpler way to achieve initial understanding of the system requirements before optimizing interference between beams.

2.6 Radio Propagation and Link Budget

Modeling the radio path for a practical environment is a complex procedure itself. There are many different abstraction levels that can be utilized depending on the scenario. Applicability of any channel model also depends on the need, i.e., what we would like

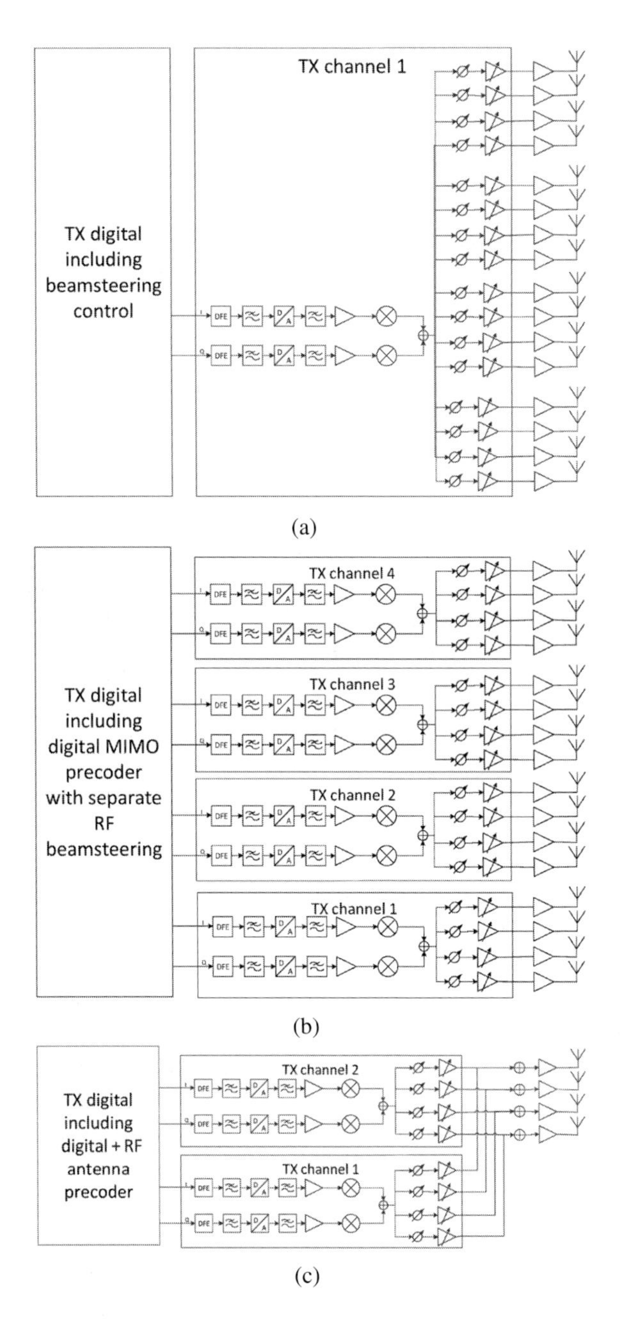

Figure 2.8 Architectures for RF beamforming: (a) RF phased array, (b) subarray-based hybrid beamformer, and (c) RF cross-connected hybrid beamformer.

to achieve with that model and whether it is appropriate for the purpose. It is a totally different task to model free space loss than to model a MIMO channel, but for proper understanding, both are valuable. However, RF performance is typically considered only against the simplest possible models, while fading channels and MIMO performance are evaluated using carefully selected power levels where RF parameters have minor impact on the performance. Such simplifications may be a reason for inaccuracy when performance is characterized over a larger dynamic range as indicated earlier in Figure 2.5. On the other hand, it is not possible to address MIMO capacity simply by evaluating beam propagation between transmitter and receiver without including phase relations in the environment. The discussion in this section considers issues and methods on how these two perspectives can be taken into account in an appropriate manner.

The first note is related to the best possible, i.e., LOS, conditions in an open area without any reflections. Path loss in LOS conditions is a function of the wavelength as

$$L_{path}(\mathrm{dB}) = 20\log_{10}\frac{4\pi d}{\lambda} \tag{2.9}$$

where d is the distance and λ is the wavelength of the carrier. The loss increases per antenna element at the slope 20 dB/dec as shown for 1 m distance in Figure 2.9. However, as a function of frequency the size of a radiating element becomes smaller. Let's assume a planar patch antenna and take the commonly used antenna distance between the elements, i.e., $\lambda/2$, as reference. If the total antenna array area is the same, it is possible to fit more antenna elements having the size of $(\lambda/2)^2$ to the same area. For constant antenna area A,

$$A = N\left(\frac{\lambda}{2}\right)^2 \tag{2.10}$$

where N is the number of antennas. Array gain to the direction of the main lobe is

$$G_{array}(\mathrm{dB}) = 10\log_{10}(N) = 10\log_{10}\frac{A}{(\lambda/2)^2}. \tag{2.11}$$

If we combine gain of a single array and path loss, we will get

$$\begin{aligned} G_{array}(\mathrm{dB}) - L_{path}(\mathrm{dB}) &= 10\log_{10}\frac{A}{(\lambda/2)^2} - 20\log_{10}\frac{4\pi d}{\lambda} \\ &= 10\log_{10}(A) - 20\log_{10}(2\pi d). \end{aligned} \tag{2.12}$$

This means that for a constant area, array gain compensates the path loss at any distance and makes propagation loss independent of the frequency. The result normalized to 1 GHz is plotted with dashed line in Figure 2.9. This result assumes that the antenna array is located only at one end of the link. With this, we can argue that beamforming using antenna arrays is an effective way to compensate path loss. However, in practice compensating the link budget calculated for a single antenna at 1 GHz an antenna array of the same area results in 900 antenna elements at 30 GHz. This is a practical problem that can be partially solved if both ends of the link are occupied with an antenna array. However, it is evident that it is not easy to achieve exactly the same link range even with directive links in beamformers. Also, individual links will become directive, which has a major impact on the radio system design. This model ignores well-known atmospheric absorption behavior that has an major impact, for example, on propagation at long range at 60 GHz frequency.

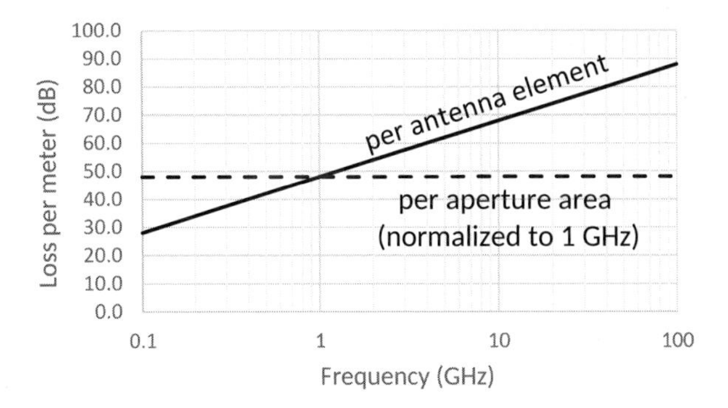

Figure 2.9 Propagation loss in line-of-sight conditions per single antenna element and for an array with constant area.

First-order range estimates can be done using link budget equation as

$$P_{RX}\,(\text{dBm}) = P_{TX} - L_{path} - L_{fade} + G_{a,TX} + G_{a,RX} \tag{2.13}$$

where P_{TX} and P_{RX} are transmitted and receiver power, L_{fade} is the fading margin, and $G_{a,TX}$ and $G_{a,RX}$ the antenna gains of the transmitter and receiver, respectively. Fading margin is a parameter that is based on the experiments in certain fading conditions and used in system analysis. Therefore, it is not suitable for generic analysis. In LOS conditions, fading is not part of the equation and is here unity i.e. 0 dB.

Let's take a 900 MHz channel bandwidth as in Table 2.4 as a reference and target for a 64-QAM transmission using a coding rate of 3/4. According to link adaptation assumption of Figure 2.2, the required minimum SNR for the transmitter and receiver are 26.0 and 25.5 dB, respectively. Maximum transmitted power in our example can be read directly from the model in Figure 2.3a, being 31.7 dBm per PA in back-off. Similarly, the minimum received power at 900 MHz bandwidth is −48 dBm as calculated and shown in Figure 2.4. If we calculate the link range in LOS conditions assuming individual antenna elements both in the transmitter and in the receiver having unity element gain (0 dBi), the longest possible communication distance for this signal using the aforementioned radio parameters with almost 80 dB of maximum path loss is 8.5 m at 27 GHz. The selected frequency is in the intersection of 5G NR bands named as n257 and n258. This shows that even with over 1 W output power from a PA, the range is limited to less than 10 m, which is not acceptable for a wide-range cellular system in general where tens or even some hundreds of meters distances would be required in a real environment. If we assume now 32 antennas for the base station and 8 for the mobile device, combined gain of the two arrays is 24 dB. In addition, the output power in a phased array can be multiplied with the number of PA, giving NTX (the number of antenna elements in the transmitter of the base station) times (15 dB) additional power compared to the single antenna case in downlink leading to ~750 m communications range in optimal conditions if both antennas are aligned properly. This example shows potential for long-range links if state-of-the-art PA components as in

[20] capable to deliver watt range output power in back-off are affordable in terms of cost, form factor, and power consumption. Such an example is modeled in this chapter. But if we look at recent, highly integrated phased arrays designed for 5G prototyping at 28 GHz without external PAs, the maximum available power in back-off for OFDM signals is less than 10 dBm [32,33]. The lower output power leads to a maximum range of less than 50 m with these antenna configurations in the best LOS scenario. Hence, tens or even hundreds of antennas per individual link are needed at least in infrastructure devices with limited power delivery per antenna with integrated PAs to achieve decent range, and especially in non–line-of-sight (NLOS) conditions as described later.

We can also map link capacity to absolute transmit and receive power levels as shown in Figure 2.10. SNR requirements for different modulations according to the link adaptation model in Figure 2.2 are mapped to maximum power of the transmitter and minimum power of the receiver at the sensitivity level based on the models in Figures 2.3 and 2.4. Based on these, it is possible to define the maximum acceptable path or channel loss for each data rate. This loss can be directly taken from the path loss as defined in the link budget or, if a more sophisticated channel model is being used, a practical way is to map the path loss of each path in the simulation to the specific data rate. This model embeds already all RF nonidealities, providing a highly abstracted approach for network-level analysis. Figure 2.10 also visualizes how the range can be enhanced with antenna arrays at both ends of the link. Cochannel interference coming from other data channels would naturally require a bit more complex approach to model the capacity at the network level.

The analysis as presented until now is including only one beam at a time in LOS conditions. Both multibeam and NLOS analysis would require more detailed channel models to estimate the impact. It is well known that radio propagation at mm-wave is not very favorable for NLOS cases. But if we would like to receive, for example, four MIMO

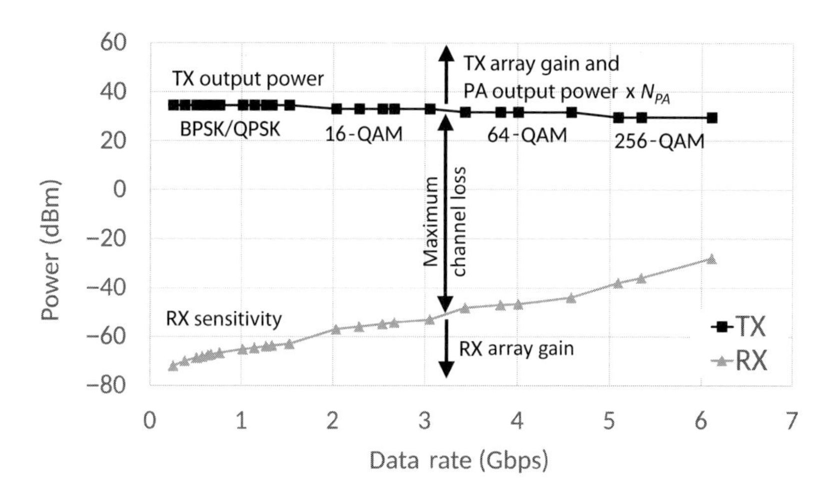

Figure 2.10 Maximum transmitted power and minimum received power according to the example in [18] for different modulations leading to throughput as given in the x-axis. (©2017 IEEE. Reprinted, with permission, from *IEEE Transactions on Antennas and Propagation*.)

streams at a time even for a single downlink or uplink connection, propagated signals must come partially either from reflected waves or, in the case of LOS, at least from two different sources if dual-polarized antennas are assumed. Propagation conditions at frequencies above 10 GHz have been discussed and studied for example in [34–36]. Based on those, it is evident that there are many scenarios where significantly more transmitted power and/or array gain would be required in case of NLOS. In the next section, an example of the modeling approach is given. For that, an accurate enough but still computationally efficient method is needed. The simulator should be able to calculate orthogonal MIMO channels based on some coding scheme. Each MIMO path can include energy from one or multiple beams coming from the same source, and therefore it is not possible to solely focus on individual beams and their reflections. For that, a proper channel model is needed.

Deterministic models based on the ray-tracing use rules of geometrical optics and model all possible propagation paths based on the environment. Such a model is complex and can be computationally heavy. It would also require complete modeling of the environment, which can be a tedious task. Therefore, as a simpler alternative, a geometry-based stochastic channel model can be applied. The propagation channel includes statistical parameters that are characterized based on the channel measurements. The approach gives also an opportunity to model antenna array separately and thus study different antenna configurations in the same environment. The quasideterministic radio channel generator (QuaDRiGa) is a publicly available model from its developers, and thus widely used for the purpose [37]. It is a suitable tool that can be utilized here as well. Mapping of the QuaDRiGa model to the real environment is somewhat possible but poses some questions about accuracy when combined with physical geometries. Recently, a new model that addresses 5G requirements of the wide frequency range and has the spatial consistency of numerous radio links has been proposed as a result of the European Union (EU)-funded METIS project [19]. Therefore, a METIS map-based channel model, which positions itself between the two other approaches, has been utilized. The results given in the next section are performed with a simulator that includes RF system modeling as presented in this chapter, a realistic scenario for antennas that are simulated using 3D antenna design software, and a physical environment where both physical space and radio channel are modeled using the METIS map-based approach implemented in Matlab.

2.7 Multiuser Multibeam Analysis Example

When analyzing multiple data channels in any MIMO system, a complex radio channel is presented as a matrix dimensioned based on the number of inputs and outputs, i.e., antennas. Conditions to find multiplexed, orthogonal paths to increase the capacity of the radio channel can be analyzed using singular value decomposition (SVD). Steps include preprocessing (or precoding) of the signal, transmission through physical radio channel including antennas and in this case also RF nonidealities, and finally postprocessing of the received signal. This is a mandatory step when we move from the analysis of individual beams to system capacity. Fundamentals of the MIMO processing can be found, for example, in [38]. Details of how SVD is applied in the modeling approach are described in [18]. Here the focus is on the scenario and findings based on that.

A large office room (30 × 16 m) is modeled using the METIS model with plasterboard walls as shown in Figure 2.11. Partitions forming the cubicles have a height of 1.2 m, and mobile devices are placed randomly over the area according to the figure. Heights of the mobile devices are also random and between 0.5 and 1.0 m. This means that not all the mobiles have an LOS connection to the base station. One base station is located at the bottom-right corner and placed at 2.0 m height pointing the beam optimally to the room. The intention is to study extremely high-speed connections for all user positions having up to four MIMO paths without dual-polarized antennas. This scenario is beyond what is expected from the first 5G systems, and the connection potential above 20 Gbps per user is simulated. However, this is with 900 MHz channel bandwidth, 256-QAM modulation, and the base station having maximum EIRP of more than 70 dBm. All of these are highly optimistic compared to the currently specified eMBB using 5G NR in 3GPP [6]. But as shown later, for example EIRP values lower than 60 dBm are sufficient even in the worst-case conditions. Mobile device density in Table 2.6 is calculated from indoor ultrahigh broadband access scenario in [39]. All 12 mobile devices are pointing the antennas optimally toward the base station. This is a somewhat optimistic view, but one can expect that future mobile devices will support multiple antenna arrays in the same device providing maximal feasible rotational flexibility as in [33]. As discussed in the earlier example, the base station has 32 and the mobile device 8 antennas, respectively. Patch antennas with 7 dBi gain per element are assumed. This is a bit optimistic value based on rather a simple simulation model. However, one should assume several decibels of element gain from a patch antenna anyway at this frequency range. These configurations

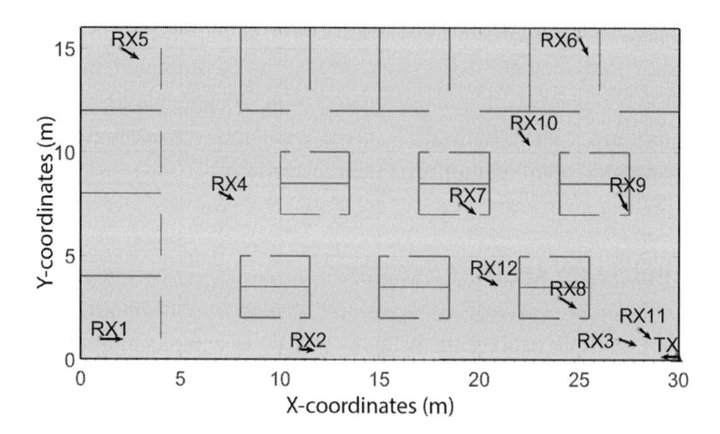

Figure 2.11 Indoor scenario of an office area with walls of the cubicles having a height of 1.2 m and made of plasterboard. The base station at bottom-right corner is located at height of 2 m [18]. (©2017 IEEE. Reprinted, with permission, from *IEEE Transactions on Antennas and Propagation.*)

Table 2.6 Scenario example of the user traffic in a dense indoor environment.

Parameter	Value	Unit
Connection density	7.5	(#/100m^2)
Activity factor	30	(%)
Area	480	(m^2)
Dimension	30 × 16	(m × m)
Number of connections in the area	36	
Active connections at a time	12	

result in total antenna array gains of 22 and 16 dB, with antenna element gain in the base station and mobile device, respectively.

The first step is to define propagation paths between the base station and mobile devices using the channel model. Results of the simulations are plotted to a polar plot both for transmitter (base station) and each receiver (mobile device) in Figure 2.12. Both azimuth and elevation planes are shown. As expected, elevation spread is small in the scenario, while directions are widely spread in the azimuth domain, which also has reflection components from walls. Already this 12-device scenario indicates the importance of angular resolution in beamforming that needs to be complemented with MIMO processing in order to separate orthogonal data channels. To support very high data rates, the target is to have four MIMO channels in each link. Absolute scale is not normalized in the figure, but it gives information on the relative strengths of the beams. If the network capacity is optimized for all connections at the same time without any hardware limitations, fully digital MIMO processing would be the choice. That would lead to a very complex matrix including 12 × 4 = 48 parallel MIMO channels for four MIMO channels per link. Also, the resulting beam pattern would become very complex and will not serve visualization of the problem the best possible way. Therefore, we will look each transmitter–receiver pair individually and calculate the beam pattern for each link at a time.

Analysis using SVD is performed for one mobile device at a time. Results of the decompositions done for two devices (numbers 5 and 12) separately are shown in Figure 2.13. They are based on the fully digital approach. However, similar results can be obtained if each stream gets an individual subarray with same size as the fully digital configuration. All antennas (32 in the base station and 8 in the mobile device) form a matrix sized as 32 × 8 elements. After decomposition four strongest (i.e., lowest loss) MIMO channels with distinct eigenvalues are then picked from the analysis. They do not directly represent any individual beams that can be implemented using a single RF beamformer but present a composite beam pattern that transmits or receives energy from several directions. The directions from Figure 2.12 that are contributing to these beam patterns are marked with the dots in Figure 2.13. The beam patterns are scaled based on the energy they receive and therefore the strongest beam points to the lowest loss path drawn at the outset sphere in Figure 2.13. That is typically the LOS component. Also it is visible that the other beam patterns form a null toward that direction in order

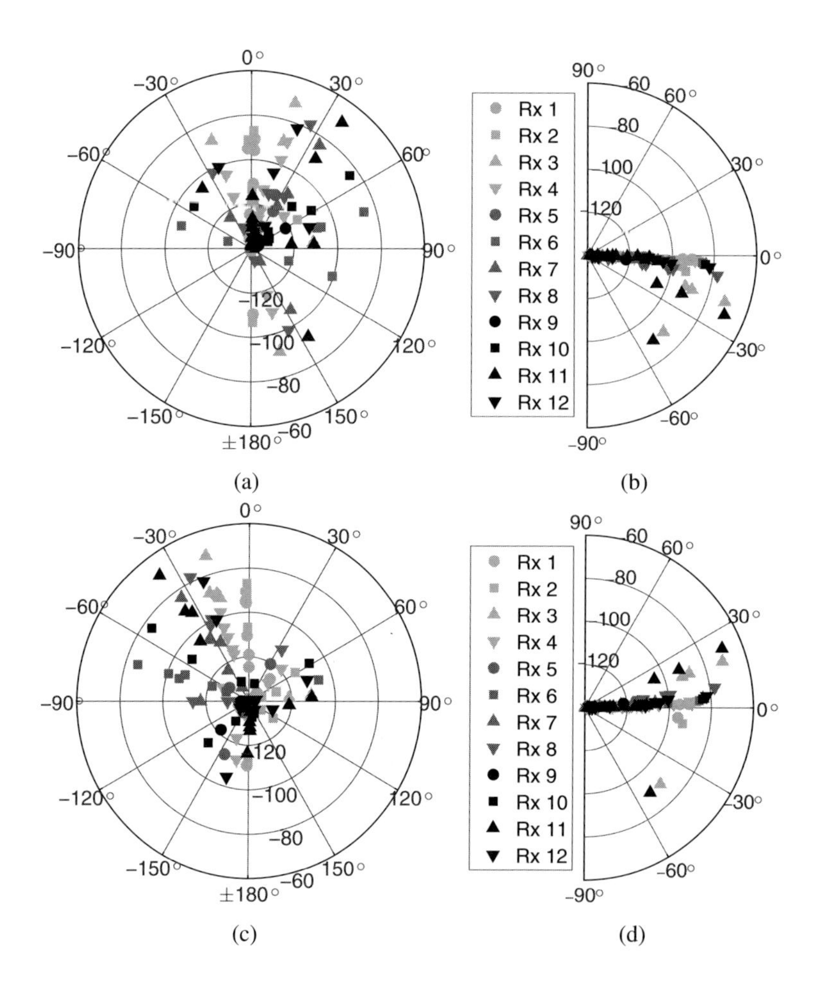

Figure 2.12 Directions of the simulated multipaths within the 80 dB dynamic range from the maxima for (a) TX azimuth, (b) TX elevation, (c) RX azimuth, and (d) RX elevation [18]. (©2017 IEEE. Reprinted, with permission, from *IEEE Transactions on Antennas and Propagation*.)

to block the energy from that component. The beam patterns that are collecting energy from other MIMO paths show some directivity, but they are more complex due to the fact that they need also improved spatial filtering properties in the beam synthesis. As shown, this is a complex process already for four MIMO channels in an mm-wave link.

Analysis with four individual subarrays with RF beamforming in a hybrid architecture would lead to a different and somewhat suboptimal result due to restrictions on how beam patterns can be constructed. We have also neglected here the effect of beam squinting, i.e. variable delay over bandwidth that is associated with phase control in wideband RF beamformers. In a fully digital architecture, this is easier to manage. However, an optimal solution from a capacity perspective using 32 antennas for four MIMO beams is definitely a suboptimal solution from a power consumption perspective. In cases where the system specification for number of antenna elements is dominated

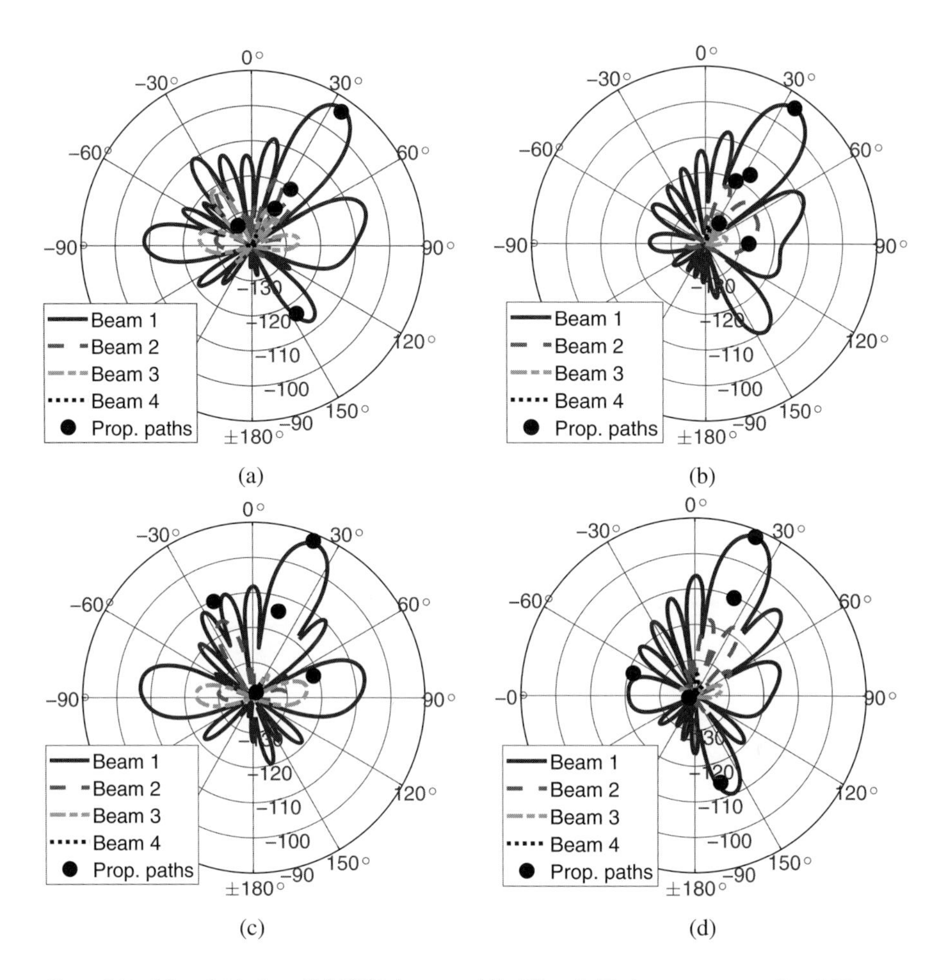

Figure 2.13 (a) and (c) show BS SVD-beams while (b) and (d) show corresponding MT SVD-beams for users 7 and 12, respectively [18]. (©2017 IEEE. Reprinted, with permission, from *IEEE Transactions on Antennas and Propagation.*)

by range requirements rather than channel capacity, a hybrid architecture is efficient in terms of system area and power. This is also a common assumption for the 5G systems operating at the mm-wave range.

In order to understand better link performance requirements from a hardware perspective, each MIMO channel based on the eigenvalue is considered as a separate beam pattern as in Figure 2.13. For all these patterns or data paths, loss of the specific path is defined in Figure 2.14. The plot shows loss of the MIMO specific paths with the fixed antenna setup for the four strongest channels (32 antennas in the base station and 8 in a mobile device, including 7 dBi element gain, with both having a single polarized configuration). Antenna element gain is bit high compared to practical values of 4 to 5 dBi, but as we include 3 dB front-end loss assumption per antenna path both in the transmitter and receiver, the variation can be embedded to that device in addition to beam alignment errors, etc., in this analysis. Free space path gain including the array

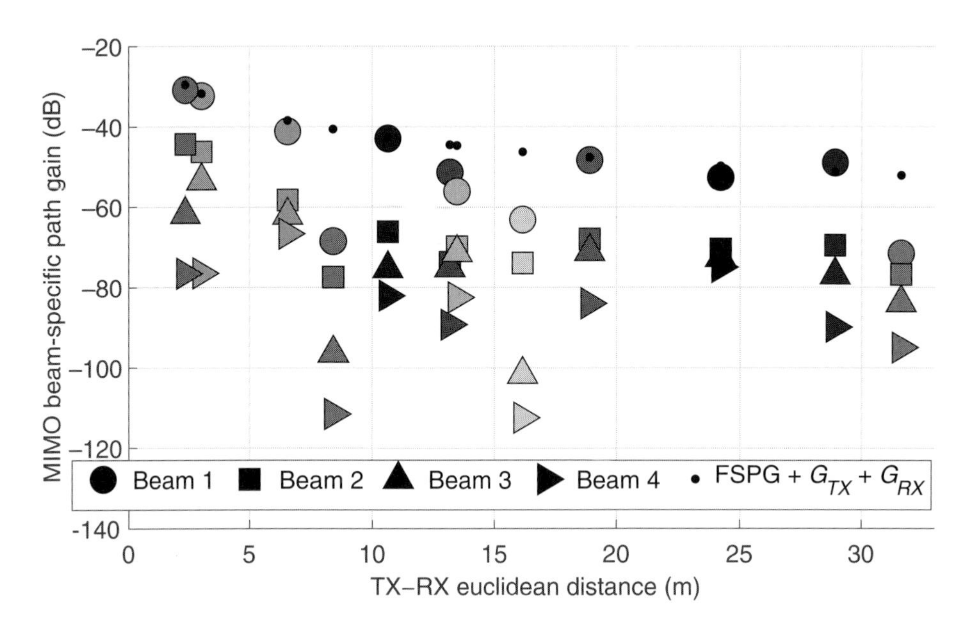

Figure 2.14 MIMO channel–based paths gains of the scenario in Figure 2.11 including the impact of array gains of the transmitter and receiver [18]. Path gains are defined for each receiver at different distances from the base station and for the four strongest MIMO channels in each case.

gains at both ends of the link are also shown as small black dots in Figure 2.14. The individual mobiles have been arranged as a function of physical distance from the base station to the x-axis. It is seen that in many cases the LOS component is detected, but there is also one case where the strongest path is more than 20 dB below the optimal LOS case even in an indoor scenario with limited furniture.

In the second step, the maximum available data rate for each MIMO channel can be defined taking RF performance constraints into account. Antenna properties are already included in the previous step but transmitted power (EIRP) and SNR/EVM of the transmitter and receiver need further consideration. Figure 2.10 for a single antenna system is modified according to this specific antenna configuration for downlink in Figure 2.15. Maximum power that can be transmitted from 32 antennas is drawn as a separate curve. That is the one that we should compare to Figure 2.14 to determine the maximum data rate that can be achieved in certain link conditions taking RF performance into account. Arrows between transmit and receive powers indicate this case. In addition, TX EIRP for 32 and RX sensitivity for 8 antennas with array gain are plotted as reference to represent all definitions of various power levels in this example. Hardware constraints from Figure 2.15 indicate that for the maximum data rate, path loss can be a bit above 70 dB when array gain is included in the propagation path and more than 100 dB for the lowest data rates described in this example.

When data from Figures 2.14 and 2.15 are combined, the total data rate for MIMO ranks 1. . . 4 can be plotted as in Figure 2.16. Rank 1 is always considered the strongest MIMO channel, and others are numbered in descending order. Almost all receivers can operate in this example at full speed with the strongest MIMO channel, but due

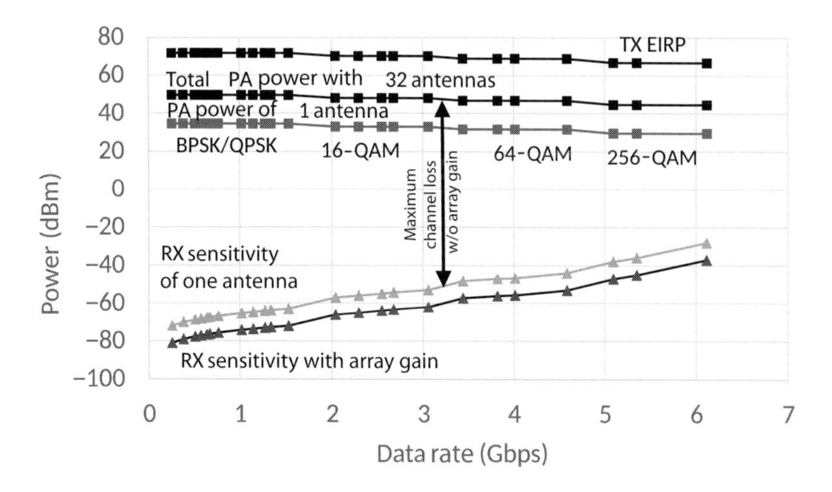

Figure 2.15 Maximum power levels and minimum sensitivities for different data rates for the case study of this chapter.

to the large spread in power levels of different MIMO paths, only one of the links can achieve almost the full speed at all four MIMO paths. Those are within 30 dB of dynamic range. There are also two cases where the fourth MIMO channel can't provide any significant addition to the data rate due to path loss that is too high for this transceiver configuration. This example shows data rates that exceed 20 Gbps. However, this is for a 900 MHz OFDM channel and 256-QAM modulation without coding. If we scale that to a 400 MHz 5G NR channel with 64-QAM and coding of 7/8, the data rate will be more than three times smaller but the range a bit longer due to smaller noise bandwidth. This clearly shows the challenge to achieve 10 Gbps data rates in practical conditions.

Because many of the paths in Figure 2.14 are actually very good channels for this platform configuration, we can finally calculate how much EIRP is needed to meet the minimum requirements for each path separately. Those are shown in Figure 2.17. We see a large spread in the required transmission powers for different MIMO channels. This leads back to discussion of the dynamic range of the transmitter and receiver. It was shown in [31] that only slight variation in power levels of different signals going through the same PA will have a major impact on the EVM where the smaller signals suffer badly from the stronger one. Even in the case when power levels are set to equal, each signal in the same RF path reduces the maximum power one PA can deliver per MIMO channel. For that reason, all channels that are transmitted through the same PA must be equalized approximately to the same level that is defined by the largest power, and the dynamic range needs to be managed in the receiver. However, as the signals are coming from different directions, even a small subarray in the receiver side will reduce dynamic range requirements of the ADCs substantially.

In this scenario, there is some headroom available even for the highest data rates, as the modeled antenna array with PAs having 9 W of peak power can deliver more than 65 dBm of EIRP power at maximum. However, this is a high level of power and

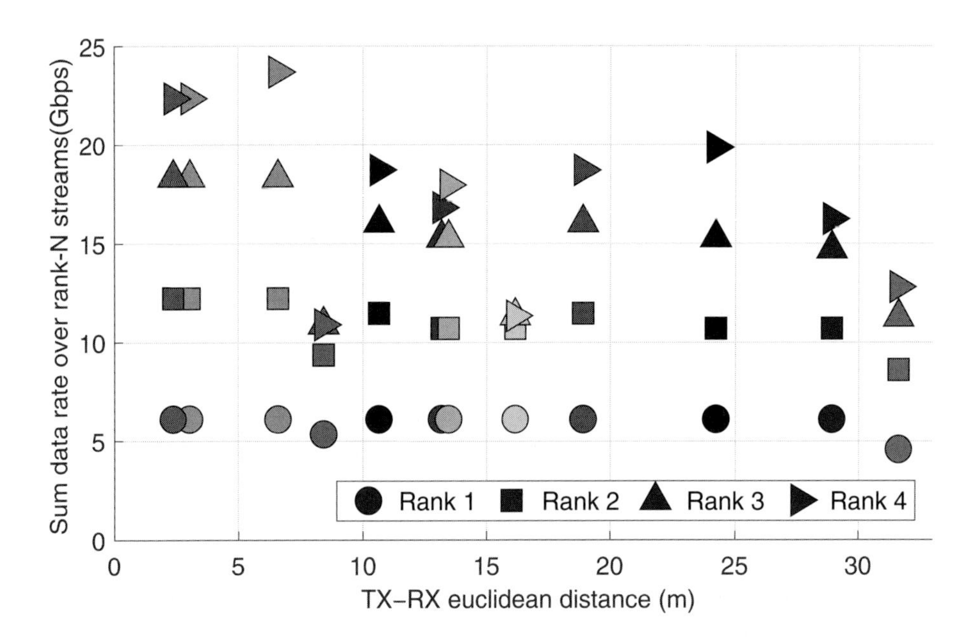

Figure 2.16 Maximum achievable data rate defined separately for each link and four MIMO channels taking RF performance into account [18]. (©2017 IEEE. Reprinted, with permission, from *IEEE Transactions on Antennas and Propagation*.)

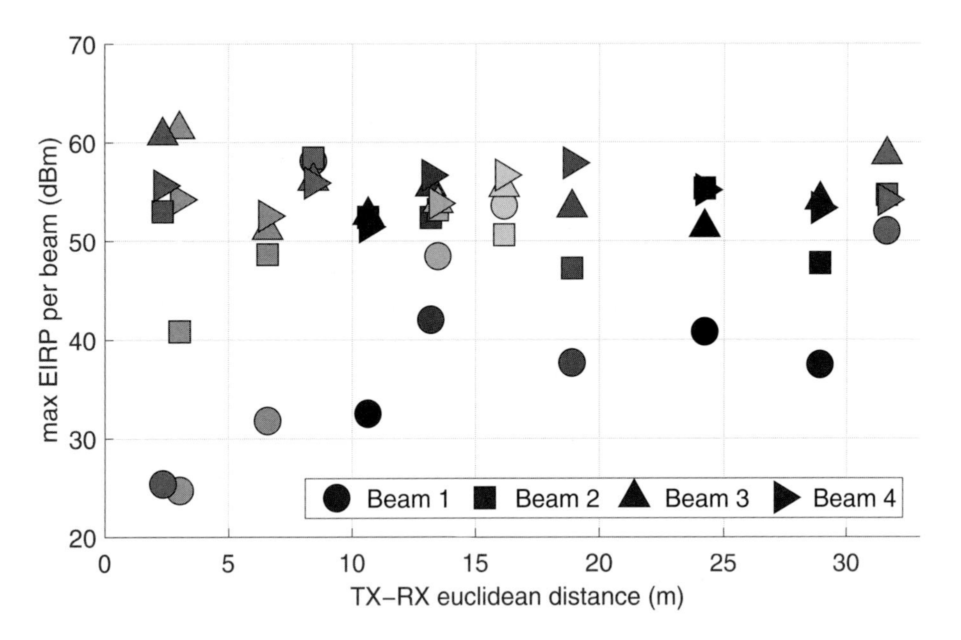

Figure 2.17 Minimum EIRP required for the transmission of each MIMO path indicating the dynamic range of the channel. SNR required for the transmission and back-off needed in electrical circuits should be included on the top of that. The dynamic range can be split between transmitter and receiver the most optimal way [18]. (©2017 IEEE. Reprinted, with permission, from *IEEE Transactions on Antennas and Propagation*.)

calculations are theoretical without considerations of regulatory power limits, thermal issues, or other physical constraints that can't be directly observed from data sheets. Therefore we may expect that the values given in Figure 2.17 are quite close to practical limits with this setup. Of course, it is possible to find other trade-offs for the number of antennas and output power of the PAs as done, for example, in the 5G prototypes described in [32] and [33].

2.8 Conclusion

Fifth-generation communications systems are rapidly evolving from research to commercial products in the coming years. Expectations are high and some of the key aspects such as mobile broadband access would require adoption of spectrum from sub-mm-wave and mm-wave regions. Designing densely populated cellular systems based on those is a new challenge to the industry. This chapter has addressed many of the challenges coming from the system requirements to RF transceivers, including both performance and architectural aspects. Radio system design from communications paradigm to integrated circuits is a complex process requiring many layers of engineering in order to achieve the set targets. In this chapter, many of those have been discussed from communications to RF specification. A system model that includes key aspects from different layers is created. It provides a highly abstracted view to different steps of the process. On the other hand, a detailed analysis can be done with a highly abstracted view. A case study of the dense indoor scenario addresses many challenges related to practical design aspects and includes a detailed model of the radio channel to visualize the complexity of the MIMO processing with beamforming and the dynamics of the radio channel at the sub-mm-wave region.

Radio system design becomes more complex in 5G systems. That is specifically coming from the need for directive communications to achieve decent range for a radio link. At the same time, spatial filtering provides new means for interference protection, although it can't be assumed as a comprehensive solution. We have seen 5G systems evolving from the first prototype systems to standards and early products in a short time frame. Still there will be many aspects that can be optimized from networks to circuits in coming versions of the standards and hardware implementations. Final optimization of the RF transceivers do not rely only on standards but on innovative architectures and thorough understanding of the cross-layer aspects in the system in the coming years.

Acknowledgments

The author would like to acknowledge Nuutti Tervo and Dr. Tommi Tuovinen for their contributions to the system model, performance analysis, many figures, and the manuscript of the paper that is basis for this chapter. Marko Leinonen deserves thanks for reviewing the manuscript and providing insights to 5G NR system specifications. Professor Matti Latva-aho has been extensively promoting 5G systems and provided me an unique opportunity to collaborate in many aspects that eventually led to the manuscript of this chapter. Professor Ari Pouttu created the concept vision for this work, and in many discussions we refined the various aspects of 5G toward a set of parameters that are close to the recent 5G NR specification. Therefore, there was no major need to retune the fundamentals but merely bring in more scalability to the established conceptual work.

I have had many discussions on 5G requirements, standards, and implementation aspects with many colleagues around the world. Special thanks are given to Dr. Sven Mattisson, Dr. Earl McCune, and Dr. Stefano Pellerano for in-depth considerations and valuable notes on the 5G systems.

This work has been financially supported by the Academy of Finland 6Genesis Flagship (grant 318927).

References

[1] IMT, "IMT vision: Framework and overall objectives of the future development of IMT for 2020 and beyond." [Online]. Available: www.itu.int/rec/R-REC-M.2083-0-201509-I/en

[2] J. Wannstrom, "LTE-Advanced." [Online]. Available: www.3gpp.org/technologies/keywords-acronyms/97-lte-advanced

[3] ISO/IEC/IEEE, *ISO/IEC/IEEE International Standard for Information Technology – Telecommunications and Information Exchange between Systems – Local and Metropolitan Area Networks – specific Requirements – Part 11: Wireless LAN Medium Access Control (MAC) and Physical Layer (PHY) Specifications Amendment 3: Enhancements for Very High Throughput in the 60 GHz Band (Adoption of IEEE Std 802.11ad-2012)*, ISO/IEC/IEEE ISO/IEC/IEEE 8802-11:2012/Amd.3:2014(E), March 2014. [Online]. Available: http://ieeexplore.ieee.org/stamp/stamp.jsp?tp=&arnumber=6774849&isnumber=6774848

[4] IEEE, *Standard for Information Technology – Telecommunications and Information Exchange between Systems Local and Metropolitan Area Networks – Specific Requirements Part 11: Wireless LAN Medium Access Control (MAC) and Physical Layer (PHY) Specifications – Amendment: Enhanced Throughput for Operation in License-Exempt Bands above 45 GHz*, IEEE Standards Association Std., 2015.

[5] *Resolution 238 (WRC-15)*, World Radiocommunication Conference Std. 2015.

[6] *3GPP Specification Series TS38*, 3GPP Std. [Online]. Available: www.3gpp.org/DynaReport/38-series.htm

[7] A. A. Zaidi, R. Baldemair, M. Andersson, S. Faxér, V. Molés-Cases, and Z. Wang, "Designing for the future: The 5G NR physical layer," *Ericsson Technology Review*, June 2017. [Online]. Available: www.ericsson.com/en/publications/ericsson-technology-review/archive/2017/designing-for-the-future-the-5g-nr-physical-layer

[8] E. McCune, "Signal design and figure of Merit for green communication links," in *2017 IEEE Radio and Wireless Symposium (RWS)*. Phoenix, IEEE, 2017, pp. 22–25.

[9] S. Andersson, L. Sundström, and S. Mattisson, "Design considerations for 5G mm-wave receivers," in *2017 Fifth International Workshop on Cloud Technologies and Energy Efficiency in Mobile Communication Networks (CLEEN)*, Turin, 2017, pp. 1–5.

[10] 3GPP, *RAN4#79 Discussion Paper R4-164226, On mm-Wave Technologies for NR*, 3GPP Std. [Online]. Available: https://portal.3gpp.org/ngppapp/CreateTdoc.aspx?mode=view&contributionId=705081

[11] 3GPP, *3GPP RAN4#80 Discussion Paper R4-166526, Discussion on BS and UE Noise Figure for mm-Waves*, 3GPP Std. [Online]. Available: https://portal.3gpp.org/ngppapp/CreateTdoc.aspx?mode=view&contributionId=722640

[12] Bluetooth Special Interest Group (SIG), *Bluetooth Core Specification v5.0*, Bluetooth SIG Std., December 2016. [Online]. Available: www.bluetooth.com/specifications/bluetooth-core-specification

[13] 3GPP, *3GPP Specification Series TS36*, 3GPP Std. [Online]. Available: www.3gpp.org/DynaReport/36-series.htm

[14] *TS 36.101, 3rd Generation Partnership Project; Technical Specification Group Radio*, 3GPP Std. [Online]. Available: www.3gpp.org/DynaReport/38-series.htm

[15] Y. Shi et al., "A 10 mm^3 inductive coupling radio for syringe-implantable smart sensor nodes," *IEEE Journal of Solid-State Circuits*, vol. 51, no. 11, pp. 2570–2583, November 2016.

[16] J. Chen, M. Berg, V. Somero, H. Y. Amin, and A. Pärssinen, "A multiple antenna system design for wearable device using theory of characteristic mode," in *12th European Conference on Antennas and Propagation (EuCAP)*, London, IET, April 2018, pp. 1–5.

[17] L. Iotti, G. LaCaille, and A. M. Niknejad, "A 12mW 70-to-100GHz mixer-first receiver front-end for mm-wave massive-MIMO arrays in 28nm CMOS," in *2018 IEEE International Solid-State Circuits Conference (ISSCC)*, IEEE, February 2018, pp. 414–416.

[18] T. Tuovinen, N. Tervo, and A. Pärssinen, "Analyzing 5G RF system performance and relation to link budget for directive MIMO," *IEEE Transactions on Antennas Propagation*, vol. 65, no. 12, pp. 6636–6645, December 2017.

[19] P. Kyösti, J. Lehtomäki, J. Medbo, and M. Latva-aho, "Map-based channel model for evaluation of 5G wireless communication systems," *IEEE Transactions on Antennas Propagation*, vol. 65, no. 12, pp. 6491–6504, December 2017.

[20] Quorvo, *27.5–31 GHz 8 W GaN Power Amplifier, TGA2595-CP datasheet*, July 2016.

[21] T. Tuovinen, N. Tervo, and A.Pärssinen, "RF system requirement analysis and simulation methods towards 5G radios using massive MIMO," in *2016 46th European Microwave Conference (EuMC)*, London, IEEE, 2016, pp. 142–145.

[22] S. Ek, T. Påhlsson, A. Carlsson, A. Axholt, A. K. Stenman, and H. Sjöland, "A 16–20 GHz LO system with 115 fs jitter for 24–30 GHz 5G in 28 nm FD-SOI CMOS," in *ESSCIRC 2017 – 43rd IEEE European Solid State Circuits Conference*, Leuven, IEEE, 2017, pp. 251–254.

[23] M. Y. Huang, T. Chi, F. Wang, T. W. Li, and H.Wang, "A 23-to-30GHz hybrid beam-forming MIMO receiver array with closed-loop multistage front-end beamformers for full-FoV dynamic and autonomous unknown signal tracking and blocker rejection," in *2018 IEEE International Solid-State Circuits Conference (ISSCC)*, San Francisco, IEEE, 2018, pp. 68–70.

[24] R. Mailloux, *Phased Array Antenna Handbook.* Artech House, 1994.

[25] N. Tervo, J. Aikio, T. Tuovinen, T. Rahkonen, and A. Pärssinen, "Digital predistortion of amplitude varying phased array utilising over-the-air combining," in *2017 IEEE MTT-S International Microwave Symposium (IMS)*, Honolulu, IEEE, 2017.

[26] L. Zhang and H. Krishnaswamy, "Arbitrary analog/RF spatial filtering for digital MIMO receiver arrays," *IEEE Journal of Solid-State Circuits*, vol. 52, no. 12, pp. 3392–3404, December 2017.

[27] A. Natarajan et al., "A fully-integrated 16-element phased-array receiver in SiGe BiCMOS for 60-GHz communications," *IEEE Journal of Solid-State Circuits*, vol. 46, no. 5, pp. 1059–1075, May 2011.

[28] E. G. Larsson, O. Edfors, F. Tufvesson, and T. L.Marzetta, "Massive MIMO for next generation wireless systems," *IEEE Communications Magazine*, vol. 52, no. 2, pp. 186–195, February 2014.

[29] S. Mondal, R. Singh, A. Hussein, and J. Paramesh, "A 25-30 GHz 8-antenna 2-stream hybrid beamforming receiver for MIMO communication," in *2017 IEEE Radio Frequency Integrated Circuits Symposium (RFIC)*, Honolulu, IEEE, 2017, pp. 112–115.

[30] X. Gao, L. Dai, S. Han, C. L. I, and R. W. Heath, "Energy-efficient hybrid analog and digital precoding for mmwave MIMO systems with large antenna Arrays," *IEEE Journal on Selected Areas in Communications*, vol. 34, no. 4, pp. 998–1009, April 2016.

[31] N. Tervo, J. Aikio, T. Tuovinen, T. Rahkonen, and A. Pärssinen, "Effects of PA nonlinearity and dynamic range in spatially multiplexed precoded MIMO systems,' in *European Wireless 2016; 22th European Wireless Conference*, Oulu, VDE, 2016, pp. 1–6.

[32] B. Sadhu et al., "A 28-GHz 32-element TRX phased-array IC with concurrent dual-polarized operation and orthogonal phase and gain control for 5G communications," *IEEE Journal of Solid-State Circuits*, vol. 52, no. 12, IEEE, December 2017, pp. 3373–3391.

[33] J. D. Dunworth et al., "A 28GHz bulk-CMOS dual-polarization phased-array transceiver with 24 channels for 5G user and basestation equipment," in *2018 IEEE International Solid-State Circuits Conference (ISSCC)*, San Francisco, February 2018, pp. 70–72.

[34] T. S. Rappaport et al., "Millimeter wave mobile communications for 5G cellular: it will work!" *IEEE Access*, vol. 1, pp. 335–349, 2013.

[35] M. K. Samimi and T. S. Rappaport, "3-D millimeter-wave statistical channel model for 5G wireless system design," *IEEE Transactions on Microwave Theory and Techniques*, vol. 64, no. 7, pp. 2207–2225, July 2016.

[36] A. Roivainen, C. F. Dias, N. Tervo, V. Hovinen, M. Sonkki, and M. Latva-aho, "Geometry-based stochastic channel model for two-story lobby environment at 10 GHz," *IEEE Transactions on Antennas and Propagation*, vol. 64, no. 9, pp. 3990–4003, September 2016.

[37] S. Jaeckel, L. Raschkowski, K. Börner, and L. Thiele, "QuaDRiGa: A 3-D multi-cell channel model with time evolution for enabling virtual field trials," *IEEE Transactions on Antennas and Propagation*, vol. 62, no. 6, pp. 3242–3256, June 2014.

[38] D. Tse and P. Viswanath, *Fundamentals of Wireless Communication*. Cambridge University Press, 2005.

[39] R. E. Hattachi and J. Erfanian, "NGMN 5G white paper," Next Generation Mobile Networks (NGMN) Alliance, February 2015. [Online]. Available: www.ngmn.org/fileadmin/ngmn/content/downloads/Technical/2015/NGMN_5G_White_Paper_V1_0.pdf

3 MU-MIMO and Massive MIMO for 5G Radios

Antonio Puglielli, Greg LaCaille, Elad Alon, Borivoje Nikolić, and Ali M. Niknejad

3.1 Spatial Processing: Untapped Potential

Wireless communications have enjoyed tremendous growth over the past two generations, both improving coverage and increasing data rates. Proliferation of mobile video and other media-rich services has created a massive increase in data consumption [1]. Current projections anticipate that mobile traffic will continue to grow rapidly into the next decade. By 2021, it is anticipated that almost two-thirds of global IP traffic will originate over a wireless connection, with 20% coming over a cellular network. Moreover, it is widely expected that new device classes, supporting augmented reality (AR) or virtual reality (VR), along with new applications such as connected vehicles and robots, will create new and even more challenging requirements for overall network traffic.

These trends reveal that, over the past decade, wireless connectivity has morphed from a luxury to a basic requirement of life throughout the world. This is a testament to rapid technological progress in the areas of wireless technology and integrated circuits. At the same time, widespread demand for ever-increasing network throughput is colliding with the fundamentals of information theory: the capabilities of wireless communication techniques have approached their theoretical limits. Since Shannon's discovery of the theory of information and communication in 1948 [2], the gap between practical systems and theoretical limits has been continuously narrowing. Advances in coding – turbo and low-density parity check (LDPC) codes [3] and modulation – Orthogonal Frequency Division Multiplexing (OFDM) [4,5] – have introduced practical algorithms that closely approach the Shannon bound. At the same time, advances in semiconductor processing have made it possible and even economical to deploy very powerful digital processing capabilities in billions of consumer devices. All told, widely available 802.11 [6] and Long Term Evolution (LTE) [7] consumer devices can communicate within a fraction of a decibel from the Shannon limit [8], meaning that these standards make near-optimal use of time and frequency resources.

Though this is a great achievement, it means that there are no simple solutions to address existing and emerging traffic demands. Previous generations of cellular networks increased data rates in one of two ways. First, wider channels were used (from 200 kHz in 2G GSM to 5 MHz in 3G WCDMA and 20 MHz in 4G LTE, even to 100 MHz in LTE-Advanced Pro). Second, infrastructure deployments were dramatically densified [9]. Today both of those techniques are running into roadblocks.

The conventional sub–6 GHz cellular bands are nearly fully allocated. At the same time, further network densification is limited by the cost and time needed to acquire backhaul connections and siting permits in millions of local jurisdictions. Compounding the challenge, crowded spectrum and dense networks naturally lead to high levels of interference [9]. For most devices on the network, realistically available speeds are limited by network interference rather than the fundamental capabilities of the standard.

This chapter will make the case for advanced spatial processing as a core component of future wireless networks. Spatially selective communication links today can be broadly classified into two main categories. First, multiple-input and multiple-output (MIMO) techniques are used in 4G and WiFi to send up to four or eight spatial streams in certain conditions. Second, mm-wave communications bands such as the 60 GHz unlicensed band use phased arrays to overcome high propagation loss. This chapter will explore how to combine and enhance both of these technologies, with the objective of designing communication links supporting 16+ simultaneous spatial streams in both <6 GHz cellular bands and new mm-wave bands.

3.2 MIMO Technology Overview

In simplest terms, MIMO technology utilizes multiple antennas at the two ends of the link in order to exploit the spatial dimension to *extend spatial capacity, increase diversity and signal-to-noise ratio (SNR), and reduce interference* [10]. Prior to MIMO technology, time (TDMA), frequency (FDMA), and orthogonal codes (CDMA) enabled spectrum sharing. For a single link, Shannon's theorem states that the only way to increase capacity has been to increase either the bandwidth or the SNR of the signal. On the other hand, utilizing the spatial domain increases capacity by offering more parallel links in space. In the 1990s, this intuition was formalized with the discovery of the multichannel capacity formula [11,12] and the early exploration of signal processing algorithms, which can make use of the spatial dimension of the environment [13–17].

Using space does not seem like an obvious option with radio waves but it is very natural in optical systems. Optical waves are also electromagnetic radiation, albeit at a much higher frequency. Infrared and optical bands (see Figure 3.1) are commonly used for light wave communication, imaging, and ranging. In these applications, very sharp beams are created by utilizing apertures that are large relative to the wavelength.

Due to the linearity of Maxwell's equations, nearly the same can be achieved at RF frequencies, provided that the aperture is electrmagnetically large. In this context, "aperture" means the area occupied by the active or passive radiating elements, which for the sake of this chapter will consist of an array of antenna elements placed in a linear one-dimensional (1D) or two-dimensional (2D) array (Figure 3.2). Intuitively, a large aperture is needed to synthesize very narrow beams since the directivity of a radiated (or received) signal is inversely proportional to the aperture. Conventional RF antenna elements are small (subwavelength) and produce nearly omnidirectional radiation patterns. As a result, signals from incoherent transceivers quickly intermingle

Figure 3.1 Spatial multiplexing using sharp beams is common at optical frequencies, and also possible at radio frequency (RF) and mm-wave frequencies using arrays or large apertures relative to the wavelength.

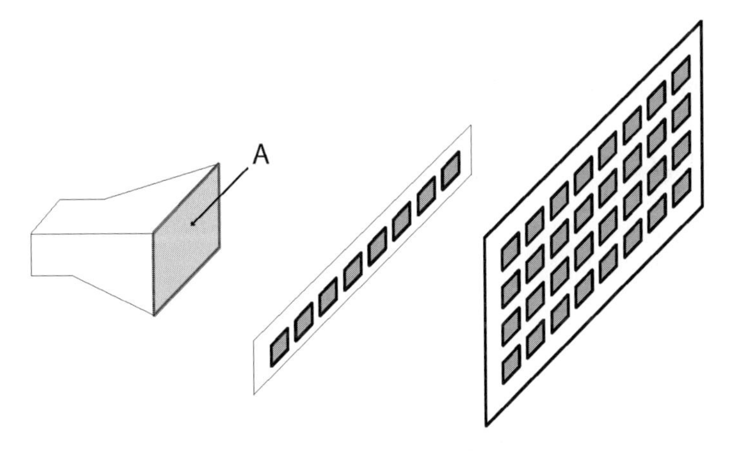

Figure 3.2 The aperture of a system can be the physical area of a directional antenna, such as a horn, or effective area of a 1D or 2D antenna array.

and interfere, making it seemingly impossible to distinguish them spatially. But if multiple transmitter or receiver antennas can be coherently excited, as shown in Figure 3.3, they have the potential opportunity to distinguish the spatial signature of signals. This system can operate as an antenna array, where cophased antenna elements synthesize an effective aperture that is many times larger than the wavelength of operation.

Under what conditions can this spatial signature be exploited? If two signals impinge on the array from the same direction (Figure 3.4), then both waves produce the same signals on each element. The antenna array provides no benefit in separating out these two signals. On the other hand, if one signal is coming in from a different angle of arrival, then the waves impart different signals on each antenna (due to the phase variation of the wavefront). This creates the opportunity to spatially distinguish the two signals.

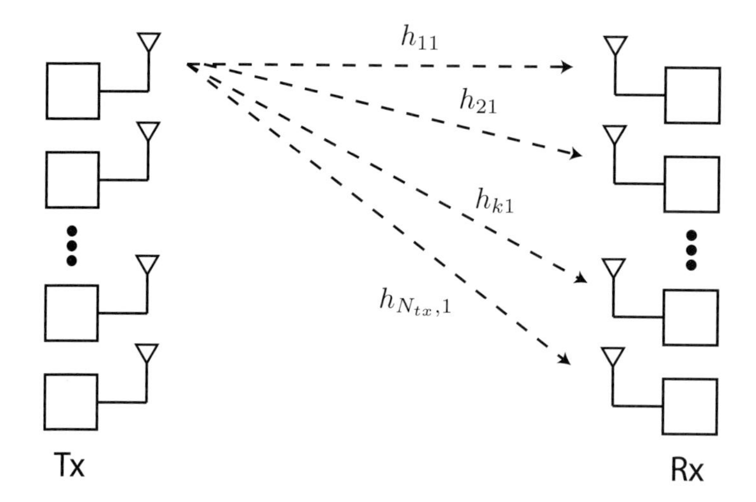

Figure 3.3 A general MIMO system consists of a number of transmitting and receiving antennas as shown.

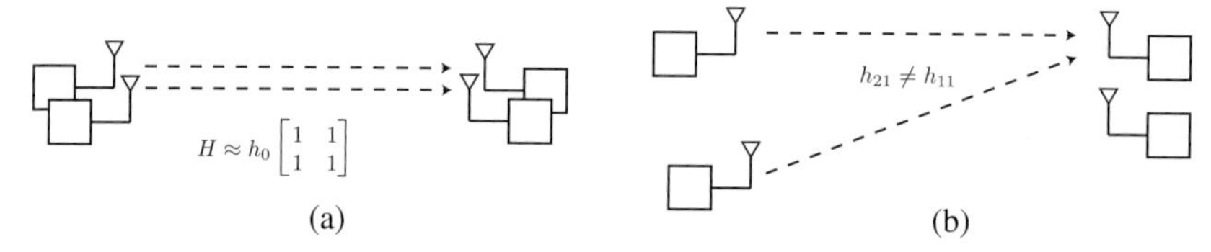

Figure 3.4 A simple 2 × 2 MIMO system cannot exploit spacial filtering when the signals arrive with (a) the same spatial signature (both LOS with very close proximity) but can distinguish (b) two signals that arrive with a different angle of arrival or different phase.

3.2.1 Spatial Multiplexing with Antenna Arrays

Even though the concept of MIMO is relatively novel, arrays of antennas have been used for spatial filtering and multiplexing for many years [18]. Let us quickly review the mathematics of phased arrays to see the connection to MIMO theory. More details of phased arrays can be found in the chapters on RF (Chapter 6) and hybrid beamforming (Chapter 2).

The motivation of traditional phased arrays is to create sharp beams of RF energy with low sidelobes. Instead of building a passive lens to focus the energy, an array of transceivers with appropriately weighted amplitude and phase is used to create a desired radiation pattern. This kind of array is widely known as a phased array since it is the phase control at each antenna (in the transmit and/or receive direction), which coherently adds signals from the desired direction of propagation of reception. As shown in Figure 3.5, the wavefront s coming from a certain direction θ impinges on a linear array, and the outputs of the array are summed together as follows

$$r(\phi) = \sum_n a_n e^{j\phi_n} s_n$$

Each element has controllable gain a_n and phase ϕ_n, which are used to synthesize the spatial response.

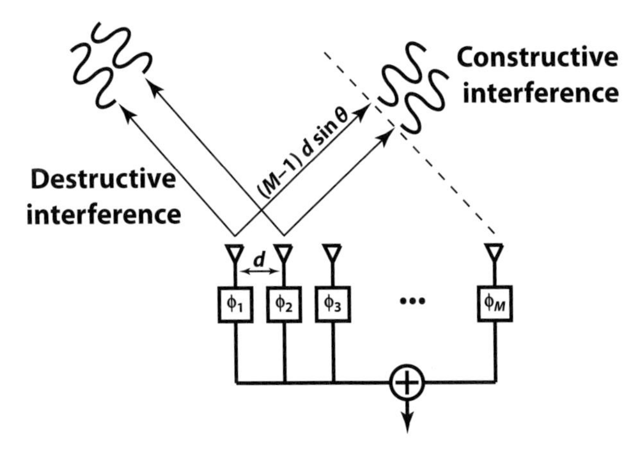

Figure 3.5 A classical linear phased array.

From the geometry of plane waves, it can be observed that the propagation delay imparts phase delay of $n \cdot kd \sin\theta$, where n is the antenna index, k is the propagation constant $k = 2\pi/\lambda$, and d is the spacing between the antenna elements. Then the received signal is given by

$$r(\phi) = \sum_n a_n e^{-jn(kd\sin(\theta)-\phi_n)} s = \sum_n a_n e^{-jn\Psi} s$$

with constructive interference occurring when $\phi_n = kd\sin(\theta)$. Note that this equation is a spatial Fourier series, or a discrete Fourier transform of the signal s along the spatial points where we sample the signal with our antenna. Fourier transform theory tells us that the spatial resolution with which signals can be distinguished is related inversely to the linear size of the array along the direction of interest.

Different antenna patterns can be synthesized by controlling a_n and ϕ_n. For a uniform array where a_n are constant, and which has linear phase profiles, the array forms the familiar pattern shown in Figure 3.6a, pointing at the desired direction. This holds true as long as the element spacing is below $\lambda/2$ to avoid spatial aliasing. If the antennas are further apart, as in Figure 3.6b, there are two main changes. The main beam is narrower, due to the fact that the aperture is larger. Also, there are multiple peaks in the array pattern (grating lobes) that arise due to spatial aliasing from undersampling the aperture. In both scenarios, the array pattern exhibits sidelobes that are somewhat weaker than the main lobe and decay with angle. Sidelobe levels are controlled by tapering the array through amplitude control. The amplitude weights a_n can be chosen in various ways (triangular or cosine patterns, for example), with exactly the same principle as Fast Fourier Transform (FFT) windowing functions in the time domain.

Finally, through linearity, phased arrays can synthesize multiple simultaneous beams simply by duplicating the phase and amplitude control elements and applying different weights to different signals of interest.

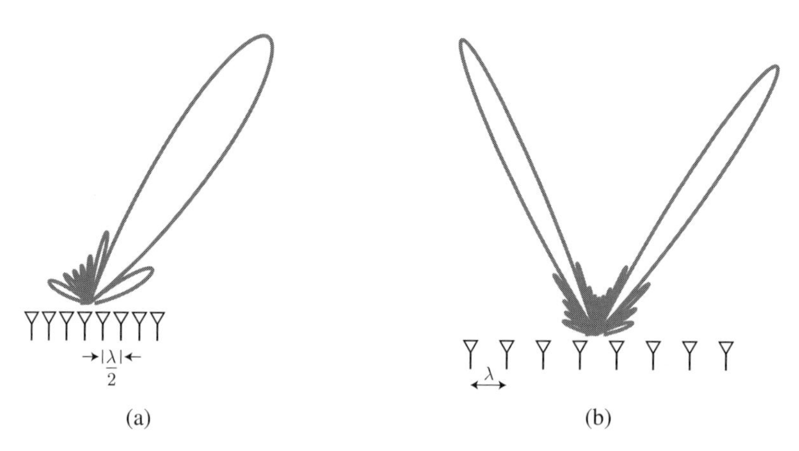

Figure 3.6 Beams formed by a phased array meeting (a) Nyquist spatial sampling versus (b) an undersampled phased array forms multiple beams due to aliasing (grating lobes).

3.2.2 MIMO: Exploiting Multipath Propagation

Traditionally, phased arrays have been designed for operation on a single direction of arrival. This results in very simple relationships for the phase and amplitude weights and for determining the sidelobe levels. How does this relate to MIMO arrays? It turns out that MIMO systems are a simple generalization of classical phased arrays.

In wireless communications, signals travel in many directions and arrive at the destination through multiple paths. If we imagine sending an impulse of energy into a transmitter, then that impulse will arrive at the receiver multiple times and at different times, since there are multiple paths from source to destination with different propagation lengths. The delay spread is a measure of the average time it takes for most of the energy to arrive at the destination. Alternatively, viewing the problem in the frequency domain, a fixed-frequency sinusoid will add constructively or destructively due to the various paths, and in general the signal will fade in some random manner with frequency or position. Since the frequency response can be obtained by taking a Fourier transform of the impulse response, either treatment is equivalent.

To handle the range of complicated propagation environments observed in communication systems, it is necessary to develop a general model for describing propagation through the environment, or channel. In general, we abstract away the specifics of any particular propagation environment and instead represent a narrowband channel by a matrix H that captures the propagation from all the transmitters to all the receivers [19]. This matrix has dimensions of number of receivers by number of transmitters. This can capture all complicated propagation effects, such as multipath, diffraction, and shadowing [20–22]. This model cleanly extends to wideband channels, where either the time-domain or frequency-domain channel response can describe the environment through its matrix impulse response or through a dispersive matrix.

The connection between classical phased arrays and MIMO arrays can be observed through the channel matrix. Classical phased arrays assume a channel matrix with a very specific form (a single direction of arrival per beam) and craft their phase and amplitude weights accordingly. By generalizing the channel matrix, we can describe arbitrarily complex environments and craft algorithms that derive optimal phase and amplitude weights to use in those environments. However, the key intuitions regarding aperture size, sidelobes, and multibeamforming translate from phased arrays.

In traditional wireless communications prior to MIMO, multipath propagation was viewed as a major detriment, requiring special techniques such as equalization in order to detect signals that are transmitted faster than the delay spread of the channel. But in MIMO communication, multipath propagation provides spatial diversity and spatial capacity that we can exploit to build more robust communication systems. Suppose that we have only a small array of transmitting and receiving antennas, small enough so that we cannot truly form sharp beams (Figure 3.3). If we assume a very narrowband modulation scheme, so that frequency selective fading can be ignored, then each antenna will receive a linear combination of the transmitted signals

$$r_i = h_{i1}t_1 + h_{i2}t_2 + h_{i3}t_3 + \cdots$$

where the coefficients h_{ij} represent the channel propagation from antenna j at the transmitter to antenna i at the receiver. The coefficients h_{ij} are the aggregate sum of the line-of-sight and multipath components due to various reflections in the environment. If we look at the output of only a single antenna, we cannot distinguish between the different t_i signals. But if we view the total summed signal from the entire array, we have (in absence of noise)

$$\vec{r} = H\vec{t}$$

where H is the matrix of channel coefficients, $\vec{r}$ is the vector of received signals, and $\vec{t}$ is the vector of transmitted signals.

If the matrix H were square, it could be simply inverted to reconstruct the individual transmitted signals. But this is an overly restrictive criterion. In practice, the link could use a different number of transmitting antennas than receiving antennas, or, even if it is square, the matrix may be ill conditioned. The rank of a matrix is a measure of the number of independent rows or columns, and is guaranteed to be bounded by the minimum of the number of rows or columns. In a wireless channel, the rank of the channel matrix describes the number of independent spatial degrees of freedom, which is at most the lesser of the number of transmitters or receivers but could be lower if the environment has degenerate propagation characteristics. For instance, a link with four transmitters and only two receivers has a matrix rank less than or equal to 2. In such a channel, only two separate spatial streams can be reliably received.

The structure of the wireless link can be clearly revealed by analyzing the singular value decomposition (SVD) of the matrix H

$$H = U\Sigma V^*$$

where Σ is a diagonal matrix, and U and V are unitary matrices. The diagonal entries of Σ are known as the singular values of the matrix H, and the number of non-zero values is equal to the rank of the matrix. Each singular value describes one orthogonal spatial degree of freedom, and the magnitude of the singular value is proportional to the gain or SNR achievable along that spatial dimension. Each mode can be excited by encoding the data along the singular vectors of the system

$$\vec{r} = H\vec{t} = U\Sigma V^*\vec{t}.$$

Multiplying the preceding equation by U^*, we have (taking advantage of the unitary properties of U and V)

$$\left(U^*\vec{r}\right) = \Sigma\left(V^*\vec{t}\right).$$

Defining a new vector of received signals and encoding the transmitted signals along the V space, we have

$$\vec{\tilde{r}} = \Sigma\vec{\tilde{t}}.$$

Since Σ is a diagonal matrix, this representation makes it obvious that the system can sustain $k = \text{rank}(H)$ independent streams. The right and left singular vectors provide a natural basis for linearly encoding and decoding, respectively, the data streams. As long as the matrix H can be estimated, the SVD technique provides at least an existence proof that it is possible to obtain optimal transmit and receive weights for MIMO communication. These weights are the MIMO generalization of the gain and phase weights in classical phased arrays.

There are some very important questions to answer with regard to MIMO. First, what determines the rank of the matrix? Second, is the complexity of the pseudoinverse computation manageable? Finally, from a practical perspective, how do we measure the channel matrix H?

3.2.3 Channel Rank

To answer the first question, it is useful to build a physical intuition for what the SVD is doing. Essentially, the SVD is identifying orthogonal pipes for sending data, based on the propagation information described in the channel matrix. Framing it in traditional antenna array terminology, the algorithm is creating antenna patterns that favor propagation along certain directions while attenuating (nulling) propagation along other directions. We can reconstruct the effective antenna pattern by taking the complex amplitude applied to each element of the array (see the sections on phased arrays).

One simple example is shown schematically in Figure 3.7, where two separate antenna patterns are synthesized, one that creates directivity in the desired direction of arrival for stream 1 and a null in the direction in stream 2, and another pattern that does the opposite. For this reason, this method of MIMO is sometimes called "zero forcing," since it tends to put a zero in the spatial transfer function in the direction of interference. With this perspective, it is clear that the matrix H encodes an image of the environment in terms of the propagation and blockage properties at the frequency of propagation.

It is therefore intuitive that the rank of the matrix should be related to the richness of the scattering environment. In free space in the absence of any reflectors, there exists

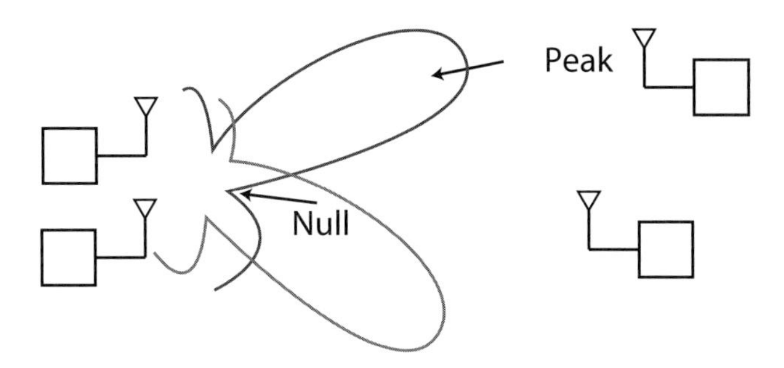

Figure 3.7 A simple 2 × 2 MIMO system orthogonalizes two beams by placing nulls in the direction of the arrival of the interference.

only a single path from the source to the destination. Under these conditions, if the antennas are colocated, there is no spatial diversity and the channel rank is just one. By introducing more reflectors and therefore new propagation paths, the channel rank is increased, *creating* more diversity and/or capacity in the channel.

It is worth mentioning that even though this discussion has considered a narrowband signal (in a flat fading channel), frequency variations can be taken into account by performing the processing in the frequency domain. This is especially convenient when using OFDM modulation schemes that naturally partition the band into subcarriers with sufficiently low bandwidth to avoid intersymbol interference.

3.2.4 Multiuser MIMO (MU-MIMO)

In outdoor environments, there is usually only a single dominant propagation path to each user, meaning that the channel rank tends to be very low even if that user is equipped with many antennas. This means that the capacity gain from conventional single-user (SU) MIMO is minimal and cellular networks have gained only modestly from adopting classic MIMO techniques.[1] A better approach is to increase the capacity of not a single link, but the aggregate links of many users by simultaneously sending each user its own data. The effective channel rank increases substantially since the users are far from each other and therefore experience highly uncorrelated propagation. Multiuser (MU) MIMO is an extension of MIMO techniques for multiple users. Generally it is assumed that each user only has one or at most very few antennas, whereas the base station can be very large (see Figure 3.8). This asymmetry is simply due to the fact that users are often mobile and both energy and size constrained.

It is easy to understand MU-MIMO as traditional MIMO where the aggregate of users forms a single larger antenna array. However, MU-MIMO must account for one key restriction that does not exist in single-user systems. Because the users cannot easily cooperate with each other, the spatial signal processing is asymmetrically partitioned

[1] In contrast, indoor WiFi links exhibit much richer propagation and have therefore benefited greatly from SU-MIMO.

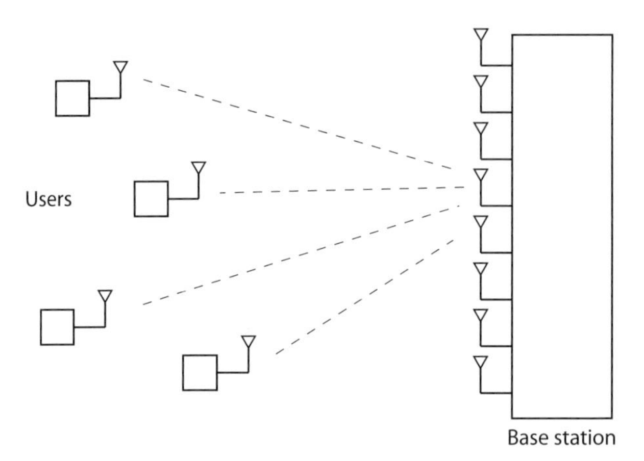

Figure 3.8 A multiuser MIMO (MU-MIMO) scenario where several users communicate simultaneously with a base station with many antennas. In massive MIMO systems, the number of antennas N is much larger than the number of user streams K, or $N \gg K$.

between base station and users. In the uplink, when the multiple users are transmitting simultaneously to the base station, the base station has (ideally) full knowledge of the channel matrix. It can form multiple beams in the users' directions, with each pattern pointing to one user and attenuating the others. In the downlink, the base station has again the full channel matrix and synthesizes transmit beams, which similarly transmit to each user without interference among the users. As a result, the users can be agnostic to any information about the other active users.

3.3 Conventional MIMO Processing

Since the discovery of MIMO theory, a significant research effort has been devoted to developing *practical* receiver and transmitter algorithms that could achieve or approach the channel capacity. This section reviews specifically those processing techniques suitable for MU-MIMO [23,24].

3.3.1 Channel Estimation

All modern wireless standards rely on coherent detection, meaning that the receiver estimates the channel gain and uses that knowledge to recover the amplitude and phase of the transmitted waveform. Channel estimation is accomplished by periodically transmitting known pilots [6]. In single-input and single-output (SISO) links, the channel estimate is used just as a demodulation reference. In MIMO systems, the channel estimate is also used to configure the spatial processing algorithm. Consequently, it is safe to assume that the receiver has available some channel state information (CSI). For example, in a MU-MIMO uplink, it is relatively straightforward for the base station to acquire an estimate of the propagation environment to all of the active users.

On the other hand, the transmitter has no obvious way to obtain CSI. Generally any transmit-side CSI is acquired by feeding back the receive-side channel estimates. Generally this overhead is considered unacceptable, so it is much more challenging to obtain good CSI at the transmitter. For instance, in a MU-MIMO downlink, it is fairly challenging to get good CSI at the base station in order for it to precancel interuser interference. There is one very important exception to this rule. If the uplink and downlink use the same frequency, the propagation environment is reciprocal – meaning that it behaves identically in both cases. This is a result of the reciprocity of electromagnetic theory. In a reciprocal uplink/downlink pair, the channel matrix, which is estimated in the uplink, can be *reused* for the downlink. This provides a very elegant solution to the challenge of obtaining transmitter CSI for a MU-MIMO downlink.

3.3.2 Linear Beamforming

The simplest spatial processing algorithm used in MIMO communication links is linear beamforming (Figure 3.9). As the name suggests, this is a conceptually straightforward extension of classical beamforming in phased arrays. User streams are transformed into antenna signals (and vice versa) through an $M \times K$ complex beamforming matrix. This

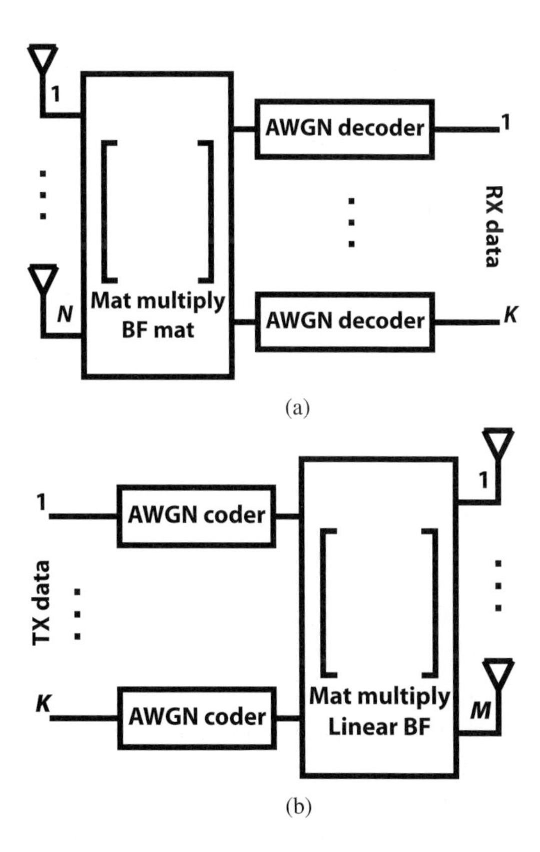

Figure 3.9 Linear beamforming at (a) RX and (b) TX.

just means that each user has a complex gain and phase weighting for each antenna. In the transmit direction, the $M \times K$ transmit beamforming matrix G_{tx} is used to compute the output voltage at every antenna by

$$y_{tx} = G_{tx} s_{tx}. \tag{3.1}$$

In the receive direction, signals are reconstructed using the $K \times N$ receive beamforming matrix G_{rx} by

$$s_{rx} = G_{rx} y_{rx}. \tag{3.2}$$

In a phased array, the beamforming weights are calculated based on the desired direction of arrival/departure, and potentially sidelobe requirements. In communication links, the estimated channel matrix is used to directly compute those weights, thus capturing all relevant information about the spatial propagation environment.

Different linear beamforming algorithms are distinguished based on the cost function they seek to minimize. The simplest flavor of linear processing is conjugate beamforming, which seeks to maximize each user's SNR independently of all other users [25]. It turns out that conventional phased-array beamforming is a special case of conjugate beamforming. Two other common cost functions seek to minimize interuser interference and to maximize the signal-to-interference-plus-noise ratio (SINR); these result in the zero-forcing (ZF) [26] and minimum mean-squared error (MMSE) [27] beamformers respectively. Intuitively, the ZF beamformer places nulls in the directions of all other users to eliminate any interference but does not consider the impact of thermal noise. In contrast, the MMSE algorithm optimally balances interference and noise.

The differences between these algorithms can be readily visualized for a line-of-sight (LOS) channel. Figure 3.10 compares the conjugate, ZF, and MMSE array patterns for user 1 of an 8×2 MIMO LOS channel. The ZF and MMSE techniques cancel user 2's interference at the expense of a wider mainlobe and increased sidelobe levels. Note that this is the same trade-off one would observe in traditional phased arrays.

3.3.3 ML and Near-ML Receivers

Linear beamforming is agnostic to any structure in the signals of interest and is therefore broadly applicable. However, higher-performance receive techniques can be used in communication links where the ultimate goal is to minimize demodulation errors. The optimal maximum a posteriori (MAP) receiver uses the received signal to determine which transmit signal was most likely to have been sent. In most cases, this is equivalent to the maximum likelihood (ML) receiver by

$$s_{rx} = \text{argmax } P(y_{rx} | H). \tag{3.3}$$

ML detection searches for the transmitted symbol, which maximizes the probability of observing what was received. This is probably optimal for all communication links, meaning that it achieves the full Shannon capacity.

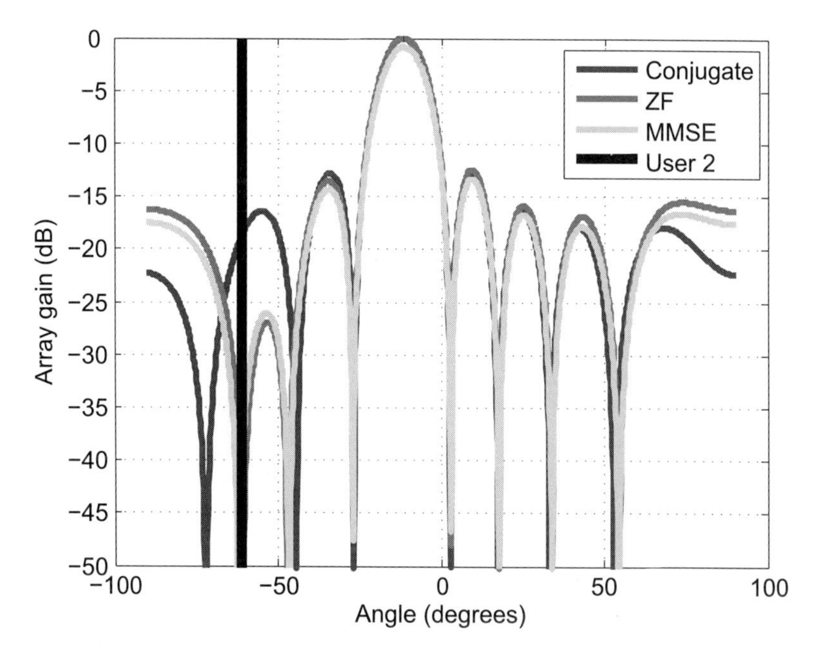

Figure 3.10 Comparison of conjugate, zero-forcing, and MMSE beam patterns for an eight-element array serving two users.

Unfortunately, true ML requires an exhaustive search over all possible transmitted sequences. For MIMO, the computational complexity is exponential with the MIMO order. To overcome this, the ML receiver can be relaxed to a linear problem by

$$s_{rx} = \operatorname{argmin} |y_{rx} - H s_{tx}|^2. \tag{3.4}$$

This forms the basis for a family of simplified approximate ML algorithms, designed to achieve near-ML performance with lower cost. The two most common are sphere decoding and K-best decoding [28,29], both of which achieve near-ML performance with complexity that is approximately cubic in the MIMO order.

3.3.4 Successive Interference Cancellation

Another popular receiver algorithm for MIMO channels is successive interference cancellation (SIC) (see Figure 3.11), which is a spatial analog of a decision feedback equalizer (DFE). The key idea is to iteratively process a single user, compute that user's contribution to the receive signal at each antenna, and subtract it out [30,31]. In this way, subsequent users experience reduced interference.

In more detail, MMSE-SIC processing proceeds as follows. First, the users are ordered by signal strength. The strongest user is spatially processed using the full MMSE beamforming matrix. Then, that user's signal is demodulated and the estimated contribution at every antenna is subtracted out. After that, the next strongest user is processed in the same way but using the reduced MMSE beamformer for users 2

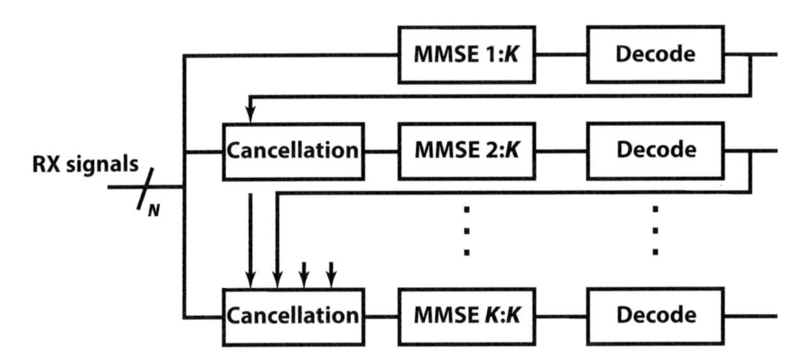

Figure 3.11 MMSE successive interference cancellation (SIC) receiver.

through K, and so on. Like the ML receiver, MMSE-SIC is theoretically optimal. As with all other decision-feedback receivers, however, it can suffer from error propagation. As such, good performance is only practically achieved in strong channel conditions; for weaker channels, sphere decoding performs better.

3.3.5 MU Downlink: Dirty Paper Coding

Of the preceding algorithms, only linear beamforming is applicable to the MU-MIMO downlink. The others are all receive-side spatial processors.

Interestingly, transmit processing must precancel the interuser interference, which can itself be predicted by the transmitter. This scenario describes the information theoretical result known as dirty paper coding (DPC), which states that if a channel is corrupted by interference that is known perfectly at the transmitter, the transmitter can code around the interference and achieve the capacity of the interference-free channel [32]. DPC itself is a theoretical result, and still requires codes to be invented for a particular scenario. One example is Tomlinson–Harashima precoding (THP), which describes a transmit-side DFE to cancel intersymbol interference [33,34]. For MIMO, a suboptimal implementation of DPC for DSL crosstalk cancellation was proposed in [35] using the QR decomposition. This technique was expanded by others [36] for MU-MIMO. In practice, these schemes suffer from very high computational complexity due to the noncausal joint processing of all users. For this reason, DPC techniques have not been used in wireless systems, and so modern standards use linear beamforming for the MU-MIMO downlink.

3.3.6 Massive MIMO: High-Order MU-MIMO

MIMO has been widely deployed in existing 802.11n/ac standards and 4G LTE. SU-MIMO capabilities were introduced very early on with MU-MIMO following shortly thereafter.

The development of 5G networks has focused on achieving very large spatial capacity gains using high-order MU-MIMO (capable of communicating with 16 or more terminals simultaneously). For comparison, existing MIMO deployments are largely limited to two or four simultaneous users in theory, and fewer in practice. Targeting such a large number of users raises entirely new system design challenges. First of all, serving a larger number of users necessarily requires creating more spatial degrees of freedom through a larger number of base station antennas. Second, the computational complexity high-order MIMO processing becomes an important consideration. The MIMO algorithms surveyed in the preceding suffer from a strong performance-complexity trade-off. Near-optimal techniques such as sphere decoding are practical in 4×2 or 4×4 MIMO links, but require excessive computational power in a 64×16 MIMO link.

Fortunately, theoretical developments in communication theory have provided an elegant approach for confronting this challenge, commonly referred to as massive MIMO. The key idea, discovered by Marzetta, is that in MU-MIMO links with a larger number of base station antennas compared to users, linear beamforming is asymptotically optimal as the number of base station antennas grows large [37–39]. In simple terms, this states that with a sufficiently large base station, we can have our cake (serve a very large number of users), and eat it too (only using the simplest form of spatial processing)!

Massive MIMO is a very promising component of 5G technologies for three reasons. First, MU-MIMO is a natural area of focus since it addresses a key problem of providing high data rates to many users while also natively and robustly providing high spatial diversity in any environment. Second, massive MIMO addresses the key challenge of providing spatially multiplexed data streams to a very large number of users and appears to scale to very large numbers of users indeed. Finally, the key ideas behind massive MIMO scale well to different frequency bands and propagation environments; in particular, the very same concepts can be easily merged with the design of analog and hybrid phased arrays at mm-wave frequencies. This is quite attractive, not least because operation in mm-wave bands already requires the use of antenna arrays to overcome propagation loss.

A number of measurement campaigns have been undertaken to determine how well massive MIMO might perform in practice, with encouraging results [40,41]. Based on this theoretical promise and experimental validation, LTE Advanced Pro (Release 15+) has introduced Full-Dimensional MIMO (FD-MIMO) [42,43]. This standard increases the base station size up to 32 or even 64 transceivers, supporting 16 or 32 spatial streams. FD-MIMO introduces new reference signal schemes to estimate this high-dimensional channel efficiently. FD-MIMO also introduces new base station hardware models to accommodate these challenging spatial multiplexing requirements.

3.4 System Architecture for Large Arrays

While massive MIMO provides an elegant way to achieve tremendous spatial multiplexing, it also requires the implementation of very high channel count base stations in an economical fashion. Studies have shown that the ratio of base station transceivers to users should be in the range of 4:8. This tells us that to serve 16 or 32 users simultaneously, base stations with 64 to 128 antennas will be required. The large number of transceiver chains leads to elevated power consumption, synchronization challenges, and signal routing constraints. These challenges have spurred research into system architectures that are specially suited for massive MIMO base stations.

It is worth pointing out that at mm-wave frequencies, the small wavelength makes it possible to implement hundreds of antenna elements in a router-size form factor. Such a system would further stretch the design challenges of practical massive MIMO architectures, on top of the existing challenges of mm-wave radio design.

3.4.1 State of the Art

Massive MIMO below 6 GHz

A number of groups in academia and industry have development massive MIMO prototypes operating in cellular bands below 6 GHz [44–49]. These testbeds all share similar characteristics: 32 to 128 elements, serving 10 to 20 users over 20 MHz of bandwidth, with a highly digital and centralized implementation. Each antenna is served by a full transceiver chain and local data converters (ADC and DAC). The digital I/Q samples are transferred over a high-speed, high-capacity backplane to and from the central processor, which is itself responsible for the entirety of the digital signal processing. This processor generally consists of one or more field programmable gate arrays (FPGAs).

The fully centralized architecture is both conceptually and practically simpler, since all data are available in one place. However, this comes at the cost of very high data throughput requirements in the backplane, and very high computational load on the central processor. For example, the Lund University testbed requires a backplane with over 450 Gbps of aggregate capacity [48]. Similarly, the Samsung prototype has 80% FPGA utilization (on a single high-end Xilinx Virtex 7-690T model) using only 20 MHz bandwidth with 12 simultaneous users.

One work has proposed a massive MIMO testbed using distributed processing [44]. In this architecture, each pair of transceivers is equipped with a small local FPGA that is responsible for performing most of the digital front-end tasks (filtering, FFT, etc.) and, crucially, distributed conjugate beamforming. As described in more detail later in this section, beamforming weights are both computed and applied locally; the interconnect only moves around the user data streams rather than the antenna I/Q samples. In this implementation, the distributed beamforming is limited only to conjugate beamforming.

Phased Arrays for mm-Wave Communications

At the same time, the past decade has witnessed the development of mm-wave CMOS technology, and today we are on the brink of commercializing of mm-wave radios. Early effort was focused on the unlicensed 60 GHz band beginning about a decade ago. More recently, as the FCC and other regulatory bodies have begun to open up other mm-wave bands (such as 24, 28, 39, and 71 GHz) for 5G deployments, there has been a concerted effort to develop phased arrays for those frequencies. Today a large number of groups have reported highly integrated phased arrays with anywhere from 4 to 32 elements on a single die, using analog beamforming to form a single beam [50–56].

Research effort has now turned to copackaging subarrays to realize improved beamforming capabilities (either narrower beams or greater number of beams) at the module level. In these systems, package- or board-level analog combining can form a small number of aggregate beams. IBM and Ericsson [57] report a single-chip solution with two cross-polarized 16-element phased arrays at 28 GHz. They combine multiple such chips on a package to form a 64-element array, which can be operated either as a 64-element one-beam array or as four independent 16-element arrays. Because the same front-end cannot form more than one beam, there is a tight trade-off between number of beams and directivity. In a similar vein, the University of California, San Diego (UCSD) [58] has demonstrated a four-element 28 GHz front-end integrated circuit (IC); eight of these ICs are combined on board to form a 32-antenna beam with 300 m range [59].

Finally, some researchers are beginning to investigate forming multiple beams from a single mm-wave array. The authors in [60] report a 60 GHz front-end with two separate four-element arrays, each forming a single beam. The two beams are digitally processed for interference nulling. North Carolina State University [61] has shown the ability to form two beams from a single four-element subarray at 60 GHz. However, the two beams cannot be independently steered; rather, the auxiliary beam direction is constrained based on where the other beam is pointed.

3.4.2 A Scalable Beamforming-Aware Array Architecture

As discussed earlier, linear beamforming is the preferred candidate for enabling aggressive spatial processing due to its favorable combination of low computational complexity along with high performance in the massive MIMO regime. However, it is not sufficient just for the algorithm have minimal complexity. It is equally necessary to ensure that the actual hardware that is deployed – consisting of radios, data converters, signal processing, data interconnect, synchronization, etc – be simple, low cost, and low power. This motivates the design of array architectures that are suitable and optimal for realizing massive MIMO systems in practice.

We can identify three main system design goals:

1. The array architecture should be modular and scalable, such that the number of antenna elements and number of beams could be easily changed in the design phase.
2. The array architecture should abstract away implementation details as much as possible, since these may differ widely based on system specifications, carrier frequency, and technology evolution.
3. The array architecture should be translatable between < 6 GHz cellular bands and high-frequency mm-wave bands.

Viewed as a full system, the massive MIMO array appears as a black box with two interfaces: on one side the physical antennas occupying the desired array aperture, and on the other side a logical and physical data port that connects to the outside network. In this light, the system architecture design boils down to two key questions. First, how should the antennas be connected to the processing element that interfaces to the data port? Second, how should the required hardware and signal processing functions be organized, ordered, and grouped?

These design goals are addressed in this section by proposing an array architecture suitable for a large range of implementation goals and scenarios [62,63]. The key conclusion is that by exploiting the natural parallelism of linear beamforming, large beamforming arrays can be readily mapped into an efficient, modular, and scalable hardware architecture.

Large Arrays Must Use Distributed Processing

Because the antennas have a physical size and spacing that is on the order of the wavelength, an array with many elements will be physically large relative to the carrier frequency. For instance, arrays operating in the low GHz range will have dimensions on the order of meters; arrays operating at 60 GHz will have dimensions on the order of tens of centimeters. At the same time, the incoming and outgoing data streams have a single physical interface to the higher layers of the network and/or application stack. As a result, a key feature of any antenna array is the dispersal and aggregation of information between the physically dispersed antenna elements and the central processor/network interface. In massive arrays, data movement is the main bottleneck and affects many aspects of the array implementation.

As a reference point, consider a fully centralized array architecture (left panel of Figure 3.12). In this architecture, all computation is performed at the central processor – in the transmit direction, this processor computes the signal for every single antenna element, while in the receive direction every receiver forwards its ADC samples for

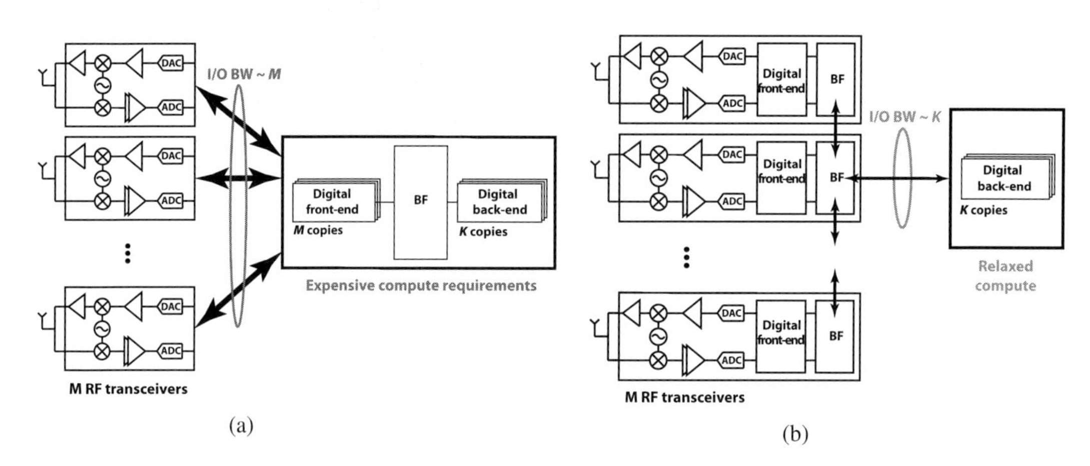

Figure 3.12 Comparison of (a) centralized and (b) distributed processing architectures for massive MIMO arrays.

processing. In this scenario, the central processor needs a total I/O bandwidth proportional to the number of antennas (M) as

$$R_{centr} = Mf_sN_b \tag{3.5}$$

where f_s is the data sampling rate and N_b is the number of bits for each sample. As an example, a 128-element array operating over 100 MHz bandwidth with 20 bits for I/Q samples would require at least 256 Gbps I/O bandwidth in the processor. This would require a complex data aggregation and routing network and would consume a large number of I/O on the processor.

Is there anything that can be done about this? The key observation is that while each antenna transmits/receives a different signal, these signals are not linearly independent. Rather, there are only K unique signals in the system, corresponding to the K user data streams. It should be possible to exploit this redundancy to reduce the dimensionality of the data interconnect and move around only K rather than M signals as

$$R_{distr} = Kf_sN_b. \tag{3.6}$$

This dimensionality reduction can be unlocked through distributed beamforming. Figure 3.13 shows how the beamforming and data distribution are implemented for uplink and downlink cases, using the distributed array architecture. Because matrix multiplication is highly parallel, the multiply and accumulate operations can be straightforwardly distributed to each remote transceiver or implemented as a reduction tree inside the data distribution network. This results in the distributed processing architecture shown in the right panel of Figure 3.12. Depending on the modulation scheme and where beamforming fits in the signal processing chain, antenna-specific signal

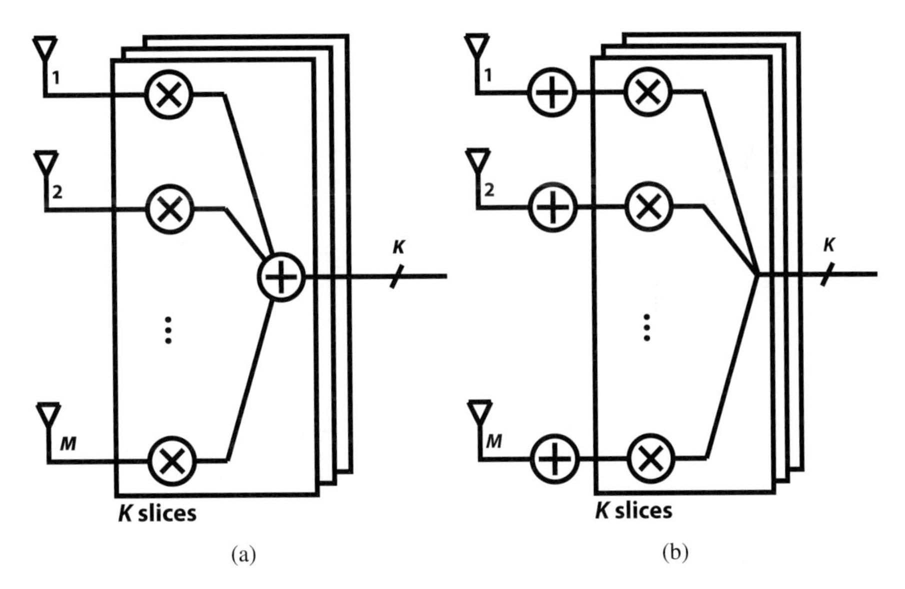

Figure 3.13 Implementation of distributed beamforming and data interconnect for (a) uplink and (b) downlink.

processing functions (denoted by a digital front-end) are also implemented locally at each transceiver.

Distributed beamforming could be performed with any beamforming matrix, regardless of which cost function is used to compute those weights. Fortunately, any linear beamforming operation can be split into a conjugate beamforming step followed by a $K \times K$ "cleanup" matrix that achieves the desired cost function – e.g., zero-forcing or MMSE. This post processing matrix has reduced dimensionality and does not affect the data movement requirements. In light of this, conjugate beamforming is a natural candidate for the distributed beamforming without imposing any performance penalty.

It is worth noting that this discussion readily extends to analog beamforming implementations. For example, RF or local oscillating (LO) phase shifting in many mm-wave arrays today is more distributed than it is centralized – the phase shifter is physically close to each antenna while a network of relatively large combiners is used to combine or split the signal power.

Long-Distance Interconnect Must Be Digital

The distributed signal processing described in the preceding subsection is suited equally well to analog and digital implementations. Paired with a digital interconnect, data are distributed and aggregated using serializer/deserializer (SerDes) lanes and digital adders. With an analog interconnect, signal distribution and summation are performed using analog splitters and combiners.

When the number of antennas *and* the number of beams is small, all-analog interconnect is preferred [57,59]. However, analog signal distribution does not scale well as the number of antennas or beams is increased. Dense and long-distance analog routing (especially at the board level) suffers from loss, crosstalk, and routing congestion. Particularly for a large number of beams, crosstalk and electromagnetic interference (EMI) management can significantly drive up the complexity and cost of the distribution network while the loss erases much of the SNR gain from beamforming.

Based on these issues, we can conclude that even at mm-wave frequencies, *long-distance* routing will favor a digital interconnect. Digital data distributed more flexibly and scalably extends to large numbers of elements and users. In practice, this means that a cluster of antennas should be coprocessed (either in analog or digital fashion, depending on the signal processing requirements), and each cluster should communicate to its neighbors and central processor with a digital interconnect. It is important to point out that there is no precise defintion of "long-distance." Rather, the trade-off between analog and digital interconnect is a function of the specifications (data rate, distance, etc.) and the available interconnect technology. However, this does suggest that all large arrays, regardless of carrier frequency, will favor an all-digital or hybrid analog–digital implementation, with hybrid dominating at high carrier frequencies.

The Array Should Be Composed of Common Modules

Thus far, we have proposed that signal processing should be implemented in a distributed fashion, with digital interconnect providing long-range communication capabilities between the distributed nodes and the central processor. This naturally suggests

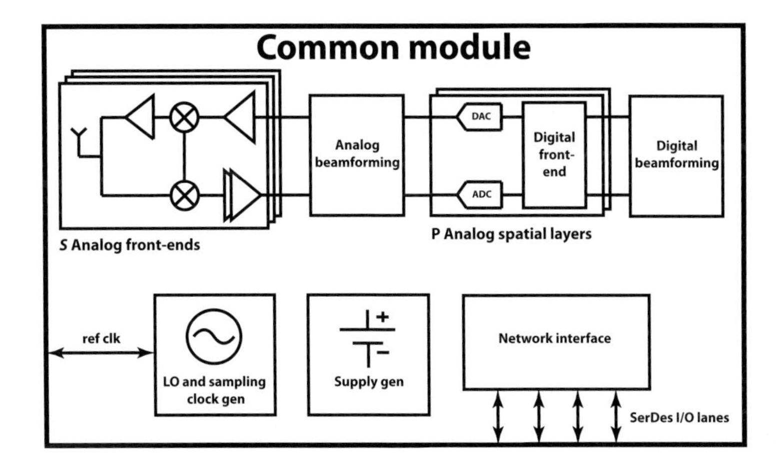

Figure 3.14 Block diagram of a generic common module for a massive MIMO array, including signal path and shared support functions.

grouping a cluster of S nearby antennas together into a subarray, implemented by a common module that encapsulates the transceivers, distributed signal processing, and analog/digital conversion (Figure 3.14).

One key benefit of this architecture is sharing auxiliary functions across multiple elements. For example, each common module can share a network interface, frequency generation, and supply generation, reducing the overhead of those functions. This also ends up creating hierarchy within the array, which significantly simplifies the design and implementation of complex systems.

How should the number of antennas per module, S, be chosen? This is largely an engineering decision, which trades off the benefits of sharing functions against the challenge of copackaging a large number of transceivers and antennas. In fact, the common module provides a logical packaging and assembly unit. Each common module could consist of one or more chips along with in-package antennas. Much of the packaging complexity comes from the area budget of a single die and the I/O and routing area on package for antenna feedlines. The level of module integration depends on the carrier frequency and bandwidth, as well as the transceivers' silicon area and other engineering considerations. At mm-wave frequencies, it is common to integrate 32 or more elements on a single die, based on silicon area, number of I/O pads, and antenna routing length. At lower carrier frequencies, generally the number of elements per die is smaller since the silicon area tends to be larger (for the passives) and the antennas are farther apart.

The module abstraction also provides a clean logical partition in the hierarchy. The implementation of the module is an engineering decision that should not impact how the overall system is put together. For example, a module could be implemented as a single mixed-signal system-on-chip, as separate analog and digital chips, or even as multiple front-end ICs with analog combining on-package and an FPGA-based digital processor. The module abstraction hides all of these implementation decisions behind a common interface.

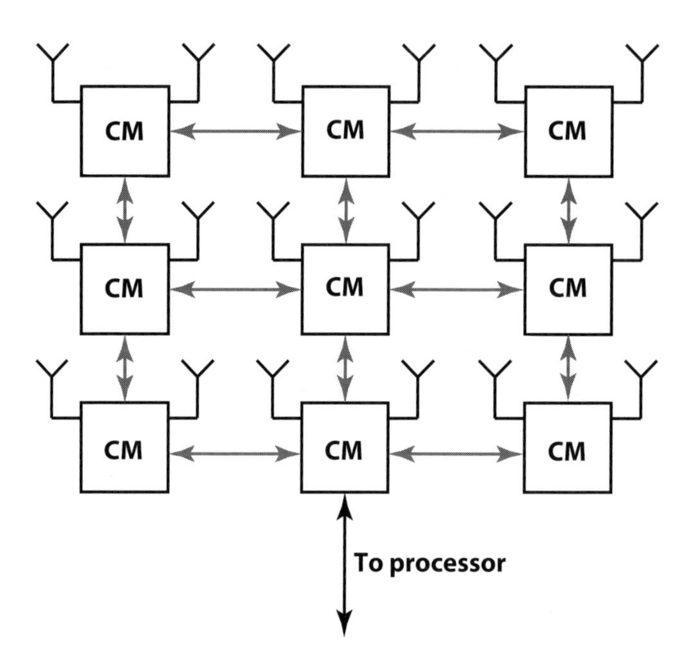

Figure 3.15 Modular and scalable implementation of a massive MIMO array using common modules.

As a result, the modular design permits simple scaling to larger number of antenna elements as depicted in Figure 3.15. Since the modules are identical, more can be added onto the interconnect without changing the fundamental way in which the array is organized. For these reasons, this distributed, modular design is the preferred design paradigm for implementing next-generation large antenna arrays.

3.5 Impairments in Large Arrays

In addition to system-level design challenges, the analysis of impairments in large arrays is somewhat different than in more conventional communication systems. The massive regime presents new challenges but also new opportunities to manage these effects.

3.5.1 Synchronization

Since beamforming comes down to applying amplitude and phase weights at each transceiver in order to point the beams, it is obvious that phase noise would corrupt the beamforming weights and lead to beam pattern error. In fact, since it is only *relative* phase shifts that affect the beam pattern, only *uncorrelated* phase noise between the front-ends affects the beam pattern. Several groups have investigated the impact of uncorrelated phase noise in massive MIMO systems [64–66].

The authors in [67] analyzed the synchronization subsystem in an OFDM massive MIMO array, consisting of a common low-frequency reference distributed to all the transceivers, an RF phased-lock loop (RF-PLL) for LO generation independently at each front-end, and an array-level carrier recovery (CR) loop using embedded pilots to track and cancel residual phase noise. The authors showed the following:

- Uncorrelated phase noise between front-ends averages with a gain of M.
- Uncorrelated phase noise below the CR bandwidth corrupts the beam pattern and leads to both intra- and interuser interference
- The optimum bandwidth in the RF-PLL is a function not only of the relative phase noise levels of the reference and the voltage-controlled oscillator (VCO), but also a function of the CR bandwidth.

In short, it was shown that there is an *optimal* level of phase noise correlation in massive arrays, and it is the job of the LO generation circuitry to manage that level of correlation. If that design objective is met, then averaging of uncorrelated phase noise presents opportunities to simplify the design specifications of the VCO and PLL. These insights apply as well to different modulation schemes and carrier frequencies.

3.5.2 Reciprocity and Channel Estimation

Channel estimation plays a critical role in any MIMO system since the channel estimates are used directly in crafting and applying the MIMO transmit/receive signal processing. Massive MIMO presents also a unique channel estimation challenge. In the uplink, each user transmits its pilots successively while each base station transceiver can estimate its channel to that user in parallel. As a result, the channel estimation burden is proportional to the number of users (K). In the downlink, the users can estimate their channels in parallel, but each base station antenna must be processed serially. This results in an overhead proportional to the number of antennas (M). Since by design $M \gg K$, the downlink channel estimation overhead is much greater.

To avoid this problem, most massive MIMO systems use reciprocity in a time-division duplex (TDD) channel. If the same carrier frequency is used in the uplink and downlink, the propagation environment is identical and therefore the uplink channel estimate can be reused in the downlink. However, although the environment itself is reciprocal, the base station front-ends are not. In order to accurately form downlink beams, the gains and phases of the transceiver front-ends must be calibrated to a relative (not absolute) reference.

One simple way to do this is to use an external antenna with over-the-air calibration. However, this is expensive and not practical in real deployments. The authors in [44] proposed a relative calibration scheme using one antenna in the array as the reference and successively transmitting and receiving calibration standards from that one antenna. This is a more attractive approach, but it is very challenging to realize good transmission between elements within a 2D array. Reciprocity calibration remains an open challenge in massive MIMO arrays.

3.5.3 Transmit Linearity

Multiuser beamforming imposes significant new constraints on transmitter design. Both amplitude weighting for beam pattern synthesis, as well as sending multiple-user data streams from a single transmitter, tend to increase the dynamic range and the peak to average power ratio (PAPR) of the transmit data. New system and circuit techniques are needed to handle this challenge.

There are three main opportunities for dealing with this requirement. First, it has been shown that in-band and out-of-band nonlinear products can be partially uncorrelated and experience some averaging over the air [68,69]. Further study is needed to understand how much gain in error vector magnitude (EVM) and adjacent channel leakage ratio (ACLR) could be obtained from this effect. Second, the central limit theorem suggests that the PAPR of the multiuser signals will remain bounded and close to that of a complex Gaussian signal. Finally, some groups have studied PAPR-aware beamforming, which constrains the beamformer design to transmit a constant envelope signal from each transmitter [70,71].

Efficient transmitter design for massive MIMO systems remains an open area of research with a number of promising avenues for improving performance and reducing complexity.

3.6 Conclusion

This chapter has provided an overview of MIMO communication, from single-user variants to multiuser variants that are being considered for 5G systems. MU-MIMO has the potential to greatly increase the aggregate spectral capacity of wireless mobile communication systems by the application of spatial multiplexing and interference cancellation. As we have seen, MU-MIMO applies equally well to both indoor and outdoor channels because the users are scattered through the environment, and spatial diversity is naturally provided by the varying angle of arrival of signals coming into the base station. When the number of antennas greatly exceeds the number of users, the so-called massive MIMO domain, the signal processing algorithms become potentially simpler to implement with practical hardware. But as we have seen, for massive MIMO to become widespread, many challenges must be addressed. In particular, processing signals from hundreds of antennas, each with a wide bandwidth (100 MHz to 1 GHz), poses several challenges from data routing and back-end signal processing. Novel techniques that perform distributed processing, combined analog/digital processing, and reduced resolution signal acquisition have the potential to make massive MIMO practical and low cost.

References

[1] Cisco, "Cisco Visual Networking Index: Forecast and methodology, 2016–2021," Tech. Rep., June 2017.

[2] C. E. Shannon, "A mathematical theory of communication," *Bell System Technical Journal*, vol. 27, no. 3, pp. 379–423, 1948. [Online]. Available: http://dx.doi.org/10.1002/j.1538-7305.1948.tb01338.x

[3] R. Gallager, "Low-density parity-check codes," *IRE Transactions on Information Theory*, vol. 8, no. 1, pp. 21–28, 1962.
[4] J. Salz and S. B. Weinstein, "Fourier transform communication system," in *Proceedings of the First ACM Symposium on Problems in the Optimization of Data Communications Systems*, ACM, 1969, pp. 99–128.
[5] M. Zimmerman and A. Kirsch, "The AN/GSC-10 (KATHRYN) variable rate data modem for HF radio," *IEEE Transactions on Communication Technology*, vol. 15, no. 2, pp. 197–204, April 1967.
[6] *Wireless LAN Media Access Control (MAC) and Physical Layer (PHY) Specifications*, IEEE Std. 802.11-2012.
[7] *Evolved Universal Terrestrial Radio Access (E-UTRA); LTE Physical Layer; General Description*, 3GPP Std. TS 36.201, 2015.
[8] A. Goldsmith, "5G and beyond: What lies ahead for wireless system design," in *Proceedings of PIMRC 2014*, 2014.
[9] J. G. Andrews, X. Zhang, G. D. Durgin, and A. K. Gupta, "Are we approaching the fundamental limits of wireless network densification?" *IEEE Communications Magazine*, vol. 54, no. 10, pp. 184–190, October 2016.
[10] A. J. Paulraj, D. A. Gore, R. U. Nabar, and H. Bolcskei, "An overview of mimo communications: A key to gigabit wireless," *Proceedings of the IEEE*, vol. 92, no. 2, pp. 198–218, February 2004.
[11] G. J. Foschini and M. J. Gans, "On limits of wireless communications in a fading environment when using multiple antennas," *Wireless Personal Communications*, vol. 6, no. 3, pp. 311–335, March 1998.
[12] I. E. Telatar, "Capacity of multi-antenna Gaussian channels," *European Transactions on Telecommunications*, vol. 10, no. 6, pp. 585–595, November–December 1999.
[13] G. J. Foschini, "Cross-polarization canceler/equalizer," US Patent, 23 December 23, 1986.
[14] J. Winters, "On the capacity of radio communication systems with diversity in a Rayleigh fading environment," *IEEE Journal on Selected Areas in Communications*, vol. 5, no. 5, pp. 871–878, June 1987.
[15] J. H. Winters, J. Salz, and R. D. Gitlin, "The impact of antenna diversity on the capacity of wireless communication systems," *IEEE Transactions on Communications*, vol. 42, no. 234, pp. 1740–1751, February 1994.
[16] A. J. Paulraj and T. Kailath, "Increasing capacity in wireless broadcast systems using distributed transmission/directional reception (DTDR)," Patent 5 345 599, September 6, 1994.
[17] G. J. Foschini, "Layered space-time architecture for wireless communication in a fading environment when using multi-element antennas," *Bell Labs Technical Journal*, vol. 1, no. 2, pp. 41–59, Autumn 1996.
[18] B. D. V. Veen and K. M. Buckley, "Beamforming: A versatile approach to spatial filtering," *IEEE ASSP Magazine*, vol. 5, no. 2, pp. 4–24, April 1988.
[19] D. Tse and P. Viswanath, *Fundamentals of Wireless Communication*. Cambridge University Press, 2005.
[20] E. Bonek, "MIMO propagation and channel modeling," in *MIMO: From Theory to Implementation*, A. Sibille, C. Oestges, and A. Zanella, Eds. Elsevier Science, 2010, pp. 27–54.
[21] H. Xu, D. Chizhik, H. Huang, and R. Valenzuela, "A generalized space-time multiple-input multiple-output (MIMO) channel model," *IEEE Transactions on Wireless Communications*, vol. 3, no. 3, pp. 966–975, May 2004.

[22] L. Schumacher, K. I. Pedersen, and P. E. Mogensen, "From antenna spacings to theoretical capacities: Guidelines for simulating MIMO systems," in *The 13th IEEE International Symposium on Personal, Indoor and Mobile Radio Communications*, Pavilhao Altantico, vol. 2, September 2002, pp. 587–592.

[23] Q. H. Spencer, C. B. Peel, A. L. Swindlehurst, and M. Haardt, "An introduction to the multi-user MIMO downlink," *IEEE Communications Magazine*, vol. 42, no. 10, pp. 60–67, October 2004.

[24] D. Gesbert, M. Kountouris, R. W. Heath. Jr., C. b. Chae, and T. Salzer, "Shifting the MIMO Paradigm," *IEEE Signal Processing Magazine*, vol. 24, no. 5, pp. 36–46, September 2007.

[25] T. K. Y. Lo, "Maximum ratio transmission," *IEEE Transactions on Communications*, vol. 47, no. 10, pp. 1458–1461, October 1999.

[26] T. Haustein, C. von Helmolt, E. Jorswieck, V. Jungnickel, and V. Pohl, "Performance of MIMO systems with channel inversion," in *Vehicular Technology Conference. IEEE 55th Vehicular Technology Conference. VTC Spring 2002 (Cat. No.02CH37367)*, vol. 1, 2002, pp. 35–39.

[27] M. Joham, W. Utschick, and J. A. Nossek, "Linear transmit processing in MIMO communications systems," *IEEE Transactions on Signal Processing*, vol. 53, no. 8, pp. 2700–2712, August 2005.

[28] Z. Guo and P. Nilsson, "Algorithm and implementation of the K-best sphere decoding for MIMO detection," *IEEE Journal on Selected Areas in Communications*, vol. 24, no. 3, pp. 491–503, March 2006.

[29] B. Hassibi and H. Vikalo, "On the sphere-decoding algorithm. I. Expected complexity," *IEEE Transactions on Signal Processing*, vol. 53, no. 8, pp. 2806–2818, August 2005.

[30] M. K. Varanasi and T. Guess, "Optimum decision feedback multiuser equalization with successive decoding achieves the total capacity of the Gaussian multiple-access channel," in *Conference Record of the 31st Asilomar Conference on Signals, Systems and Computers (Cat. No.97CB36136)*, vol. 2, November 1997, pp. 1405–1409.

[31] P. W. Wolniansky, G. J. Foschini, G. D. Golden, and R. A. Valenzuela, "V-BLAST: An architecture for realizing very high data rates over the rich-scattering wireless channel," in 1998 URSI International Symposium on Signals, Systems, and Electronics, 1998 (ISSSE 98), September 1998, pp. 295–300.

[32] M. Costa, "Writing on dirty paper (Corresp.)," *IEEE Transactions on Information Theory*, vol. 29, no. 3, pp. 439–441, May 1983.

[33] M. Tomlinson, "New automatic equaliser employing modulo arithmetic," *Electronics Letters*, vol. 7, no. 5, pp. 138–139, March 1971.

[34] H. Harashima and H. Miyakawa, "Matched-transmission technique for channels with intersymbol interference," *IEEE Transactions on Communications*, vol. 20, no. 4, pp. 774–780, August 1972.

[35] G. Ginis and J. M. Cioffi, "A multi-user precoding scheme achieving crosstalk cancellation with application to DSL systems," in *Conference Record of the 34th Asilomar Conference on Signals, Systems and Computers (Cat. No.00CH37154)*, vol. 2, October 2000, pp. 1627–1631.

[36] G. Caire and S. Shamai, "On the achievable throughput of a multiantenna Gaussian broadcast channel," *IEEE Transactions on Information Theory*, vol. 49, no. 7, pp. 1691–1706, July 2003.

[37] T. L. Marzetta, "Noncooperative cellular wireless with unlimited numbers of base station Antennas," *IEEE Transactions on Wireless Communications*, vol. 9, no. 11, pp. 3590–3600, November 2010.
[38] F. Rusek, D. Persson, B. K. Lau, et al., "Scaling up MIMO: Opportunities and challenges with very large arrays," *IEEE Signal Processing Magazine*, vol. 30, no. 1, pp. 40–60, January 2013.
[39] E. G. Larsson, O. Edfors, F. Tufvesson, and T. L. Marzetta, "Massive MIMO for next generation wireless systems," *IEEE Communications Magazine*, vol. 52, no. 2, pp. 186–195, February 2014.
[40] X. Gao, F. Tufvesson, O. Edfors, and F. Rusek, "Measured propagation characteristics for very-large MIMO at 2.6 GHz," in *2012 Conference Record of the 46th Asilomar Conference on Signals, Systems and Computers (ASILOMAR)*, 2012, pp. 295–299.
[41] X. Gao, O. Edfors, F. Rusek, and F. Tufvesson, "Massive MIMO performance evaluation based on measured propagation data," *IEEE Transactions on Wireless Communications*, vol. 14, no. 7, pp. 3899–3911, July 2015.
[42] H. Ji, Y. Kim, J. Lee, et al., "Overview of full-dimension MIMO in LTE-advanced pro," *IEEE Communications Magazine*, vol. 55, no. 2, pp. 176–184, February 2017.
[43] Y. Kim, H. Ji, J. Lee, et al., "Full dimension mimo (FD-MIMO): The next evolution of MIMO in LTE systems," *IEEE Wireless Communications*, vol. 21, no. 2, pp. 26–33, April 2014.
[44] C. Shepard, H. Yu, N. Anand, et al., "Argos: Practical many-antenna base stations," in *Proceedings of the 18th Annual International Conference on Mobile Computing and Networking*, ser. Mobicom '12. ACM, 2012, pp. 53–64.
[45] H. Suzuki, S. Hajime, K. Rodney, et al., "Highly spectrally efficient Ngara Rural Wireless Broadband Access Demonstrator," in *2012 International Symposium on Communications and Information Technologies (ISCIT)*, 2012.
[46] C. Shepard, H. Yu, and L. Zhong, "ArgosV2: A flexible many-antenna research platform," in *Proceedings of the 19th Annual International Conference on Mobile Computing & Networking*, ser. MobiCom '13, ACM, 2013, pp. 163–166.
[47] "5G massive MIMO testbed: From theory to reality – national instruments," Online. Available: www.ni.com/white-paper/52382/en/
[48] J. Vieira, S. Malkowsky, K. Nieman, et al., "A flexible 100-antenna testbed for massive MIMO," in *2014 IEEE Globecom Workshops (GC Wkshps.)*, December 2014, pp. 287–293.
[49] G. Xu, Y. Li, J. Yuan, et al., "Full dimension MIMO (FD-MIMO): Demonstrating commercial feasibility," *IEEE Journal on Selected Areas in Communications*, vol. 35, no. 8, pp. 1876–1886, August 2017.
[50] A. Valdes-Garcia, S. T. Nicolson, J. W. Lai, et al., "A fully integrated 16-element phased-array transmitter in SiGe BiCMOS for 60-GHz communications," *IEEE Journal of Solid-State Circuits*, vol. 45, no. 12, pp. 2757–2773, December 2010.
[51] S. Emami, R. F. Wiser, E. Ali, et al., "A 60GHz CMOS phased-array transceiver pair for multi-Gb/s wireless communications," in *2011 IEEE International Solid-State Circuits Conference*, February 2011, pp. 164–166.
[52] A. Natarajan, S. K. Reynolds, M. D. Tsai, et al., "A fully-integrated 16-element phased-array receiver in SiGe BiCMOS for 60-GHz communications," *IEEE Journal of Solid-State Circuits*, vol. 46, no. 5, pp. 1059–1075, May 2011.

[53] K. Okada, N. Li, K. Matsushita, K. Bunsen, et al., "A 60-GHz 16QAM/8PSK/QPSK/BPSK direct-conversion transceiver for IEEE802.15.3c," *IEEE J. Solid-State Circuits*, vol. 46, no. 12, pp. 2988–3004, December 2011.

[54] N. Saito, T. Tsukizawa, N. Shirakata, et al., "A fully integrated 60-GHz CMOS transceiver chipset based on WiGig/IEEE 802.11ad with Built-in self calibration for mobile usage," *IEEE Journal of Solid-State Circuits*, vol. 48, no. 12, pp. 3146–3159, December 2013.

[55] M. Boers, B. Afshar, I. Vassiliou, et al., "A 16TX/16RX 60 GHz 802.11ad chipset with single coaxial interface and polarization diversity," *IEEE Journal of Solid-State Circuits*, vol. 49, no. 12, pp. 3031–3045, December 2014.

[56] H. T. Kim, B. S. Park, S. M. Oh, et al., "A 28GHz CMOS direct conversion transceiver with packaged antenna arrays for 5G cellular system," in *2017 IEEE Radio Frequency Integrated Circuits Symposium (RFIC)*, June 2017, pp. 69–72.

[57] B. Sadhu, Y. Tousi, J. Hallin, et al., "A 28GHz 32-element phased-array transceiver IC with concurrent dual polarized beams and 1.4 degree beam-steering resolution for 5G communication," in *2017 IEEE International Solid-State Circuits Conference (ISSCC)*, February 2017, pp. 128–129.

[58] U. Kodak and G. M. Rebeiz, "Bi-directional flip-chip 28 GHz phased-array core-chip in 45nm CMOS SOI for high-efficiency high-linearity 5G systems," in *2017 IEEE Radio Frequency Integrated Circuits Symposium (RFIC)*, June 2017, pp. 61–64.

[59] K. Kibaroglu, M. Sayginer, and G. M. Rebeiz, "An ultra low-cost 32-element 28 GHz phased-array transceiver with 41 dBm EIRP and 1.0-1.6 Gbps 16-QAM link at 300 meters," in *2017 IEEE Radio Frequency Integrated Circuits Symposium (RFIC)*, June 2017, pp. 73–76.

[60] K. Takinami, N. Shirakata, K. Tanaka, et al., "A 60GHz wireless transceiver employing hybrid analog/digital beamforming with interference suppression for multiuser gigabit/s radio access," in *2015 Symposium on VLSI Circuits (VLSI Circuits)*, June 2015, pp. C306–C307.

[61] Y. S. Yeh, B. Walker, E. Balboni, and B. Floyd, "A 28-GHz phased-array receiver front end with dual-vector distributed beamforming," *IEEE Journal of Solid-State Circuits*, vol. 52, no. 5, pp. 1230–1244, May 2017.

[62] A. Puglielli, N. Narevsky, P. Lu, et al., "A scalable massive MIMO array architecture based on common modules," in *2015 IEEE International Conference on Communication Workshop (ICCW)*, June 2015, pp. 1310–1315.

[63] A. Puglielli, A. Townley, G. LaCaille, et al., "Design of energy- and cost-efficient massive MIMO arrays," *Proceedings of the IEEE*, vol. 104, no. 3, pp. 586–606, March 2016.

[64] T. Höhne and V. Ranki, "Phase noise in beamforming," *IEEE Transactions on Wireless Communications*, vol. 9, no. 12, pp. 3682–3689, December 2010.

[65] A. Pitarokoilis, S. K. Mohammed, and E. G. Larsson, "Uplink performance of time-reversal MRC in massive MIMO systems subject to phase noise," *IEEE Transactions on Wireless Communications*, vol. 14, no. 2, pp. 711–723, February 2015.

[66] M. R. Khanzadi, G. Durisi, and T. Eriksson, "Capacity of SIMO and MISO phase-noise channels with common/separate oscillators," *IEEE Transactions on Communications*, vol. 63, no. 9, pp. 3218–3231, September 2015.

[67] A. Puglielli, G. LaCaille, A. M. Niknejad, G. Wright, B. Nikolić, and E. Alon, "Phase noise scaling and tracking in OFDM multi-user beamforming arrays," in *2016 IEEE International Conference on Communications (ICC)*, May 2016, pp. 1–6.

[68] C. Mollen, U. Gustavsson, T. Eriksson, and E. G. Larsson, "Out-of-band radiation measure for MIMO arrays with beamformed transmission," in *2016 IEEE International Conference on Communications (ICC)*, May 2016, pp. 1–6.

[69] C. Mollen, E. G. Larsson, U. Gustavsson, T. Eriksson, and R. W. Heath, "Out-of-band radiation from large antenna arrays," *IEEE Communications Magazine*, vol. 56, no. 4, pp. 196–203, April 2018.

[70] S. K. Mohammed and E. G. Larsson, "Per-antenna constant envelope precoding for large multi-user MIMO systems," *IEEE Transactions on Communications*, vol. 61, no. 3, pp. 1059–1071, March 2013.

[71] C. Studer and E. G. Larsson, "PAR-aware large-scale multi-user MIMO-OFDM downlink," *IEEE Journal on Selected Areas in Communications*, vol. 31, no. 2, pp. 303–313, February 2013.

4 RF and Millimeter-Wave Full-Duplex Wireless for 5G and Beyond

Harish Krishnaswamy and Tolga Dinc

Full-duplex wireless, where transmitters and receivers operate on the same frequency band at the same time, have the potential to double network capacity at the physical layer while offering numerous other benefits at higher layers of the network. The main challenge with full duplex wireless is the transmitter self-interference, which lies on top of the received signal and can be a billion to a trillion times (90–120 dB) more powerful. Over the last five years, there has been significant research progress within the systems community as well as the Radio Frequency Integrated Circuit (RFIC) community on techniques for transmitter self-interference cancellation (SIC) across different domains, from antenna to radio frequency (RF) and analog to digital. This chapter will review recent research progress, and will conclude with a discussion of outstanding challenges and opportunities.

4.1 Overview of Full-Duplex

Demand for wireless capacity has exploded in recent years with increasing use of bandwidth (BW)–hungry media applications in smartphones, tablets, and smart TVs and keeps growing exponentially every year. As a result, a thousandfold increase in data traffic is projected over the next 10 years along with tenfold to hundredfold increase in data rates and ultra-low network latency [1]. Solutions for delivering the thousandfold increase in capacity fall into three fundamental groups: reducing cell size (especially in urban settings in where the data demand is higher), allocating more spectrum, and improving spectral efficiency [2]. Low RF frequencies (sub-6 GHz) are plagued with limited available spectrum and interference issues, therefore smaller cells at these frequencies are unlikely to address the total capacity demand [3]. On the other hand, millimeter-wave (mm-wave) frequencies (over 24 GHz) offer multi-GHz channel bandwidths and allow integration of interference mitigation techniques (e.g., phased arrays) in smaller form factors. Hence, mm-wave beamforming has recently gained significant research attention as one of the most promising solutions to deliver the data traffic demands of the 5G era [4–7].

Same-channel full-duplex (or in-band full-duplex), e.g., simultaneous data transmission and reception on the same frequency channel, is another technology has drawn significant interest in the recent years and will be referred to as just full-duplex (FD) in the rest of this chapter. FD can theoretically double the spectral efficiency in the

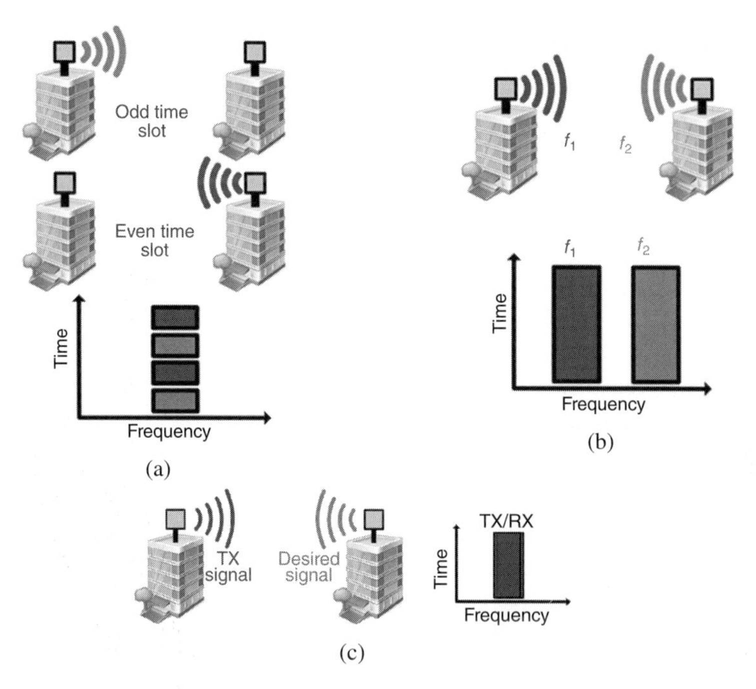

Figure 4.1 Full-duplex operation versus traditional duplexing schemes. (a) Time-division duplexing scheme, (b) Frequency-division duplexing, and (c) same-channel full-duplex, which enables simultaneous transmission and reception on the same frequency channel.

physical layer over traditional duplexing schemes such as time division duplex (TDD) or frequency division duplex (FDD) (Figure 4.1). Compared to FDD, which can support simultaneous transmission and reception using surface acoustic wave (SAW) or bulk acoustic wave (BAW) filters but over two different frequency channels, same-channel FD can use available spectral resources to the fullest extent for simultaneous transmission and reception in each channels, doubling the spectral efficiency.[1] In addition, FD offers many other benefits in higher layers such as increase in access-layer throughput, collision avoidance, solving the hidden node problem, and low latency [10]. However, self-interference (SI) from transmitter (TX) to its own receiver (RX) poses a fundamental challenge as it can be can be more than a billion times stronger than the desired RX signal, depending on the application. A total SI suppression of 90 dB or more must be achieved across multiple domains – antenna, RF/analog and digital – to suppress the SI below the receiver noise floor, enabling full-duplex operation.

System-level demonstrations leveraging off-the-shelf components (e.g. [11,12]) have established the feasibility of full-duplex operation. However, self-interference suppression techniques proposed in these works are not compatible for small-form factor integrated circuit implementations. More recently, complementary metal-oxide semiconductor (CMOS) integrated circuits (ICs) for full-duplex applications have been

[1] When evaluating the capacity gain achieved through FD in real-world scenarios, factors such as asymmetric uplink and downlink and interference between cells must be considered [8,9].

demonstrated at low RF frequencies [13–17], as well as at mm-wave frequencies [18–22]. This chapter will review these research efforts on FD integrated radios spanning FD antenna interfaces, RF, and analog and digital SIC techniques. A special focus will be put on the millimeter-wave full-duplex technology, which can potentially offer the dual benefits of wide BWs and improved spectral efficiency.

4.2 Millimeter-Wave Full-Duplex Applications

There are many potential future applications for mm-wave full-duplex, applications ranging from vehicular radars, 5G small cells, and mm-wave backhaul to virtual/augmented reality.

4.2.1 Millimeter-Wave Backhaul

Rapid growth of mobile data traffic with densification of small cells will bring massive capacity pressure on the backhaul, which is a somewhat less addressed bottleneck of the overall system. Cost-effective backhauling schemes with fiberlike throughput and low latency are essential to connect 5G small-cell base stations to other 5G base stations and the network [23]. Fiber is hard and costly to deploy, prohibiting its deployment in every 5G small cell [24]. On the other hand, mm-wave communication (e.g., E-band backhaul) offers a flexible and cost-effective candidate for 5G backhaul with fiberlike throughput. Additionally, there has been interest in using the 60 GHz unlicensed band for backhauling over shorter distances between densely deployed small cells (Figure 4.2a) [25]. Simultaneous uplink and downlink is essential to reduce latency in such networks, and E-band uses two different frequency bands for that reason (effectively providing frequency division duplexing between 71–76 GHz and 81–86 GHz using waveguide diplexers). Full-duplex would enable such operation in a single frequency band, enabling aggregation of two mm-wave bands for increased capacity.

4.2.2 Milimeter-Wave FD Relaying

One of the main issues of mm-wave wireless communication is higher propagation loss compared to low RF frequencies. The wireless link range is limited due to high channel losses at high frequencies. Relay nodes between sources and destinations would help to extend link range as well as improve link margin [27], enabling robust connection in severe conditions. However, traditional relay nodes use frequency-division or time-division duplexing, resulting in poor spectral efficiency or unwanted latency in the network, respectively. Figure 4.2b depicts mm-wave FD relay nodes that can be used to extend the wireless link range, with significant spectral efficiency and latency improvement over existing half-duplex (HD) counterparts [27]. Such FD mm-wave relays can be implemented using separate TX and RX antennas or a single shared antenna (Figure 4.2b). For the latter, a millimeter-wave circulator would be preferred for low-loss operation while preserving channel reciprocity.

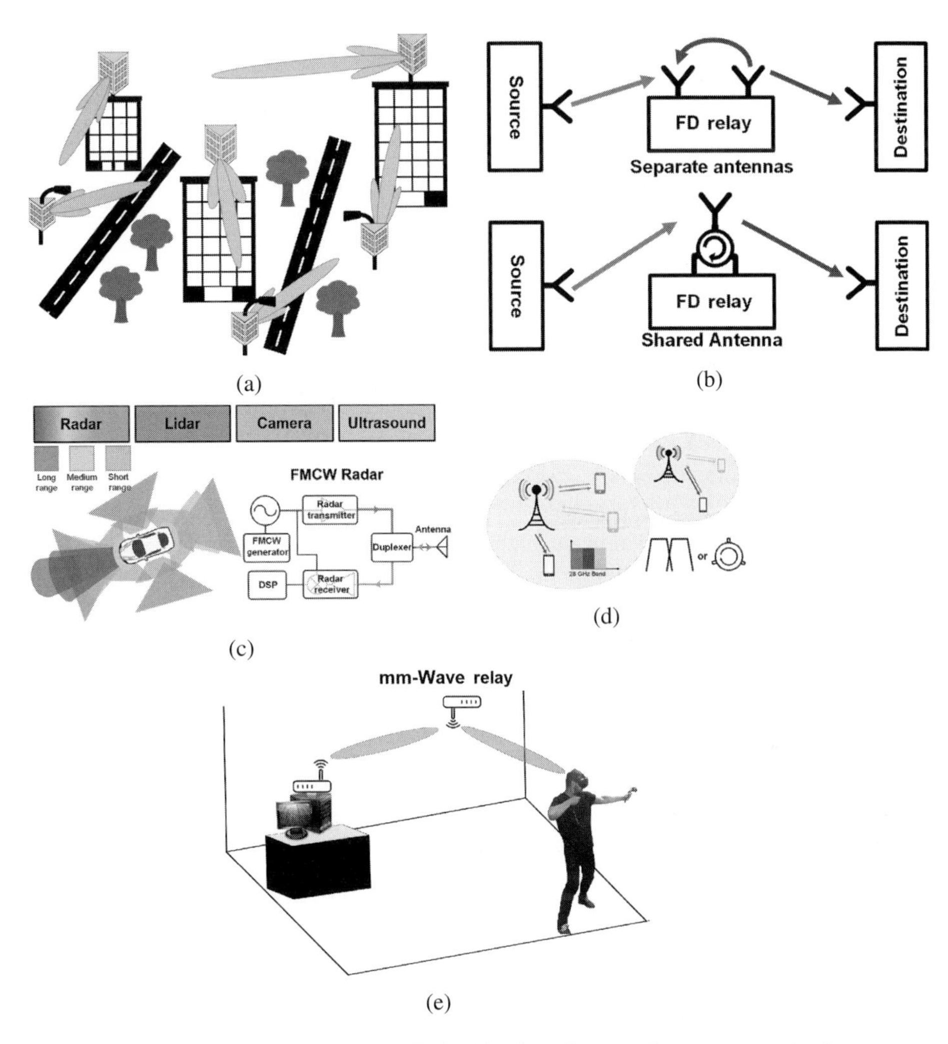

Figure 4.2 (a) A mm-wave phased-array FD backhaul mesh network can support simultaneous same-channel up and down links, improving spectral efficiency and latency. (b) mm-wave FD relays with TX and RX antenna pairs or shared antenna with a mm-wave circulator. (c) SI suppression techniques developed in the context of mm-wave FD are also applicable to automotive radars (e.g., a low-loss, noise, and high-isolation duplexer). (d) mm-wave 5G small-cell base stations enabled by a circulator eliminates the need for high-quality diplexers. (e) mm-wave FD can extend the range of untethered wireless VR/AR headsets with reduced latency. Adapted from [26].

4.2.3 Millimeter-Wave Vehicular Radar

TX-to-RX SI is referred to as spillover in frequency-modulated continuous wave (FMCW) automotive radars and is one of the remaining issues waiting to be solved. Spillover can result from limited isolation in the antenna interface, and from strong reflections arising from very close objects such as the vehicle's bumper or fascia. The spillover at the RX input can be much stronger than the reflected powers by distant objects that the radar aims to detect [28] and must be suppressed to prevent saturation of the receiver. SI suppression techniques developed in the context of mm-wave FD is also applicable to solving the spillover problem in automotive radars. A fully integrated low-loss circulator with high isolation would replace lossy passive shared antenna interfaces such as hybrids (which have a theoretical loss of 3 dB, typically 4 dB), which are widely used at monostatic radars to share a single antenna between the TX and RX.

4.2.4 5G Small-Cell Base Stations

Millimeter-wave small-cell 5G base stations will communicate with multiple users simultaneously in uplink and downlink in the adjacent channels. This would necessiate high-quality mm-wave diplexers, which are bulky. Alternatively, the techniques developed in the context of FD could be adopted to isolate the transmitter and receiver. For example, a fully integrated low-loss, high-isolation circulator with high-power handling capability could replace high-quality mm-wave diplexers to share a single antenna between TX and RX.

4.2.5 Virtual Reality (VR)/Augmented Reality (AR) Headsets

Head-mounted displays projects high-quality video (e.g., 2,160 × 1,200 resolution) to each eye at a high frame rate (e.g., 90 Hz) [29] to create a virtual reality experience by projecting. To prevent an adverse user experience (e.g., VR sickness), the very high data rate video streams (approaching 20 Gbps for emerging VR headsets) need to be delivered with very low latency (less than 5 ms) [30]. For a smooth VR experience, a huge amount of data has to be sent back and forth between the computer, the headset, and the positional tracker, requiring bidirectional communication [29]. Millimeter-wave full-duplex wireless links can be a promising solution to cut the cord of VR headsets as they have the potential of delivering high-speed data with low latency. Additionally, the range of wireless VR headsets can be extended through mm-wave full-duplex relays, improving the user's mobilility [31].

4.3 Full-Duplex Challenge and System Considerations

Figure 4.3 depicts the self-interference problem in a typical full-duplex wireless radio. Self-interference arises from the inherent coupling in typical antenna interfaces as well as environmental reflections and consists of a leakage of the main transmitted signal and nonlinear distortions of the transmitter as well as the TX-side local oscillator (LO) phase noise. SI should be suppressed sufficiently below the receiver noise floor (P_n) to enable reception of the weak desired signal. Therefore, assuming a 6 dB margin for SI suppression below the noise floor, a total SI suppression of P_{TX}-P_n + 6 dB must be achieved in antenna, RF/analog, and digital domains where P_{TX} is the transmitter output power. For example, an FD transceiver at 60 GHz with a typical transmit power P_{TX} of +14 dBm, the thermal noise $N_{th} = 10\log(kTB)$, RX noise figure NF_{RX} of 5 dB, and channel BW of 2.16 GHz, a margin $M = 6$ dB would require a self-interference cancellation SIC of

$$\begin{aligned} SIC &= P_{TX} - (N_{th} + 10\log BW + NF_{RX}) + M \\ &= +14\,\text{dBm} - (-174\,\text{dBm/Hz} + 10\log(2.16\,\text{GHz}) + 5) + 6\,\text{dB} \\ &= 96\,\text{dB} \end{aligned} \tag{4.1}$$

For low-RF FD transceivers, the total SIC requirement can be even higher, especially for long-range wireless communication. An low RF FD link with +20 dBm TX average output power, 20 MHz signal BW, and RX noise figure of 4 dB would require 120 dB SIC. Such a high level of SI suppression can only be achieved by enhancing the transmit–receive (T/R) isolation at the antenna as well as performing SIC in the antenna, RF/analog, and digital domains.

Assuming an ADC with an effective number of bits (ENOB) between 8 and 12 bits and an effective dynamic range DR of 6.02 × (ENOB-2), resulting in a DR between 36 and 60 dB, is used in an mm-wave/RF FD link, then the remaining 60 dB SI suppression must be achieved by the antenna interface and the RF canceller in Figure 4.3. While partitioning this 60 dB between the antenna and RF domain cancellers, the RX noise figure (NF_{RX}) degradation due to the RF canceller's noise should be taken into account. If G_{SIC} and C_{RX} are the RF SI canceller gain and the coupling coefficient of the coupler at the receiver input, respectively, and assuming C_{RX} is weak, the noise factor of the FD transceiver (F_{total}) shown in Figure 4.3 can be derived as

$$F_{total} = F_{RX} + (F_{SIC} - 1)\,G_{SIC}C_{RX} \tag{4.2}$$

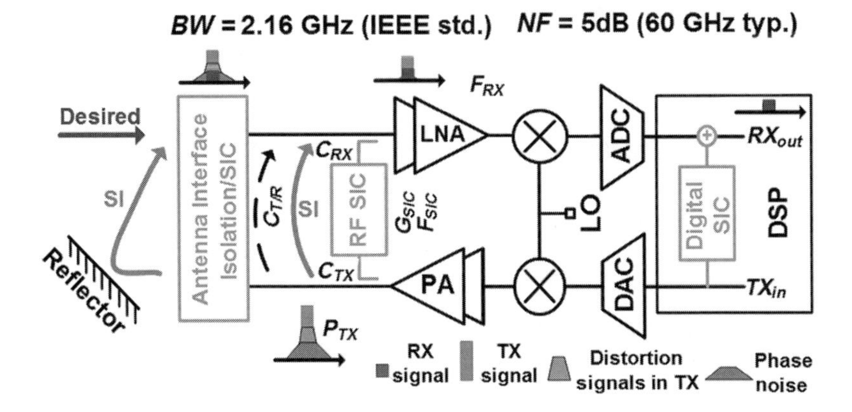

Figure 4.3 Self-interference issues in full-duplex radios. SI must be suppressed in the antenna, RF/analog, and digital domains to prevent receiver sensitivity degradation.

where F_{SIC} and F_{RX} are the noise factor of the RF SI canceller and the receiver, respectively. The last term in (4.2) represents the amount of RX NF degradation due to the RF SI canceller. For perfect SI cancellation, the magnitude of the transfer function through the antenna interface, $C_{T/R}(1 - C_{TX})$, where $C_{T/R}$ and C_{TX} are the coupling coefficient from the TX output to RX input and the coupling coefficient of the coupler at the power amplifier output, respectively, and the RF canceller, $C_{TX}G_{SIC}C_{RX}$, must be equal. This simplifies (4.2) as

$$F_{total} = F_{RX} + (F_{SIC} - 1)\left(\frac{C_{T/R}(1 - C_{TX})}{C_{TX}}\right). \tag{4.3}$$

From (4.3) we conclude that a lower $C_{T/R}$ reduces the NF degradation due to the RF canceller as it allows a lower C_{RX} in (4.2), motivating the need for a higher SI suppression in the antenna domain. Additionally, (4.3) reveals a trade-off between the receiver NF and the transmitter efficiency degradations due to the RF SI canceller as a lower C_{TX} increases the receiver NF degradation but reduces the transmitter efficiency degradation. It is noteworthy that a higher antenna interface isolation or antenna-domain SIC eases this trade-off. For a given PA output power, *SNR* degradation ΔSNR, compared to its half-duplex counterpart can be written (in dB scale) as

$$\Delta SNR = 10\log F_{total} - 10\log F_{RX} - 10\log(1 - C_{TX}) \tag{4.4}$$

Assuming an *NF* of 10 dB for the RF canceller ($F_{SIC} = 10$), ΔSNR versus $10\log(C_{TX})$ is plotted in Figure 4.4a for different T/R isolation levels at the antenna interface. To keep the *SNR* degradation negligibly small, more than 40 dB isolation must be achieved in the antenna interface. Note that for a low C_{TX} (e.g., smaller than -20 dB for $10\log(C_{T/R})$ < -40 dB), the TX efficiency degradation, namely the last term in (4.4), is extremely small so that ΔSNR is governed by the receiver noise figure degradation. On the other hand, as C_{TX} becomes larger, ΔSNR increases due to the TX efficiency degradation as more power is stolen away by the RF canceller, reducing the transmitter power going into the antenna interface.

Higher SI suppression in the antenna interface also relaxes the linearity requirement on the receiver and the RF canceller, leading to lower power consumption. Achieving low $C_{T/R}$ also opens up the possibility of performing RF SIC after some low-noise amplification, further reducing the NF degradation due to the RF canceller. The spurious-free dynamic range (*SFDR*) requirement of the receiver in dB scale is given by

$$SFDR = P_{TX} + 10\log C_{T/R}(1 - C_{TX}) - SIC_{RF} - P_n \tag{4.5}$$

and reduces with decreasing $C_{T/R}$. Lower $C_{T/R}$ also implies that more loss can be tolerated in the RF canceller and allows designing an all passive RF canceller with higher linearity so that the third-order distortion products generated by the RF canceller (not shown in Figure 4.3) fall below P_n. The *IIP3* requirement on the RF SI canceller in Figure 4.3 can be expressed as

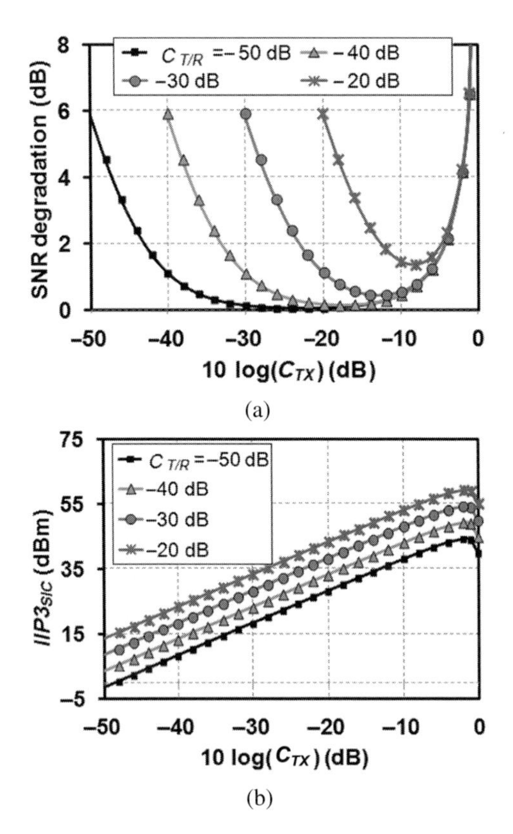

Figure 4.4 (a) *SNR* degradation versus the transmit side coupling coefficient assuming $10\log(F_{RX}) = 5\,\text{dB}$ and $10\log(F_{SIC}) = 10\,\text{dB}$. *SNR* degradation becomes negligible for 40 dB T/R isolation at the antenna interface. (b) Required *IIP3* for the RF SI canceller versus the C_{TX}. $IIP3_{SIC}$ is 18 dBm for $C_{T/R} = -50\,\text{dB}$ and $C_{TX} = -30\,\text{dB}$.

$$IIP3_{SIC} = \frac{3P_{TX} - P_n + 10\log C_{T/R}(1 - C_{TX}) + 20\log C_{TX}}{2}. \tag{4.6}$$

Equation (4.6) is plotted versus C_{TX} for different $C_{T/R}$ levels in Figure 4.4b. The RF canceller must have an $IIP3_{SIC}$ higher than +45 dBm for $10\log(C_{T/R}) = -30\,\text{dB}$ and $10\log(C_{TX}) = -10\,\text{dB}$ whereas an $IIP3_{SIC}$ of +18 dBm is required for $10\log(C_{T/R}) = -50\,\text{dB}$ and $10\log(C_{TX}) = -30\,\text{dB}$. The latter is more manageable, motivating the need for an isolation higher than 50 dB in the antenna interface. Alternatively, if the RF canceller is not linear enough, the distortion products can be estimated and canceled in the digital canceller [11]. However, this may be power inefficient depending on the required computational complexity, especially at mm-wave systems.

A common limitation for both interfaces is the reflections from nearby objects. A nearby reflector creates another interference path from the transmitter to the receiver for both cases. The effect of the environment on the T/R isolation is not predictable during the design process and varies in the field. In order to combat the scattering from environment, the SI suppression technique within the antenna interface must be reconfigurable.

4.4 Self-Interference Suppression Techniques

There have been numerous research efforts on integrated SI interference suppression techniques and FD radios in the recent years. In general, these efforts can be divided into two main categories as TX-to-RX isolation enhancement (especially at the antenna interface) and SI cancellation, which involves nulling of the SI signal by adding its inverse replica. This section will review the state-of-the art SI suppression techniques in the antenna, RF, and analog domains.

4.4.1 Antenna Suppression

As revealed by the FD system analysis in the previous section, SI suppression within the antenna interface is crucial for enabling full-duplex operation. It relaxes the dynamic range requirements of the receiver blocks (RF, analog, and digital) as well as the RF/analog and digital SI cancellation circuits. Other than providing a high TX-to-RX isolation in the excess of 30–40 dB, FD antenna interfaces must support wide SI suppression BWs of emerging standards, be reconfigurable to combat reflections from the environment (which can be frequency dependent and time varying) and preserve the TX/RX antenna patterns while suppressing the SI.

Figure 4.5 shows various SI suppression techniques in the antenna domain, which can be divided into two main groups. The techniques in the first group targets reducing the initial inherent coupling between the TX and RX ports. The simplest method would be using separate antennas for TX and RX as shown in Figure 4.5a, but this requires prohibitively large separation between the T/R antenna pairs [32]. For example, a simple free-space path loss–based model for isolation reveals that 8λ spacing is required for around 40 dB isolation. This corresponds to 2.4 m and 4 cm at 1 and 60 GHz, respectively, which clearly is not an area-efficient approach and thus not suitable for small-form-factor FD radios. Other approaches in this group include separating the TX and RX antennas in polarization (Figure 4.5b) [33,34] and shadowing the near fields of TX and RX antennas [35]. Although these techniques are capable of providing good isolation over wide BWs, they are typically static approaches that cannot respond to a changing electromagnetic environment.

The second category consists of the techniques that essentially perform self-interference cancellation in the antenna domain and will be referred to as antenna cancellation. Generally speaking, SIC in the antenna domain is achieved either with use of an auxiliary antenna or an auxiliary coupling path between the TX and RX antennas/ports to create an inverse replica of the main coupling signal. For example, Figure 4.5c shows a two-TX and a single-RX antenna system in which the RX antenna's

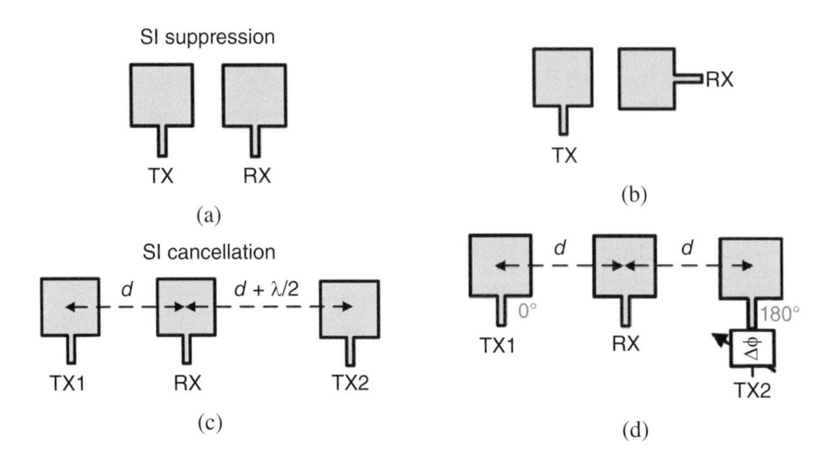

Figure 4.5 SI suppression techniques in the antenna domain can be divided into two main categories. The first group targets enhancing the TX-to-RX isolation through use of (a) spatial separation and (b) orthogonal polarization for TX and RX antennas. The second group is based on performing self-interference cancellation in the antenna domain using 180° (c) spatial phase shift and (d) RF phase shift.

distance from the TX antennas differs by $\lambda/2$ at the operating frequency [32]. Similarly, in Figure 4.5d, the TX antennas are fed 180° out of phase and the RX antenna is placed in symmetrically between the TX antennas to cause destructive interference [36]. This out of phase feeding technique is extended to create wideband multi-element arrays consisting of multiple transmitting elements and a centrally located receiving antenna in [37–39]. These cancellation techniques, in general, are vulnerable to manufacturing and antenna excitation tolerances (phase and amplitude imbalance). Additionally, they may create an undesirable effect on the far-field antenna pattern if a simultaneous optimization for far-field and antenna cancellation is not performed.

There have also been research efforts on reconfigurable antenna cancellation techniques that can potentially combat variable scattering from the environment. In [40,41], authors propose placing tunable resonant baffles between antenna elements for tunable FD operation, and they provide 40 dB isolation over 12 and 55 MHz SIC BW, respectively.

The aforementioned antenna cancellation techniques are mainly demonstrated at low RF frequencies. More recently, there has been research progress on mm-wave antenna interfaces as well. Two particularly interesting antenna cancellation methods achieving wideband suppression are depicted in Figure 4.6. A compact wideband reconfigurable polarization-based antenna cancellation technique is demonstrated in [19], exploiting the polarization to embed SIC within the antenna. As depicted in Figure 4.6a, it uses a co-located TX/RX antenna pair with orthogonal polarizations to improve initial TX-to-RX isolation. Additionally, an auxiliary port that is copolarized with the TX antenna is embedded on the RX antenna and terminated with a reflective on-chip RLC termination with variable R and C. This auxiliary port creates an indirect port between the TX and RX port, and the RLC termination conditions the signal in this path to create an inverse replica of the SI at the RX port, achieving SIC.

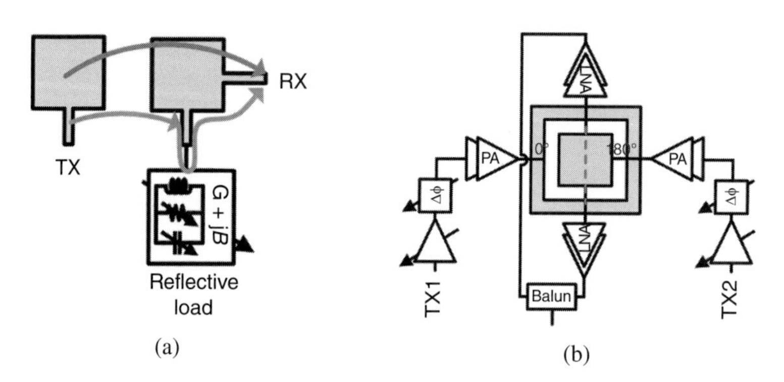

Figure 4.6 State-of-the-art mm-wave antenna cancellation techniques: (a) reconfigurable polarization-based antenna cancellation adapted from [19], and (b) multifeed SIC antenna, adapted from [22].

Due to the electronically programmable nature of this reflective termination, SIC can be reconfigured to respond to a changing electromagnetic (EM) environment. The polarization-based antenna cancellation technique provides more than 50 dB isolation over 8 GHz BW centered at around 60 GHz (in simulation). A 4.6 GHz printed circuit board (PCB)–based prototype in [42] was also demonstrated with 50 dB isolation over 300 MHz BW.

Figure 4.6b depicts a more recently proposed multi-feed SIC antenna, providing a TX-to-RX isolation higher than 35 dB over 15 GHz BW in a single antenna footprint. It consists of a four-feed on-chip antenna implemented on high-resistivity silicon, supporting TX and RX radiations with orthogonal polarizations. The two TX feeds are driven 180° out of phase, and the RX feeds are placed symmetrically between the TX feeds, causing destructive interference at the RX ports. Based on the same principle depicted in Figure 4.5d, this method also relies on perfect symmetry and differential excitations. Therefore, it is inherently frequency independent in theory, providing a wideband SIC without any frequency-tuning element. In practice, it requires phase shifters and variable gain amplifiers in the TX paths to compensate amplitude and phase imbalances between the two TX excitations.

4.4.2 Shared Antenna Interface

The high complexity of antenna suppression techniques due to employing either multiple antennas or multiple ports makes expansion of these techniques to FD multiple-input, multiple-output (MIMO) challenging. Compact FD antenna interfaces sharing a single antenna between the TX and RX ports would be more readily compatible with MIMO, and would also ease channel reciprocity calibration. Additionally, for mobile FD applications, a shared antenna interface is necessitated by form-factor constraints. However, achieving a CMOS-compatible, small form-factor, low-loss FD antenna interface with low noise, high isolation, linearity, and power handling is a significant challenge, irrespective of operation frequency. Among possible choices, ferrite circulators

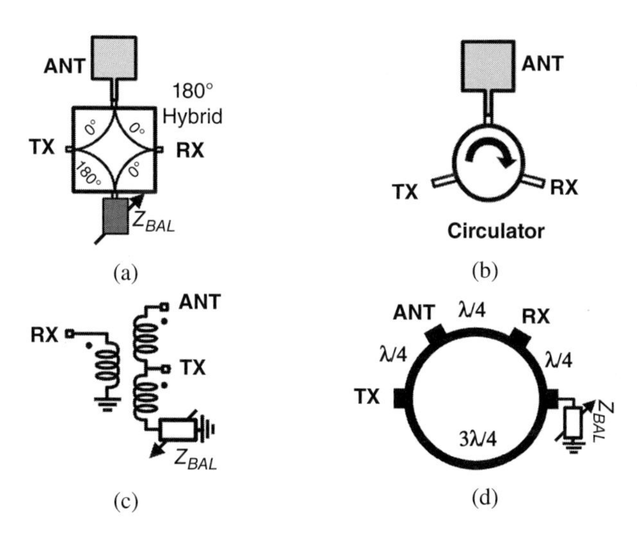

Figure 4.7 Two different shared antenna interfaces for FD operation: (a) three-port passive networks such as electrical balance duplexer, (b) circulator. The EBD concept is based on a 180° hybrid terminated with an antenna balancing termination. The hybrid can be implemented based on (c) transformer or (d) directional couplers such as a rat-race coupler. Adapted from [43].

that break reciprocity to circulate the signal only in a single direction (e.g., clockwise direction in Figure 4.7a) are bulky and are not CMOS-compatible. Therefore, passive reciprocal shared-antenna interfaces, such as the EBD [44–48], are widely explored as on-chip solutions. Figure 4.7b shows a generic view of EBD concept which is based on a 180° hybrid terminated with a passive termination for antenna impedance balancing. Ideally, the balancing network mimics the antenna impedance so that the SI at the RX port due to imperfect matching at the antenna port is canceled. This concept was initially used in telephone networks, but recently regained interest in regard to RF and mm-wave as a shared-antenna interface. The RF implementations are generally implemented using transformer-based hybrids as in Figure 4.7c for small-form factor, whereas a transmission line–based hybrid such as a rat-race coupler (Figure 4.7d) is possible at mm-wave frequencies. The balancing networks are generally implemented with fixed on-chip inductors and switched capacitor and/or resistor banks. The state-of-the-art EBDs provide more than 50 dB isolation over relatively wide BWs (typically 10 to 15 % fractional BWs) with good power handling and linearity. For example, recently, an EBD with more than 50 dB isolation over >200 MHz BW with a power handling of +27 dBm and TX-to-ANT *IIP3* of more than +70 dBm was demonstrated in 180 nm silicon on insulator (SOI) CMOS [48], using device stacking techniques. However, EBDs suffer from a 3 dB fundamental loss (typically around 4 dB at RF and mm-wave frequencies) due to the fact that a three-port passive network cannot be reciprocal, lossless, and matched at all ports at the same time. This 3 dB fundamental loss can be avoided by breaking reciprocity.

Reciprocity can be broken by materials with asymmetric permittivity or permeability tensors, nonlinearity, or time variance. Traditionally, asymmetric permeability tensors are achieved in magnetic materials such as ferrites under external biasing magnetic field. However, ferrites are bulky as they need biasing magnets and are not compatible with IC fabrication processes. An alternative is exploiting an asymmetric port-to-port behavior of active devices [50,51], but this approach is limited by linearity and noise performance. Nonlinearity-based approaches provide nonreciprocity over only certain signal power levels (e.g., larger than 15 dBm input power in [52]). Therefore, in recent years, breaking reciprocity through time variance, specifically spatio-temporal modulation of the constitutive parameters, has gained significant research attention. Early approaches have focused on permittivity as the modulated constitutive parameter [21,53–55]. In the RF domain, permittivity modulation is achieved using varactors, which has a limited modulation index (C_{max}/C_{min} ratio is typically ranging from 2 to 4 in CMOS), resulting in a trade-off among loss, size, bandwidth, and linearity.

On the other hand, conductivity in semiconductors can be varied with a large modulation index over a wide range of frequencies (on-off conductance ratios of CMOS transistors can be as high as 10^3–10^5 [56]). Recently, switch-based spatio-temporal conductivity modulation methods have been explored in silicon [43,57–60] and are depicted in Figure 4.8b. In essence, these approaches include two sets of transistor switches on either end of a delay medium. The switches toggle between on and off states with a periodic modulation signal. Lossless phase nonreciprocity is achieved by synchronizing the modulation clocks of the two sets of switches with an appropriate delay. Figure 4.8a shows an N-path filter-based approach in where the signal is periodically commutated through a bank of capacitors by switching the two sets of transistors in a synchronized manner. The switches are modulated through nonoverlapping clocks with 12.5 % duty cycle and a relative delay of 90° is introduced between them to break phase reciprocity, providing +90° and −90° phase shift in the forward and reverse directions. This structure essentially realizes an ultra-compact gyrator, a fundamental nonreciprocal component postulated by Tellegen in [61] that provides a nonreciprocal phase difference of π between forward and reverse directions. In [57], the gyrator is placed in a $3\lambda/4$ transmission line loop to convert phase nonreciprocity to nonreciprocal wave propagation as depicted in Figure 4.8c. A circulator can be realized by placing three ports that are $\lambda/4$ away from each other. The first passive CMOS circulator is demonstrated at 750 MHz with an insertion loss of 1.7 dB, isolation higher than 40 dB [57].

In principle, a similar approach (wrapping a $3\lambda/4$ transmission line loop around a gyrator) can be adopted to design a mm-wave circulator as well. However, an N-path-based gyrator cannot be expanded to mm-waves due to its stringent clocking requirements. More recently, inspired by N-path-based nonreciprocity, a switched transmission line–based approach is demonstrated as shown in Figure 4.8b. This approach consists of two sets of fully balanced (Gilbert quad) switches on either end of a differential transmission line delay. The switches are modulated through 50% duty cycle periodic square pulses, and the transmission line delay is equal to one quarter of the modulation period ($T_m/4$). The modulation of the right switches is delayed with respect to those on the left by the same amount ($T_m/4$) to break phase reciprocity. In the forward direction,

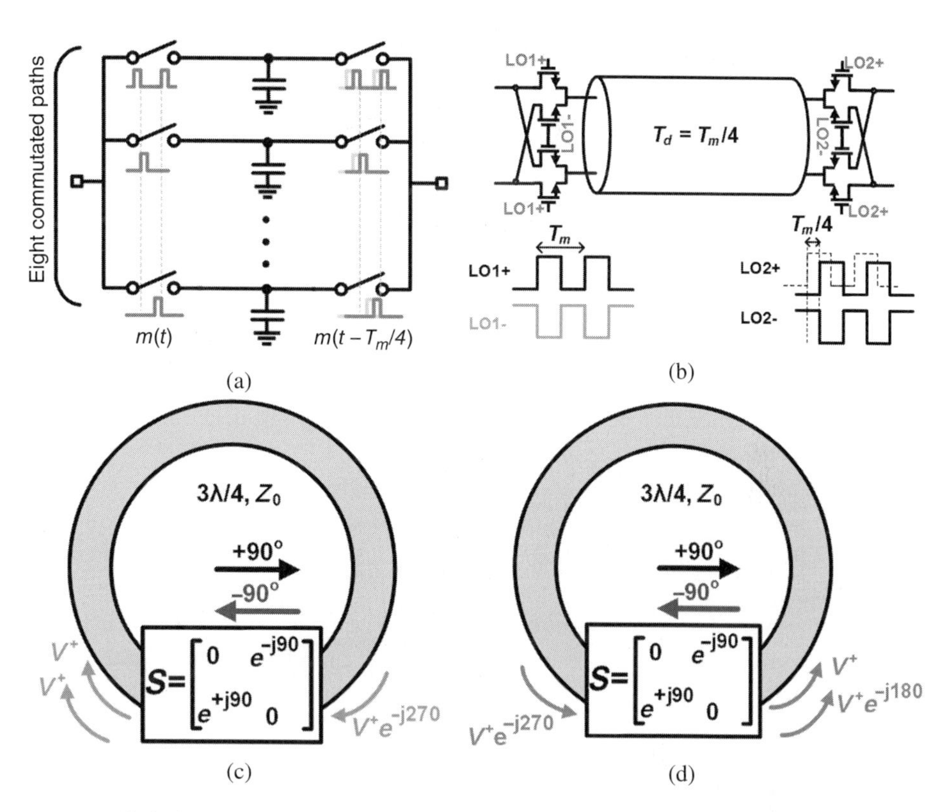

Figure 4.8 Switch-based spatio-temporal conductivity modulation breaks reciprocity on CMOS. In the essence, these approaches include two sets of transistor switches on either end of a delay medium. (a) N-path-based gyrator. (b) Switched transmission line–based gyrator with theoretically infinite BW. (c) Nonreciprocal wave propagation can achieved by wrapping up a $3\lambda/4$ transmission line loop around a gyrator. Adapted from [43] and [49]. (©2017 IEEE. Reprinted, with permission, from *IEEE Journal of Solid-State Circuits.*)

the input signal passes to output after a delay of $T_m/4$, whereas in the reverse direction it experiences a delay of $T_m/4$ and an additional sign flip, thus behaving as an ideal passive lossless gyrator over theoretically infinite BW. In this structure, a modulation frequency that is arbitrarily lower than the operation frequency can be used, enabling mm-wave operation. In reality, the modulation frequency should be chosen based on a trade-off among the loss, delay line length, and power consumption. In [43], the modulation was performed at one-third of the operating frequency (8.33 GHz) to realize a 25 GHz passive circulator in 45 nm SOI CMOS. It achieves an insertion loss of 3.3 dB with a 1 dB BW of 4.6 GHz, isolation higher than 18.3 dB (limited by the measurement setup) over the same BW, and an IP1dB higher than +21 dBm. More recently, the same wideband gyrator concept was used in a 1 GHz passive circulator in 180 nm SOI CMOS to demonstrate watt-level power handling (30.7 dBm) and >+50 dBm TX-to-ANT *IIP3* [62] along with a loss-free and inductor-free antenna balancing technique.

4.4.3 Integrated Low-RF FD Radios

Numerous FD CMOS receivers and transceivers with a variety of RF/analog SIC techniques have been reported in the recent years. This section will review the state-of-the-art silicon FD radios employing a combination of SI suppression techniques.

The SI channel usually has a large group delay (nanosecond-scale at RF frequencies) due to the frequency-selective behavior of the antenna interface as well as the environmental reflections. Conventional RF cancellers based on frequency-flat amplitude and phase cannot emulate the group delay of the SI channel over a wide frequency range, resulting in a limited RF SIC BW. As illustrated in Figure 4.9a, the frequency-selective SI channel response can be mimicked using a continuous-time finite impulse response (FIR) filter that performs equalization of the group delay and group delay variation in the SI channel. In fact, earlier PCB-based FD radios [11] employed transmission line–based time delay and attenuators to achieve SIC in the RF domain. However, it is not area efficient to integrate nanosecond-scale delay lines on silicon. Therefore, new RF SIC techniques with compact form factors are required to enable wideband FD RF transceivers.

A frequency domain equalization (FDE) approach was proposed in [64] to replace the conventional time-domain delay–based FIR cancellers, addressing the challenge regarding the integrated wideband RF SIC. Shown in Figure 4.9b, this approach employs second-order reconfigurable bandpass filters (BPFs) with amplitude and phase control, essentially performing FDE, to achieve wideband RF SIC. In general, an RF canceller with $2N$ degrees of freedom can be used to achieve perfect SIC at N different frequencies, resulting in wideband cancellation. There are multiple ways to use the $2N$ degrees of freedom. The magnitude and phase response of the SI channel can be synthesized at N different frequencies or the magnitude and phase as well as their sloped can be mimicked at $N/2$ frequencies. In Figure 4.9b, each second-order bandpass filter bank has four degrees of freedom (amplitude, phase, quality factor, and center frequency of the BPF), mimicking not only the magnitude and phase of the SI channel but also their slopes at a single frequency point. Using a bank of independently controlled filters further enhances the SIC BW by enabling such mimicking at multiple frequency points. Nanosecond-scale delays are enabled by implementing the baseband filters as the two-port N-path-based Gm-C filters with embedded variable attenuation and phase shifting, which allows four degrees of freedom (gain, phase, quality factor, and center frequency). A 0.8–1.4 GHz FD receiver using the FDE was demonstrated with >20 dB SIC over 25 MHz BW (using a narrowband antenna pair with peak group delay of 8 ns as the antenna interface).

Figure 4.10 shows another CMOS transceiver architecture that was reported for small-form factor short-range FD systems [66]. This architecture taps the TX signal close to the antenna and uses a second down-converter to add a phase and amplitude adjusted copy in the analog/BB domain. Phase-shift, amplitude scaling, and I/Q down-conversion functions are combined in a vector-modulator (VM) passive mixer. The RX and SI cancellation paths are based on passive mixer-first architecture for superior in-band linearity performance, lowering the SI-induced distortion, which

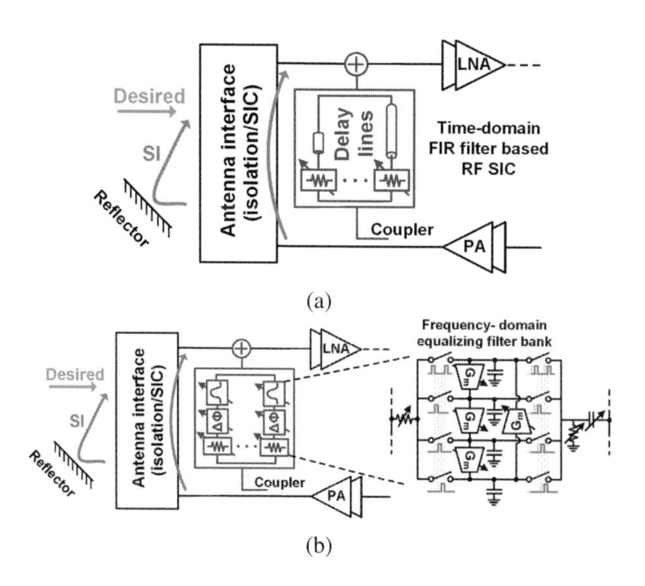

Figure 4.9 RF domain SIC (a) based on time-domain equalization with delay lines (not amenable for on-chip integration at low-RF frequencies). (b) Frequency-domain equalization employs a bank of second-order N-path based Gm-C filters to achieve wideband RF SIC on-chip. Adapted from [63].

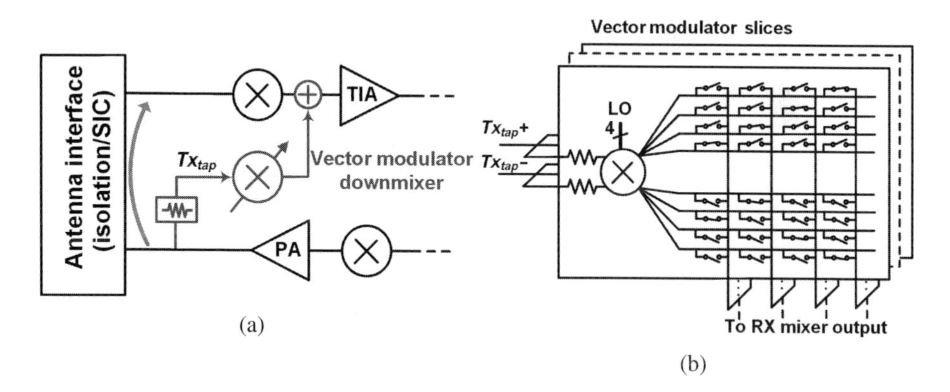

Figure 4.10 (a) SI-cancelling CMOS transceiver based on vector-modulator. (b) The VM is a 31-slice version of the main receiver, each slice followed by static phase rotator switches. The VM diverts SI currents through linear passive networks before amplification. Adapted from [65].

would increase the complexity of the digital SIC. The VM mixer is designed as a sliced version of the four-phase RX mixer, and each slice is followed by multiplexer switches to achieve VM functionality. A 2.5 GHz transceiver that employs a VM with 32-by-32 phase/amplitude constellation points demonstrated 27 dB SIC over 16.25 MHz BW. However, this comes at the expense of 3–5 dB degradation in the RX NF.

More recently, a transceiver architecture with dual-path CMOS SIC was proposed in [17] for long-range wireless FD applications. Figure 4.11a shows a simplified block

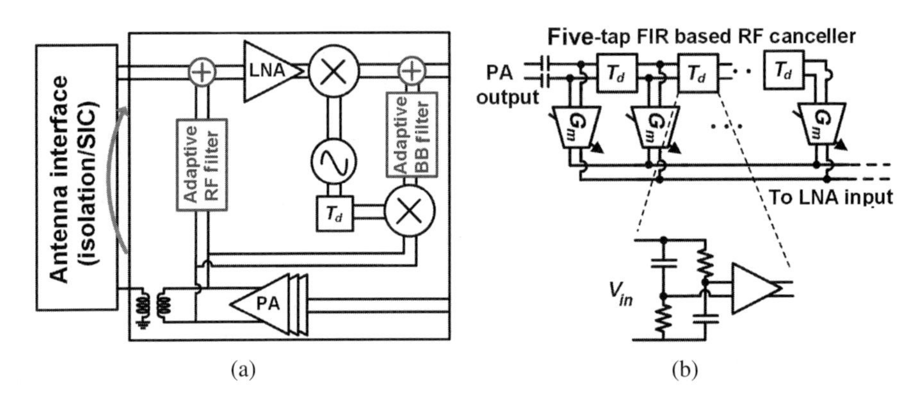

Figure 4.11 (a) Dual-injection SIC transceiver architecture for long-range FD wireless. The SIC was performed through RF and BB cancellers that are based on continuous-time FIR filters. (b) Five-taps continuous FIR filter where each tap includes a true-time-delay element, buffer, and seven-bits VGA. Adapted from [68].

diagram of the transceiver. It employs RF and baseband (BB) cancellers that are based on continuous-time FIR filters to create an inverse time-domain response of the SI channel. The RF canceller is implemented as a five-taps FIR filter where each tap includes a true-time-delay element, buffer, and seven-bit variable gain amplifier (VGA), shown in Figure 4.11b. The RF true-time-delay elements are realized as RC-CR all-pass filters, allowing compact integration. The VGA is designed as inverter-based Gm stages. Both RF and BB cancellers tap from the TX output to capture the TX path imperfections (noise and distortions) as well. The number of RF canceller taps was determined based on the trade-offs among the SIC BW, the RF NF degradation, and power consumption [67]. Increasing the number of the RF filter taps would mimic the SI channel more closely, enabling wider SIC BWs. However, this increases the RX NF degradation due to the RF cancellation path as well as the RF SIC power consumption. The authors found that five taps was optimum to enable an SIC BW of around 40 MHz [67]. The BB canceller is implemented as 14-tap FIR filter based on Gm-C all-pass true-time-delay elements, providing 10 ns delay. A 40 nm prototype demonstrated more than 50 dB SIC over 42 MHz BW centered at 1.96 GHz with an RX NF degradation of less than 1.55 dB.

The FD radio ICs covered in this section so far assumes a duplexer or antenna pair (typically off-chip) would be available as an antenna interface. The FD receivers and transceivers incorporating duplexing functionality on silicon was also explored in the recent years. For example, targeting short-range FD applications with relatively relaxed SIC requirements, a wideband FD transceiver based on BB duplexing low-noise amplifiers (LNAs) was presented in [13] (Figure 4.12a). It uses a passive mixer-first architecture to move the duplexing operation from RF to BB. A fully differential baseband duplexing amplifier was realized based on the well-known RF noise-cancelling LNA topology but with the addition of a load sharing technique as shown in Figure 4.12b. Under the same design considerations to achieve noise

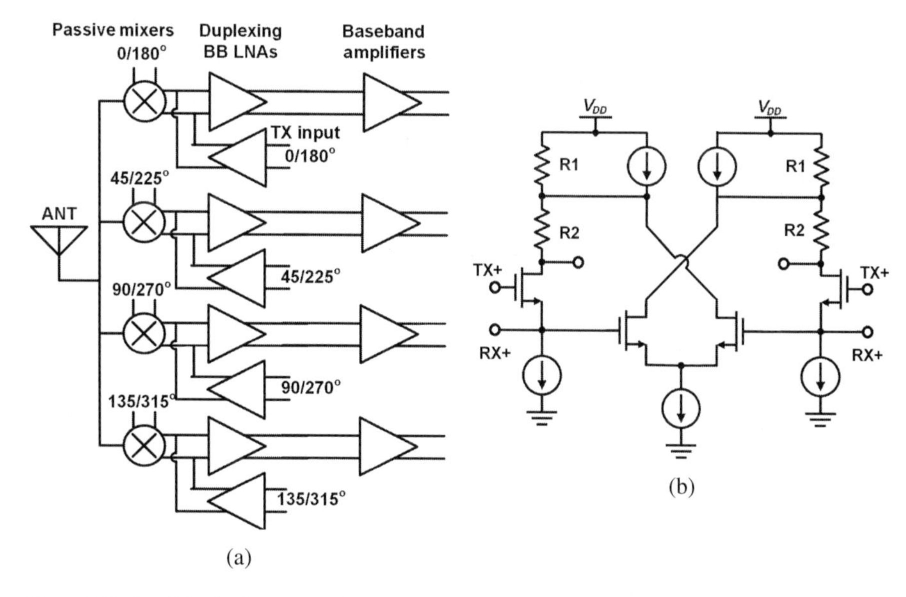

Figure 4.12 (a) Full-duplex transceiver based on a mixer-first architecture with tunable baseband duplexing LNAs. The bidirectional transparency of the passive mixer moves the duplexing function to baseband. (b) Fully differential duplexing LNA concept. Adapted from [13].

cancellations, the BB duplexing LNAs perform multiple functions such as buffering the BB TX signals into the mixer, providing impedance matching, and amplifying low noise of the desired RX signals while rejecting the SI from the transmitter. Additionally, to maintain the SIC across a wide frequency range, a complex feed-forward network was integrated. Covering a frequency range of 0.1–1.5 GHz, an FD transceiver prototype enables short-range FD operation with a TX-to-RX isolation of more than 33.5 dB up to a TX output power of −17.3 dBm and assumes another 50 dB isolation in the digital domain.

Another CMOS FD receiver IC incorporating a shared antenna interface is reported in [69], integrating the magnetic-free N-path-based passive circulator discussed in the previous section. Figure 4.13a shows the overall architecture of the FD receiver IC operating between 610 and 850 MHz. In this work, the reciprocal $3\lambda/4$ transmission line of the magnetic-free N-path-based circulator was implemented as the left-handed artificial transmission line with on-chip series capacitors and off-chip shunt inductors. The N-path-based gyrator is placed next to the RX port (Port 3) as shown in Figure 4.13b to enhance the TX-to-ANT linearity. Placing the N-path filter-based gyrator next to the RX port ensures low voltage swing on both of its ends. In addition to the magnetic-free passive circulator, an analog BB SI canceller is also featured. The analog BB canceller taps from the TX BB and injects a phase and amplitude adjusted cancellation signal at the RX BB. A total SI suppression of 42 dB was reported across the circulator and analog BB cancellers over 12 MHz BW. The authors also studied digital SIC–based on nonlinear delays in Matlab to cancel not only the SI but also the SI-induced IM3 distortion. In conjuction with digital SIC, 85 dB total SI suppression was demonstrated. Despite the high linearity of the circulator, this FD RX was only able to handle up to a TX power of −7 dBm due to the limited circulator isolation and LNA linearity. Additionally, it suffered from a high NF of 10.9 dB under SIC.

More recently, as illustrated in Figure 4.13c, the N-path-filter-based circulator concept was extended to a circulator-receiver in [16], which combines circulating and down-mixing functions, eliminating the RX LNA/low-noise transconductance amplifier (LNTA) and mixer of the conventional receiver. Furthermore, inspired by electrical balance duplexers, an on-chip balancing network is built into the circulator architecture to track ANT impedance, improving the TX-to-RX (Port1 to Port3) isolation. This structure is essentially a zero–intermediate frequency (IF) mixer-first RX embedded within the circulator. It does not have to provide a 50 Ω RX input impedance since

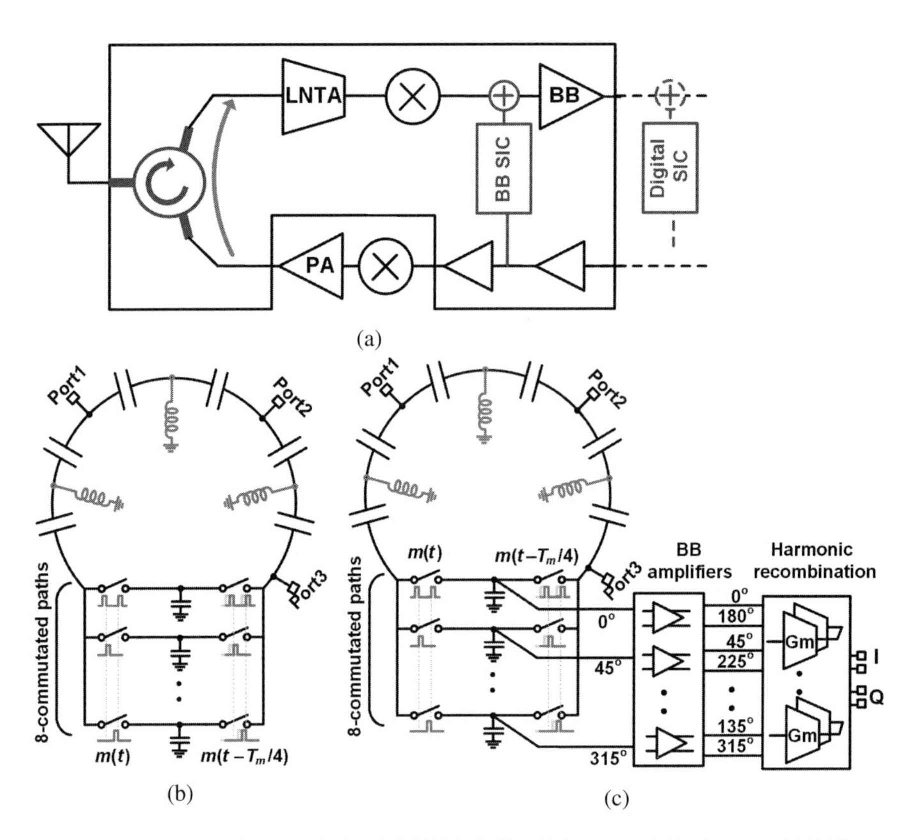

Figure 4.13 (a) Block diagram of the CMOS FD RX with integrated circulator and BB SI canceller. (b) High-linearity circulator is built by placing the RX port right at the N-path-based gyrator and miniaturized by using left-handed artificial transmission lines. Adapted from [49]. (c) Highly linear integrated magnetic-free circulator-receiver merging the passive N-path-based circulator with a down-converting mixer. Adapted from [16].

the reflected power at the RX port can circulate while the ANT port is still matched. Additional 6 dB voltage gain can be achieved by leaving the RX port open. These new concepts incorporated around the circulator results in an FD receiver architecture which has lower power consumption and NF compared to [69] and higher TX power handling capability. The circulator receiver achieves 40 dB SI suppression over 20 MHz BW and can handle up to TX power of +8 dBm. It also achieves 8 dB NF under SIC. At +8 dBm TX power, 80 dB overall SI suppression was achieved in conjunction with digital cancellation.

4.4.4 Integrated Millimeter-Wave FD Radios

Figure 4.14 shows the fully integrated 60 GHz direct-conversion transceiver reported in 45 nm SOI CMOS for full-duplex wireless applications. The FD operation is enabled by the polarization-based wideband reconfigurable antenna cancellation technique, which was covered earlier. The implementation of this technique at 60 GHz is illustrated in Figure 4.14a. First, the colocated TX and RX antenna pair with orthogonal polarizations improves the initial isolation to 32–36 dB over 54–66 GHz (in simulation). As described earlier, an auxiliary port is introduced on the RX antenna and is terminated with a high-order reconfigurable reflective termination that is integrated on the IC to maintain a high level of SIC in the face of a changing EM environment. To further suppress the residual SI, a second RF canceller from the TX output to the LNA output with >30 dB gain and $>360^\circ$ phase control is integrated. Figure 4.14b shows the architecture of the 45 nm SOI CMOS 60 GHz FD transceiver IC [19]. It is a zero-IF binary phase shift keying (BPSK) transceiver consisting of five main parts: an on-PCB TX/RX antenna pair with the polarization-based antenna cancellation, transmitter, receiver, a second RF canceller, and LO distribution. In conjunction with digital SIC in Matlab, the FD transceiver provides almost 80 dB total SI suppression over 1 GHz BW.

A simple FD link over a distance of nearly 1 m was demonstrated in [19]. Figure 4.14 shows the link setup using a 100 MHz offset continuous-wave signal and a 1 Gbps BPSK as the desired and the transmitted SI, respectively. Without any SI cancellation, the RX output is dominated by the 1 Gbps SI, whereas the antenna and RF SIC allows discerning the desired signal. Digital cancellation in Matlab further suppresses the SI so that the received signal is even cleaner and exhibits a signal-to-interference-noise-and-distortion ratio of 7.2 dB.

A 64 GHz FD transceiver front-end is reported in a 45 nm CMOS SOI process [22] (Figure 4.16). It consists of an on-chip four-feed SIC slot-loop antenna, two parallel TX paths, two parallel LNA paths, and an all-passive canceller. Similar to [19], an RF canceller is featured from the TX output to LNA output to suppress the residual SI after the antenna interface. The RF canceller includes reflection-type phase shifters and attenuators, providing $>360^\circ$ phase and ≈ 40 dB attenuation range. More than 60 dB SIC in antenna and RF domains were demonstrated over 63–65 GHz, 65–66 GHz, and 71.7–72.3 GHz. Additionally, an FD link between two FD front-ends (using external up

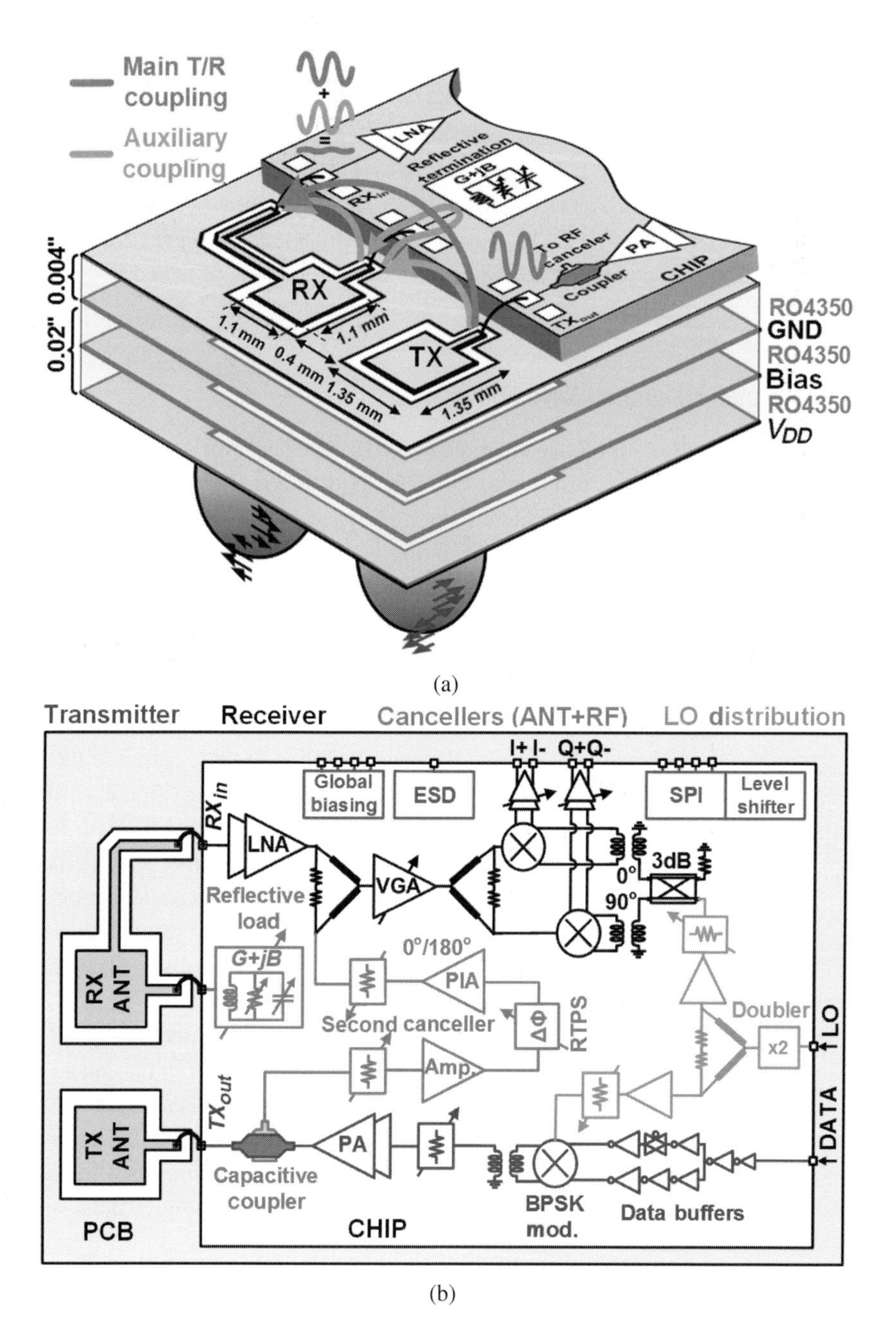

Figure 4.14 Fully integrated 60 GHz FD transceiver featuring polarization-based reconfigurable antenna cancellation and RF SIC: (a) 3D implementation view of the polarization-based antenna SIC concept and (b) architecture. Adapted from [19]. (©2016 IEEE. Reprinted, with permission, from *IEEE Journal of Solid-State Circuits*.)

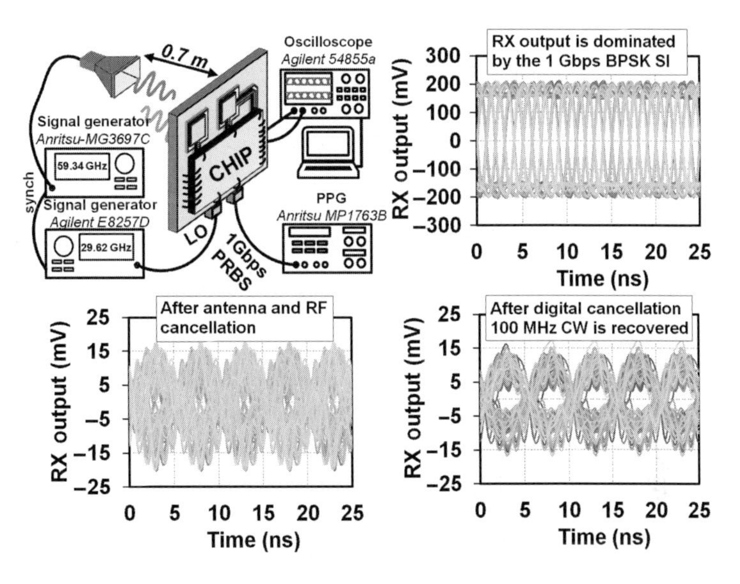

Figure 4.15 A simple 60 GHz FD link setup and demonstration. Adapted from [19]. (©2016 IEEE. Reprinted, with permission, from *IEEE Journal of Solid-State Circuits*.)

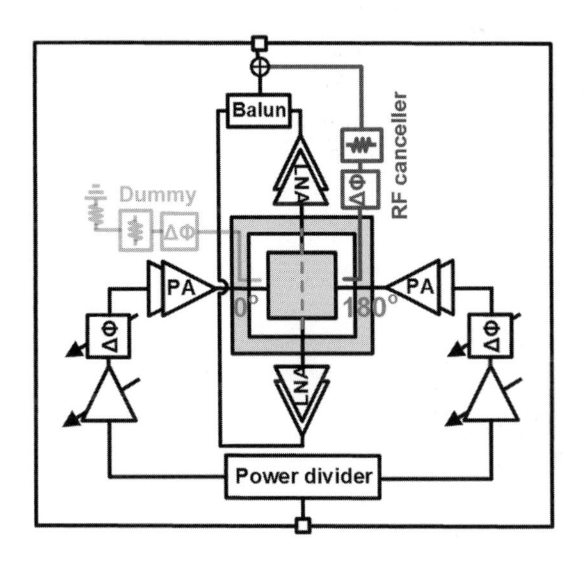

Figure 4.16 A 64 GHz FD front-end with an on-chip four-feed SIC antenna and all-passive passive canceller. Adapted from [22].

and down conversion chains) was demonstrated over 0.5 m using 4 Gbps 16-QAM and 3 Gbps 64-QAM.

4.5 Conclusion

While tremendous research progress has been made in the last few years, there are many problems remaining that need to be solved before FD can become a reality [63]. So far, all the reported integrated FD radios feature single-element receivers or transceivers that target solving the SI problem within a single wireless node. Even if they solve the SI problem, interference from other FD nodes would hamper the FD operation. Beamforming is an excellent solution for interference management at RF frequencies. Furthermore, beamforming is essential for mm-wave transceivers for robust non-light-of-side operation. Therefore, incorporation of FD in large-scale phased-array transceivers is an inevitable direction toward revealing the true benefits offered by FD. To this end, efficient SIC techniques and compact antenna interfaces are required to manage SI from any TX element to any RX element with minimal complexity. At the time of writing this chapter, the first IC-based effort on combining phased-array beamforming with FD has yet to be presented [70].

An even harder open research problem is the extension of IC-based SIC concepts to MIMO transceivers in which every TX element would transmit totally uncorrelated data streams (in contrast to uniformly phase-shifted data streams transmitted in phased arrays). In a brute force implementation, complexity would scale quadratically with the number of MIMO elements, which is impractical to sustain in terms of area and power consumption in large-scale MIMO systems. Therefore, new antenna and RF SIC techniques that scale with a low complexity close to the optimal possible (scales linearly with number of elements) are required.

References

[1] J. Zander and P. Mähönen, "Riding the data tsunami in the cloud: Myths and challenges in future wireless access," *IEEE Communications Magazine*, vol. 51, no. 3, pp. 145–151, March 2013.

[2] J. G. Andrews, S. Buzzi, W. Choi, et al., "What will 5G be?" *IEEE Journal on Selected Areas in Communications*, vol. 32, no. 6, pp. 1065–1082, June 2014.

[3] Y. Zhu, Z. Zhang, Z. Marzi, et al., "Demystifying 60GHz outdoor picocells," in *Proceedings of the 20th Annual International Conference on Mobile Computing and Networking*, ACM, 2014, pp. 5–16.

[4] S. Zihir, O. D. Gurbuz, A. Kar-Roy, S. Raman, and G. M. Rebeiz, "60-GHz 64- and 256-elements wafer-scale phased-array transmitters using full-reticle and subreticle stitching techniques," *IEEE Transactions on Microwave Theory and Techniques*, vol. 64, no. 12, pp. 4701–4719, December 2016.

[5] B. Sadhu, Y. Tousi, J. Hallin, et al., "A 28GHz 32-element phased-array transceiver IC with concurrent dual polarized beams and 1.4 degree beam-steering resolution for 5G communication," in *2017 IEEE International Solid-State Circuits Conference (ISSCC)*, February 2017, pp. 128–129.

[6] U. Kodak and G. M. Rebeiz, "Bi-directional flip-chip 28 GHz phased-array core-chip in 45nm CMOS SOI for high-efficiency high-linearity 5G systems," in *2017 IEEE Radio Frequency Integrated Circuits Symposium (RFIC)*, June 2017, pp. 61–64.

[7] H. T. Kim, B. S. Park, S. M. Oh, et al., "A 28GHz CMOS direct conversion transceiver with packaged antenna arrays for 5G cellular system," in *2017 IEEE Radio Frequency Integrated Circuits Symposium (RFIC)*, June 2017, pp. 69–72.

[8] P. M. N. H. Mahmood, G. Berardinelli, and F. Frederiksen, "Throughput analysis of full duplex communication with asymmetric traffic in small cell systems," in *International Conference on Wireless and Mobile Communications*, IEEE, October 2015.

[9] J. Marasevic and G. Zussman, "On the capacity regions of single-channel and multi-channel full-duplex links," in *Proceedings of ACM MobiHoc'16*, July 2016, pp. 241–250.

[10] A. Sabharwal, P. Schniter, D. Guo, D. Bliss, S. Rangarajan, and R. Wichman, "In-band full-duplex wireless: Challenges and opportunities," *IEEE Journal on Selected Areas in Communications*, vol. 32, no. 9, pp. 1637–1652, September 2014.

[11] D. Bharadia, E. McMilin, and S. Katti, "Full duplex radios," *SIGCOMM Computer Communication Review*, vol. 43, no. 4, pp. 375–386, August 2013. [Online]. Available: http://doi.acm.org/10.1145/2534169.2486033

[12] E. Everett, A. Sahai, and A. Sabharwal, "Passive self-interference suppression for full-duplex infrastructure nodes," *CoRR*, vol. abs/1302.2185, 2013. [Online]. Available: http://arxiv.org/abs/1302.2185

[13] D. Yang, H. Yuksel, and A. Molnar, "A wideband highly integrated and widely tunable transceiver for in-band full-duplex communication," *IEEE Journal of Solid-State Circuits*, vol. 50, no. 5, pp. 1189–1202, May 2015.

[14] D.-J. van den Broek, E. Klumperink, and B. Nauta, "A self-interference-cancelling receiver for in-band full-duplex wireless with low distortion under cancellation of strong TX leakage," in *IEEE International Solid-State Circuits Conference*, IEEE, February 2015, pp. 1–3.

[15] J. Zhou, T.-H. Chuang, T. Dinc, and H. Krishnaswamy, "19.1 Receiver with >20MHz bandwidth self-interference cancellation suitable for FDD, co-existence and full-duplex applications," in *2015 IEEE International Solid-State Circuits Conference – (ISSCC) Digest of Technical Papers*, February 2015, pp. 1–3

[16] N. Reiskarimian, M. B. Dastjerdi, J. Zhou, and H. Krishnaswamy, "Highly-linear integrated magnetic-free circulator-receiver for full-duplex wireless," in *IEEE International Solid-State Circuits Conference (ISSCC)*, IEEE, February 2017, pp. 316–317.

[17] T. Zhang, A. Najafi, C. Su, and J. C. Rudell, "A 1.7-to-2.2GHz full-duplex transceiver system with 50dB self-interference cancellation over 42MHz bandwidth," in *2017 IEEE International Solid-State Circuits Conference (ISSCC)*, IEEE, February 2017, pp. 314–315.

[18] T. Dinc and H. Krishnaswamy, "A 28GHz magnetic-free non-reciprocal passive CMOS circulator based on spatio-temporal conductance modulation," in *2017 IEEE International Solid-State Circuits Conference (ISSCC)*, IEEE, February 2017, pp. 294–295.

[19] T. Dinc, A. Chakrabarti, and H. Krishnaswamy, "A 60 GHz CMOS full-duplex transceiver and link with polarization-based antenna and RF cancellation," *IEEE Journal of Solid-State Circuits*, vol. 51, no. 5, pp. 1125–1140, May 2016.

[20] C. Lu, M. K. Matters-Kammerer, A. Zamanifekri, A. B. Smolders, and P. G. M. Baltus, "A millimeter-wave tunable hybrid-transformer-based circular polarization duplexer with

sequentially-rotated antennas," *IEEE Transactions on Microwave Theory and Techniques*, vol. 64, no. 1, pp. 166–177, January 2016.

[21] C. Yang and P. Gui, "85-110 GHz CMOS tunable nonreciprocal transmission line with 45 dB isolation for wideband transceivers," in *2017 IEEE Radio Frequency Integrated Circuits Symposium (RFIC)*, June 2017, pp. 284–287.

[22] T. Chi, J. S. Park, S. Li, and H. Wang, "A 64GHz full-duplex transceiver front-end with an on-chip multi-feed self-interference-canceling antenna and an all-passive canceler supporting 4Gb/s modulation in one antenna footprint," in *2018 IEEE International Solid-State Circuits Conference*, February 2018, pp. 76–78.

[23] R. Taori and A. Sridharan, "Point-to-multipoint in-band mmwave backhaul for 5G networks," *IEEE Communications Magazine*, vol. 53, no. 1, pp. 195–201, January 2015.

[24] "Crucial economics for mobile data backhaul: An analyss of the total cost of ownership of point-to-point, point-to-multipoint, and fiber options," Senza Fili Consulting, 2012. [Online] Available: https://cbnl.com/sites/all/files/userfiles/files/CB-002070-DC-LATEST.pdf

[25] "EdgeHaul millimeter wave small cell backhaul system," March 2015. [Online]. Available: www.interdigital.com/presentations/edgehaul-millimeter-wave-small-cell-backhaul-system

[26] T. Dinc and H. Krishnaswamy, "Millimeter-wave full-duplex wireless: Applications, antenna interfaces and systems," in *2017 IEEE Custom Integrated Circuits Conference*, April 2017, pp. 1–8.

[27] Z. Wei, X. Zhu, S. Sun, Y. Huang, A. Al-Tahmeesschi, and Y. Jiang, "Energy-efficiency of millimeter-wave full-duplex relaying systems: Challenges and solutions," *IEEE Access*, vol. 4, pp. 4848–4860, 2016.

[28] D. Guermandi, Q. Shi, A. Medra, et al., "A 79GHz binary phase-modulated continuous-wave radar transceiver with TX-to-RX spillover cancellation in 28nm CMOS," in *2015 IEEE International Solid-State Circuits Conference (ISSCC) Digest of Technical Papers*, IEEE, February 2015, pp. 1–3.

[29] "How Oculus Rift works: Everything you need to know about the VR sensation," March 2016. [Online]. Available: www.wareable.com/oculus-rift/how-oculus-rift-works

[30] "IEEE 802.11TGay use cases," September 2015. [Online]. Available: https://mentor.ieee.org/802.11/dcn/15/11-15-0625-03-00ay-ieee-802-11-tgay-usage-scenarios.pptx

[31] O. Abari, D. Bharadia, A. Duffield, and D. Katabi, "Cutting the cord in virtual reality," in *Proceedings of the 15th ACM Workshop on Hot Topics in Networks*, ser. HotNets '16, ACM, 2016, pp. 162–168. [Online]. Available: http://doi.acm.org/10.1145/3005745.3005770

[32] J. I. Choi, M. Jain, K. Srinivasan, P. Levis, and S. Katti, "Achieving single channel, full duplex wireless communication," in *Proceedings of the Sixteenth Annual International Conference on Mobile Computing and Networking*, ser. MobiCom '10, ACM, 2010, pp. 1–12. [Online]. Available: http://doi.acm.org/10.1145/1859995.1859997

[33] B. Debaillie, D.-J. van den Broek, C. Lavin, et al., "Analog/RF solutions enabling compact full-duplex radios," *IEEE Journal on Selected Areas in Communications*, vol. 32, no. 9, pp. 1662–1673, September 2014.

[34] A. Khandani, "Two-way (true full-duplex) wireless," in *Canadian Workshop on Information Theory (CWIT)*, IEEE, June 2013, pp. 33–38.

[35] E. Yetisir, C.-C. Chen, and J. Volakis, "Low-profile UWB 2-port antenna with high isolation," *IEEE Antennas and Wireless Propagation Letters*, vol. 13, pp. 55–58, 2014.

[36] E. Aryafar, M. A. Khojastepour, K. Sundaresan, S. Rangarajan, and M. Chiang, "MIDU: Enabling MIMO full duplex," in *Proceedings of the 18th Annual International Conference on Mobile Computing and Networking*, ser. Mobicom '12, ACM, 2012, pp. 257–268. [Online]. Available: http://doi.acm.org/10.1145/2348543.2348576

[37] T. Snow, C. Fulton, and W. Chappell, "Transmit-receive duplexing using digital beamforming system to cancel self-interference," *IEEE Transactions on Microwave Theory and Techniques*, vol. 59, no. 12, pp. 3494–3503, December 2011.

[38] K. Kolodziej, P. Hurst, A. Fenn, and L. Parad, "Ring array antenna with optimized beamformer for simultaneous transmit and receive," in *IEEE Antennas and Propagation Society International Symposium (APSURSI)*, IEEE, July 2012, pp. 1–2.

[39] W. Moulder, B. Perry, and J. Herd, "Wideband antenna array for simultaneous transmit and receive (STAR) applications," in *IEEE Antennas and Propagation Society International Symposium (APSURSI)*, IEEE, July 2014, pp. 243–244.

[40] A. Wegener and W. Chappell, "High isolation in antenna arrays for simultaneous transmit and receive," in *IEEE International Symposium on Phased Array Systems Technology*, IEEE, October 2013, pp. 593–597.

[41] A. Wegener, "Broadband near-field filters for simultaneous transmit and receive in a small two-dimensional array," in *IEEE Proceedings of International Microwave Symposium*, IEEE, June 2014, pp. 1–3.

[42] T. Dinc and H. Krishnaswamy, "A T/R antenna pair with polarization-based reconfigurable wideband self-interference cancellation for simultaneous transmit and receive," in *IEEE Proceedings of International Microwave Symposium*, IEEE, May 2015, pp. 1–4.

[43] T. Dinc, A. Nagulu, and H. Krishnaswamy, "A millimeter-wave non-magnetic passive SOI CMOS circulator based on spatio-temporal conductivity modulation," *IEEE Journal of Solid-State Circuits*, vol. 52, no. 4, pp. 3276–3292, December 2017.

[44] H. Darabi, A. Mirzaei, and M. Mikhemar, "Highly integrated and tunable RF front ends for reconfigurable multiband transceivers: A tutorial," *IEEE Transactions on Circuits and Systems I: Regular Papers*, vol. 58, no. 9, pp. 2038–2050, September 2011.

[45] M. Mikhemar, H. Darabi, and A. A. Abidi, "A multiband RF antenna duplexer on CMOS: Design and performance," *IEEE Journal of Solid-State Circuits*, vol. 48, no. 9, pp. 2067–2077, September 2013.

[46] M. Elkholy, M. Mikhemar, H. Darabi, and K. Entesari, "Low-loss integrated passive CMOS electrical balance duplexers with single-ended LNA," *IEEE Transactions on Microwave Theory and Techniques*, vol. 64, no. 5, pp. 1544–1559, May 2016.

[47] S. H. Abdelhalem, P. S. Gudem, and L. E. Larson, "Tunable CMOS integrated duplexer with antenna impedance tracking and high isolation in the transmit and receive bands," *IEEE Transactions on Microwave Theory and Techniques*, vol. 62, no. 9, pp. 2092–2104, September 2014.

[48] B. van Liempd, B. Hershberg, S. Ariumi, et al., "A +70-dBm IIP3 electrical-balance duplexer for highly integrated tunable front-ends," *IEEE Transactions on Microwave Theory and Techniques*, vol. 64, no. 12, pp. 4274–4286, December 2016.

[49] N. Reiskarimian, M. B. Dastjerdi, J. Zhou, and H. Krishnaswamy, "Analysis and design of commutation-based circulator-receivers for integrated full-duplex wireless," *IEEE Journal of Solid-State Circuits*, vol. 53, no. 8, pp. 2190–2201, August 2018.

[50] H. S. Wu, C. W. Wang, and C. K. C. Tzuang, "CMOS active quasi-circulator with dual transmission gains incorporating feedforward technique at K-Band," *IEEE Transactions on Microwave Theory and Techniques*, vol. 58, no. 8, pp. 2084–2091, August 2010.

[51] J.-F. Chang, J.-C. Kao, Y.-H. Lin, and H. Wang, "Design and analysis of 24-GHz active isolator and quasi-circulator," *IEEE Transactions on Microwave Theory and Techniques*, vol. 63, no. 8, pp. 2638–2649, August 2015.

[52] Ahmed M. Mahmoud, Arthur R. Davoyan, and Nader Engheta, "All-passive nonreciprocal metastructure," *Nature Communications*, vol. 6, no. 8359, July 2015.

[53] S. Qin, Q. Xu, and Y. E. Wang, "Nonreciprocal components with distributedly modulated capacitors," *IEEE Transactions on Microwave Theory and Techniques*, vol. 62, no. 10, pp. 2260–2272, October 2014.

[54] N. A. Estep, D. L. Sounas, J. Soric, and A. Alu, "Magnetic-free non-reciprocity and isolation based on parametrically modulated coupled-resonator loops," *Nature Physics*, vol. 10, pp. 923–927, 2014.

[55] N. A. Estep, D. L. Sounas, and A. Alu, "Magnetless microwave circulators based on spatiotemporally modulated rings of coupled resonators," *IEEE Transactions on Microwave Theory and Techniques*, vol. 64, no. 2, pp. 502–518, February 2016.

[56] S. Tyagi, C. Auth, P. Bai, et al., "An advanced low power, high performance, strained channel 65nm technology," in *IEEE International Electron Devices Meeting, 2005. IEDM Technical Digest*, December 2005, pp. 245–247.

[57] N. Reiskarimian and H. Krishnaswamy, "Magnetic-free non-reciprocity based on staggered commutation," *Nature Communications*, vol. 7, no. 11217, April 2016.

[58] T. Dinc, M. Tymchenko, A. Nagulu, D. Sounas, A. Alù, and H. Krishnaswamy, "Synchronized conductivity modulation to realize broadband lossless magnetic-free non-reciprocity," in *Nature Communications*, vol. 8, no. 10, October 2017.

[59] M. M. Biedka, R. Zhu, Q. M. Xu, and Y. E. Wang, "Ultra-wide band non-reciprocity through sequentially-switched delay lines," *Scientific Reports*, vol. 7, 2017.

[60] J. Krol and S. Gong, "A non-magnetic gyrator utilizing switched delay lines," in *IEEE European Microwave Conference*, IEEE, 2017, pp. 452–455.

[61] B. D. H. Tellegen, "The gyrator a new electric network element," in *Philips Res. Rep.*, April 1948, pp. 81–101.

[62] A. Nagulu, A. Alù, and H. Krishnaswamy, "Fully-integrated non-magnetic 180nm SOI circulator with >1W P1dB, >+50dBm IIP3 and high isolation across 1.85 VSWR," in *2018 IEEE RFIC*, June 2018, pp. 104–107.

[63] J. Zhou, N. Reiskarimian, J. Diakonikolas, et al., "Integrated full duplex radios," *IEEE Communications Magazine*, vol. 55, no. 4, pp. 142–151, April 2017.

[64] J. Zhou, T.-H. Chuang, T. Dinc, and H. Krishnaswamy, "Integrated wideband self-interference cancellation in the RF domain for FDD and full-duplex wireless," *IEEE Journal of Solid-State Circuits*, vol. 50, no. 12, pp. 3015–3031, 2015.

[65] D.-J. van den Broek, "CMOS front-end techniques for in-band full-duplex radio," Ph.D. dissertation, University of Twente, Enschede, Netherlands, 2017.

[66] D. J. van den Broek, E. A. M. Klumperink, and B. Nauta, "An in-band full-duplex radio receiver with a passive vector modulator downmixer for self-interference cancellation," *IEEE Journal of Solid-State Circuits*, vol. 50, no. 12, pp. 3003–3014, December 2015.

[67] T. Zhang, C. Su, A. Najafi, and J. C. Rudell, "Wideband dual-injection path self-interference cancellation architecture for full-duplex transceivers," *IEEE Journal of Solid-State Circuits*, vol. PP, no. 99, pp. 1–14, 2018.

[68] T. Zhang, "Integrated wideband self-interference cancellation techniques for FDD and full-duplex wireless communication," Ph.D. dissertation, University of Washington, Seattle, 2017.

[69] N. Reiskarimian, J. Zhou, and H. Krishnaswamy, "A CMOS Passive LPTV nonmagnetic circulator and its application in a full-duplex receiver," *IEEE Journal of Solid-State Circuits*, vol. 52, no. 5, pp. 1358–1372, May 2017.

[70] M. B. Dastjerdi, N. Reiskarimian, T. Chen, G. Zussman, and H. Krishnaswamy, "Full duplex circulator-receiver phased array employing self-interference cancellation via beamforming," in *2018 IEEE RFIC*, June 2018, pp. 108–111.

5 Flexible Integrated Architectures for Frequency Division Duplex Communication

Lucas A. Calderin, Sameet Ramakrishnan, Elad Alon, Borivoje Nikolić, and Ali M. Niknejad

5.1 Introduction

The expansion of wireless systems and applications has led to a proliferation of wireless bands. While there is a broad industry adoption of multimode, multistandard radio transceiver integrated circuits (ICs), support for multiband operation has been provided through off-chip components, where the discrete components and filters are used to provide interference rejection at the antenna interface. In particular, the largest interferer a receiver experiences is often a transmit signal from within the same system. Radios for multiple standards on a single device may operate simultaneously at closely spaced frequency bands. In the most challenging case, the Long Term Evolution (LTE) standard has led to the adoption of 35 frequency division duplex (FDD) bands, in which the transmitter and receiver operate simultaneously in separate frequency bands over the same antenna. A duplexer is a three-port device, pictured in Figure 5.1, which interfaces the transmitter and receiver to the antenna while mitigating transmit to receive interference. Frequency flexible duplexing techniques attempt to integrate the duplexer's functionality directly onto the transceiver chip, using techniques independent of the transmit and receive frequencies, spacings, or bandwidths. The duplexer's performance is evaluated by the specifications described in this section.

Transmit to receive (TX-RX) isolation is one of the critical metrics. This isolation must be provided while simultaneously minimizing the RX band insertion loss. As the duplexer is matched to 50 ohms on both sides, any loss directly degrades the signal level while maintaining a constant noise level, thus adding dB for dB to the receiver's noise figure. The duplexer should also minimize the amount of loss from the power amplifier output to the antenna input, referred to as TX band insertion loss. As the TX output power level can be on the order of a watt, even a few decibels result in a significant wasted absolute power. For example, 2 dB of loss on a 1 watt TX signal corresponds to 350 mW of lost power, enough to supply approximately seven receive chains. Additionally, there has been recent research and commercial interest in integrated CMOS power amplifiers (PA) to lower system cost. These CMOS PAs are limited in their ability to deliver high power as compared to non-CMOS counterparts, due to the limited supply voltage and breakdown of the CMOS process. Any loss in the duplexer must be compensated by producing higher output power at the transceiver chip. This can have a superlinear power penalty, due to the need for cascoding or other circuit techniques that reduce the core PA efficiency.

Table 5.1 Example design specifications from the LTE standard.

Parameter	Value	
Channel BW	<20	MHz
Duplex spacing	30–700	MHz
Min (spacing/BW)	2	
TX peak power	23	dBm
RX noise figure	<15	dB
RX out-of-band (OOB) blocker P_{1-dB} (nonspec)	~0	dBm

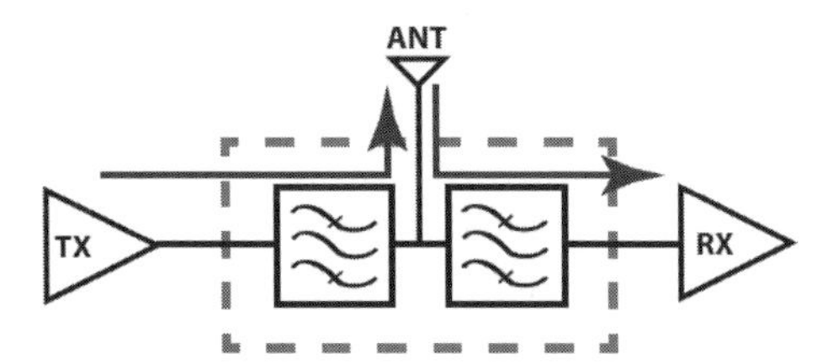

Figure 5.1 Duplexer functionality.

The LTE standard is taken as a representative example to guide the specifications of a frequency-flexible duplexing system. The relevant standard-level specifications are summarized in Table 5.1. The challenge is due to the large dynamic range difference (23 dBm TX power vs. −100 dBm RX sensitivity) between the transmitter and receiver, which must be filtered within a very sharp stop band. In the worst case for the LTE standard, the filter must reach the stop band within twice the signal bandwidth (BW). This is fundamentally why narrowband, discrete, high-Q components must be used, and integrated frequency-agnostic duplexers do not exist. These LTE specs correspond to TX-RX isolation of 45–66 dB, with RX insertion loss of 2 dB and TX insertion loss of 1 dB.

5.2 Approaches to Self-Interference Cancellation

5.2.1 Hybrids

Hybrids are passive networks capable of isolating TX and RX, by forcing the TX signal to appear as a common mode for the RX port. Constructed with coupled inductor networks, the common mode signal on one side of the transformer will not leak to the other side, neglecting interwinding capacitance. A balancing impedance, shown in Figure 5.2, is required for hybrids to operate properly, where the voltage swing present on these tunable impedance nodes normally sets the linearity of the full system, allowing for high-power TX rejection [1–7]. A major drawback for hybrids is the direct trade-off between insertion losses on the TX and RX paths, fundamentally limited to a sum

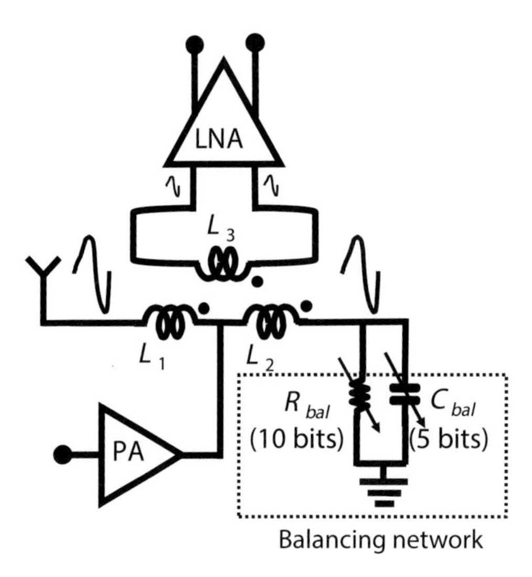

Figure 5.2 Hybrid with balancing impedance [2].

of 6 dB, such that $IL_{TX} = 6\text{ dB} - IL_{RX}$ [8]. Practical instantiations of this technique normally result in 4 dB loss for both the TX and RX paths, a significant penalty for the transceiver.

5.2.2 Active Cancellation

To offset the insertion loss penalty of hybrids, another technique is to generate a high-fidelity replica of the TX signal, either by coupling a portion of the interferer's signal directly or by using the interferer's baseband data. This signal can then be fed through a replica path that matches the TX-to-RX interference propagation path, and used to cancel the TX-to-RX coupling at the RX port.

Several techniques directly couple a portion of the transmit signal [9–11], and use a bank of analog filters to generate a replica of the TX-to-RX inference path. The output of these filters are then subtracted at the input of the RX, as shown in Figure 5.3a. This technique has the advantage that any nonlinearity present in the transmitter is inherently captured in the cancellation signal. However, the power coupled into the replica network directly adds to insertion loss on the PA. The main disadvantage is the linearity requirements for the replica path components, which directly set the maximum TX interference that can be canceled. Furthermore, because the replica path is created using analog components, there is a limited bandwidth over which the filtration can be adjusted to match the TX-to-RX inference propagation path, making it challenging to achieve high cancellation over a wide bandwidth.

Finally, mixed signal techniques that reproduce the interference signal by using baseband data and a digital-to-analog converter (DAC) have been used in the Ethernet domain [12] to cancel interference from multiple simultaneous network streams.

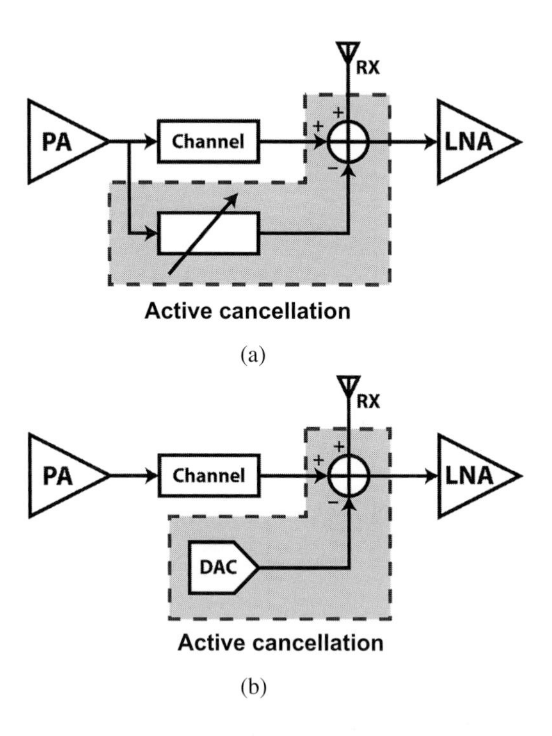

Figure 5.3 Active cancellation methods: (a) with channel replica and (b) with DAC.

This technique is illustrated in Figure 5.3b and has the advantage that no power is removed from the transmit path. Additionally, the filtering needed to replicate the TX-to-RX interference propagation path can be applied in the digital domain, widening the bandwidth over which active cancellation is effective.

5.3 System Concept and Architecture

5.3.1 Conceptual Overview

Consider a PA connected directly to an antenna; a large voltage swing is imposed upon the the antenna node, and a large current flows through the circuit. Next, let's place a current source in series with the PA, from the negative PA terminal to the ground, as shown in Figure 5.4. If this current source sinks the same current as the the initial configuration with only the antenna, then it would not have a voltage swing across it. Moreover, the PA would not notice a difference in the network and the output power would remain unchanged. Due to the zero voltage condition across the current source, any impedance may be placed in shunt with the current source without producing a voltage swing, and without impacting the TX output power, depicted in Figure 5.5. This architecture can be used to place an LNA in shunt with the current source, leading to no TX-RX interference and, in principle, no TX efficiency degradation.

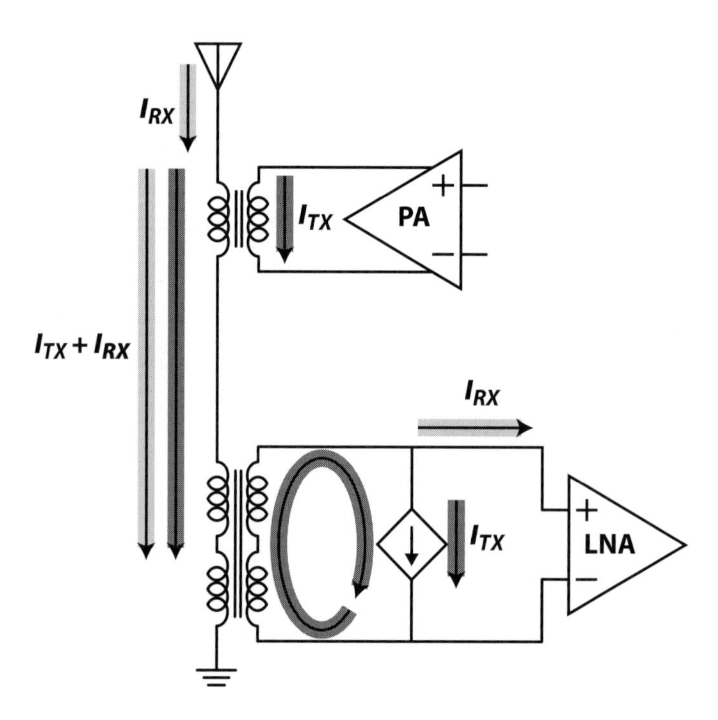

Figure 5.4 Top-level conceptual diagram of cancellation architecture.

Note that the dual of this circuit utilizing a series voltage source is also possible, but is significantly harder to implement in current technologies. While a differential shunt current source can easily be produced, a floating series voltage source is not as simple. Secondly, the voltage source must cancel the large antenna swing, a major challenge in low-voltage CMOS processes. The current source canceling architecture produces ideally zero voltage swing across the canceller regardless of the TX output power. Because of this virtual short, scaling up the current for cancellation simply involves increasing transistor width, or adding more unit cells.

The cancellation source could be purely analog, where the TX signal is sensed, filtered, and distorted to match the TX-to-RX interference path, or could be mixed-signal, where digital filtering is performed on the TX data and a DAC is used to produce the cancellation signal. For treating a wide array of interference path nonidealities, the flexibility of processing TX data in the digital domain is preferred. An additional advantage to the mixed-signal approach is the decoupling of canceller linearity and the signal-to-noise ratio (SNR). In the case of a purely analog canceller, a high-power input signal to the canceller causes compression, while a low input signal degrades replica output SNR. In selecting the number of bits for the cancellation DAC, the residual error signal will be bounded to $\pm 1\text{LSB}_{DAC}$, regardless of the TX power. Therefore, the mean power of the TX residual will be equal to $P_{TX} - 6N_{Bits,DAC}$, shown in Figure 5.5c.

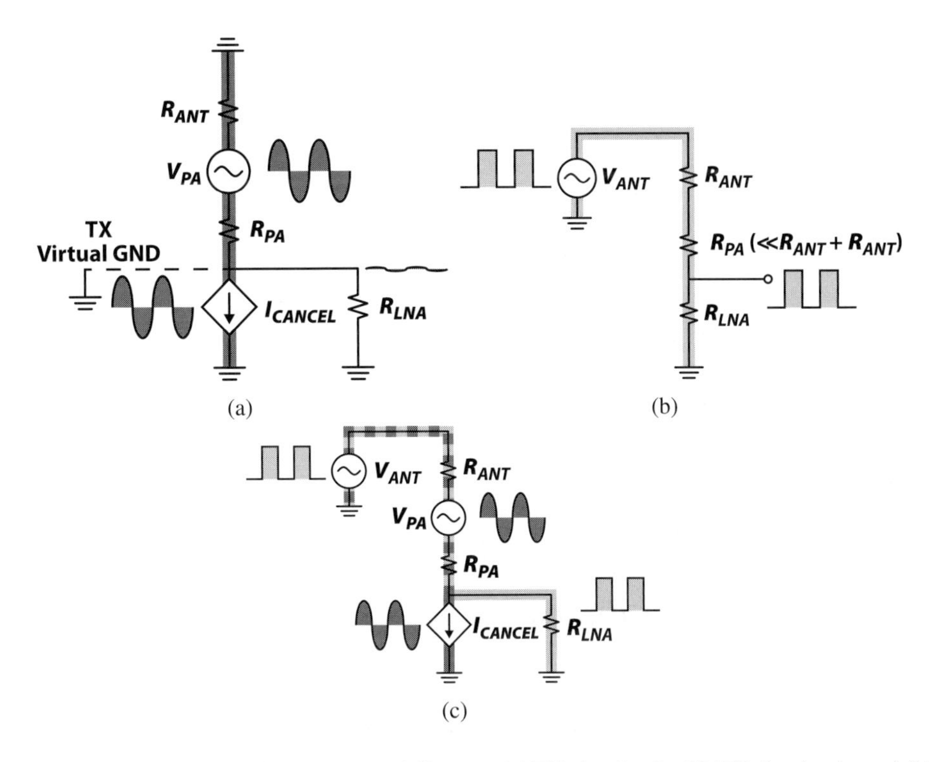

Figure 5.5 Breakdown of the conceptual diagram: (a) TX signal only, (b) RX signal only, and (b) simultaneous TX/RX.

5.4 System Implementation Considerations

This section focuses on implementation details of the current DAC cancellation architecture, and goes into depth of the benefits and drawbacks of the presented cancellation method.

5.4.1 DAC Power Consumption

Because this is an active cancellation methodology, it is very important to characterize the DAC's power consumption relative to the TX as a function of output power. While somewhat counterintuitive, it is true that the DAC need only cancel the TX *current*, and not its full power. Since TX current rises with the square root of TX power, it can be deduced that the DAC power consumption would also rise with the square root of TX power. For a more rigorous analysis of power, a topology first needs to be chosen for the DAC. Given the desire to create a floating differential current source, it makes sense to create the DAC unit cells using a current-steering differential pair, with its supply connected to the RX balun center tap. Using a tail current source and hard switching on the DAC unit cell input makes sense from a linearity perspective and has noise advantages that will be shown later in this section. The DAC supply voltage is primarily dictated by headroom requirements, and is independent of TX peak power to first order.

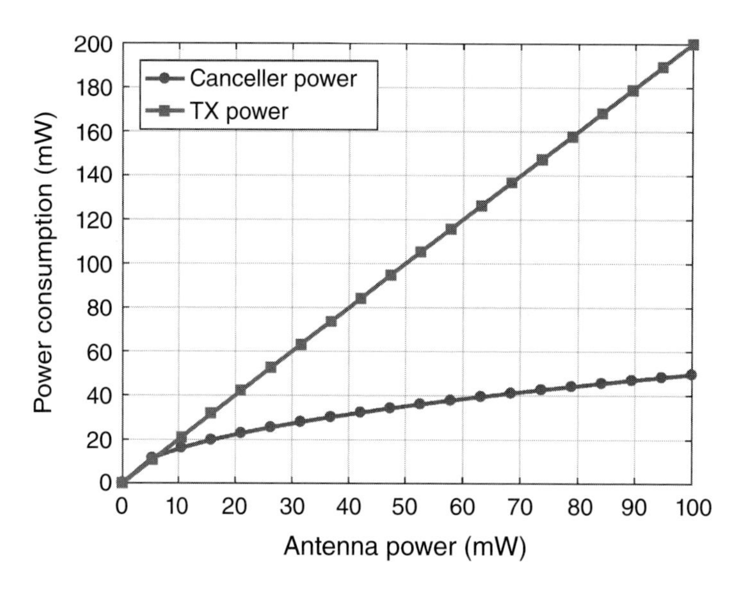

Figure 5.6 Power consumption of canceller versus TX.

For a TX power of P_{TX}, an antenna impedance of R_{Ant}, and an RX balun turns ratio of N_{Turns}, the sinusoidal current amplitude flowing through a short at the RX input port is equal to

$$I_{TX,RX} = \sqrt{\frac{2P_{TX}}{R_{Ant}} \frac{1}{N_{Turns}}}. \tag{5.1}$$

The hard-switched square waveform of the DAC has a differential amplitude of $I_{Tail}/2$. This 50% duty-cycle square wave has a fundamental sinusoidal amplitude of $\frac{2}{\pi} I_{Tail}$. Therefore, power consumed by the DAC to cancel a TX power output power level of P_{TX} is

$$P_{DAC} = V_{DD} \frac{\pi}{N_{Turns}} \sqrt{\frac{P_{TX}}{2R_{Ant}}}. \tag{5.2}$$

To present a practical case, a 2:1 turns ratio balun, 50 Ω antenna, and 1 volt DAC supply is chosen. The power consumption of the cancellation DAC is plotted against the power consumption of a PA with 50% efficiency in Figure 5.6. At +20 dBm, the power consumption of the PA is 4 × that of the DAC, due to the $\sqrt{P_{TX}}$ dependency.

5.4.2 System Thermal Noise

It is very important that the duplexing system minimizes the impact to RX noise figure, as noise figure directly impacts receive distance for a given TX power. Along with noise produced in the RX chain, there are three other sources within the transceiver: TX thermal and phase noise, which leak through the interference path to the RX; and cancellation DAC noise, which is injected directly at the RX input. The RX and TX

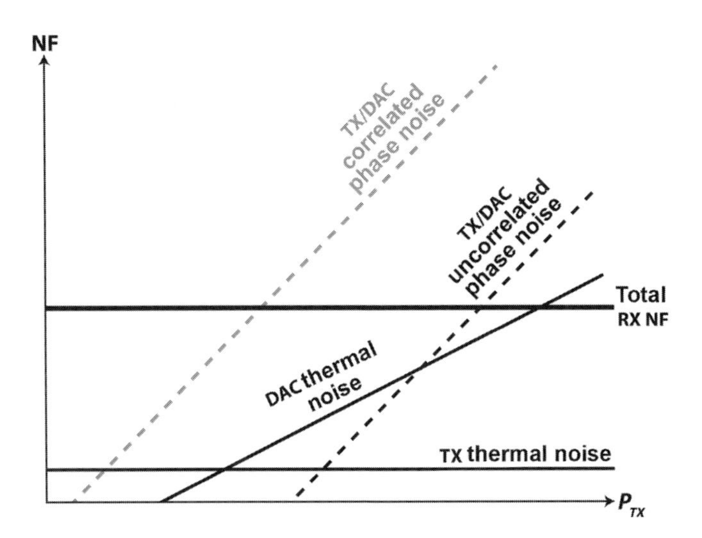

Figure 5.7 Significant transceiver noise contributions.

thermal noise are independent of the TX power and form the base sensitivity of the network, while the DAC thermal noise and TX phase noise grow with TX power. This increase in desensitization sets the practical limit on TX output power, as shown in Figure 5.7, where the relative magnitudes and growth rates of each noise source are illustrated. TX thermal noise is negligible, so the majority of design effort should be focused on minimizing the DAC thermal noise and the TX phase noise.

A final source of noise that is worth commenting on is RX LO phase noise. Through reciprocal mixing, TX interference of the RX band is mixed by the phase noise of the RX LO, spreading its energy to the RX baseband. This source produces negligible desensitization in this architecture because the TX interference is subtracted before the RX LO.

TX Thermal Noise

As stated earlier, the TX must present a low output impedance for low RX insertion loss. If a switching power amplifier is used as the TX, the PA can be thought of as a passive linear time-varying network, where the real part of the TX output impedance sets the voltage noise level. The noise figure due to TX and RX is shown in (5.3), where $\overline{v^2}_{RX,n}$ is the input-referred voltage noise of the RX, and $\overline{v^2}_{Ant,n}$ is the antenna noise referred to the input of the RX. Requiring the RX insertion loss due to the TX to be small necessitates that the real part of the TX output impedance, $R_{TX} \ll R_{Ant}$, meaning the noise figure penalty is similarly small. Shown in Figure 5.8, a simulation using transformer and PA parameters from [13], the noise figure due to the TX alone is small, impacting the total noise figure by <1 dB.

$$F = 1 + \frac{R_{TX}}{R_{Ant}} + \frac{\overline{v^2}_{RX,n}}{\overline{v^2}_{Ant,n}} \tag{5.3}$$

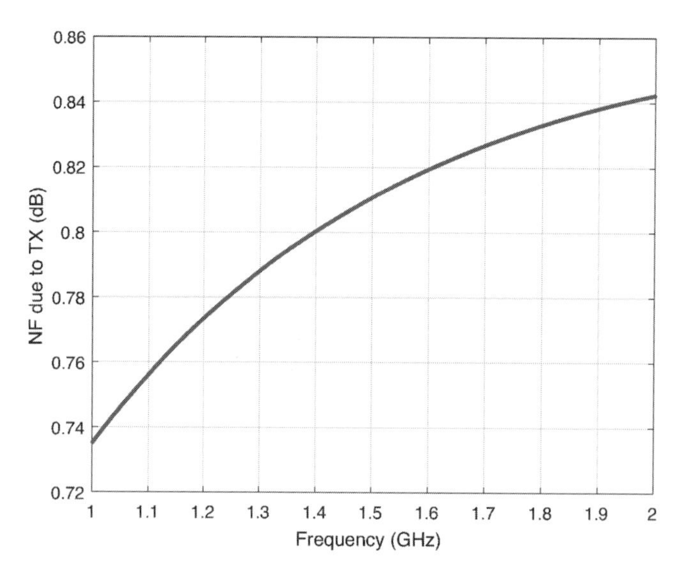

Figure 5.8 Noise figure due to TX only.

The effects of TX thermal noise added to the system and the loss due to the TX output impedance are one and the same, and should not be considered as independent degradations. Using the Friis cascade noise figure expression (5.7) to determine overall noise figure

$$\overline{v^2}_{Ant,TX,n} = 4kT\,(R_{Ant} + R_{TX}) \tag{5.4}$$

$$G_{A,TX} = \frac{R_{Ant}}{R_{Ant} + R_{TX}} \tag{5.5}$$

$$F_{TX} = 1 + \frac{R_{TX}}{R_{Ant}} \tag{5.6}$$

$$F_{Total} = 1 + (F_{TX} - 1) + \frac{\left(1 + \frac{\overline{v^2}_{RX,n}}{\overline{v^2}_{Ant,TX,n}}\right) - 1}{G_{A,TX}} \tag{5.7}$$

The total noise figure is exactly the same as taking into account only the effect of the noise voltage of the TX, as shown in (5.3).

TX Phase Noise

Phase noise from the TX that falls into the Rx band and leaks into the receiver is a very strong RX desensitization mechanism for the canceller system. This is due to the high output power of the TX and the fact that TX interference falling into the RX band due to phase noise increases decibel for decibel with TX power. There is a highly effective way of mitigating this effect, as explained in [14]. If the phase noise profiles of the TX and DAC are identical, filtering of the DAC to match the TX leakage signal will also cause cancellation of the TX phase noise. This condition of matching phase noise profiles

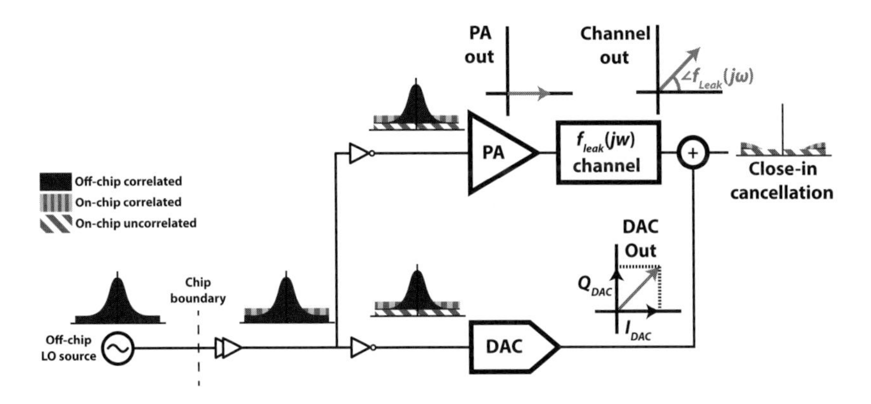

Figure 5.9 Feedforward cancellation of TX/DAC shared phase noise.

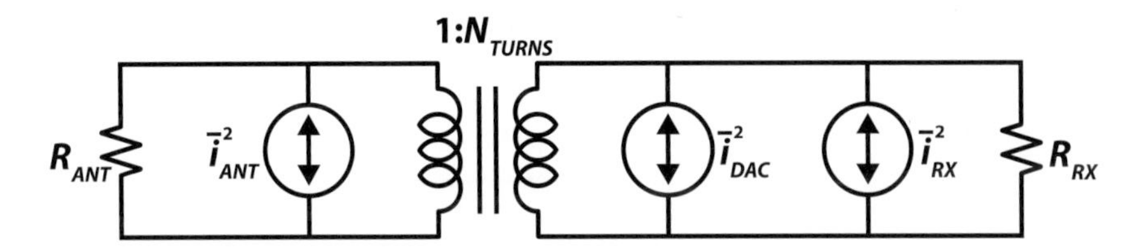

Figure 5.10 DAC noise network.

suggests that the TX and DAC LO chains share as many elements as possible so as to not introduce uncorrelated phase noise. This also suggests that polar implementations of the DAC and TX are not desired, since separate phase interpolators in the LO chains produce significant unshared noise.

This phase noise cancellation mechanism suggests yet another design choice, illustrated in Figure 5.9. A high amount of phase noise cancellation is present only over the frequency range where the TX interference path has only small variation in amplitude and phase. If the TX and RX are collocated on the chip and if there is no external isolation, the interference path is roughly constant over a wide bandwidth. A faraway TX or the use of an off-chip filter will create significant variation and severely limit cancellation bandwidth.

DAC Thermal Noise

To isolate the DAC's effect on RX noise figure, analyze Figure 5.10, where the TX is assumed to be zero output impedance, contributing zero thermal noise power.

$$F_{RX,DAC} = F_{RX} + N^2 \frac{\bar{i}^2_{n,DAC}}{\bar{i}^2_{n,Ant}} \tag{5.8}$$

It is clear from this equation that the DAC current noise adds directly to the RX noise figure. Therefore, it is important to accurately model this noise current. A general model for the DAC output noise current as a function of the TX leakage signal can be created

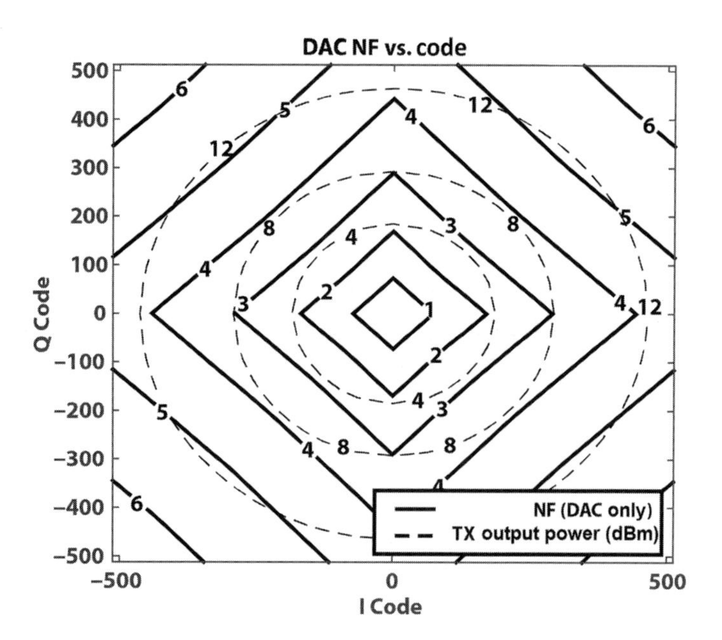

Figure 5.11 DAC noise figure versus TX power.

by considering the DAC as a noisy tail device connected to a noiseless mixer driven by a square wave. The vast majority of RF current DACs conform to this model due to their construction as a tail transistor with hard-driven switches.

A full analysis of the noise contribution of the DAC vs. code is very involved, and interested readers are pointed to [15] for more detail. Some general intuitive statements can be made, however. The tail sources of each unit cell are uncorrelated with one another, so in the case of a polar DAC with only thermometer cells, the total DAC output noise rises with $\sqrt{P_{TX}}$. The full expression in this case, in terms of v_{Ov}, the tail overdrive voltage and γ, the transistor noise coefficient, is reproduced in (5.9).

$$\frac{\bar{i}^2_{n,Total}}{\Delta f} = 8kT\gamma \frac{\sqrt{2\frac{P_{TX}}{R_{Ant}}}}{\pi N v_{Ov}} \tag{5.9}$$

The noise figure due to the thermal noise of the DAC only is equal to

$$F = 1 + \gamma \sqrt{\frac{8}{\pi} P_{TX} R_{Ant}} \frac{N_{Turns}}{v_{Ov}}. \tag{5.10}$$

For the more involved Cartesian case, a mapping of complex DAC code to output noise figure for a TX power of >12 dBm is shown in Figure 5.11 without derivation. An important point of note on the graph is that the output noise level is not just a function of the DAC current amplitude, but also a function of current phase.

5.5 System Degradation

It is worthwhile to benchmark the performance of this active cancellation system against the hybrid, as that is a duplexing architecture that can also handle high TX power. The insertion losses for TX and RX are typically 4 dB each in practical implementations of the hybrid, raising the RX noise figure (NF) by 4 dB and reducing PA efficiency by 60%.

In the active cancellation network, noise added by the DAC and TX factors into NF increase. The balun loss for the PA and extra power consumption from the cancellation path effectively reduce PA efficiency. The digital filtering on the Tx to Rx replica path can be conservatively estimated for an LTE-type system at 8 taps at 200 MS/s, with 10-bit coefficients. According to [16], the power consumption for this filter would be around 10 mW in a 65 nm process. Digital predistortion of the DAC is achieved through a lookup table, estimated to cost 15 mW in power. There is additional power consumed to run the adaptation algorithms to change these filters and lookup tables as the network or other nonidealities change. However, if the dynamics of the channel are far slower than the data rate, the power consumption of these digital algorithms can be amortized over a very large operation time, making their average power consumption negligible. In Figure 5.12a, degradation of the TX efficiency due to these mechanisms is plotted with class-A and class-B DAC back-off for a modulated data signal with 6 dB peak-to-average power ratio (PAPR) (Figure 5.2).

Compared with an 8 dB combined loss from the hybrid, the replica DAC gives better performance than the hybrid for approximately +5 dBm to above +20 dBm TX power.

Table 5.2 TX and RX specifications used to compare the performance of the proposed scheme compared to a hybrid.

(a) TX loss parameters.

Parameter	Value
TX average back-off (dB)	6
TX average PAE (%)	25
Digital filter power (mW)	10
Canceller DPD power (mW)	15
Insertion loss (IL) from RX winding (dB)	−0.35
DAC supply voltage (V)	1

(b) RX NF parameters.

Parameter	Value
RX NF (dB)	2.5
RX XFMR IL (dB)	1
RX XFMR N_{Turns}	2
R_{TX} (ω)	7
DAC V_{ov} (mV)	800
Uncorrelated phase noise (dBc/Hz)	−190

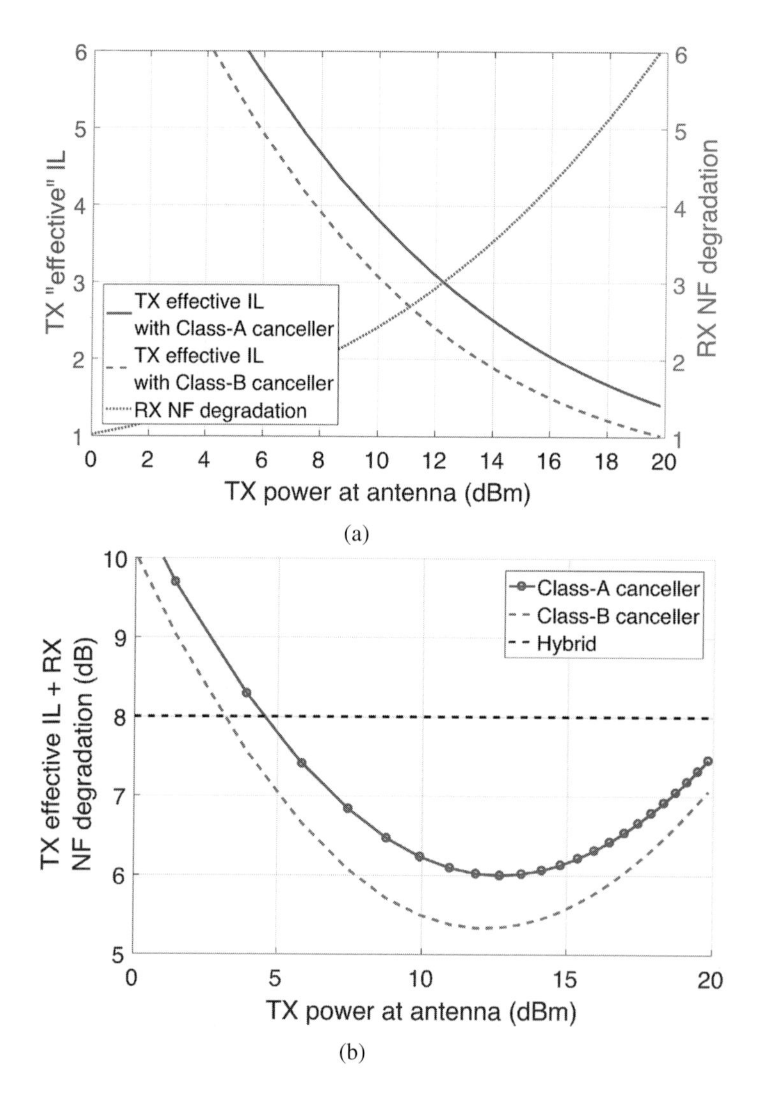

Figure 5.12 TX and RX degradation. (a) TX/RX degradation versus output power and (b) total degradation comparison with hybrid.

Additionally, the tunability of the active cancellation solution over a wide range of antenna voltage standing wave ratio (VSWR) and TX-to-RX interference paths make this a more attractive solution than the hybrid.

5.6 Transmitter

A low, code-independent PA output impedance lowers insertion loss in the series configuration, and prevents mixing of the RX signal with the TX when outputting modulated data. The switched-capacitor power amplifier [17] architecture satisfies all criteria for this active cancellation system.

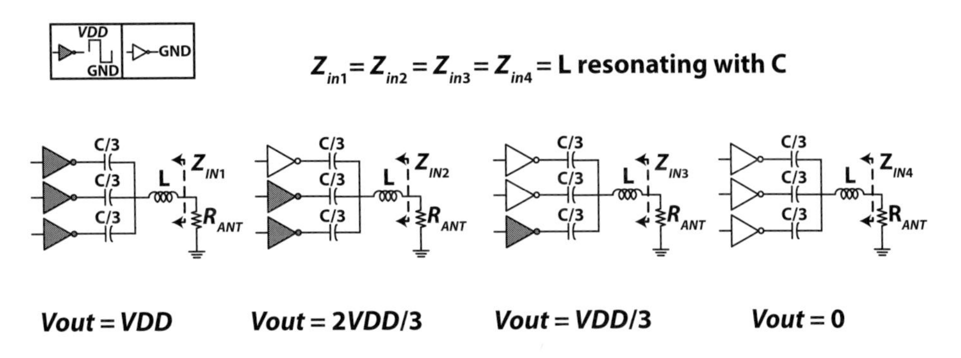

Figure 5.13 Operation of switched-capacitor power amplifier [18].

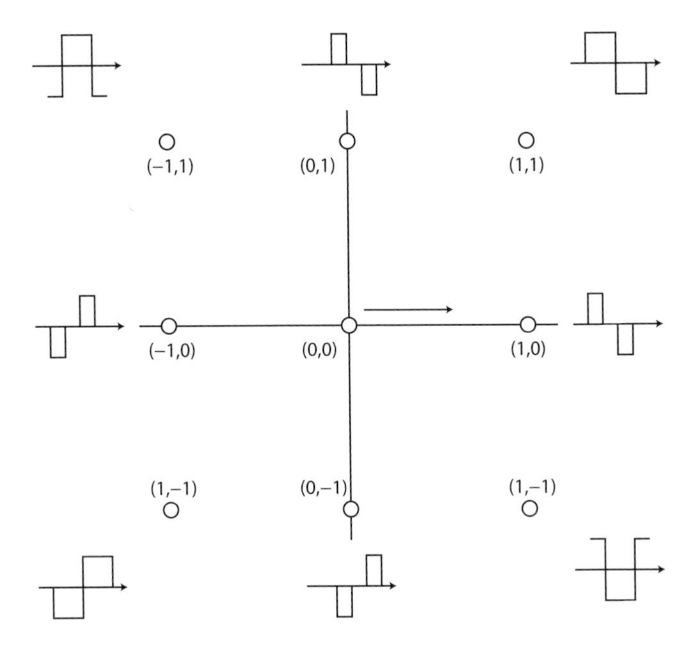

Figure 5.14 Unit cell output constellation.

Shown in Figure 5.13, the PA acts as a code-dependent voltage divider, exhibiting high linearity and identical output impedance across the full output range.

Sizing and matching network optimization for this architecture can be found in [19]. The switched capacitor power amplifier (SCPA) can be implemented using a Cartesian I/Q cell-sharing technique, where each enabled unit cell can output any combination of ± 1 and $\pm j$ using pulse width modulation, where the $I = 0$ and $Q = 0$ waveforms are 25% duty cycle, and the $|I| = |Q|$ waveforms are 50%, shown in Figure 5.14. For the same peak output power, this method uses the same area as a polar PA and $\sqrt{2}$ less area than a conventional Cartesian PA having separate I and Q unit cells. Consider the peak power case of the I/Q cell-sharing architecture, where all unit cells are enabled and output with 50% duty cycle. This is indistinguishable from the polar PA

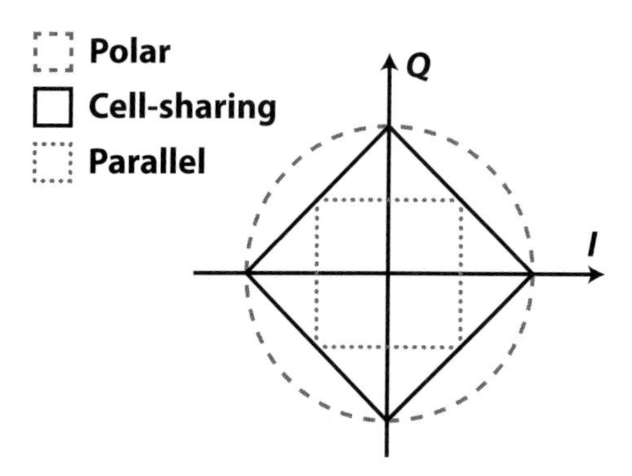

Figure 5.15 Available constellation regions for PA architectures.

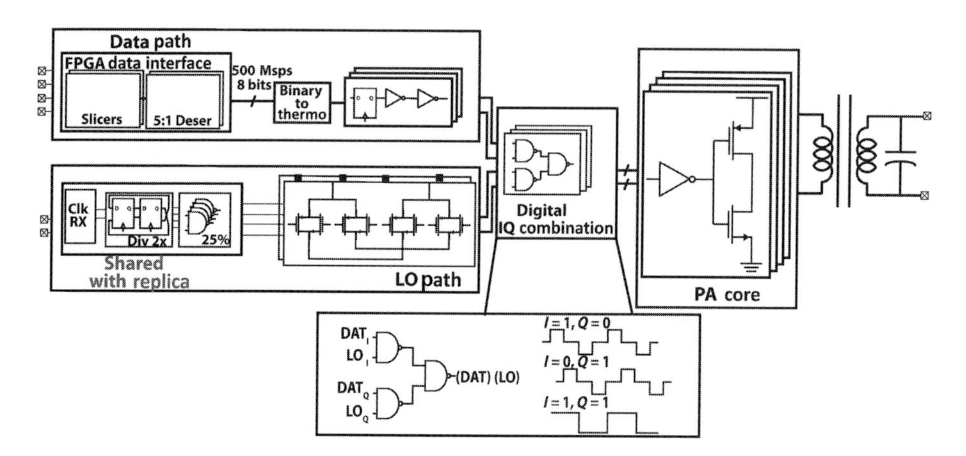

Figure 5.16 TX top level.

at maximum power with the same area, since both output 50% duty cycle waveforms in this scenario. For $I = 0$ or $Q = 0$ cases, the maximum output power is half that of the polar case. The achievable constellation region is a rhombus inscribing the polar architecture's circle, illustrated in black in Figure 5.15.

The PA core was integrated along with the matching network balun, LO distribution, data deserializers, and retiming circuitry, shown in the top-level schematic of Figure 5.16.

5.7 Cancellation DAC Design

5.7.1 DAC Linearity

While it may first appear that the cancellation DAC must be highly linear in order to provide large TX cancellation, it is actually found that as long as the DAC has a sufficient number of physical bits, its required effective number of bits is far lower.

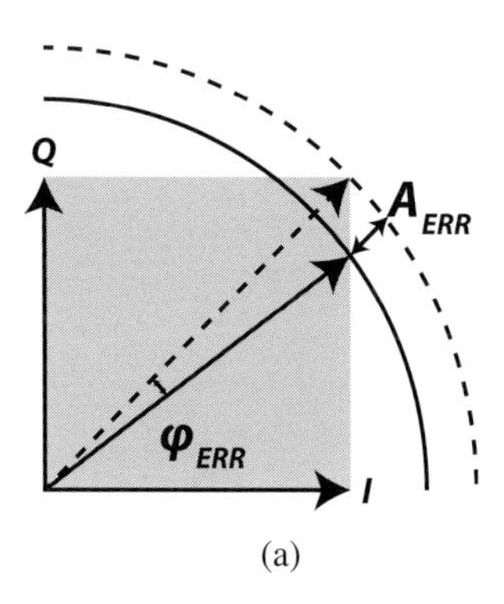

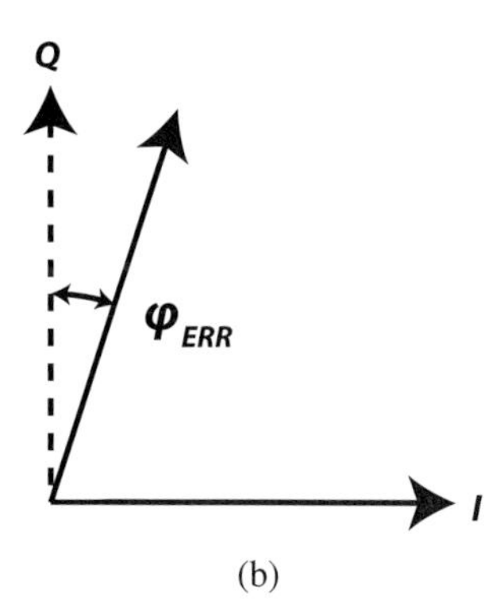

Figure 5.17 Definitions of DAC static mismatch. (a) Quadrature phase mismatch definition and (b) I/Q mismatch definition.

Because the cancellation DAC outputs a complex current signal, integral nonlinearity (INL)/differential nonlinearity (DNL) are insufficient to define the full range of possible nonlinearities over codes. For a complex-output DAC, there is quadrature skew, illustrated in Figure 5.17a, where the I and Q signals are either different amplitudes, have a phase relationship other than 90°, or both. Next, there is summation skew, where equal-code I and Q signals may be the same amplitude and correct phase offset, but their summation has an amplitude different from $\sqrt{2}$, or a different angle from 45°. In this work's implementation of the cancellation DAC, I and Q have the same amplitude for the same code to first order.

The analysis of DAC nonlinearities is somewhat involved, so various results are given without derivation and more details can be found in [15]. Two main points can be made about the effect of these nonlinearities. First, using nonlinear predistortion makes TX interference cancellation highly robust to nonlinearities compared to a simple FIR. Second, even with predistortion, different DAC nonlinearities affect cancellation differently, and interestingly enough, interference rejection is a strong function of DAC segmentation at a fixed number of physical bits, shown in Figure 5.18.

5.7.2 DAC Thermal Noise Cancellation

In receivers, thermal noise reduction is routinely performed, creating multiple paths where the noise of a device interferes destructively while the signal interferes constructively [20–22]. While this is a common trait of state-of-the-art receivers and analog devices requiring very high sensitivity, thermal noise mitigation methods are generally not employed for transmitters or DACs due to the large signal levels they produce. In DACs especially, the quantization noise floor is far higher than the thermal noise floor. Take, for example, a transmitter capable of outputting +20 dBm on a 50 Ω load. In order for a thermal noise floor increase of 10 dB to negatively affect the signal-to-quantum-noise ratio (SQNR) of a Nyquist DAC by 3 dB, 18 bits

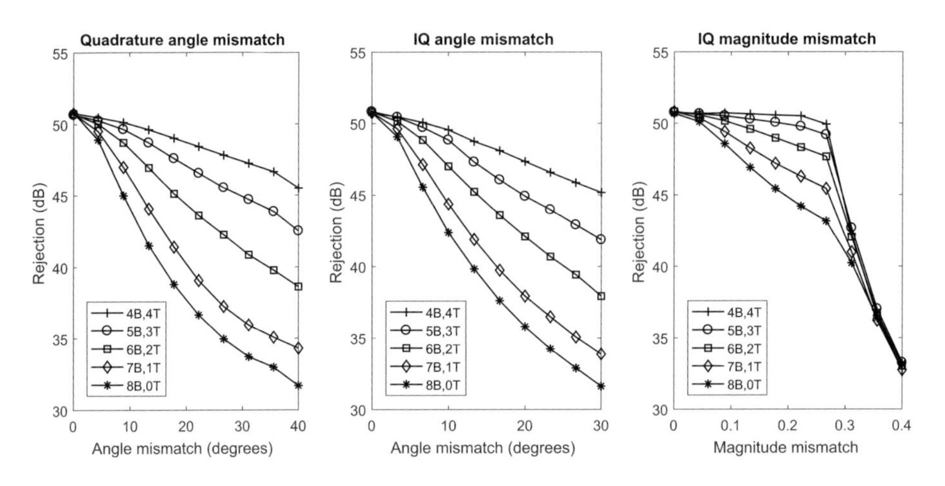

Figure 5.18 Rejection vs. DAC segmentation for I/Q summation nonlinearity.

would be required. Even if an oversampling ratio of 100 were used, 15 bits would still be required. This is far above the requirements for the TX, but a 10 dB RX noise floor increase due to a cancellation DAC would heavily impact overall system performance. Note that in the LTE standard, a 15 dB noise figure limit is used for FDD [23]. Taking inspiration from feedback techniques to improve spectral efficiency in voltage-controlled oscillators (VCOs) requiring high spectral purity, [24,25], the cancellation DAC tail source noise can be mitigated with a topology shown in Figure 5.19.

Two nodes where the DAC tail current can be sensed separately from interference are the RX balun center tap and the source of the unit cell tail device. A baseband feedback loop from the center tap to the tail gates reduces noise unconverted by the DAC LO, while a $2F_{LO}$ inductive degeneration on the tail sources reduces noise downconverted by the DAC LO. If the DAC is Class-A, the current through center tap and source is independent of the DAC signal. The DAC was given a dedicated supply to minimize supply noise on the center tap node, and an isolated DAC ground pin was used to minimize interference around $2F_{TX}$.

In both cases, the current noise is converted to a voltage through a resonant impedance and is fed back using the DAC aggregate G_m. Because of the high bias current of the DAC, this G_m is very large (300 mS), creating a large loop gain for moderate source and center tap impedances. While this DAC G_m is constant, the effective noise reduction is code dependent because of noise injected into the loop from inactive cells. This effect is most prominent at low codes is not an issue because at lower DAC codes, the RX is the dominant source of desensitization.

A simplified model of the DAC and baseband noise feedback can be used to analyze the impact of noise. The switching quad is represented as a single cascode transistor because there is always a path from the tail to the center tap, independent of data and the LO phase. The only distinction between unit cells in this case is which cells are active (the tail outputs to the differential RX input) or inactive (tail shunted to

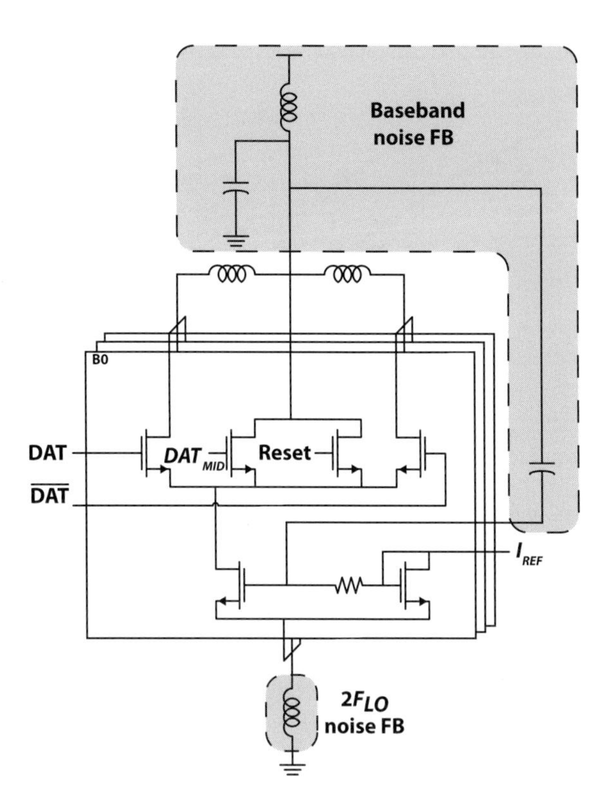

Figure 5.19 DAC with noise feedback highlighted.

center tap). All active cells are lumped into a single-tail transistor of transconductance $G_{m,A}$, and all inactive cells are lumped into a tail with $G_{m,I}$. In this simplification, it is clear that the active and inactive devices are simply diode connected to the center tap node. I_A is the total active tail current signal, i_A^2 and i_I^2 are the active and inactive tail device noise sources, respectively. A further simplification can be made where the diode-connected inactive transistor is replaced with a resistor and the inactive current noise is kept in the same position. Intuitively, since both the center tap impedance and the disabled cells are connected to small signal ground, and it doesn't matter what current flows through them, they can be put in parallel. This reduction in the effective center tap impedance implies that, even ignoring the current noise injected by the inactive devices, the current noise rejection is code dependent.

The feedback noise current normalized by nonfeedback current as a function of replica DAC code is plotted in Figure 5.20, assuming a fully thermometer DAC and $G_m Z_{CT} = 2$. In this plot, the output noise is normalized by the noise without any feedback, assuming that all cells are active. The maximum noise variance with feedback is less than one-third of the maximum without, and after approximately half the cells are active, the output noise reduces with increasing code because the attenuation rises with n^2, while the noise power rises with n.

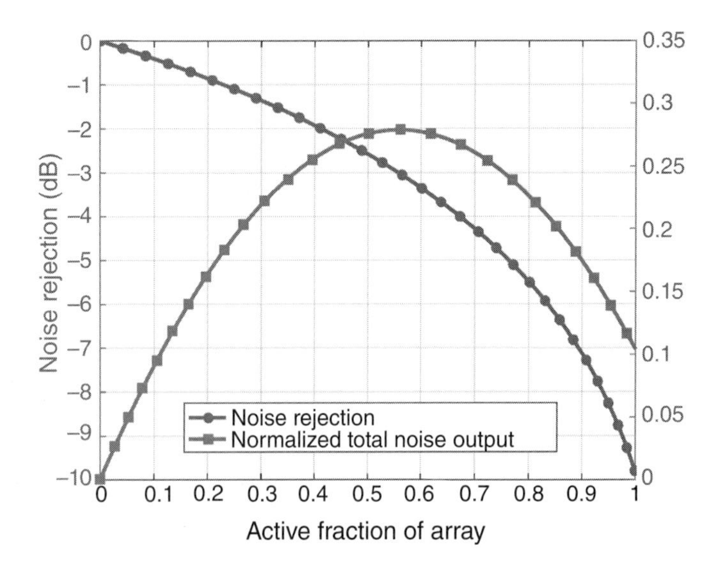

Figure 5.20 Feedback rejection and normalized output noise.

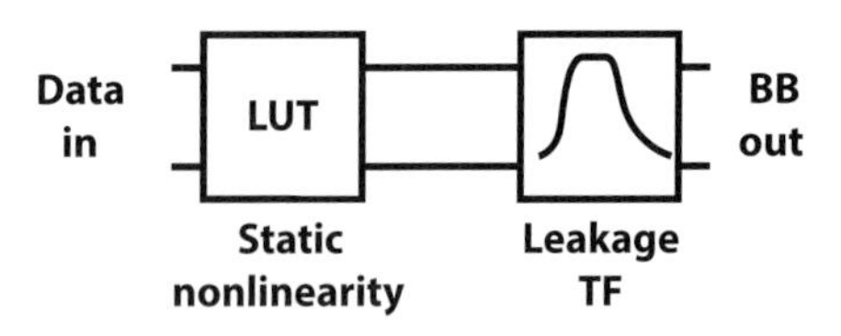

Figure 5.21 DAC model.

5.8 Quantization Noise Measurements

While the analog cancellation prevents the RX chain from compressing during simultaneous TX output, desensitization from TX and DAC quantization noise is still an issue. Unlike thermal noise, though, the DAC's quantization noise is deterministic. Accordingly, rather than specifying number of bits to keep the quantization noise below the receivers noise floor, the DAC's resolution can be chosen to ensure the quantization noise floor does not compress the receiver. The DAC's quantization noise can then be subtracted in the digital domain. In this section, the calibration process for the transceiver digital cancellation is shown, along with results.

Shown in Figure 5.21 is the full-complexity model used for cancellation. This model contains both static and dynamic nonlinearities as well as multiple linear memory stages. While in the course of the work's development, each part of this model has been adapted, it will be shown that major simplifications can be made to this model.

5.8.1 Channel Memory

A measurement of a long pulse response shows that the vast majority of memory in the complex baseband channel is due to components external to the chip (see Figure 5.22).

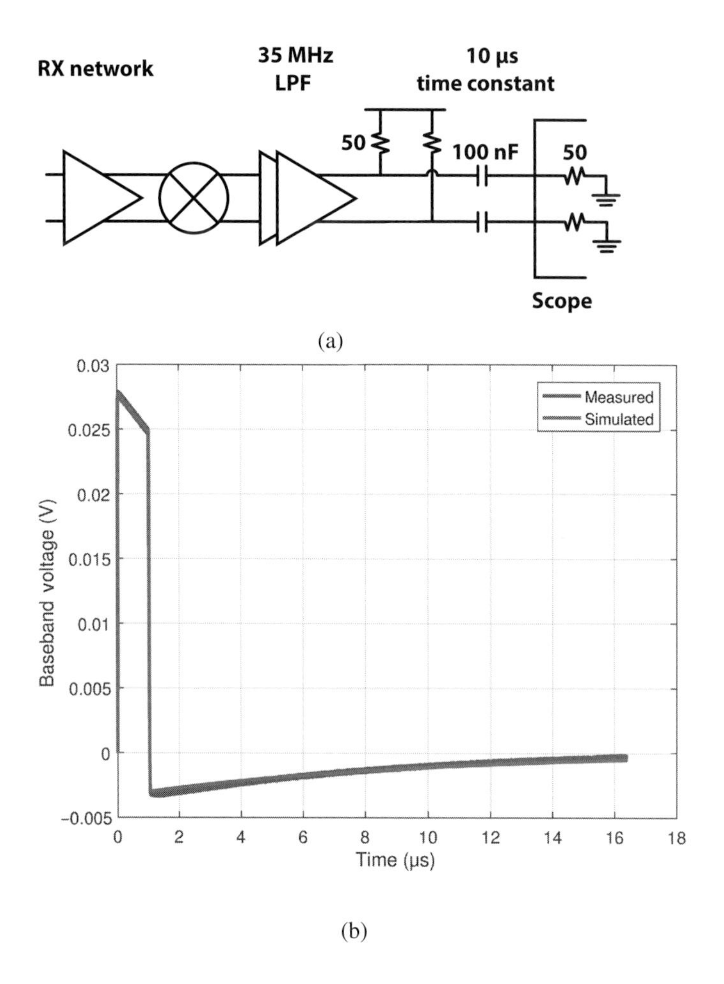

Figure 5.22 (a) Baseband channel and (b) effective baseband pulse response (of 1 μs).

This is mainly due to the fact that the internal channel response has such a wide bandwidth (>1 GHz) owing to the direct connection between TX and RX with no external isolation. Corroborating this measurement is the TX phase noise cancellation measurement from Section 5.4.2, which shows a channel bandwidth much higher than the data bandwidth. In the rest of the digital cancellation work, this channel is simplified by removing the 100 nF coupling capacitors from the system, allowing the internal channel to be ignored.

In the case where the channel response cannot be ignored, an iterative procedure to model the TX leakage pulse response can be used. The constellation measurement and calibration procedure is detailed in Figure 5.23 and consists of transmitting a data packet containing every point in the desired constellation, then using the measured pulse response of the channel to subtract the intersymbol interference (ISI) from the measured constellation sequence. After multiple iterations, the pulse response is deconvolved from the measured sequence, leading to a refined constellation in Figure 5.24b, which has symmetric characteristics and clearly shows some amount of DNL, which are both indicative of a correct calibration procedure.

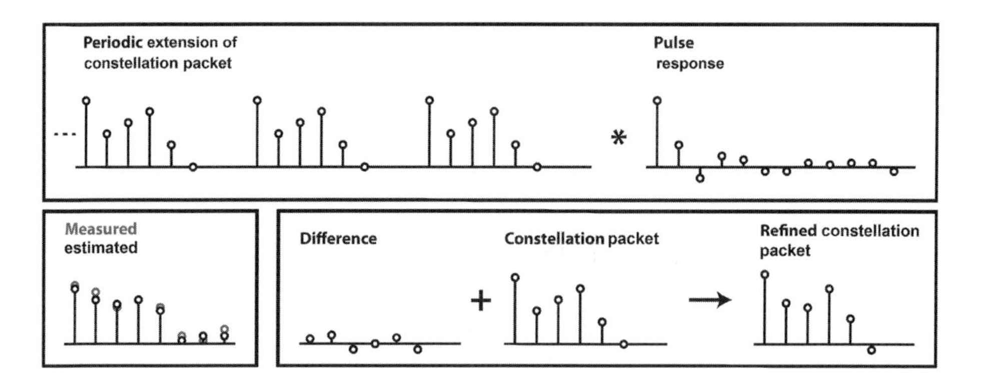

Figure 5.23 Constellation refinement procedure.

5.8.2 PA Dynamic Nonlinearity

The PA can be modeled as a tunable capacitive voltage divider connected to its supply. When this supply has some ripple, $V_{DD}t$, the ripple directly modulates the output waveform $O_{PA}(t) = \frac{V_{DD}(t)}{V_{DD,nominal}} O_{PA,ideal}(t)$. An illustration of this effect is shown in Figure 5.25.

In the case of a PA with an isolated supply, the ripple amplitude is a strong function of the PA output power due to series resistance in the supply network. In the presence of a series resistance on the supply, the voltage ripple is a linear function of the current draw. In the case of a switched-capacitor PA, the current draw is proportional to $\sqrt{P_{TX}}$, or $|C_{TX}|$, where $C_{TX} = I_{TX} + jQ_{TX}$. This produces a second-order nonlinearity that contains any further nonlinear or memory elements that are part of the supply network. Figure 5.26 shows a measurement of the pulse response of the PA with its accompanying supply response. These are directly on top of one another, so measurement, prediction, or reduction of supply ripple is paramount to proper TX digital cancellation.

The voltage supply ripple is a function of the baseband PA code and also directly modulates the output waveform, so this effect is independent of the frequency of the PA. Realistically, there is significant memory on the power supply node. The presence of memory in the supply impedance does not prevent a purely baseband analysis of this nonlinearity, but it can no longer be simply modeled by a constellation distortion. To observe the effect of memory on the supply node, a 64 μs step applied in Figure 5.26, where it is clear that there is a multiplicative effect of the supply ripple on the unit step waveform, and that the supply ripple has a large time constant, approximately 10 μs, affecting hundreds of symbols before settling. This very long time constant is due to the large decoupling capacitors on the PA supply. While removing them would make the supply ripple shorter, making this nonlinearity closer to memoryless, it would add significant supply noise to the output of the PA, adding wideband interference to the RX band. Therefore, this memory effect should be modeled rather than removed.

A constant data step creates a step increase in current, but so does a modulated sequence with constant average power, as shown in Figure 5.27. The same initial droop can be seen as in the unit step response, followed by a relatively constant period as the

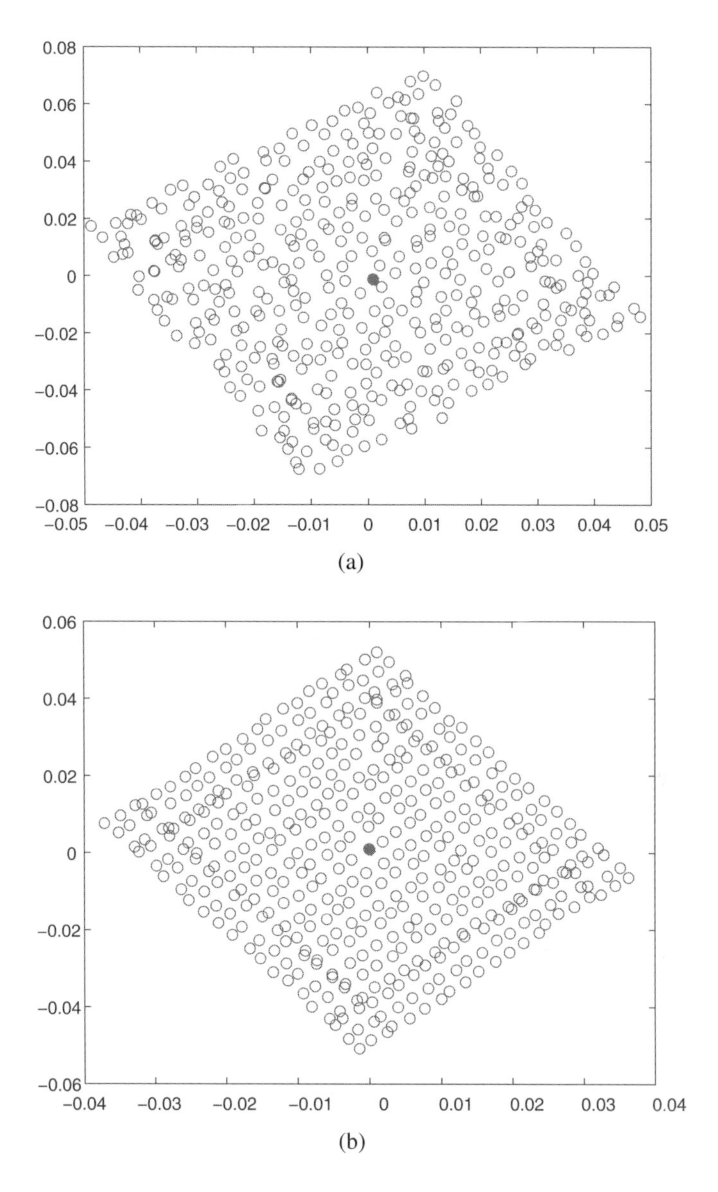

Figure 5.24 Comparison of measured constellations (a) before and (b) after channel deconvolution.

data are changed but the average power is kept constant. Most important to note is that the relatively constant period has a very low variance, meaning that signals of consistent average power over their packet length present minor issues with regard to prediction and cancellation.

If the level of supply decoupling is not adequate, there are two approaches one can use to model this dynamic nonlinearity. The full PA baseband model is shown in Figure 5.28. A model can be made of the off-chip power supply network (at the level of precision required, there is no benefit to considering the on-chip passive network, which has far higher bandwidth), which could include both passive devices

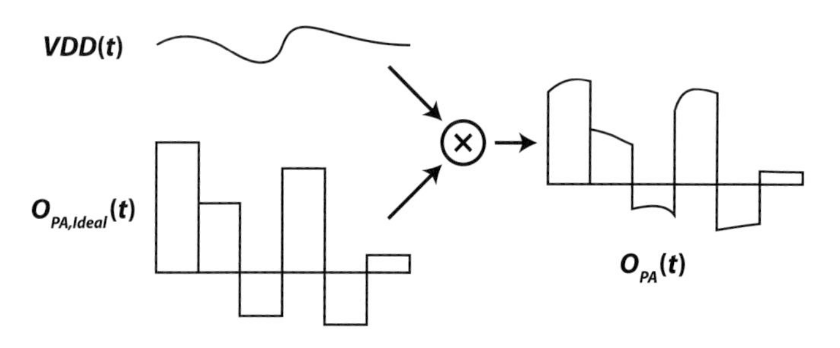

Figure 5.25 TX passes supply ripple to output.

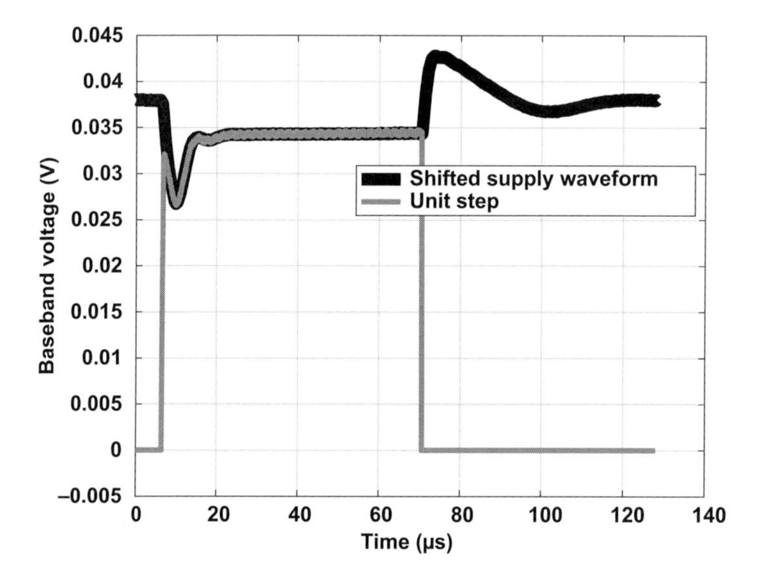

Figure 5.26 TX unit step compared with supply ripple.

and nonlinear models of active devices, such as the low-dropout regulator (LDO) supplying the PA. This model, coupled with a model of the current draw of the PA given code, would provide an estimate of the power supply ripple as a function of the data sequence.

Another approach is to simply measure the power supply ripple. This can be done off-chip by probing the PA power supply pin, or a dedicated low-bandwidth power supply measurement device could be implemented on-chip. Measuring the power supply ripple allows for far lower model complexity, a strong advantage. One disadvantage is the fact that the ripple cannot be predicted, therefore this could not be used for predistortion unless an iterative adaptation loop is used.

In this work, measuring the supply ripple off-chip was used. The same procedure for refining the constellation, shown in Figure 5.23, is performed, but the supply ripple is recorded while the constellation packet is output. Iteratively using this method with the new model provides far better matching in the constellations at higher codes, as shown in Figure 5.29. The small variation in initial constellation points is almost entirely due

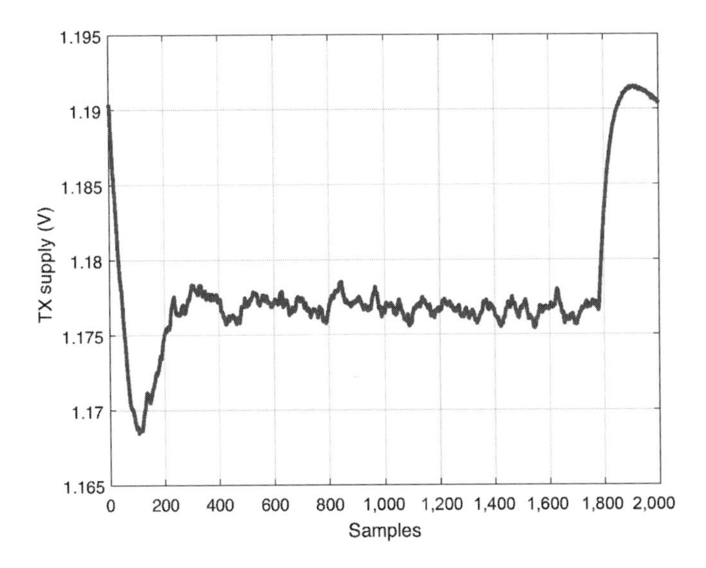

Figure 5.27 TX supply, arbitrary sequence with dead time.

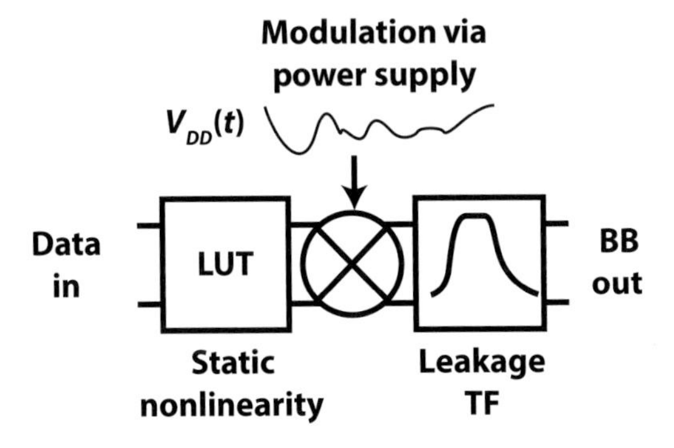

Figure 5.28 PA model including supply modulation.

to the supply ripple differences. Using this technique, 30 dB of additional rejection of the TX and DAC's quantization noise and nonlinearities can be provided in the digital domain.

5.9 Measurement Results

Can this active cancellation architecture truly reduce the TX-to-RX interference? To verify, chips were measured for their performance on the two key system metrics: TX-Rx isolation, and RX noise degradation.

In the first test, the TX and RX are operated at four different frequency offsets: fully overlapping (0 MHz), 40, 80, and 120 MHz spacing. As the cancellation system

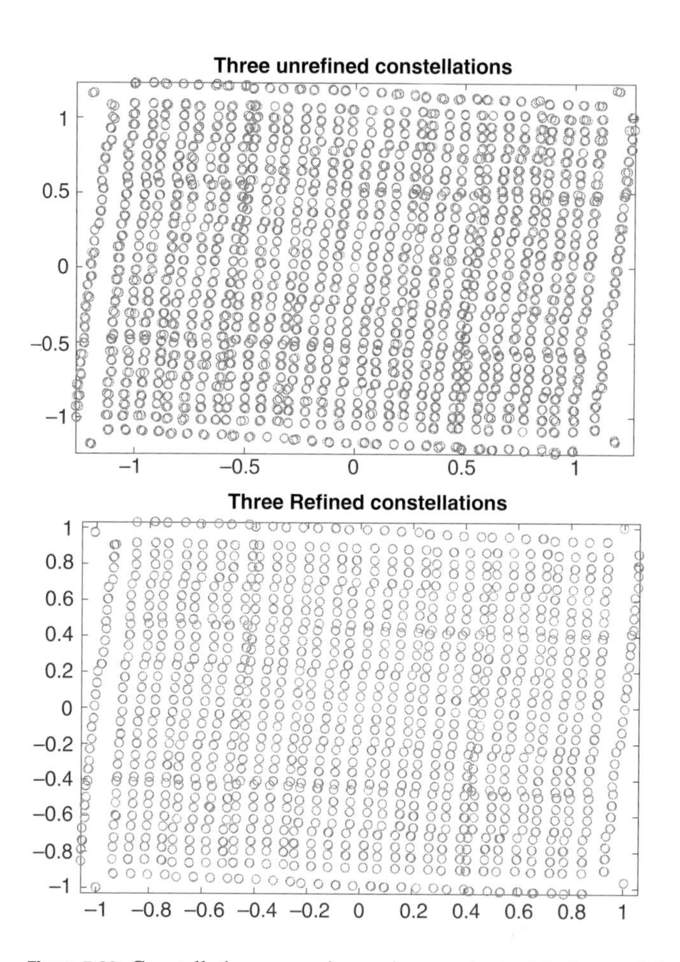

Figure 5.29 Constellation comparison using supply modulation model.

is effective until the TX interference begins to desensitize the receiver, receiver gain compression is measured as TX power is increased. The results are given in Figure 5.30. The transceiver can receive with <1 dB RX gain compression for up to a +18 dBm TX signal.

The TX signal and the residual TX-to-RX interference after cancellation are plotted in Figure 5.31a. Across TX output power, the postcancellation TX interference remains constant, confirming that the least significant bit (LSB) of the DAC sets degree of cancellation. Note that this scheme provides a fixed interference level, unlike conventional active cancellation or duplexers, which offer a fixed isolation. Gain compression due to unfiltered transmitter harmonics limit the linearity of this system.

The oversampling ratio *OSR* on the DAC and TX data is increased from 3.125× to 6.25× to 12.5×, to measure its impact on the residual TX interference at the receiver. For each doubling of the oversampling ratio, a 3 dB decrease is seen in the residual TX inference power in the 20 MHz band, apparent in Figure 5.32. This is because the DAC quantization noise, which sets the postcancellation residual, is spread out over a

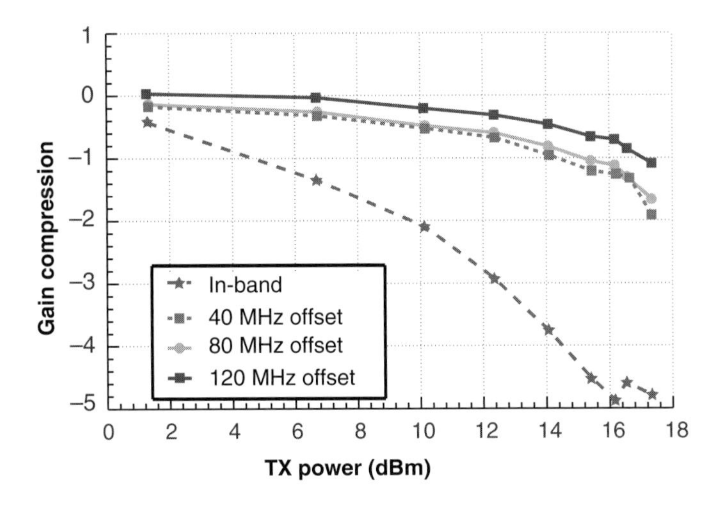

Figure 5.30 Gain compression vs. TX power.

wider bandwidth as oversampling is increased. Utilizing 9 bits of DAC dynamic range ($ENOB$) and 12.5× data oversampling, a maximum cancellation ΔP_{Cancel} of 64 dB was measured. This matches up very well with the expected cancellation of

$$\Delta P_{Cancel,\mathrm{dB}} = 6ENOB + OSR_{\mathrm{dB}} = \\ 6 \cdot 9 + 10\log_{10} 12.5 = 65\,\mathrm{dB}. \quad (5.11)$$

To determine the RF bandwidth over which cancellation is effective, the TX and RX LO frequency was swept with a fixed TX output power of 0 dBm. At each center frequency point, the DAC data was readapted to optimally match the frequency response of the TX-to-RX interference. Results are shown in Figure 5.33. The residual power remained constant over the TX frequency sweep from 1 to 1.8 GHz due to the high bandwidth of the virtual ground current subtraction node. The limited bandwidth of the RX matching network does not affect the bandwidth of cancellation. It simply changes the amplitude/phase shift of the TX current, which is compensated by adapting the input code to the cancellation DAC.

A major advantage of the cancellation DAC over purely analog cancellation architectures is its ability to maintain the aforementioned isolation as the TX-to-RX interference path varies. This strength is shown by measuring TX-to-RX isolation over antenna VSWR. In this measurement, with setup shown in Figure 5.34a, an antenna tuner, in the form of a sliding short in parallel with a 50 Ω calibration standard, was used to vary the antenna impedance up to 5:1 VSWR. For all points along the sweep range, >50dB of cancellation was achieved after the DAC input is readapted to match the interference path, shown in Figure 5.34b.

The impact of the simultaneous TX/RX operation on the receiver noise figure must also be characterized. The chip had a 15 dB noise figure for +16 dBm TX output power canceled as 40 MHz duplex spacing (7.6 dB nominal NF). This noise figure is broken

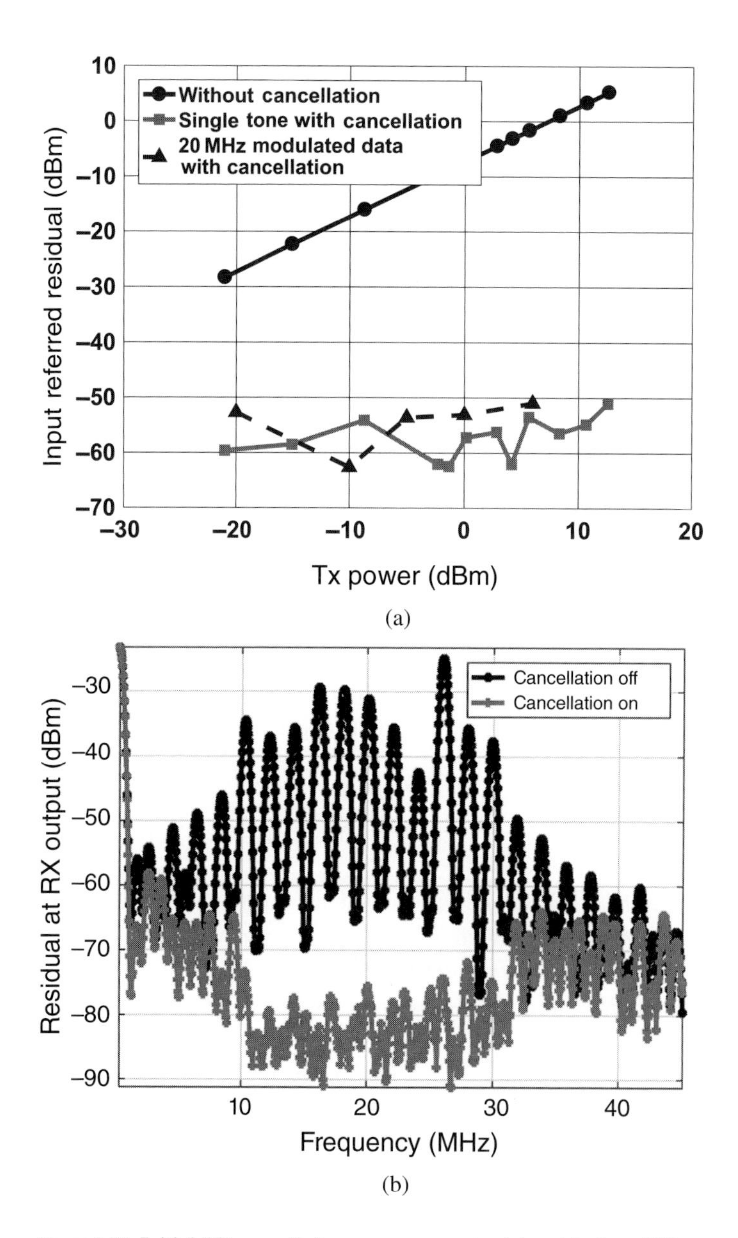

Figure 5.31 Initial TX cancellation measurements: (a) residual vs. TX power and (b) spectrum of modulated data cancellation.

down into its constituent components – TX thermal noise, RX noise, TX phase noise, and DAC thermal noise – shown in Figure 5.35.

To test the efficacy of DAC thermal noise cancellation, measurements were performed at low TX-RX offset so that thermal noise would be visible due to the large degree of phase noise cancellation at low offsets. Up to 3 dB DAC thermal noise reduction was found, with a 1 dB bandwidth of 2 MHz. As the measurements are performed at low

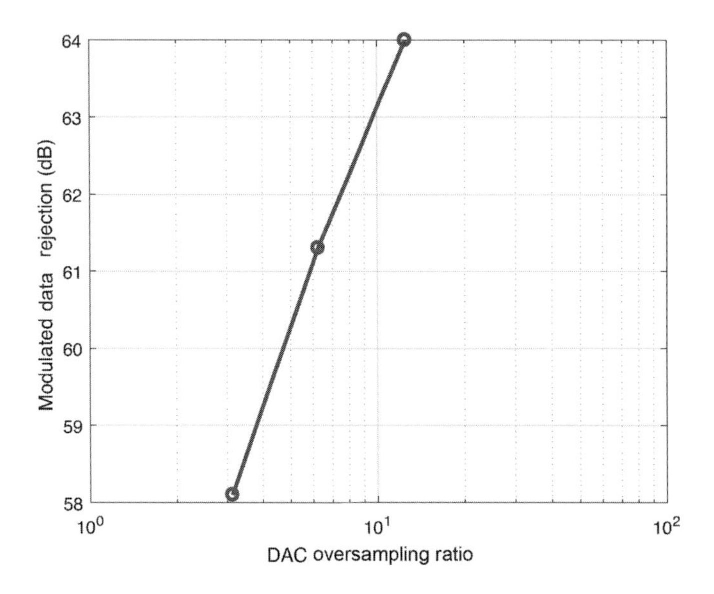

Figure 5.32 TX rejection vs. DAC oversampling ratio.

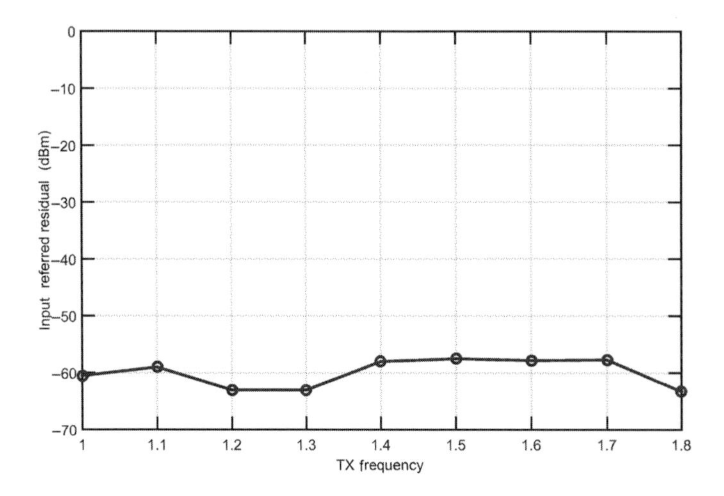

Figure 5.33 Residual vs. TX frequency.

offset, the bandwidth is smaller than it would be for the same Q at 40 MHz offset, where the bandwidth would be 15 MHz.

To characterize phase noise cancellation in isolation, phase noise of varying bandwidth was injected into the TX LO input by using an external noise source with a filter. Power combining this noise signal with the LO and feeding it into the limiting buffer chain creates an LO with phase noise because the amplitude is unchanged at the output of the limiter [26]. This injected noise is much higher power than the intrinsic noise of the system, allowing accurate verification of phase noise cancellation, diagrammed in Figure 5.36. First, 100 MHz bandwidth noise was injected to characterize phase noise

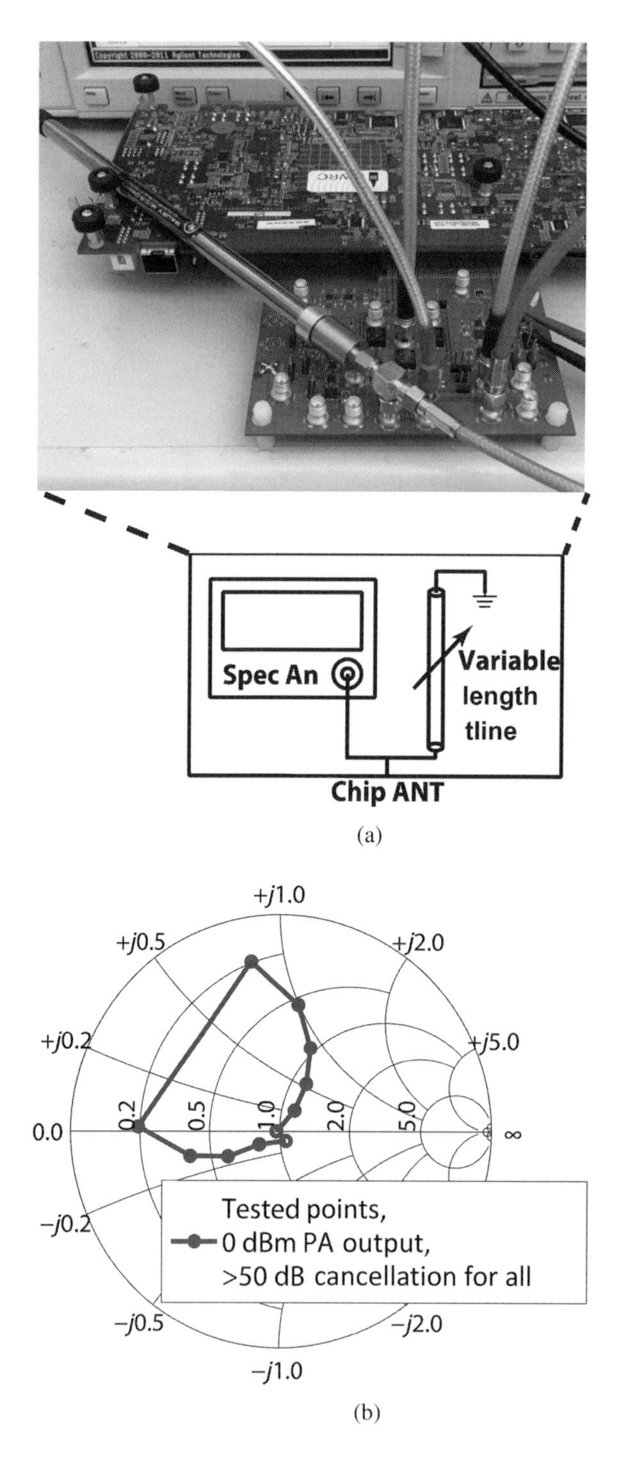

Figure 5.34 VSWR (a) test setup and (b) results (sweep over VSWR, up to 5:1).

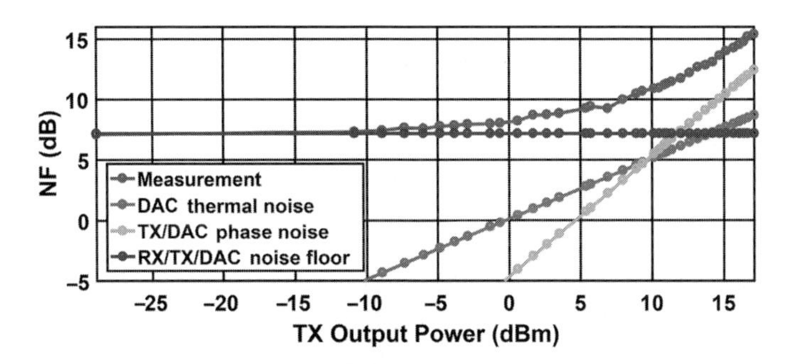

Figure 5.35 Breakdown of noise vs. cancellation power.

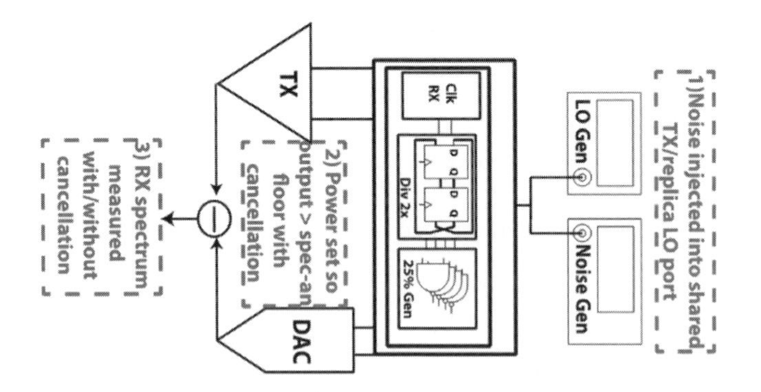

Figure 5.36 Test setup for phase noise cancellation measurement.

propagation without the effect of noise folding due to higher harmonics. The result, shown in red in Figure 5.37, follows the same contour of the cancellation of single tone spurs added to the LO. For a duplex spacing of 40 MHz, the closest of the LTE FDD bands, 20 dB of phase noise cancellation is observed. Next, the TX and RX were isolated from one another using a shunt connection to ground between the TX and RX baluns, and a meter-long cable was connected between the two. This emulates a channel with large group delay. The significant reduction in phase noise cancellation bandwidth, shown in Figure 5.38, is due to the large frequency-dependent phase shift of the cable. Additionally, it can be seen that past 40 MHz, the phase noise level with cancellation on is larger than the phase noise with cancellation disabled due to the fact that the correlated phase noise between the TX and DAC adds constructively once the phase shift between the two is large enough.

If wideband noise is injected into the TX LO, shown in black in Figure 5.37, cancellation falls off and levels out to a lower value due to noise at higher harmonics of the LO folding down with a different phase shift. The mechanism causing this phase noise cancellation limitation is detailed in [19]. A summary of this detailed analysis is that noise injected at some higher harmonics is folded down with an opposite phase shift between I and Q LO, as shown in Figure 5.37. Because the DAC uses I and Q weighting to replicate the TX's filtering, the oppositely phase-shifted noise is not canceled, limiting overall phase noise cancellation. The test setup for Figure 5.37 is

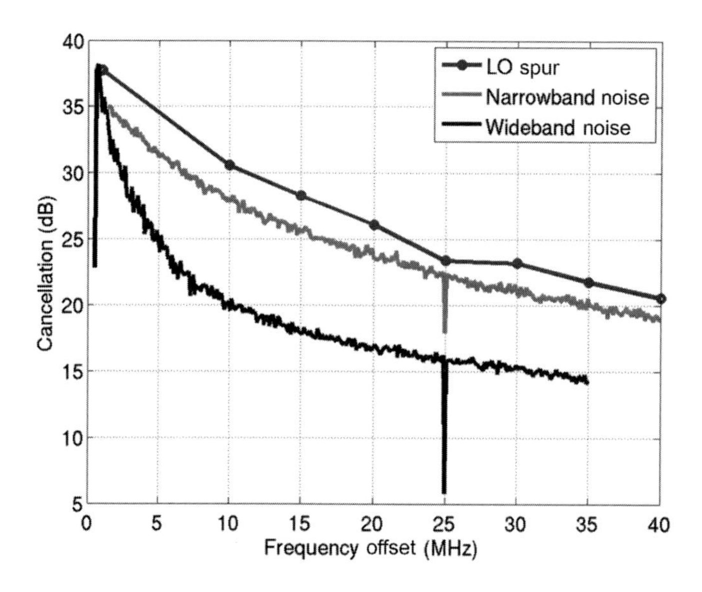

Figure 5.37 Phase noise cancellation measurement with single-tone, narrowband, and wideband injection.

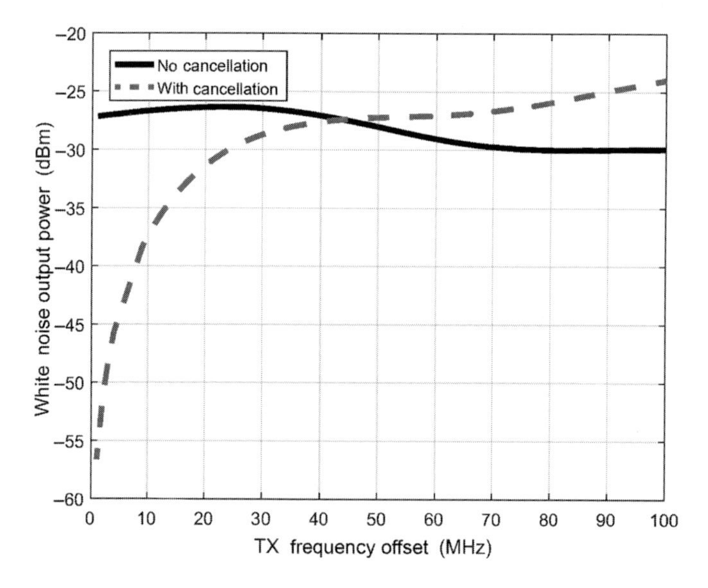

Figure 5.38 Phase noise cancellation with 1 m cable.

shown in Figure 5.39, where phase noise spurs are injected at higher harmonics and the phase difference between the downconverted spurs of codes C and jC are measured. Note that the higher levels of cancellation close to the TX frequency in Figure 5.37 are due to the Lorentzian phase noise spectrum of the source LO, where close-in phase noise dominates over the far-out white phase noise, obscuring cancellation limits due to folding at small frequency offsets.

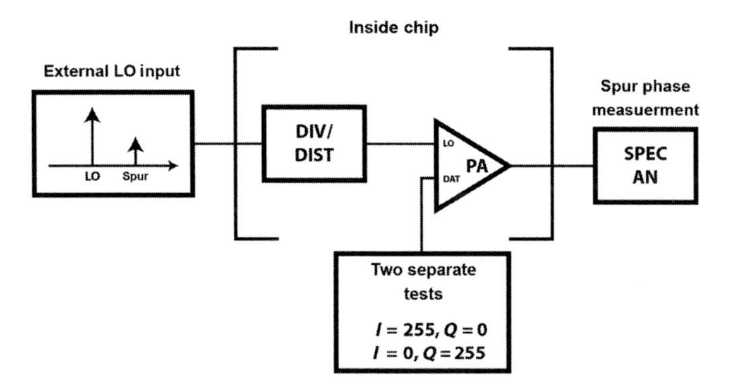

Figure 5.39 Test setup for phase noise folding measurement.

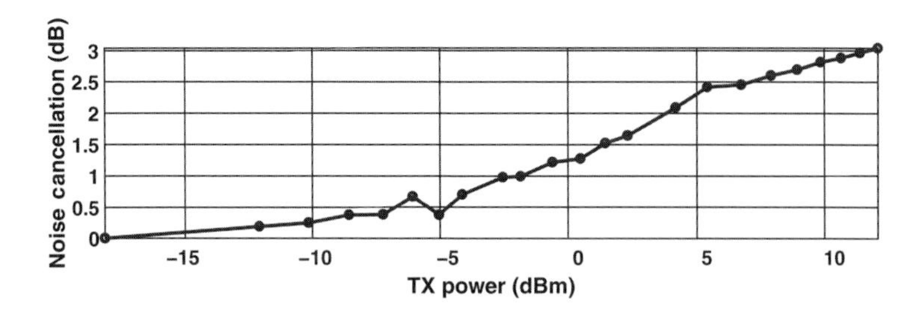

Figure 5.40 Measured thermal noise cancellation.

Two resonant feedback networks are used to reduce the tail thermal noise at F_{Duplex} and $2F_{TX} + F_{Duplex}$, shown in Figure 5.19. Resonant networks are used to provide a high impedance for frequencies of interest, while not injecting significant noise from the passive components. As the resonance around $2F_{TX}$ is more sensitive to parasitics, the $2F_{TX}$ impedance was placed on the tail source, and the baseband resonance fed back to the tail gate. The $2F_{TX}$ resonance is synthesized with a printed circuit board (PCB) via network, while the baseband resonance uses discrete passives. At low TX powers, the majority of the noise that flows through the center tap impedance and is fed back is due to disabled cells, offsetting the feedback cancellation. However, at low TX powers, the NF is dominated by the RX chain, not DAC thermal noise. The DAC thermal noise is most dominant at $>$+10 dBm FD operation with external isolation. In this regime, peak thermal noise cancellation of 3 dB is measured at +12 dBm, shown in Figure 5.40. The 1 dB BW is 2.25 MHz, corresponding to 15 MHz BW at $F_{Duplex} = 40$ MHz with the same resonant Q.

5.10 Conclusion

This chapter has described the architecture and a practical implementation of an effective active cancellation scheme that uses a replica current DAC in shunt with the receiver input in order to handle large transmit power of +18 dBm with receiver compression of less than 1 dB. Moreover, the shunt current is shown to consume significantly less power than the transmitter, due to the fact that transmit current alone, and not the transmit power, is canceled. The main sources of noise figure degradation due to the cancellation architecture are found to be phase noise from the transmitter and thermal noise from the DAC. Cancellation methods for both sources are detailed, along with cancellation of quantization noise from the transmitter and DAC, and supported by the prototype measurements. The presented architecture shows promise for use in integrated duplexers with narrow spacings between transmit and receive bands.

References

[1] M. Mikhemar, H. Darabi, and A. Abidi, "A tunable integrated duplexer with 50dB isolation in 40nm CMOS," in *2009 IEEE International Solid-State Circuits Conference: Digest of Technical Papers*, IEEE, February 2009, pp. 386–387, 387a.

[2] M. Mikhemar, H. Darabi, and A. Abidi, "An on-chip wideband and low-loss duplexer for 3G/4G CMOS radios," in *2010 Symposium on VLSI Circuits*, June 2010, pp. 129–130.

[3] J. G. Kim, S. Ko, S. Jeon, J. W. Park, and S. Hong, "Balanced topology to cancel Tx leakage in CW radar," *IEEE Microwave and Wireless Components Letters*, vol. 14, September 2004, pp. 443–445.

[4] T. Zhang, A. R. Suvarna, V. Bhagavatula, and J. C. Rudell, "An integrated CMOS passive transmitter leakage suppression technique for FDD radios," *2014 Radio Frequency Integrated Circuits Symposium*, IEEE, pp. 43–46. 2014.

[5] S. H. Abdelhalem, P. S. Gudem, and L. E. Larson, "Hybrid transformer-based tunable differential duplexer in a 90-nm CMOS process," *IEEE Transactions of Microwave Theory and Techniques*, vol. 61, no. 3, pp. 1316–1326, 2013.

[6] B. van Liempd et al., "A +70dBm IIP3 single-ended electrical-balance duplexer in 0.18um SOI CMOS," in *2015 IEEE International Solid-State Circuits Conference (ISSCC) Digest of Technical Papers,* IEEE, pp. 1–3, 2015.

[7] K. D. Chu, M. Katanbaf, T. Zhang, C. Su, and J. C. Rudell, "A broadband and deep-TX self-interference cancellation technique for full-duplex and frequency-domain-duplex transceiver applications," in *2018 IEEE International Solid-State Circuits Conference (ISSCC)*, February 2018, pp. 170–172.

[8] M. Mikhemar, "Interference cancellation in software-defined CMOS receivers," Ph.D. dissertation, University of California, Los Angeles, 2010.

[9] J. Zhou, T. H. Chuang, T. Dinc, and H. Krishnaswamy, "19.1 receiver with 20MHz bandwidth self-interference cancellation suitable for FDD, co-existence and full-duplex applications," in *2015 IEEE International Solid-State Circuits Conference (ISSCC) Digest of Technical Papers*, February 2015, pp. 1–3.

[10] J. Zhou, P. R. Kinget, and H. Krishnaswamy, "20.6 A blocker-resilient wideband receiver with low-noise active two-point cancellation of 0dBm TX leakage and TX noise in RX band for FDD/co-existence," in *2014 IEEE International Solid-State Circuits Conference Digest of Technical Papers (ISSCC)*, IEEE, February 2014, pp. 352–353.

[11] D. J. van den Broek, E. A. M. Klumperink, and B. Nauta, "19.2 A self-interference-cancelling receiver for in-band full-duplex wireless with low distortion under cancellation

of strong TX leakage," in *2015 IEEE International Solid-State Circuits Conference (ISSCC) Digest of Technical Papers*, IEEE, February 2015, pp. 1–3.

[12] T. Lee and B. Razavi, "A 125-MHz mixed-signal echo canceller for Gigabit Ethernet on copper wire," *IEEE Journal of Solid-State Circuits*, vol. 36, no. 3, pp. 366–373, March 2001.

[13] L. Calderin, S. Ramakrishnan, A. Puglielli, E. Alon, B. Nikolić, and A. M. Niknejad, "Analysis and design of integrated active cancellation transceiver for frequency division duplex systems," *IEEE Journal of Solid-State Circuits*, vol. 52, no. 8, pp. 2038–2054, August 2017.

[14] D. J. van den Broek, E. A. M. Klumperink, and B. Nauta, "A self-interference cancelling front-end for in-band full-duplex wireless and its phase noise performance," in *2015 IEEE Radio Frequency Integrated Circuits Symposium (RFIC)*, IEEE, May 2015, pp. 75–78.

[15] L. Calderin, "Flexible integrated architectures for frequency division duplex communication," Ph.D. dissertation, University of California, Berkeley, 2017.

[16] F. Sheikh, "Power-performance tradeoffs in ASICS for next generation wireless datapaths," Ph.D. dissertation, University of California, Berkeley, 2008.

[17] S. Yoo, J. S. Walling, E. C. Woo, B. Jann, and D. J. Allstot, "A switched-capacitor RF power amplifier," *IEEE Journal of Solid-State Circuits*, vol. 46, no. 12, pp. 2977–2987, 2011.

[18] V. Vorapipat, C. Levy, and P. Asbeck, "A wideband voltage mode Doherty power amplifier," in *2016 IEEE Radio Frequency Integrated Circuits Symposium (RFIC)*, IEEE, May 2016, pp. 266–269.

[19] S. Ramakrishnan, "Design of integrated full-duplex wireless transceivers," Ph.D. dissertation, University of California, Berkeley, 2016.

[20] D. Murphy, H. Darabi, A. Abidi, et al., "A blocker-tolerant, noise-cancelling receiver suitable for wideband wireless applications," *IEEE Journal of Solid-State Circuits*, vol. 47, no. 12, pp. 2943–2963, December 2012.

[21] F. Bruccoleri, E. A. M. Klumperink, and B. Nauta, "Wide-band CMOS low-noise amplifier exploiting thermal noise canceling," *IEEE Journal of Solid-State Circuits*, vol. 39, no. 2, pp. 275–282, February 2004.

[22] S. C. Blaakmeer, E. A. M. Klumperink, D. M. W. Leenaerts, and B. Nauta, "The Blixer, a wideband balun-LNA-I/Q-Mixer topology," *IEEE Journal of Solid-State Circuits*, vol. 43, no. 12, pp. 2706–2715, December 2008.

[23] 3GPP, "LTE specifications." [Online]. Available: www.3gpp.org/specifications

[24] E. Hegazi, H. Sjoland, and A. A. Abidi, "A filtering technique to lower LC oscillator phase noise," *IEEE Journal of Solid-State Circuits*, vol. 36, no. 12, pp. 1921–1930, December 2001.

[25] D. Murphy, H. Darabi, and H. Wu, "25.3 A VCO with implicit common-mode resonance," in *2015 IEEE International Solid-State Circuits Conference (ISSCC) Digest of Technical Papers*, February 2015, pp. 1–3.

[26] A. Hajimiri and T. H. Lee, "A general theory of phase noise in electrical oscillators," *IEEE Journal of Solid-State Circuits*, vol. 33, no. 2, pp. 179–194, February 1998.

6 Scalable RF and Millimeter-Wave Multibeam Approaches

Arun Natarajan

6.1 Large-Scale Phased and MIMO Arrays

Large-scale arrays at radio-frequency (RF) and millimeter-wave (mm-wave) have emerged as a promising candidate to address peak data rate, last mile, and wireless backhaul applications for 5G and beyond-5G links. Single/multibeam phased arrays as well as full-aperture multiple-input and multiple output (MIMO) transceivers are of interest for such applications [1–3]. Millimeter-wave multibeam arrays have been demonstrated for last-mile Gb/s links. Massive MIMO techniques that leverage large MIMO array sizes (with respect to number of users) also demand cost-effective transceivers with tens to thousands of elements. Scalable phased and MIMO arrays based on tiling unit cells present a cost-effective approach for flexible array deployments.

Scalable Arrays Using Tiled Unit Cells: Developing a scalable array by tiling unit cells involves balancing cost and complexity trade-offs associated with integrated circuit (IC) development and fabrication and the cost/complexity associated with packages that must route large number of RF, intermediate frequency (IF), or baseband signals. A single-element IC approach leads to high power consumption for every-element local oscilllator (LO) synchronization and impedance-matched drivers. At low RF frequencies (<10 GHz), large antenna spacing is offset by low-packaging losses making single-element unit cells potentially feasible. At these frequencies, the interface to the antenna can also be considered independent of the IC unit cell. However, at mm-wave frequencies (>30 GHz), physically short antenna spacings ($\sim\lambda/2 < 5$ mm), packaging losses, and manufacturing challenges with impedance-controlled multilayer packaging interconnect imply that unit cells with multiple elements as well as the antennas and IC antenna interfaces are preferable.

Scalable Array Challenges: For array TX (RX), a unit cell that contains N-elements must distribute (combine) the input signal to (from) each of the N-elements while providing variable phase-shift and variable-gain functionality in each element. Since IF signal distribution is preferable to RF signal distribution in the package, the unit cell may include frequency translation. However, this requires phase locking between multiple unit cells, which in turn requires LO or lower-frequency reference distribution. As mentioned earlier, at mm-wave frequencies, the unit cell must be envisioned while also considering the interface between the IC and antennas.

The need for large-scale arrays for 5G networks has motivated scalable, integrated phased-array architectures based on RF and/or LO-path phase-shifting at frequencies from 6 GHz to beyond 100 GHz [4–8]. The scalable low-IF 6–18 GHz array RX in [6] incorporates two RX, with each RX capable of providing two outputs with independent variable-phase shift and variable-gain for multibeam arrays. The IC includes a phase-locked loop (PLL) that operates from a 50 MHz reference enabling phase-locking between multiple ICs. The sub-100 MHz IF and reference (REF) frequencies simplify multi-IC packaging; however, an N-element array requires N such ICs increasing packaging complexity. In [9], a scalable 2 × 2 unit cell is presented to achieve a 32-element 28 GHz array where each element in the four-element TRX unit cell includes low-noise amplifier (LNA), power amplifier (PA), and phase shifters. A 4:1 combiner/distribution network is also included in the unit cell. However, the signal distribution is still at RF, and additional ICs are required for signal combining and distribution to the phased-array unit cell.

In general, digital beamforming arrays that incorporate A/D and D/A conversion in each unit cell considerably reduce RF, IF, or analog baseband signal distribution and reduce sensitivity to packaging when the array is scaled to a larger number of elements. The potential for such a digital-intensive scalable array has been demonstrated in S-band using monolithic microwave integrated circuits (MMICs) and commercial-off-the-shelf (COTS) components in [10,11]. However, the absence of any spatial filtering prior to the analog-to-digital converter (ADC) implies that in-band or cochannel interferers (CCI) are present at each ADC input, resulting in nonlinearities. Therefore, ADC power consumption and signal-to-noise-and-distortion ratio (SNDR) remain key bottlenecks for scalable digital beamforming (DBF)/MIMO transceivers. State-of-the-art ADC performance is reflected in an ADC achieving ~55 dB SNDR for ~20 MHz bandwidth using ~8 mW in [12]. Increasing SNDR by 30 dB in each element to account for CCI is equivalent to a 5-bit increase in the effective number of bits (ENOB), and hence ADC power consumption will increase by 32 times to as much as 1,000 times for thermal-noise limited ADCs [13]. Furthermore, technology scaling provides fairly slow energy-efficiency improvements for ADCs with high ENOB [13,14]. Therefore, reducing ADC dynamic range and output data rate is critical for the feasibility of scalable MIMO/DBF arrays. Analog MIMO beamforming to provide spatial filtering prior to the ADC has been investigated for cognitive radio [15] with results reinforcing ADC dynamic range advantages of spatial filtering. MIMO simulations with three elements in the TX and RX in [15] predict a reduction of more than 3 bits in required ADC ENOB. In addition, it must be noted however that digital-IO can lead to high-power consumption in wideband arrays, limiting array size (8 bits I and Q at 1GS/s, implies 16 Gb/s, translating to 160 mW/IC assuming a 10 pJ/bit serial link efficiency).

6.2 Reconfigurable Spatial Filtering

Emerging 5G applications rely on dense spectrum reuse to increase network capacity, leading to increased in-band interference or CCI in receivers. While phased arrays/MIMO arrays enable spatial filtering of CCI, DBF in such arrays is desirable for reconfigurable, concurrent multiple beams. However, the absence of analog spatial filtering results in high ADC dynamic range requirements to tolerate CCI/jammers. This has led to interest in development of notched arrays with spatiospectral notching of jammers/interferers in RF/analog prior to the ADC and DBF. Blocker suppression at RF and IF is critical to address intermodulation products between blocker and desired signal. In the following, we discuss techniques to achieve reconfigurable spatiospectral notch filtering (SNF) at RF/IF using N-path and MIMO filtering techniques.

6.2.1 MIMO Spatial Filtering at RF

Spatial filtering for MIMO systems can be viewed from an architectural level as implementing spatial notches in antenna pattern at each element at specific frequencies

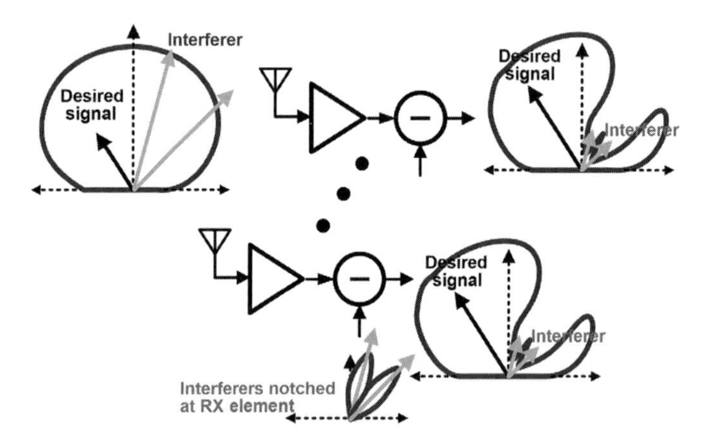

Figure 6.1 Spatial cancelling/notch approach for MIMO arrays.

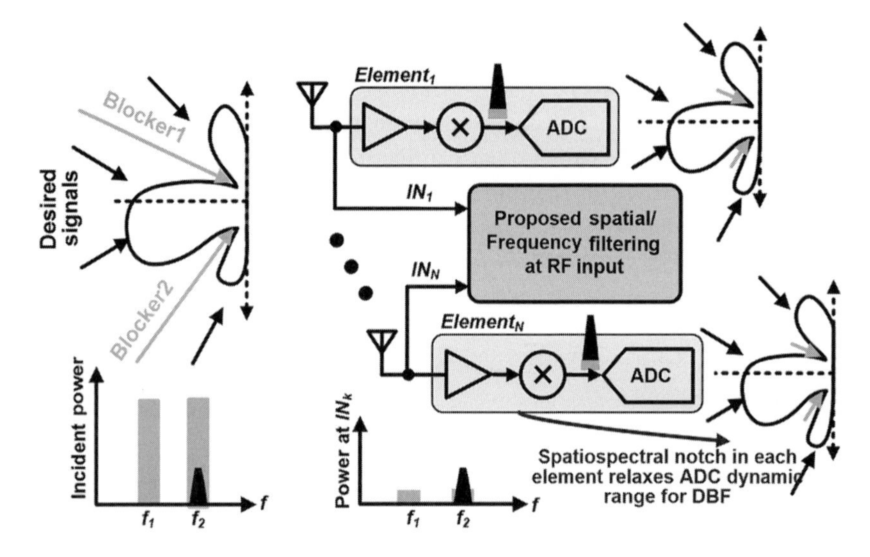

Figure 6.2 N-path filter at RF input for an N-element array [16]. (©2017 IEEE. Reprinted, with permission, from *2017 IEEE International Microwave Symposium.*)

(Figure 6.1). The angle of incidence (AoI) and the frequency of the notches must be broadly reconfigurable to address dynamic interferer scenarios.

A scalable, reconfigurable RX architecture for parallel spatiospectral notch filtering (PSNF) is presented in [16] (Figure 6.2). This N-path filter-based approach, outlined in the next section, allows scalable concurrent rejection of blockers at independent frequencies/AoI at each antenna input in a DBF array.

6.3 N-Path Spatiospectral Filtering

Frequency-translated filtering using the impedance translation property of N-path passive mixers promises highly selective filtering around a tunable LO-defined frequency [17–23]. N-path mixer-based tunable bandpass and bandstop filters driven by nonoverlapping clock pulses (NOPs) have been theoretically analyzed in [19,18,21]. Integrated N-path filters provide tunable center frequency and high linearity, and, with higher-order filtering, provide high selectivity to mitigate large out-of-band blockers [20,24–27]. For example, combined bandstop and bandpass filtered approaches have demonstrated ~10–13 dBm blocker tolerance, which is suitable for tunable surface acoustic wave (SAW)-less receiver applications [27–29].

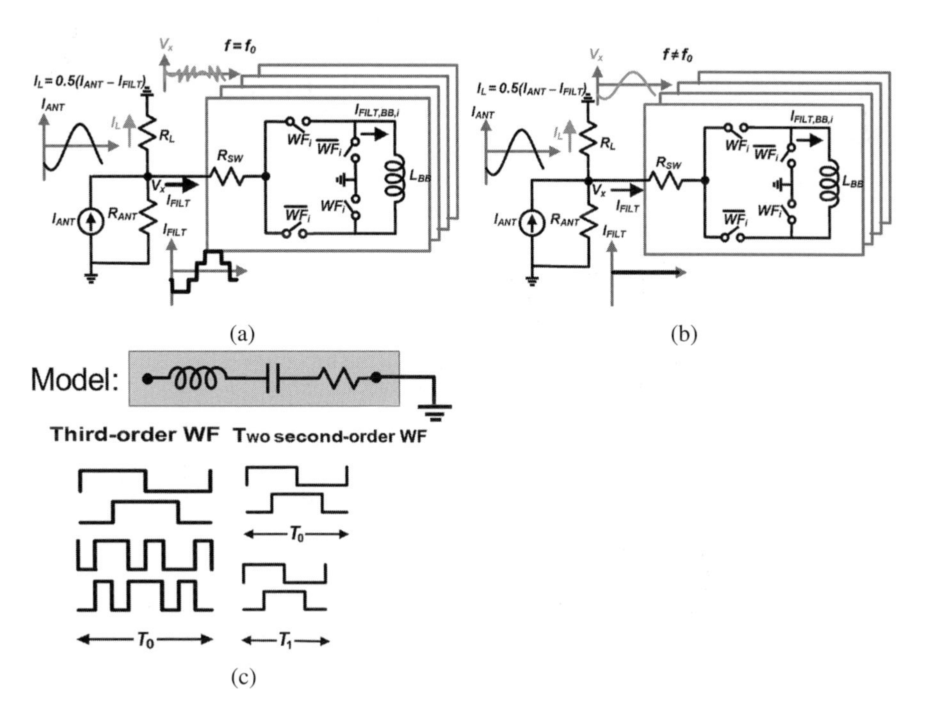

Figure 6.3 Bandstop filter using Walsh function sequence mixing and impedance translation of passive mixers. Filter response corresponds to the case when the input signal is (a) harmonically related to and (b) not harmonically related to WF-seq running at frequency f_0. (c) WF-seq driving N-path correlators.

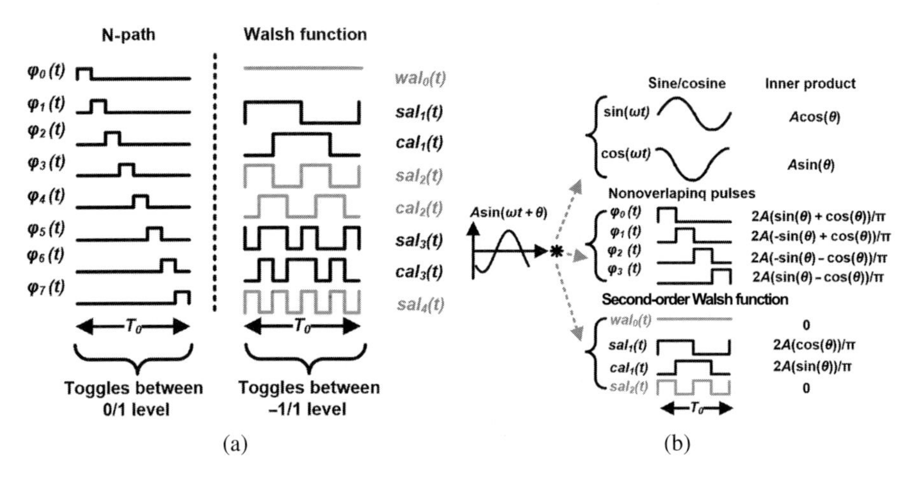

Figure 6.4 Walsh function sequences orthogonal and like nonoverlapping pulses; increasing WF order leads to better signal approximation; functions are restricted to ± 1.

The proposed shunt filter approach is shown in Figure 6.3, where N-path switches are driven by Walsh function sequences (WF-seq) instead of NOP. Notably, Walsh function sequences, like the Fourier series, are well known as a complete orthogonal basis system to represent a signal (sal(i) and cal(i) in Figure 6.4b) [30]. WF-seq present the following benefits: (i) similar to N-phase NOP LO, WF-seq orthogonality implies that passive mixers driven by WF-seq can be connected together at RF without scaling and that notch depth is increased with higher-order WF-seq; (ii) unlike NOP, each correlator in Figure 6.4 is always connected to the RF port with the WF-seq approach ensuring a current path for the baseband current; (iii) since the correlation for a sinusoid input with some WF-seq (wal(0), cal(2), sal(2), sal(4) – gray in Figure 6.4) results in zero output if both have the same period T_0, and correlation with those sequences is not required; and (iv) harmonic properties of the WF-seq filter are also equivalent to NOP N-path filters.

Each set of switches and baseband gyrator/capacitor can be considered to be a correlator that senses RF voltage and returns current based on the projection of the RF input voltage on a basis function determined by the mixer switches. An input signal that leads to low baseband voltage on $C_{BB,K}$ (in blue) leads to small baseband current, $I_{BB,K}$, and small RF current, I_{FILT}. This translates to a high PSNF input impedance and hence no filtering. However, a blocker signal that is correlated with the mixer switching signals leads to a nonzero voltage on $C_{BB,K}$ (in red) and hence to gyrator output currents that are upconverted with I_{FILT} following I_{ANT}. This creates a low RF impedance, attenuating RF input voltage (Figure 6.3).

Each PSNF element in Figure 6.5 consists of four correlators in this implementation. Spatiospectral filtering can be achieved by connecting capacitors corresponding to one set of correlators across elements (Figure 6.5). For example, if we assume that the WF-seq are in-phase in all elements, a blocker with broadside AoI results in constructive addition on $C_{BB,K}$ (Figure 6.5). This leads to currents, $I_{BB,K}$ at all gyrator outputs, causing low RF impedance and blocker notch filtering at all element inputs. If the desired signal AoI results in null voltage on $C_{BB,K}$, and hence null gyrator output, all RF inputs

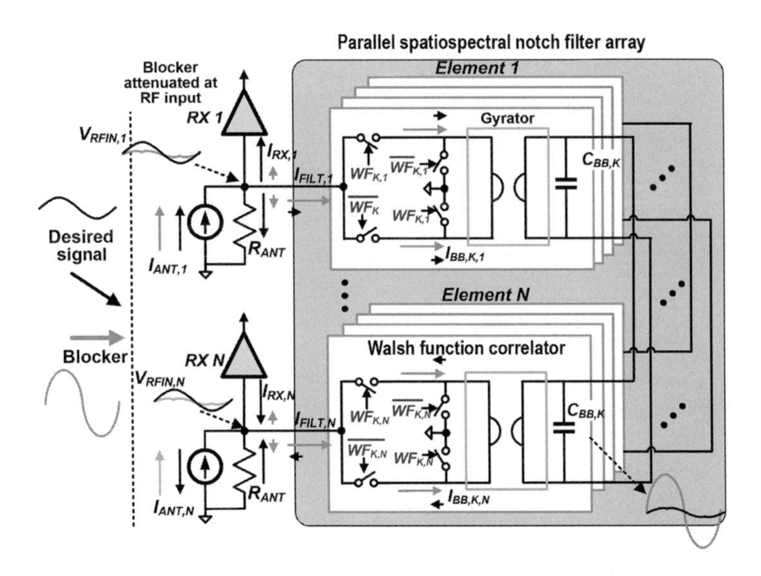

Figure 6.5 PSNF array using Walsh function sequence mixing and impedance translation of passive mixers [16]. (©2017 IEEE. Reprinted, with permission, from *2017 IEEE International Microwave Symposium*.)

see a high PSNF impedance, and the desired signal is unaffected at all elements. The AoI corresponding to notch filtering can be steered by changing the relative phase of the WF-seq applied to correlators in each element.

Importantly, since the correlators in the PSNF correlate input voltage and return current, the approach is not affected by overlap between WF-seq across correlators. This is unlike N-path filters with NOP in [18] that correlate current and return voltage, making them sensitive to overlap. The PSNF's insensitivity to overlap between basis functions allows arbitrary WF-seq to be applied at each correlator. For instance, if single-frequency/AoI third-order WF-seq (Figure 6.3a) are applied to the four correlators in each array element, a single frequency/AoI notch filter is created that is equivalent to an eight-phase NOP filter. On the other hand, a concurrent dual-frequency/AoI notch can be achieved by applying two second-order WF-seq at two independent frequencies (Figure 6.3a) to the four correlators in each element. Additionally, the gyrator capacitors in the PSNF approach also capture the blocker signal for subsequent feed-forward cancellation (FFC).

The CMOS four-element, four-correlator notch filter implemented in [16] is shown in Figure 6.6. A gain-boosted N-path RX [31] is included at one element output to demonstrate an RX following the PSNF. Since a G_{M1} following the mixer, as in Figure 6.5, leads to high flicker noise or capacitive switching losses, a translational approach is adopted with G_{M1} preceding the mixer. The four-element PSNF is implemented in 65 nm CMOS and operates from 0.3 to 1.4 GHz. Figure 6.7a and 6.7b show the measured S11 and Figure 6.7c and 6.7d show the measured S21 for one PSNF input/output pair with third-order WF-seq for single frequency and two second-order

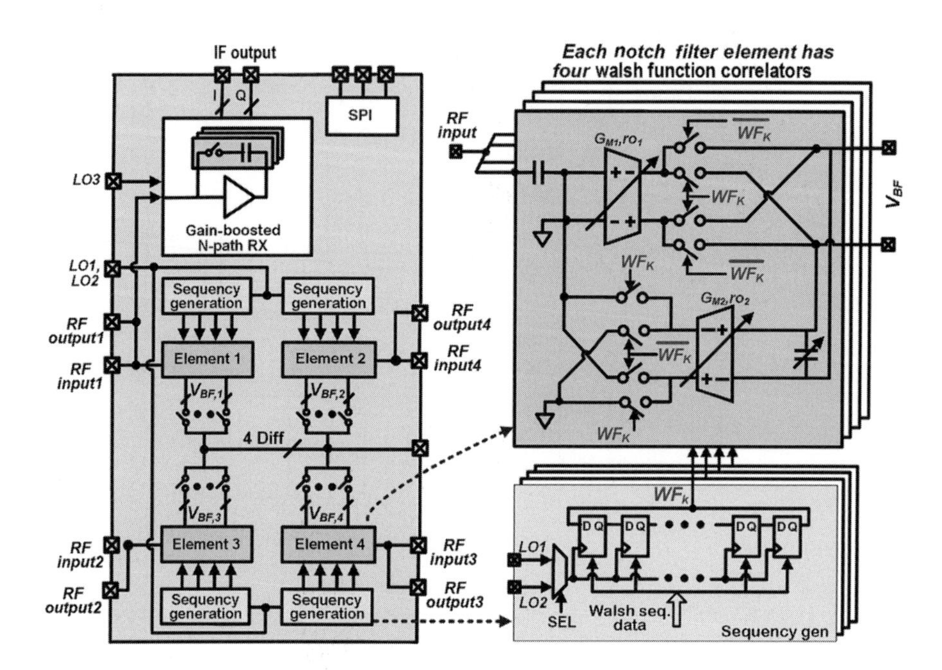

Figure 6.6 Schematic of 65 nm CMOS implementation of a four-element PSNF with four correlators in each element operating from 0.3 to 1.4 GHz [16]. (©2017 IEEE. Reprinted, with permission, from 2017 *IEEE International Microwave Symposium.*)

WF-seq for concurrent dual-frequency tunable notch. The PSNF achieves 14 and 20 dB spectral notch depth for second- and third-order WF-seq respectively. The measured four-element PSNF array factor at each output is shown across AoI/frequency for two frequency and relative phase-shift settings of WF-seq, demonstrating concurrent dual-frequency/AoI spatiospectral notch filtering (Figure 6.8a and 6.8b).

6.4 Scalable mm-Wave Packaging

As described in Section 6.1, integration in CMOS/SiGe, with excellent yields and matching between elements, makes large-scale reconfigurable mm-wave arrays feasible [5,32,33]. Scalable arrays with tiled unit cells have been demonstrated as a path toward achieving arrays with hundreds of elements [34,35]. While silicon ICs can achieve high yields, the mm-wave interface between the IC and package is challenging for a large number of mm-wave IO.

Antenna-in-package: Antenna-in-package (AiP) approaches have been demonstrated for multielement mm-wave ICs using low-temperature cofired ceramics (LTCC) and multilayer organic (MLO) laminates [35–38]. In this case, elements on the IC are interfaced to single-polarization or dual-polarization antennas on the package. This requires routing of impedance-controlled lines and vias in a compact area on a package while minimizing routing losses. For example, >1.5 dB interconnect loss in [38] leads to overall efficiency of 63% even if 90% antennas efficiency is assumed. Element-to-

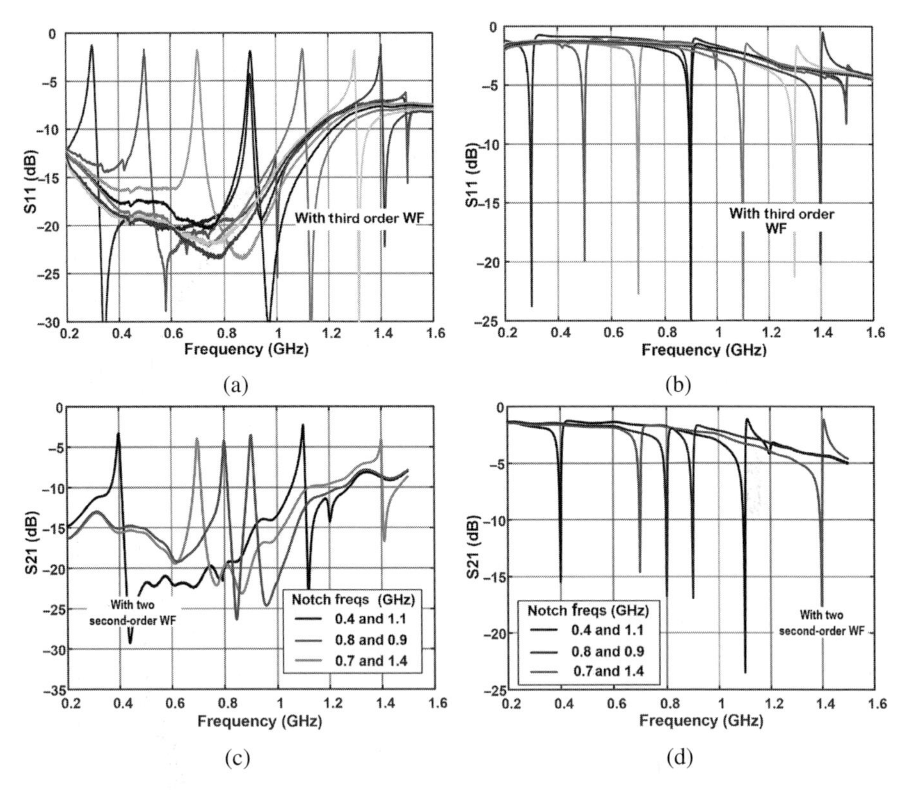

Figure 6.7 Measured PSNF-element S11 with (a) third-order WF-seq for single notch and (b) two second-order WF-seq for two notch frequencies; (c) measured S21 from one of the RF inputs to RF output with third-order WF-seq; and (d) two second-order WF-seq [16]. (©2017 IEEE. Reprinted, with permission, from *2017 IEEE International Microwave Symposium.*)

element variations must also be minimized (±1 dB have been achieved). In [38], four 16-element ICs are combined together to acheive a 64-element unit cell, which is tiled to achieve a large array with a fill factor of 64% at W-band. A larger array with >300 elements in the W-band has also been reported (with smaller fill factors) with lower routing complexity by separating TX and RX antennas [35]. Even as silicon integration makes a tiled approach to large-scale mm-wave arrays practical, packaging complexity and cost have motivated research on alternative approaches to AiP for low-cost arrays.

Antenna Cointegration: On-chip antennas are attractive for such mm-wave arrays if comparable system performance can be achieved. However, silicon substrate has high dielectric constant (∼11.7) and often low resistivity (∼10 Ω-cm), leading to low efficiency if electromagnetic (EM) energy is confined inside the substrate by the antenna. While an on-chip ground plane can isolate the substrate, antennatoground plane distance is limited to 9–15 μm, leading to poor radiation efficiency and narrow bandwidths. On-chip antenna performance can be improved by adding a superstrate on top of the antenna [39] or by proximity coupling to an antenna on a superstrate [40]. A wafer-scale array has been demonstrated with custom lithography in [4] that improves antenna efficiency.

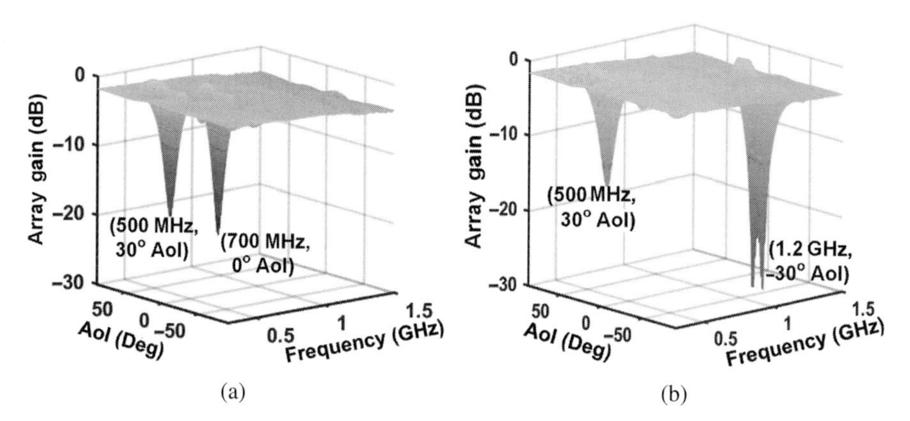

Figure 6.8 Measured array gain for four-element array for two settings demonstrating concurrent dual frequency/AoI notch filtering [16]. (©2017 IEEE. Reprinted, with permission, from *2017 IEEE International Microwave Symposium.*)

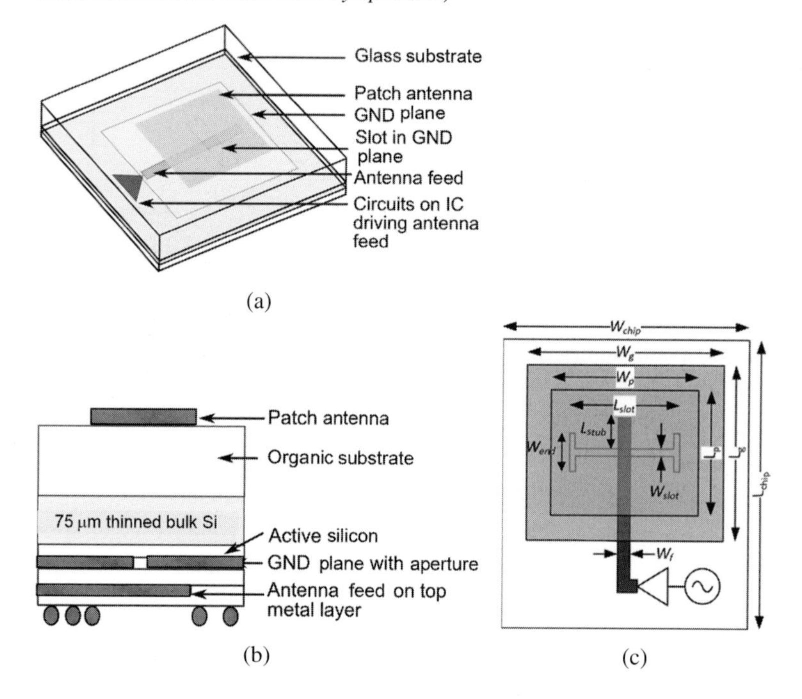

Figure 6.9 (a) Wafer-scale approach for antenna-IC co-integration using aperture-coupled feed on IC, (b) side view of antenna–IC structure stack-up, and (c) top view of structure

Wafer-scale-compatible antenna-IC cointegration approaches can significantly simplify mm-wave packaging and test by eliminating mm-wave I/O to/from the IC. However, they must achieve ~50% efficiency in order to be competitive with overall efficiencies achieved with state-of-the-art MLO and LTCC packaging.

Figure 6.9 shows a wafer-scale-compatible antenna cointegration scheme in [41] that relies on aperture-coupling between the on-chip feed and the patch antenna on a substrate that is bonded to a thinned silicon IC. The thick top-metal layer on the IC is used for antenna feed, and the accompanying ground plane is created using the lower metal layers. CMOS lithography allows the creation of a precise slot in the ground plane that aperturecouples the feed to a patch antenna on the substrate without the need for any off-chip via that conducts mm-wave signals mm-wave signals. The silicon die (or wafer) with circuits is thinned to reduce loss. The patch antenna metalization is created on the substrate, and the die (or wafer) and substrate can be bonded together using well-established adhesive techniques. From an antenna performance perspective, this technique preserves all the benefits of aperture coupling – wide bandwidth as well as isolation between the antenna layer and feed line layer – which allows for transmission line (t-line) structures to be created without interfering with antenna performance. Bandwidth enhancement techniques such as stacked aperture-coupled patches [42] are also potentially feasible.

As shown in Figure 6.9, design variables include patch and slot dimensions (L_p, W_p, W_{slot}, L_{slot}, and W_{end}), on-chip ground plane size (W_g and L_g), feed structure parameters (L_{stub}), and substrate parameters (ϵ_{LCP}, h_{LCP}, and h_{si}).

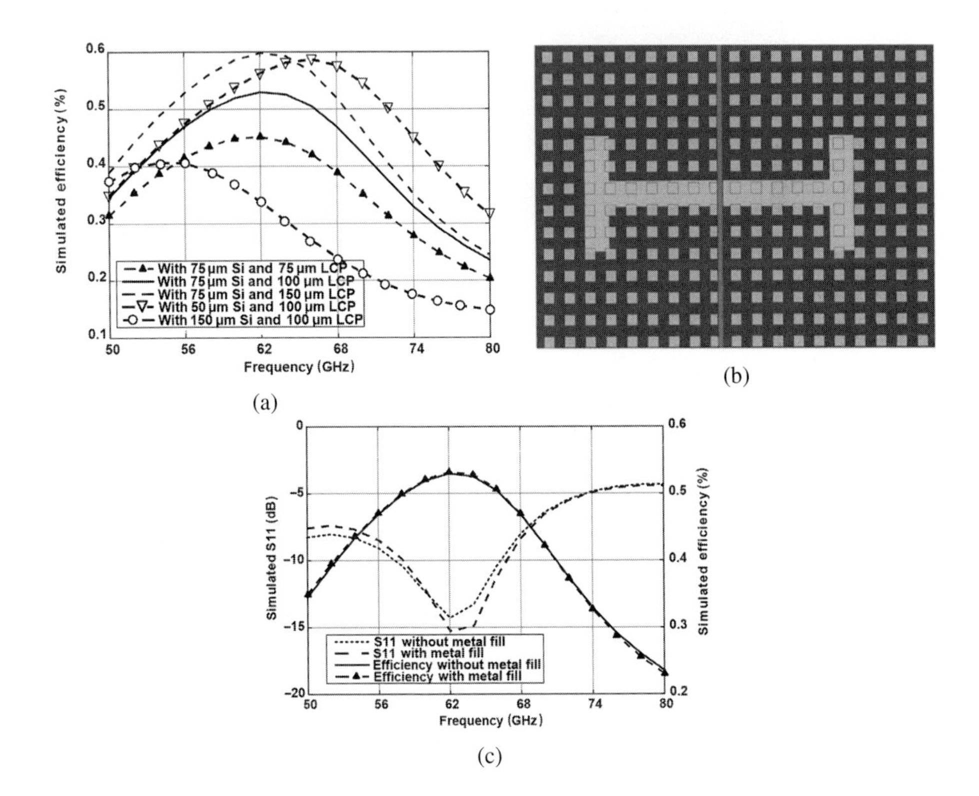

Figure 6.10 (a) Efficiency variation with different LCP and silicon thickness, (b) feedline and slot with metal fill, (c) impact of metal fill on S11 and efficiency [41]. (©2017 IEEE. Reprinted, with permission, from *2017 IEEE International Microwave Symposium.*)

Since wafer-scale compatibility is targeted, the patch substrate material must have low dielectric constant, low loss at mm-wave, and silicon-compatible coefficient of thermal expansion. Liquid crystal polymer (LCP) has been identified as a potential low-cost, high-performance, mm-wave substrate [43] and is selected in this work. Notably, similar performance is also achieved in simulation with quartz as the substrate. LCP has a low dielectric constant ($\epsilon_{LCP} \approx 3.1$) and low loss at mm-wave ($\tan \delta_{LCP} \approx 0.003$), which is comparable to LTCC. While increasing LCP substrate thickness can initially lead to higher radiation efficiency and bandwidth, a very thick substrate leads to lower efficiency due to surface-wave loss.

The impact of silicon and LCP thickness is shown in Figure 6.10 – reducing silicon thickness improves efficiency, e.g., changing thickness from 150 to 50 μm improves efficiency from 38% to >50%. While thinning silicon dies can lead to reliability challenges, 3D IC integration has motivated research into die thinning and bonding techniques for robust packaging. In the rototype, silicon thickness of unit ~75 μm is selected to balance

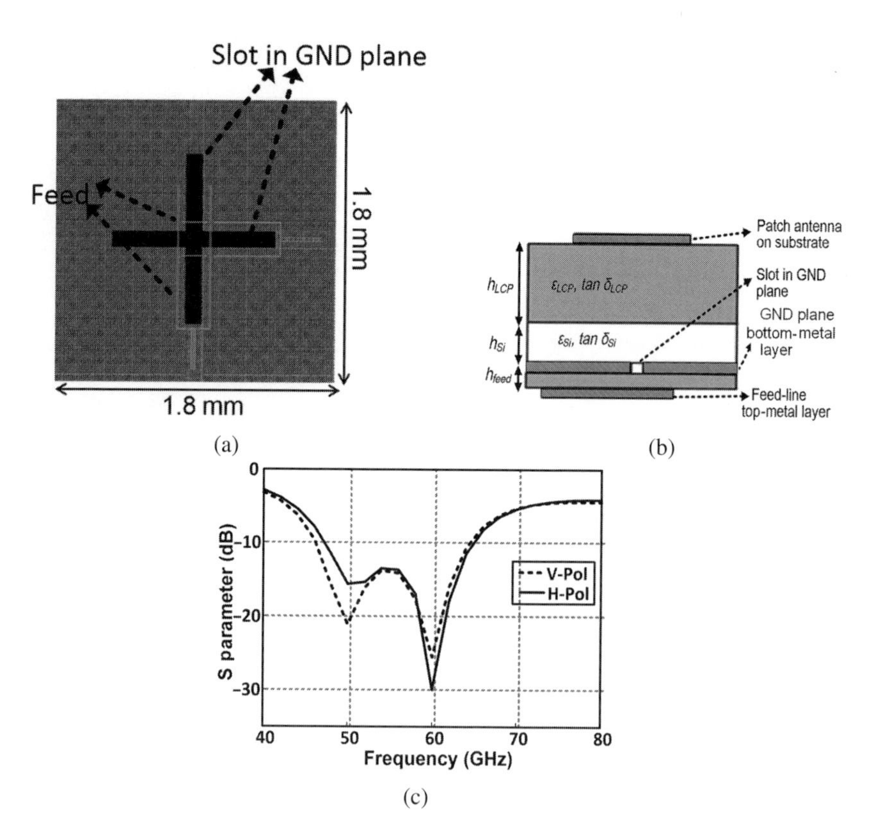

Figure 6.11 Antenna cointegration with dual-polarization on-chip feeds and slot aperture-coupled to the antenna on LCP through the backside of the die. (a) Top view of slot and feed. (b) Cross-section view showing die and LCP stackup. The feed lines are on the top metal, while ground plane with the orthogonal slots are on the bottom-metal layer. (c) A 60 GHz dual-polarization RXFE with antenna feed and slot coupled to patch on backside of the IC and cross-polarization cancellation.

efficiency with ease of chip handling for packaging. Similarly, increasing LCP thickness from unit ~75 μm to unit ~150 μm can increase efficiency as well. Readily-available LCP material with 100 μm thickness (Rogers 3850) is used, leading to 52% efficiency and 9 GHz bandwidth in simulation (Figure 6.10a).

The approach can be extended to a dual-polarization (dual-pol) TX/RX with orthogonal slots and dual feeds [44]. The area inefficiency of the ground plane in [41] can be addressed by implementing t-line-based circuits in the ground plane area. Given the aperture-coupled approach, the ground plane separates the antenna and the t-line-based

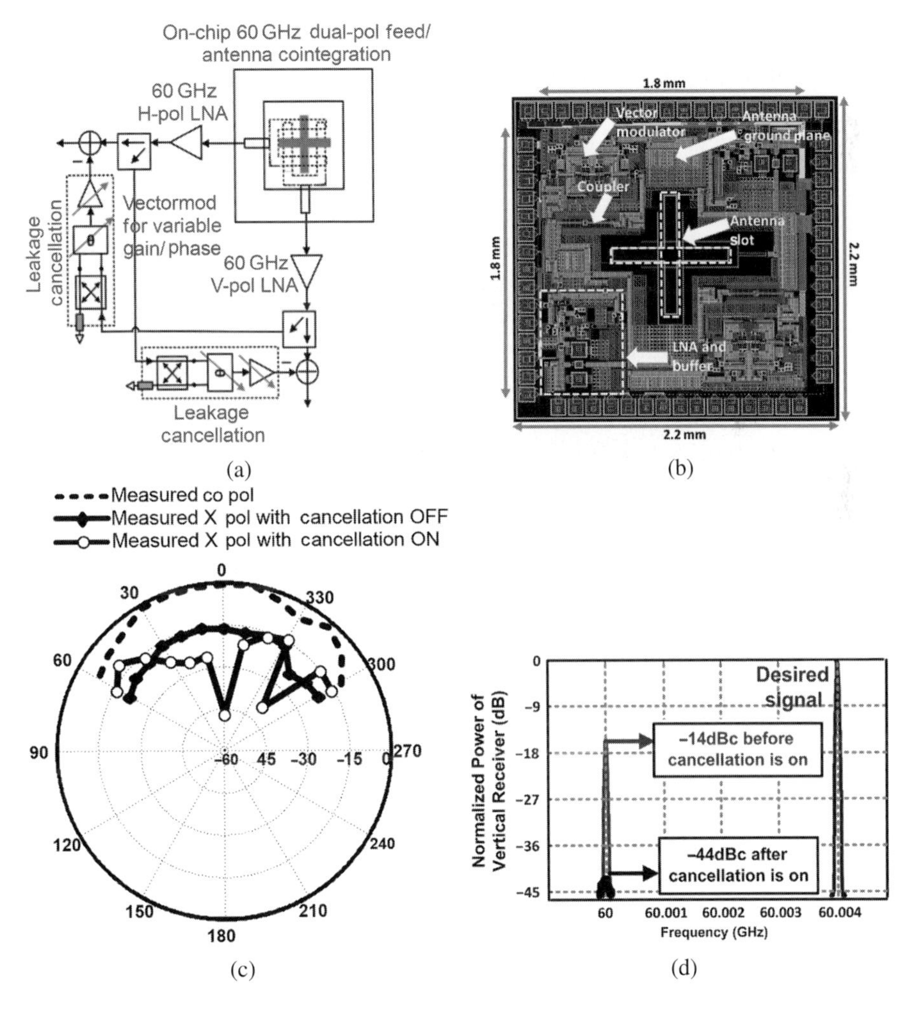

Figure 6.12 (a) 60 GHz dual-polarization RX architecture with dual-polarization antenna/feed cointegration and cross-polarization leakage cancellation at RF. (b) layout demonstrating use of t-line based circuits to reuse antenna ground plane area. Aperture-coupled approach ensures t-line circuits separated from antenna by ground plane. (c) Measured antenna pattern with and without cross-polarization cancellation at LNA output. (d) Cross-polarization levels up to −44 dBc after cancellation after packaging using the stackup in Figure 6.11b.

circuits (Figure 6.9), allowing the ground plane area to be reused for circuits, provided interconnect and matching networks are implemented using t-lines. Integrating the feed network on the ICs enables architectures that leverage the ability to integrate large-scale complex circuits on silicon.

Based on this approach, a dual-polarization antenna feed driving a dual-polarization 60 GHz RX front-end (RXFE) with cross-polarization cancellation is shown in Figure 6.11a and 6.11b. In order to achieve 60 GHz dual-polarization operation, orthogonal slots are designed in the ground plane with slot length of 800 μm and slot width of 90 μm. A forked dual-polarization feed structure, as shown in Figure 6.11a, is designed for each antenna with width and length to ensure 50 Ω input impedance for each feed. Figure 6.11c shows simulated dual-antenna S-parameters at 60 GHz demonstrating good matching at 60 GHz for both V- and H-polarization. An overall efficiency of ~50% and 2.7 dBi gain at mm-wave are simulated assuming 75 μm silicon thickness and 175 μm LCP thickness for the stackup in Figure 6.11b.

As described earlier, the antenna ground plane area is unutilized in the single-polarization feed structure described in [41]. In the aperture-coupled approach, the t-lines with the signal layer on the top-metal layer are separated from the antenna by the ground plane. Therefore, on-chip t-line networks can be designed that reuse the ground plane. Active devices can be considered to be placed in slots in the ground plane (1.8 mm × 1.8 mm) that are relatively small compared to the wavelength. Therefore, circuits can potentially be integrated in the ground plane if grounded coplanar waveguide-based t-line networks are used. Figure 6.12a and 6.12b show a dual-polarization antenna cointegrated 60 GHz RX implemented in the TowerJazz SBC18 process, where two LNAs and cross-polarization cancellation circuits are implemented in the ground-plane area, leading to efficient antenna utilization as shown in Figure 6.12b. Measurements in Figure 6.12c and 6.12d demonstrate the feasibility of this approach for achieving dual-polarization RX/TX with wafer-scale compatible antenna cointegration at mm-wave frequencies.

References

[1] V. Jungnickel, K. Manolakis, W. Zirwas, et al., "The role of small cells, coordinated multipoint, and massive MIMO in 5G," *IEEE Communications Magazine*, vol. 52, no. 5, pp. 44–51, May 2014.

[2] S. Rangan, T. Rappaport, and E. Erkip, "Millimeter-wave cellular wireless networks: Potentials and challenges," *Proceedings of the IEEE*, vol. 102, no. 3, pp. 366–385, March 2014.

[3] J. Andrews, S. Buzzi, W. Choi, et al., "What will 5G be?" *IEEE Journal on Selected Areas in Communications*, vol. 32, no. 6, pp. 1065–1082, June 2014.

[4] S. Zihir, O. Gurbuz, A. Karroy, S. Raman, and G. Rebeiz, "A 60 GHz 64-element wafer-scale phased-array with full-reticle design," in *IEEE MTT-S International Microwave Symposium (IMS)*, IEEE, 2015.

[5] A. Natarajan, A. Valdes-Garcia, B. Sadhu, S. Reynolds, and B. Parker, "W-band dual-polarization phased-array transceiver front-end in SiGe BiCMOS," *IEEE Transactions on Microwave Theory and Techniques*, vol. 63, no. 6, pp. 1989–2002, June 2015.

[6] S. Jeon, Y.-J. Wang, H. Wang, et al., "A scalable 6-to-18 GHz concurrent dual-band quad-beam phased-array receiver in CMOS," *IEEE Journal of Solid-State Circuits*, vol. 43, no. 12, pp. 2660–2673, 2008.

[7] X. Gu, A. Valdes-Garcia, A. Natarajan, B. Sadhu, D. Liu, and S. K. Reynolds, "W-band scalable phased arrays for imaging and communications," *IEEE Communications Magazine*, vol. 53, no. 4, pp. 196–204, 2015.

[8] X. Guan, H. Hashemi, and A. Hajimiri, "A fully-integrated 24-GHz eight-element phased-array receiver in silicon," *IEEE Journal of Solid-State Circuits*, vol. 39, no. 12, pp. 2311–2320, December 2004.

[9] K. Kibaroglu, M. Sayginer, and G. M. Rebeiz, "A low-cost scalable 32-element 28-GHz phased array transceiver for 5G communication links based on a 2 × 2 beamformer flip-chip unit cell," *IEEE Journal of Solid-State Circuits*, vol. 53, no. 5, pp. 1260–1274, May 2018.

[10] W. Chappell and C. Fulton, "Digital Array Radar panel development," in *IEEE International Symposium on Phased Array Systems and Technology*, October 2010, pp. 50–60.

[11] W. H. Weedon, "Phased array digital beamforming hardware development at Applied Radar," in *IEEE International Symposium on Phased Array Systems and Technology*, IEEE, October 2010, pp. 854–859.

[12] M. Andersson, M. Anderson, L. Sundstrom, S. Mattisson, and P. Andreani, "A filtering ΔΣ ADC for LTE and beyond," *IEEE Journal of Solid-State Circuits*, vol. 49, no. 7, pp. 1535–1547, July 2014.

[13] B. Murmann, "A/D converter trends: Power dissipation, scaling and digitally assisted architectures," in *2008 IEEE Custom Integrated Circuits Conference*, 2008, pp. 105–112.

[14] B. E. Jonsson, "On CMOS scaling and A/D-converter performance," in *IEEE 2010 NORCHIP*, IEEE, 2010, pp. 1–4.

[15] J. H. Van den Heuvel, J.-P. M. Linnartz, P. G. Baltus, and D. Cabric, "Full MIMO spatial filtering approach for dynamic range reduction in wideband cognitive radios," *IEEE Transactions on Circuits and Systems I: Regular Papers*, vol. 59, no. 11, pp. 2761–2773, 2012.

[16] A. Agrawal and A. Natarajan, "A concurrent dual-frequency/angle-of-incidence spatio-spectral notch filter using Walsh function passive sequence mixers," in *2017 IEEE International Microwave Symposium (IMS)*, IEEE, June 2017, pp. 1606–1609.

[17] C. Andrews and A. C. Molnar, "Implications of passive mixer transparency for impedance matching and noise figure in passive mixer-first receivers," *IEEE Transactions on Circuits and Systems I: Regular Papers*, vol. 57, no. 12, pp. 3092–3103, December 2010.

[18] A. Ghaffari, E. A. Klumperink, M. C. Soer, and B. Nauta, "Tunable high-Q N-path band-pass filters: Modeling and verification," *IEEE Journal of Solid-State Circuits*, vol. 46, no. 5, pp. 998–1010, 2011.

[19] A. Ghaffari, E. A. M. Klumperink, and B. Nauta, "Tunable N-path notch filters for blocker suppression: Modeling and verification," *IEEE Journal of Solid-State Circuits*, vol. 48, no. 6, pp. 1370–1382, June 2013.

[20] M. Darvishi, R. van der Zee, E. A. Klumperink, and B. Nauta, "Widely tunable 4th order switched g-c band-pass filter based on n-path filters," *IEEE Journal of Solid-State Circuits*, vol. 47, no. 12, pp. 3105–3119, 2012.

[21] A. Mirzaei, H. Darabi, J. C. Leete, and Y. Chang, "Analysis and optimization of direct-conversion receivers with 25% duty-cycle current-driven passive mixers," *IEEE Transactions on Circuits and Syst. I: Regular Papers*, vol. 57, no. 9, pp. 2353–2366, September 2010.

[22] H. Darabi, A. Mirzaei, and M. Mikhemar, "Highly integrated and tunable RF front ends for reconfigurable multiband transceivers: A tutorial," *IEEE Transactions on Circuits and Systems I: Regular Papers*, vol. 58, no. 9, pp. 2038–2050, September 2011.

[23] E. A. Klumperink, H. J. Westerveld, and B. Nauta, "N-path filters and mixer-first receivers: A review," in *Proceedings of IEEE Custom Integrated Circuits Conference (CICC)*, IEEE, 2017, pp. 1–8.

[24] N. Reiskarimian and H. Krishnaswamy, "Design of all-passive higher-order CMOS N-path filters," in *Proceedings of the IEEE Radio Frequency Integrated Circuits Symposium (RFIC)*, IEEE, May 2015, pp. 83–86.

[25] M. Darvishi, R. van der Zee, and B. Nauta, "Design of active N-path filters," *IEEE Journal of Solid-State Circuits*, vol. 48, no. 12, pp. 2962–2976, December 2013.

[26] Y. Lien, E. Klumperink, B. Tenbroek, J. Strange, and B. Nauta, "A mixer-first receiver with enhanced selectivity by capacitive positive feedback achieving +39dBm IIP3 <3db noise figure for SAW-less LTE radio," in *Proceedings of the IEEE Radio Frequency Integrated Circuits Symposium (RFIC)*, June 2017, IEEE, pp. 280–283.

[27] Y. Lien, E. Klumperink, B. Tenbroek, J. Strange, and B. Nauta, "A high-linearity CMOS receiver achieving +44dBm IIP3 and +13dBm B1dB for SAW-less LTE radio," in *IEEE International Solid-State Circuits Conference (ISSCC)*, IEEEE, February 2017, pp. 412–413.

[28] C. K. Luo, P. S. Gudem, and J. F. Buckwalter, "A 0.2-3.6-GHz 10-dBm B1dB 29-dBm IIP3 tunable filter for transmit leakage suppression in SAW-less 3G/4G FDD receivers," *IEEE Transactions on Microwave Theory and Techniques*, vol. 63, no. 10, pp. 3514–3524, October 2015.

[29] C. K. Luo, P. S. Gudem, and J. F. Buckwalter, "A 0.4âĂŞ6-GHz 17-dBm B1dB 36-dBm IIP3 channel-selecting low-noise Amplifier for SAW-less 3G/4G FDD diversity receivers," *IEEE Transactions on Microwave Theory and Techniques*, vol. 64, no. 4, pp. 1110–1121, February 2016.

[30] H. F. Harmuth, "Applications of Walsh functions in communications," *IEEE Spectrum*, vol. 6, no. 11, pp. 82–91, November 1969.

[31] Z. Lin, P. l. Mak, and R. P. Martins, "A 0.028mm^2 11mW single-mixing blocker-tolerant receiver with double-RF N-path filtering, S11 centering, +13dBm OB-IIP3 and 1.5-to-2.9dB NF," in *IEEE International Solid-State Circuits Conference (ISSCC) Digest of Technical Papers*, IEEE, February 2015, pp. 1–3.

[32] F. Golcuk, T. Kanar, and G. Rebeiz, "A 90-100-GHz 4×4 SiGe BiCMOS polarimetric transmit/receive phased array with simultaneous receive-beams capabilities," *IEEE Transactions on Microwave Theory and Techniques*, vol. 61, no. 8, pp. 3099–3114, August 2013.

[33] M. Boers et al., "A 16TX/16RX 60GHz 802.11ad chipset with single coaxial interface and polarization diversity," in *2014 IEEE International Solid-State Circuit Conference Digest of Technical Papers (ISSCC)*, IEEE, February 2014, pp. 344–345.

[34] T. Sowlati, S. Sarkar, B. Perumana, et al., "A 60GHz 144-element phased-array transceiver with 51dbm maximum eirp and ±60 beam steering for backhaul application," in *2018 IEEE International Solid-State Circuits Conference (ISSCC)*, IEEE, 2018, pp. 66–68.

[35] S. Shahramian, M. Holyoak, A. Singh, B. J. Farahani, and Y. Baeyens, "A fully integrated scalable W-band phased-array module with integrated antennas, self-alignment and self-test," in *2018 IEEE International Solid-State Circuits Conference (ISSCC)*, IEEE, 2018, pp. 74–76.

[36] D. G. Kam, D. Liu, A. Natarajan, S. Reynolds, H.-C. Chen, and B. Floyd, "Ltcc packages with embedded phased-array antennas for 60 GHz communications," *IEEE Microwave and Wireless Components Letters*, vol. 21, no. 3, pp. 142–144, March 2011.

[37] A. Balankutty, S. Pellerano, T. Kamgaing, K. Tantwai, and Y. Palaskas, "A 12-element 60GHz CMOS phased array transmitter on LTCC package with integrated antennas," in *Solid State Circuits Conference (A-SSCC), 2011 IEEE Asian*, November, pp. 273–276.

[38] X. Gu, D. Liu, C. Baks, et al., "A compact 4-chip package with 64 embedded dual-polarization antennas for W-band phased-array transceivers," in *2014 Electronic Components and Technology Conference (ECTC)*, May 2014, pp. 1650–1655.

[39] J. M. Edwards and G. M. Rebeiz, "High-efficiency elliptical slot antennas with quartz superstrates for silicon RFICs," *IEEE Transactions on Antennas and Propagation*, vol. 60, no. 11, pp. 5010–5020, November 2012.

[40] Y.-C. Ou and G. Rebeiz, "On-chip slot-ring and high-gain horn antennas for millimeter-wave wafer-scale silicon systems," *IEEE Transactions on Microwave Theory and Techniques*, vol. 59, no. 8, pp. 1963–1972, August 2011.

[41] Y. Liu, A. Agrawal, and A. Natarajan, "Millimeter-wave IC-antenna cointegration for integrated transmitters and receivers," *IEEE Antennas and Wireless Propagation Letters*, vol. 15, pp. 1848–1852, 2016.

[42] S. Targonski, R. Waterhouse, and D. Pozar, "Design of wide-band aperture-stacked patch microstrip antennas," *IEEE Transactions on Antennas and Propagation*, vol. 46, no. 9, pp. 1245–1251, September 1998.

[43] D. C. Thompson et al., "Characterization of liquid crystal polymer (LCP) material and transmission lines on LCP from 30 to 110 GHz," *IEEE Transactions on Microwave Theory and Techniques*, pp. 1343–1352, April 2004.

[44] Y. Liu and A. Natarajan, "60 GHz concurrent dual-polarization RX front-end in SiGe with antenna-IC co-integration," in *2017 IEEE Bipolar Components Technology Meeting (BCTM)*, IEEE, October 2017, pp. 42–45.

7 Millimeter-Wave Radar SoC Integration in CMOS

Piet Wambacq, Davide Guermandi, André Bourdoux, and Jan Craninckx

Abstract

The interest in compact and low-power automotive radar sensors toward fully autonomous vehicles pushes research in 79 GHz radar. Further, many indoor sensing applications are emerging, going from people detection over vital signs monitoring to gesture recognition. This chapter discusses the integration of a mm-wave radar transceiver and then treats more in detail the design of a 28 nm complementary metal-oxide semiconductor (CMOS) fully integrated 79 GHz multiple-input and multiple output (MIMO) radar system on a chip (SoC) using continuous-wave phase modulation.

7.1 Introduction

A radar (RAdioDetection And Ranging) is a position and velocity sensor that uses electromagnetic waves. A signal is transmitted by the transmit part of the radar system, then it "hits" an object, which reflects a signal that is detected by the receiver part of the radar system. Developed originally for military applications in the first half of the previous century, it found its way to many other application fields, from which the automotive is a very important one. Autonomous driving vehicles that will be on the road in the next years, if not already today, can only be enabled by dozens of heterogeneous sensors to make them aware of their surroundings [1]. A key role in the sensing will be taken by mm-wave radars thanks to their inherent capability of measuring distance and speed at the same time and robustness to light and environmental conditions. For these reasons, mm-wave radars become part of driving assistance packages offered in high- and mid-end cars. Different radar types are needed for long range (hundreds of meters straight ahead of the vehicle), medium range (around 75 m from the four vehicle corners), and short range (creating a detection zone around the vehicle). Medium- and short-range radar will require phased array or MIMO capabilities to detect not only distance and velocity of the objects but also their angle with respect to the vehicle [2,3]. At the same time, the growing demand of positioning technologies for smart home and, in general, environment- and person-aware systems such as surveillance, indoor location, vital sign monitoring, and gesture recognition opens new market opportunities for compact and low-cost radar sensors like [4] and [5].

Operating a radar at mm-wave frequencies has several advantages, related to the physical quantities that are to be detected: range, velocity, and angle:

- As shown in the next sections, the range resolution, which measures how objects can be distinguished from each other, is inversely proportional to the modulation bandwidth (BW). A large bandwidth can be realized more easily around a carrier frequency at mm-wave frequencies than at lower frequencies, since it results in a lower fractional bandwidth.
- Velocity detection is based on the Doppler effect: when an object moves, the frequency of the reflected signal is different from the frequency of the transmitted signal. This frequency difference is called Doppler shift and it is historically used in radars, as described in [6]; it is positive if the target moves toward the radar and negative if it moves away. As this Doppler shift is proportional to the carrier frequency and the target speed component in the direction of the radar, operation at higher frequencies gives the potential of a higher resolution to determine the velocity of moving objects. This, however, means that for a given velocity resolution, the radar acquisition time is smaller.
- The angular resolution improves when the wavelength is small compared to the antenna size. High-resolution antenna arrays can be realized in a more compact way at higher frequencies.

These advantages, combined with the ability of modern silicon technologies to work at mm-wave frequencies have led to compact radar modules that are based on silicon chips. In this chapter, we focus on millimeter-wave radar suitable for both for automotive short-range radar (SRR) and indoor sensing. For SRR automotive applications, the frequency band from 77 to 81 GHz (in short, 79 GHz) has been allocated in Europe and Japan and will probably also be allocated in USA, Canada, and most other countries. These SRR systems provide a detection range of a few tens of meters all around the car, using many sensors on the car body. The 79 GHz SRR has specific requirements: first, it has a wide bandwidth (up to 4 GHz), which in turn makes it harder to implement; second, since many sensors will be installed around the body of the car, low-cost implementation is crucial; and third, because of the very large expected quantities and the amount of signal processing, single-chip implementation in bulk CMOS processes is very attractive, as opposed to SiGe for the RF and CMOS for the digital parts.

The choice of the waveform is critical since this drives both the radar performance and its implementation complexity. In pulse-based radar systems, energy is transmitted over short pulses. This requires high peak powers at the transmit side, which are not compatible with CMOS integration. Therefore, this approach is not popular in automotive system. Implementations in CMOS, however, have been reported at mm-wave by [7], for very short-range applications.

On the other hand, transmitting a continuous wave increases the output energy for a given supply. If combined with constant-envelope modulation (phase or frequency modulation), the transmitter efficiency can be maximized while the receiver front-end is relatively simple. In addition, a continuous-wave modulation allows, with intensive digital computation, to increase processing gain and extract Doppler information. This requires operations such as correlations, accumulations, and fast Fourier transforms (FFTs) which in heavily downscaled CMOS can be performed with a low-power consumption.

Both frequency-modulated continuous wave (FMCW) and phase-modulated continuous wave (PMCW) have been successfully demonstrated both in CMOS and SiGe [4,8–12]. Most of them do not integrate the mm-wave parts together with the necessary digital signal processing. In this chapter, after a high-level discussion of FMCW and PMCW and a comparison between the two waveforms, a fully integrated 2×2 MIMO PMCW radar SoC that combines mm-wave CMOS front-ends and frequency synthesis with high-speed application-specific digital processing is discussed. The digital part is discussed more in detail in [13] and [14]. With the 77 and 79 GHz bands exclusively assigned worldwide for automotive radars, the presented radar concept can be used (by redesign of the RX and TX front-Ends for the desired frequency band) for high-resolution, single-chip, radar-based indoor/outdoor sensing, as part of a more comprehensive Internet of Things (IoT) scenario.

The outline of this book chapter is as follows. First FMCW and PMCW waveforms are introduced in Sections 7.2 and 7.3, respectively, with a comparison in Section 7.4; then the derivation of the radar SoC specification is given in Section 7.5, with attention to MIMO PMCW radar, analyzed in Section 7.6. Circuit insights are given in Section 7.7. Finally, experimental results are shown in Section 7.8, followed by conclusions.

7.2 Frequency-Modulated Continuous-Wave Radar

FMCW, described in [15], is the most widely used waveform in automotive radar. Worldwide, a large know-how has been built on FMCW radar implementations. The most widely used modulation is with a sawtooth wave (see Figure 7.1): here the frequency increases linearly over time. When it has reached its maximum value, it returns quickly to its minimum value and the increase restarts. Due to its analogy with the audio frequency of a bird chirp, it is called a chirp waveform. The useful part of the signal is the so-called up-chirp, where frequency increases over time at a constant rate. The down-chirp is not used. In an FMCW radar, the received signal that has been reflected by an object at distance R is a delayed version of the transmitted chirp signal. When mixing the received signal with the signal that is actually transmitted, a beat frequency Δf occurs at the mixer output. From Figure 7.1, we find

$$\Delta f = BW \cdot \frac{\Delta t}{T_{chirp}} \tag{7.1}$$

in which T_{chirp} is the duration of the up-chirp, BW is the bandwidth that is spanned by the chirp, and Δt is the time that is needed for a wave to travel from the transmitter to the object and then back to the receive section of the radar. In this time, a distance of $2R$ is bridged at the speed of light c such that

$$\Delta t = 2 \cdot \frac{R}{c}. \tag{7.2}$$

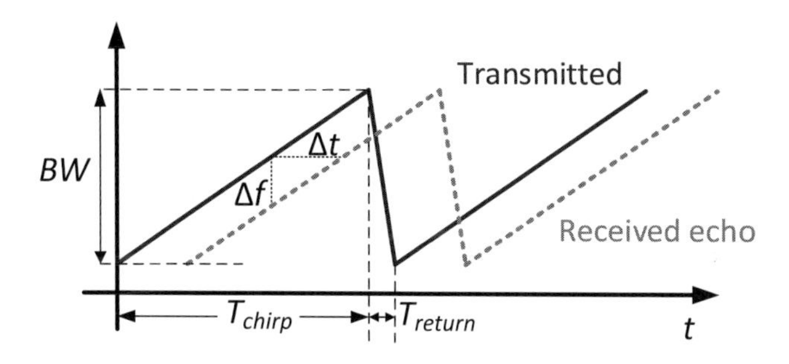

Figure 7.1 Chirp waveform.

In this way, the range can be derived from the value of Δf.

$$R = \Delta f \cdot \frac{T_{chirp}}{BW} \cdot \frac{c}{2} \tag{7.3}$$

Using then $1/T_{chirp}$ for Δf in (7.3) yields the range resolution R_{res}:

$$R_{res} = \frac{c}{2 \cdot BW} \tag{7.4}$$

For example, a 10 cm range resolution requires a 1.5 GHz bandwidth BW.

For a target velocity component in the direction to the radar v and the starting carrier frequency f_C of the chip, the Doppler shift f_D is computed as

$$f_D = 2 \cdot f_C \frac{v}{c - v} \approx 2 \cdot f_C \frac{v}{c} = 2 \cdot \frac{v}{\lambda} \tag{7.5}$$

resulting in a received frequency slightly different from the transmitted one for target speed much lower than the speed of light.

It is important to observe that both the range shift Δf and the Doppler shift f_D are detected in the frequency domain and should be discriminated. Two possible methods are described in [16]. For "slow" chirps, both frequency shifts Δf and f_D are found in the kHz range for common automotive distances and speeds: the separation can be achieved varying the slope between two successive chirps; this results in two different beat frequencies from which Δf and f_D can be computed. The second solution is to use "fast" chirps (chirps with a higher slope): in this case, the frequency shift due to the target Δf is found at much higher frequency with respect to the Doppler shift f_D, and the two can be separated with subsequent FFTs.

Using fast chirps, the value of Δf is determined in the digital part of the receiver via an FFT. Within one chirp period T_{chirp}, a set of M time-domain samples is taken for this FFT. The frequency resolution of this FFT corresponds to the inverse of the chirp duration. Repeating the chirp N times yields an $N \times M$ matrix of time-domain samples; then an FFT of N points, taken over one row of this matrix, reveals the Doppler information from which the speed of the object can be determined. In summary, determining range and velocity of an object requires two FFT operations. The FMCW waveform is used in several commercial car radar systems. Chirp durations range from 10 to 100 μs while bandwidths are from several hundreds of megahertz to a few gigahertz.

An important problem in continuous-wave radar is the leakage of the transmitted signal directly into the receive part of the radar, as discussed in [17]. Since both TX and RX operate continuously in the same band, power that leaks from the transmitter, for example due to limited antenna isolation, is received by the RX. At the output of the down-conversion mixer in an FMCW receiver, this leakage, referred to as TX-to-RX spillover or self-interference, gives rise to a DC signal, or at very low frequency due to the short distance between TX and RX antennas. As the spillover can be several orders of magnitude stronger in power than the signal received from targeted objects, care must be taken in the design of the low-noise amplifier (LNA) and the down-conversion mixer to avoid saturation by the spillover signal. The down-converted spillover can be suppressed after the mixer with a high-pass filter, chopping techniques, etc., which involves low-frequency analog signal processing. These techniques are also useful to reduce flicker noise, which is always a point of attention in CMOS implementations.

7.3 Phase-Modulated Continuous-Wave Radar

PMCW radars represent an interesting alternative to FMCW radars for high-resolution short- and medium-range applications. Differently from FMCW, they do not need a linear frequency ramp to determine time of flight, which is instead measured by parallel correlations as described in [18]. As a consequence, the architecture is intrinsically digitally intensive and better tailored for advanced CMOS technologies such as 28 nm. The block diagram and the operating principle of a PMCW radar are shown in Figure 7.2. A binary sequence with good periodic autocorrelation property and length L_C, such as an m-sequence [19], is used to binary phase modulate a (mm-wave) local oscillator (LO). The modulated signal is then amplified and radiated over the air. Reflected signals are received and down-converted by the same (quadrature) LO to result in a superposition of delayed copies of the same sequence with delays equal to the time of flight. After digitization, parallel correlators fed with delayed copies of the transmitted sequence (delays $\tau = kT_C$ with $k = [1:L_C]$) seek nonzero correlation, meaning targets corresponding to that specific time of flight, hence distance. The range resolution R_{res} is given by the distance traveled by the electromagnetic wave at the speed of light c in one symbol duration divided by 2 as

$$R_{res} = \frac{c}{2 \cdot F_C} \tag{7.6}$$

in which F_C is the bit (or symbol) rate. $F_C = 2\,\text{Gbps}$ corresponds to a resolution of 7.5 cm. Clearly, (7.6) resembles to the resolution equation of FMCW (7.4).

Outputs of the correlators are called "range gates" or "range bins"; they represent the quantized distance from the radar antennas with resolution R_{res}, e.g., a target at 1 m distance is found at range gate 13 = round(1/0.075). Just as with FMCW, the Doppler shift can be used to measure the speed of the target since it will appear as a frequency offset at the RX baseband. For a target velocity component in the direction to

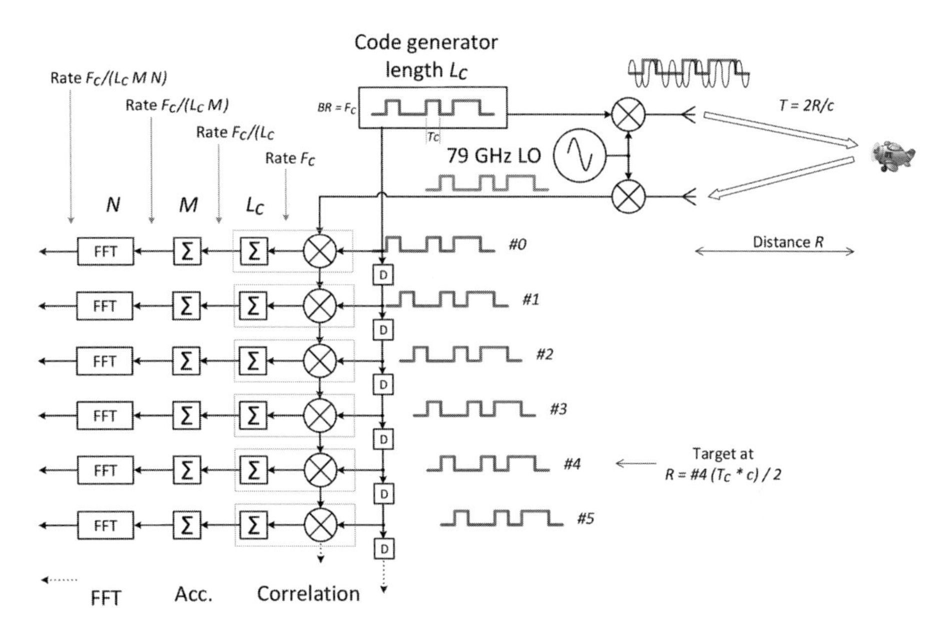

Figure 7.2 PMCW radar block diagram and operating principle. Correlator # 4 detects the presence of the target with a nonzero output [14]. (©2016 IEEE. Reprinted, with permission, from *2016 IEEE Asian Solid-State Circuits Conference [A-SSCC]*.)

the radar v and to begin f_C the carrier frequency, the Doppler shift f_D is computed again with

$$f_D = 2 \cdot f_C \frac{v}{c - v} \approx 2 \cdot f_C \frac{v}{c} = 2 \cdot \frac{v}{\lambda}. \tag{7.7}$$

To increase the signal-to-noise ratio (SNR), correlation results are coherently accumulated M times. A further FFT of N accumulated complex values (I and Q) is used to compute Doppler shifts and further increase SNR. The length of the observation time, being equal to $L_C \cdot M \cdot N$, determines the Doppler resolution. The dwell time is computed as

$$T_{DW} = \frac{L_C \cdot M \cdot N}{F_C}. \tag{7.8}$$

Assuming the noise to be uncorrelated between samples, the overall processing gain of correlation, accumulation, and FFT is

$$G_P = 10 \log_{10}(L_C \cdot M \cdot N). \tag{7.9}$$

TX-to-RX spillover: just as with FMCW, the TX-to-RX spillover requires careful attention and, if possible, mitigation as proposed in [20]. An antenna isolation greater than 40–45 dB cannot be guaranteed within the accuracy of the electromagnetic (EM) simulation and while keeping the antenna distance within the limit of a compact module. The TX-to-RX spillover acts no differently than a very large target at a short distance; the system is inherently able to discriminate the spillover from the targets as long as it has enough dynamic range to handle it. However, the TX noise also leaks at the RX input and may affect the noise floor if not low enough.

7.4 Comparison between FMCW and PMCW

It is clear that the whole automotive radar industry clearly favors FMCW because of existing know-how, intellectual property (IP), and implementations in long-range radars at 76.5 GHz and short-range radars around 79 GHz. PMCW is not widely used for automotive applications. It has, however, a lot of advantages that could not be exploited until recently. This is changing with the advent of fast low-power analog-to-digital converters (ADCs). The novelty of PMCW could be turned into an advantage since there are less IP and patents blocking access to the market.

7.4.1 Sensitivity to Phase Noise and Flicker Noise

Simulations in demanding scenarios with phase noise show that, when the integrated phase noise is not too high (better than −18 dBc), both systems are fairly equivalent. At higher phase noise, PMCW suffers slightly from range sidelobe degradation whereas FMCW suffers slightly from a worse range behavior at long range. Both systems show similar low sensitivity to flicker noise because their intermediate frequency (IF) bandwidth is large ($\gg$1 MHz). In FMCW, flicker noise my cause false alarms in the first range gates.

7.4.2 TX Orthogonality for MIMO Radar

The use of binary symbols in PMCW brings along a significant advantage for MIMO radars. Indeed, MIMO radars require perfectly orthogonal waveforms on the different TX antennas if they are transmitting simultaneously, which is desirable for fast illumination. As is well known from communication, this orthogonality is achievable with binary codes. On the contrary, orthogonality is more difficult to achieve with FMCW.

7.4.3 Interference Robustness

Interference robustness between different radars is easier to achieve with PMCW since different sequences can be attributed to different radars and/or the radar can pseudorandomly change sequences (code hopping). This is harder to achieve with FMCW systems since they have to exploit up- or down-slopes together with different slope durations (which changes the waveform properties).

7.4.4 IF Bandwidth and ADC

The most often mentioned advantage of FMCW systems is the low IF bandwidth. Indeed, the mixing at RX with the FM slope achieves a kind of time (or range) to frequency conversion that heavily reduces the IF bandwidth. Although correct, this statement requires two comments:

1. In slow-sloped FMCW systems, the IF bandwidth is small ($\leq$ 1 MHz) and ambiguity results between range-induced frequency shift and motion-induced Doppler shift. Disambiguation and deghosting are daunting when more than two or three targets are present. This is avoided in advanced fast-sloped FMCW systems in which the frequency shift due to Doppler at the highest velocity is much lower than the frequency shift due to range, even at the shortest range. This requires much higher IF bandwidths (typically 100 MHz).
2. The analog processing gain of FMCW results in a higher dynamic range of the signal before the ADC, whereas it has been shown in [21] that a low resolution is sufficient for PMCW radars.

7.4.5 TX-to-RX Spillover

As previously discussed, the TX-to-RX spillover affects both PMCW and FMCW systems. It manifests itself as a strong target at range 0 and it can potentially saturate the receiver. For FMCW, however, the spillover can be removed right after the downconversion mixer in the receiver with techniques similar to DC offset compensation. For PMCW, this spillover removal is more complicated because of its wideband nature.

7.4.6 Waveform Generation and Linearity

The waveform generation of PMCW is very simple: no upconversion is needed, and the biphase modulation implementation is straightforward. The FMCW waveform does not require upconversion if the voltage-controlled oscillator (VCO) is swept over the RF bandwidth. This requires, however, a delicate linearization loop to avoid degradation of the range resolution. An alternative is to upconvert a very linear slope generated at low frequencies.

7.4.7 Other Aspects

The use of biphase modulation in PMCW makes it possible to use the radar hardware for communications. One way to do this is to embed low-rate communication into the radar waveform itself (as in GPS signals). The FMCW system could do this also by using up- and down-slopes to encode bits (0 and 1). The PMCW radar could stop the ranging for a short period and switch to high-rate (up to 2 Gsymbols/s [Gsps]) communication. This is clearly a plus point for PMCW.

7.5 Link Budget for a PMCW Radar

Main specifications and design parameters of the short-range radar described in this chapter are summarized in Table 7.1.

Table 7.1 Summary of radar system target specification.

Parameter	Value	Unit
Maximum range	30	m
Range resolution	7.5	cm
Minimum detectable target	−8	dBm^2
Carrier frequency	79	GHz
Chip rate	1.975	Gsps
MIMO	2×2[a]	
	4×4[b]	
TX output power	10	dBm
RX noise figure	10	dB

[a] Single chip.
[b] Combination of two chips.

7.5.1 Link Budget for Single-Antenna TX and RX and MIMO Systems

For a single antenna system (SISO), the radar receiver SNR for a given target at distance R is computed with the radar equation as

$$SNR_{SISO} = \frac{P_{TX} \cdot G_{TX} \cdot G_{RX} \cdot G_P \cdot \lambda^2 \cdot \sigma}{(4\pi)^3 \cdot R^4 \cdot k_B \cdot T \cdot B_N \cdot NF_{RX} \cdot L_{TX} \cdot L_{RX}} \tag{7.10}$$

in which P_{TX} is the transmitted power, G_{TX} and G_{RX} are the antenna gain at the TX and RX side respectively, G_P is the processing gain defined in (7.9), and $k_B \cdot T \cdot B_N \cdot NF_{RX}$ is the receiver input-referred noise power over its noise bandwidth B_N. Further, L_{TX}, L_{RX} are the implementation losses on TX and RX side, for example due to the loss in the connection to and from the antenna. Finally, σ is the radar cross section (RCS). This quantity, expressed in decibel square meters, indicates how detectable an object is with radar. A large RCS indicates that an object can be detected more easily. The RCS depends on the size, the shape, and the material of the target. The radar equation (7.10) resembles the link budget/SNR of a communication radio link with the notable difference that the signal strength, and hence the SNR, decreases with the fourth power of the distance, rather than with the second power.

A binary phase shift keying (BPSK) modulation with bit rate F_C results in a power spectral density shaped as $\text{sinc}^2(\pi f/F_C)$ centered around the carrier frequency. From (7.6), we know that higher rates correspond to finer resolutions. With $F_C \approx 2$ Gbps, the main lobe occupies 4 GHz centered around 79 GHz with nulls at ± 2 GHz, the maximum bandwidth allowed around 79 GHz by European Telecommunications Standards Institute (ETSI) ETSI regulations as defined in [22]. To maximize SNR, an RX baseband bandwidth of $F_C/2 \approx 1$ GHz is chosen, resulting in a noise bandwidth B_N of $1\,\text{GHz} \cdot 2(\pi/2) \approx 3.14$ GHz for a single-pole baseband filter. After downconversion and filtering, the received signal is the superposition of delayed versions of the sequence at bit rate F_C. With a baseband bandwidth over 1 GHz, even small oversampling factors as 3 or 4 would result in a very high sampling rate for the ADC. However, since the sampling clock is perfectly synchronous with the received sequence, being generated from the same PLL, the information in the received signal can be recovered by sampling with only one sample per bit at F_C, similarly to serial link with a perfect clock recovery, and oversampling is not needed. As an alternative to immediate digitization, parallel analog correlations between the received signal and the transmitter sequence can be considered, followed by lower-speed and higher-resolution ADCs. This solution is, however, impractical for a large number of parallel correlations (see Figure 7.2) that are more efficiently computed in the digital domain.

With an antenna gain of $G_{TX} = G_{RX} \approx 3$ dBi, a TX output power P_{TX} around 10 dBm like in [23], an estimated NF around 10 dB that also includes TX and RX losses (L_{TX} and L_{RX}), the SNR_{SISO} from (7.10) is as low as -64 dB for a target radar cross section of -8 dBm^2 (corresponding to a person according to [24]) 30 m away from the radar before the processing gain G_P. A 2×2 MIMO can increase the SNR to -58 dB and 4×4 MIMO to -52 dB according to (7.13) that will be introduce in Section 7.6, meaning at least 62 dB of processing gain is needed to boost the target SNR > 10 dB for detection. This can be achieved with $L_C = 511$, $M = 232$, and $N = 64$ ($G_p \approx 68.8$ dB from (7.9)). From (7.8), the dwell time is $T_{DW} \approx 3.84$ ms. With 64 FFT points (N), the Doppler resolution is $D_{res} = 1/3.84\ \text{ms} \approx 260$ Hz with a maximum unambiguous Doppler shift of $N/2 \cdot D_{res} \approx 8.33$ kHz. From (7.5), a velocity resolution of 0.49 m/s $\approx$ 1.8 km/h and an unambiguous velocity of 15.8 m/s $\approx$ 57 km/h are found. Considering the combination of receiver chain front-end, variable gain amplifier (VGA), and ADC, it is possible to outline a gain partitioning. As previously discussed, with a transmitted power of 10 dBm at the antenna, a spillover as high as -30 dBm is found at the RX input. This is most likely the highest receiver input signal even if large targets are found at a short distance. With a full-scale level corresponding to 0 dBm at the ADC input, corresponding to a voltage amplitude of 310 mV if the signal were sinusoidal, a gain of 30 dB is required in the RX chain; assuming 2 dB attenuation from the antenna to the die, the gain can be separated as 20 dB in the front-end and 12 dB in the baseband. With the effective antenna isolation not precisely known, variable gain must be added in the RX chain to ensure an optimal ADC loading under different circumstances. If the TX-to-RX spillover is higher than expected, the gain should be reduced in the LNA to avoid saturation in the mixer (at the price of a noise figure degradation); if it is lower, extra gain can be added in the baseband for a better ADC loading. With the RX NF = 10 dB and RX Gain = 30 dB, the thermal noise at the ADC input can be written as

$$N_{ADC,in} = K_B \cdot T + NF + G_{RX} + 10\log_{10}(B_N) \approx -39\ \text{dBm}. \tag{7.11}$$

With the ADC full scale (FS) of 0 dBm and a 7-bit resolution, the quantization noise is

$$N_{ADC,q} \approx FS - (6.02 \cdot N_b + 1.76) \approx -43.9\ \text{dBm}. \tag{7.12}$$

The thermal noise is just above the quantization noise and high enough to toggle a few least significant bits (LSBs) in the ADC and randomize it so to benefit from processing gain. This calculation is done in the presence of only the maximum TX-to-RX spillover and the minimum detectable target, which is an extreme case; in a more realistic scenario, more targets will be present, thus increasing the variability of the signal at the ADC input, leading to a more pronounced randomization of the quantization noise. More bits in the ADC would improve the overall SNR at the cost of adding complexity to a 2 Gsps ADC and increase the power consumption of the digital data path because of the wider data words.

7.5.2 LO Phase Noise

In a communication system, TX and RX unavoidably use different oscillators and PLLs locked on different crystals. Therefore, the uncorrelated phase noise of both the noise of TX and RX oscillators, will add up, thus, will add, thus decreasing SNR. In a radar, however, TX and RX use the same local oscillator, meaning that receiver and transmitter LO phase noise are actually correlated. For the sake of simplicity, the phase noise is assumed to be limited by the bandwidth of the frequency synthesizer at 1 MHz offset from the carrier. This means that any reflection received with a two-way propagation delay much shorter than 1 μs will see the same phase fluctuation in the transmitter and receiver, being then insensitive to phase noise. For a target at 30 m the time of flight is ~0.2 μs, small enough to neglect the effect of the phase noise.

7.6 MIMO Techniques for PMCW Radars

Similarly to communication, multiple receivers and transmitters can be used to improve the SNR as discussed in [25]. When N_{TX} and N_{RX} are the numbers of transmitters and receivers respectively, SNR is increased as

$$SNR_{MIMO} = SNR_{SISO} \cdot N_{TX} \cdot N_{RX} \tag{7.13}$$

if the observation time (dwell time T_{DW}) is kept constant.

To realize an $N_{TX} \times N_{RX}$ MIMO virtual array, each RX should discriminate the signal transmitted from each TX separately without any interference from others, creating $N_{TX} \cdot N_{RX}$ independent channels [25]. The transmitted sequences from the different TX should therefore have zero cross-correlation or, using signal processing terminology, should be orthogonal. This section will review two different solutions to make the TX antenna signals orthogonal by code design [26]. Techniques such as frequency division or time division are applicable as in communication; orthogonal coding, however, allows all the TX to operate simultaneously on the full available bandwidth.

Notice that the benefit of multiple transmitters in MIMO is not squared as in the beamforming case, as reported in [27]. However, it must be considered that MIMO gets the full scene in one illumination whereas beamforming gets information only in the beamformed direction. Therefore, several beams must be generated sequentially, say N_{TX} beams to cover the full scene with beamforming. By coherently combining MIMO signals during the same time it takes to generate the N_{TX} beams, MIMO gets an extra coherent gain of $\approx 10\log_{10}(N_{TX})$, and both techniques achieve the same overall gain over the full scene.

7.6.1 TX Orthogonality by Sequence Engineering: Different TX Use Different Sequences

In order to recover reflections from each TX separately, each RX should correlate in N_{TX} parallel correlators with the N_{TX} orthogonal (zero cross-correlation) sequences used in the TX. Nonzero cross-correlation between the N_{TX} different sequences would result in range sidelobes, either in the form of an increased floor or in high correlation values ("ghost" targets) for distances at which no target is found. As reported in [26], however, known sequence families do not have zero cross-correlation and relying on sequence cross-correlation will lead to poor radar performance, with isolation between different channels in the MIMO virtual array in the order of only 30 dB.

7.6.2 TX Orthogonality with Outer Code: All TX Use the Same Sequence

An alternative approach is to use the same sequence for all TX and to use an outer code to make the TX orthogonal to each other. The outer code needs to have zero cross-correlation for zero delay, like for example Walsh–Hadamard (or simply Hadamard) codes [28]. The length of the outer code should match the number of TX N_{TX}. The modulation scheme of four TX using an outer coding is illustrated in Figure 7.3 together with the Hadamard matrix of order 4. For each TX, the sequence "S" is repeated M times and the block is repeated N_{TX} times with a possible sign inversion as indicated in the corresponding rows of the Hadamard matrix. This produces an aggregate packet of $N_{TX} \cdot M \cdot L_C$ samples. At the receive side, as depicted in Figure 7.4 for a single RX, $N_{TX} \cdot M \cdot L_C$ symbols are correlated with the original sequence "S" to produce $N_{TX} \cdot M$ samples. The M correlations relative to each of the four columns of the Hadamard matrix are accumulated into N_{TX} separate accumulators and eventually combined (added/subtracted) according to the Hadamard matrix. The final result is N_{TX} samples relative to the N_{TX} TX, computed in each RX.

7.6.3 Comparison of the Two Approaches and Implementation

Implementationwise, the MIMO with outer code is simpler because only one correlator is needed per RX and the outer code processing is small size and performed at low rate, while the MIMO with different sequences per TX requires N_{TX} full-length correlators in each RX. In high Doppler, however, the MIMO with outer code has a factor N_{TX} penalty with respect to the case of different sequences, with the dwell time increased by the factor N_{TX}. The radar system described in this chapter is designed to work either with independent sequences for each TX/RX pair as independent radars, or with outer code up to 4×4 MIMO. Only one correlator bank is placed in each receiver, meaning that MIMO operation can only be achieved by using outer codes and not by orthogonal coding.

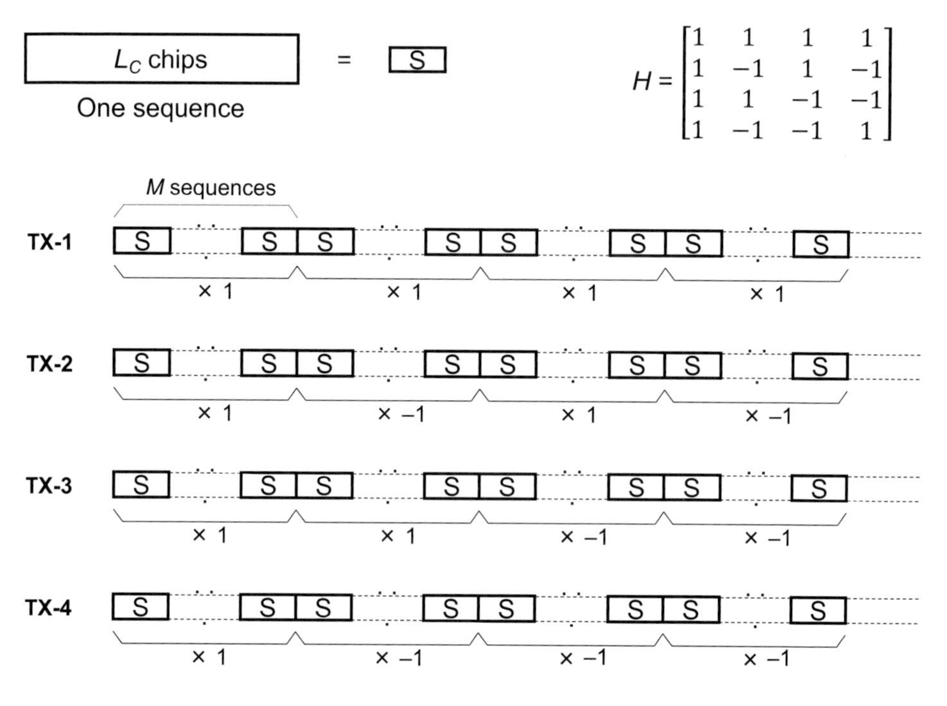

Figure 7.3 TX MIMO signal generation with an outer code (same sequence "S" on all TX antennas but with an outer code) for four TX antennas [13]. (©2017 IEEE. Reprinted, with permission, from *2017 IEEE Journal of Solid-State Circuits.*)

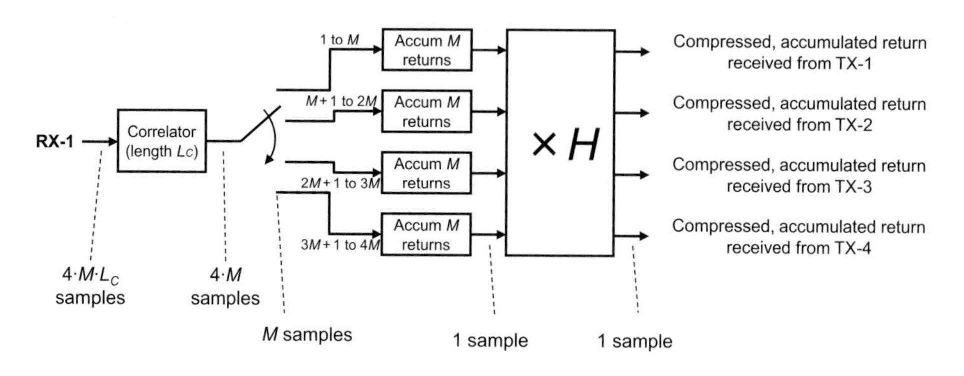

Figure 7.4 RX antenna MIMO processing for one range gate.

7.7 Analog and Millimeter-Wave Circuits

An overall block diagram of the SoC is illustrated in Figure 7.5. Two identical antenna paths share the same frequency synthesis PLL used both for LO and clock generation. In this section, the fundamental analog building blocks are described and analyzed.

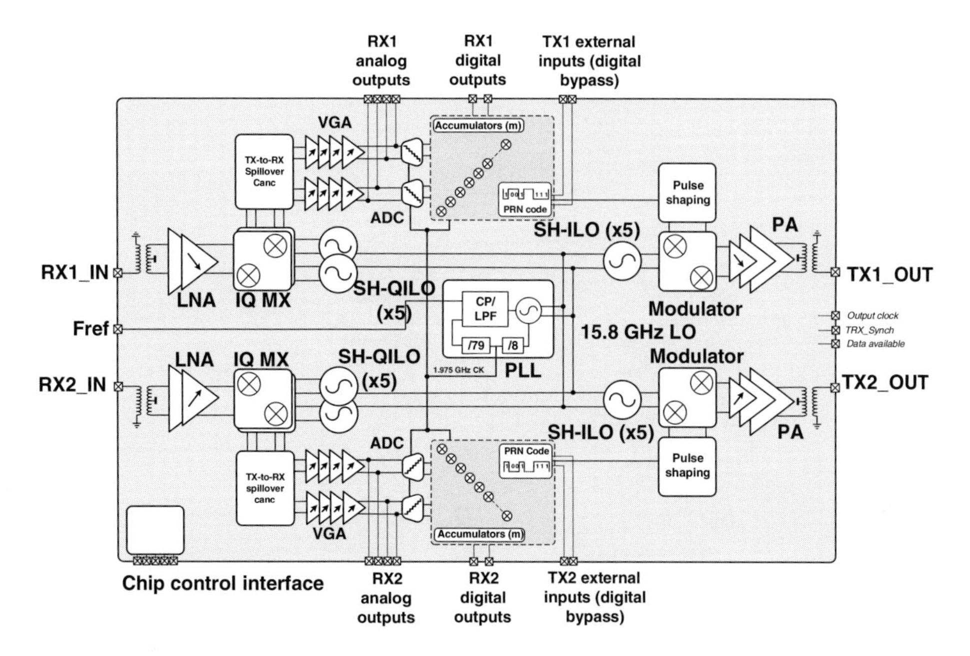

Figure 7.5 IC block diagram [13]. (©2017 IEEE. Reprinted, with permission, from 2017 *IEEE Journal of Solid-State Circuits.*)

7.7.1 Frequency Generation

Subharmonic Injection-Locked Oscillators

As described in Section 7.5.2, the TX and RX should share the same local oscillator to mitigate the impact of phase noise on the received SNR. A single frequency synthesizer is therefore required, and the local oscillator must be distributed to the two TX and the two RX front-ends.

Similarly to [20] and [29], the mm-wave carrier is not generated by a fundamental oscillator but by frequency multipliers starting from a lower-frequency VCO. This has the benefit to reduce the power consumption of the LO distribution and to avoid pulling between the TX outputs and the VCO. Moreover, at lower oscillation frequency, a higher tuning range and often a better quality factor can be achieved due to the lower impact of parasitics on the overall tank capacitance.

Subharmonic injection-locked oscillators (SH-ILO) are used as $\times$ 5 frequency multipliers because of their capability to operate at mm-wave without degrading the phase noise, at least inside their locking range. The schematic of the subharmonic quadrature injection-locked oscillators (SHQ-ILO) is shown in Figure 7.6. It consists of two SH-ILOs injected by quadrature signals at one-fifth of the desired frequency, which in this case is 15.8 GHz. As the quadrature is preserved when multiplying with an odd frequency number; the resulting fifth harmonic at 79 GHz is also available in quadrature format. An N-type metal-oxide-semiconductor (NMOS)-only ring oscillator between the outputs improves the coupling and the quadrature accuracy. The injected signal is amplified with CMOS inverters rather than tuned amplifiers as in [29]. In this 28 nm

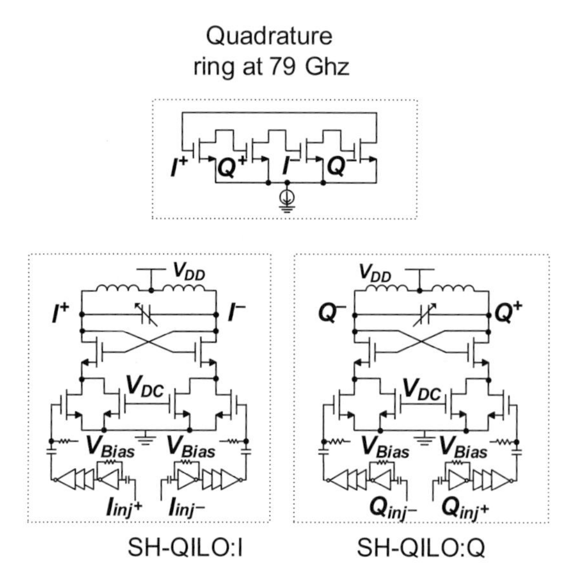

Figure 7.6 Sub-harmonic quadrature injection-locked oscillator schematic [20]. (©2015 IEEE. Reprinted, with permission, from *2015 IEEE International Solid-State Circuits Conference (ISSCC)* Digest of Technical Papers.)

CMOS technology, the use of CMOS inverters at 15.8 GHz is possible. The quasisquare wave generated by the inverter chain enhances the harmonic signal level, which results in a stronger fifth harmonic signal at the injection node, improving the locking range as shown in [30]. To center the locking range around the 15.8 GHz input frequency, a binary weighted 4-bit capacitor bank is used to adjust this locking range.

PLL and VCO

The SH-ILOs described in the preceding section are locked on a 15.8 GHz LO (79 GHz/5) generated by an on-chip integer-N PLL, illustrated in Figure 7.7. The same PLL also generates the clock for the ADC and the digital core at $F_C = 15.8/8 = 1.975$ GHz. The divider-by-eight in the feedback loop, realized as a cascade of three CML dividers, guarantees that the clock phase is locked on the 25 MHz reference input, regardless of the divider initial state.

A class-B LC VCO with tail inductor filtering, shown in Figure 7.8, is chosen for $1/f^3$ phase noise minimization as proposed in [31], together with a very large tail current device (3,072/0.8 µm). In the bias part of the VCO, a current mirror is used with noise filtering in between the diode-connected transistor and the current source of the VCO. To reduce the voltage drop on the filter resistor due to gate leakage current, thick-oxide devices are used in the current mirror. A 6-bit capacitor bank is used for coarse tuning while an NMOS varactor is used by the PLL loop. The capacitor bank is realized with switched metal-oxide-metal (MOM) capacitors that are binary weighted with an LSB of 5.7 fF. This capacitance resonates with a two-turn inductor of 350 pH.

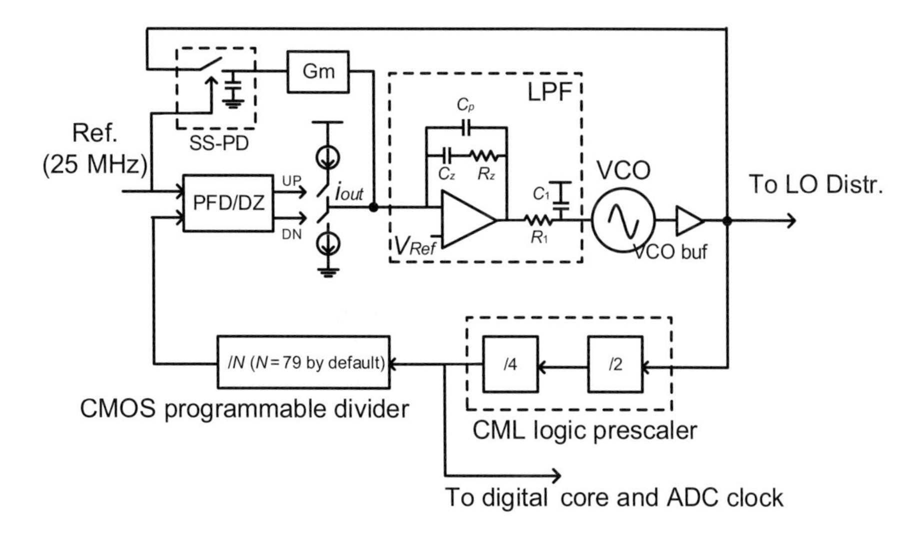

Figure 7.7 Phase-locked loop block diagram [13]. (©2017 IEEE. Reprinted, with permission, from 2017 *IEEE Journal of Solid-State Circuits.*)

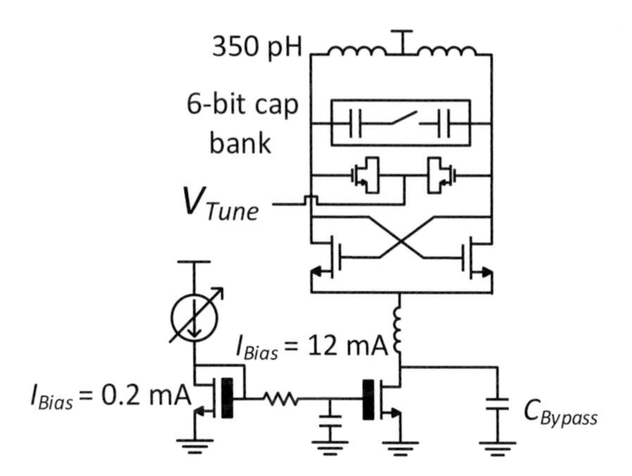

Figure 7.8 15.8 GHz VCO schematic [13]. (©2017 IEEE. Reprinted, with permission, from 2017 *IEEE Journal of Solid-State Circuits.*)

To reduce the PLL integrated phase noise, a subsampling loop, introduced by [32], can be enabled to bypass the divider and the charge pump. This is particularly important if two chips are used together to realize a MIMO 4 × 4 system; in that case, in fact, only the phase noise coming from the common reference will be correlated and the phase noise cannot be neglected even for low offset frequencies and close targets, as previously discussed.

LO Distribution

The 15.8 GHz PLL output is distributed to the two transmitters and two receivers (see Figure 7.9) with differential signaling. Tuned LC buffers are used where the signal is split in two, first from the PLL to the two antenna paths and then again inside the antenna

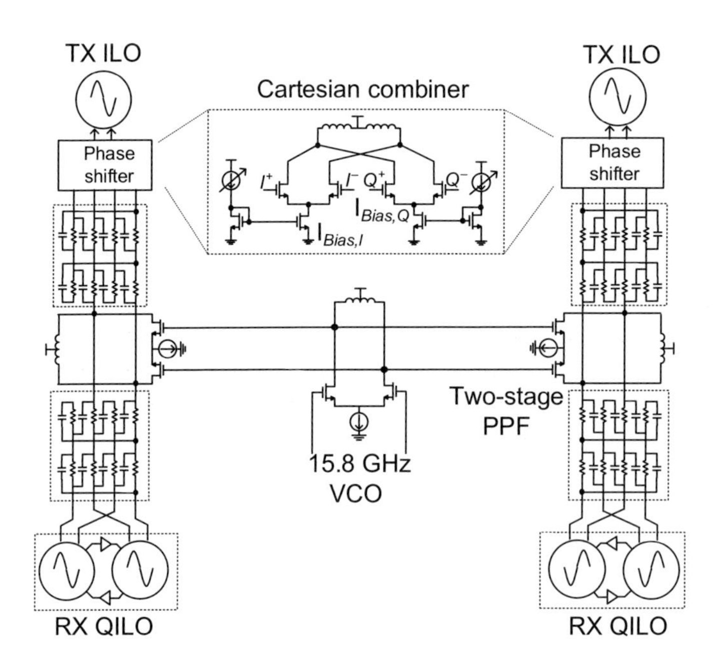

Figure 7.9 LO distribution schematic [13]. (©2017 IEEE. Reprinted, with permission, from 2017 *IEEE Journal of Solid-State Circuits*.)

path between TX and RX. Buffers are designed to overcome the attenuation of the traces. A bypass input for an external 15.8 GHz signal is also foreseen.

For the two receive paths, a quadrature LO signal is needed. This is obtained starting from a two-stage polyphase filter that takes the differential 15.8 GHz signal as an input, from which a quadrature signal is generated, which is injected into the SH-QILO (see Figure 7.6). The same polyphase filters, followed by a Cartesian combiner, are used in the transmit path to realize an LO phase shifter. In this way, the absolute phase of each transmitter can be tuned to align the phases of the different transmitters over mismatches and local variations for optimal beam radiation direction.

Transmitter

The transmitter is realized as a BPSK modulator followed by a three-stage PA, similar to the ones shown in [22,23]. A three-stage PA drives the antenna with a saturated power larger than 10 dBm at 79 GHz. Each stage of the PA is a push-pull NMOS common-source amplifier with C_{gd} neutralization to improve stability. The width of the transistors is progressively larger as they are closer to the output. Minimum channel length is used for all transistors. The modulator and the different PA stages are coupled by means of transformers; the final transformer also serves as balun and electrostatic discharge (ESD) protection toward the antenna. Sidelobes of the BPSK modulation outside the specified band are suppressed by a combination of filtering and harmonic rejection on the sequence in the modulator, as in [22,23].

Receiver

The RX front-end is realized with a two-stage LNA followed by an active Gilbert cell mixer. The front-end is followed by a baseband VGA, as shown in Figure 7.10. The gain of the LNA and mixer are 18 and 2 dB, respectively. The VGA has a gain from 10 to 45 dB in 28 steps. An active Gilbert cell mixer is preferred over a passive mixer, as for the latter it is difficult to realize a good virtual ground at its output over the wide baseband bandwidth. Moreover, an active mixer is less sensitive to the LO amplitude, which can vary considerably across the ILO locking range. On the other hand, an active mixer is more noisy, especially in the $1/f$ noise region, demanding more LNA gain to reduce the influence of $1/f$ noise. With 18 dB LNA gain and 2 dB mixer gain, the mixer should be designed to handle -10 dBm at its output. Digitally controlled current sources (current DACs or, briefly, iDACs) in parallel with the resistive load are used for offset compensation.

The two differential LNA stages are based on a push-pull NMOS structure with C_{gd} neutralization. Transformers are used at the LNA input (as balun and ESD protection)

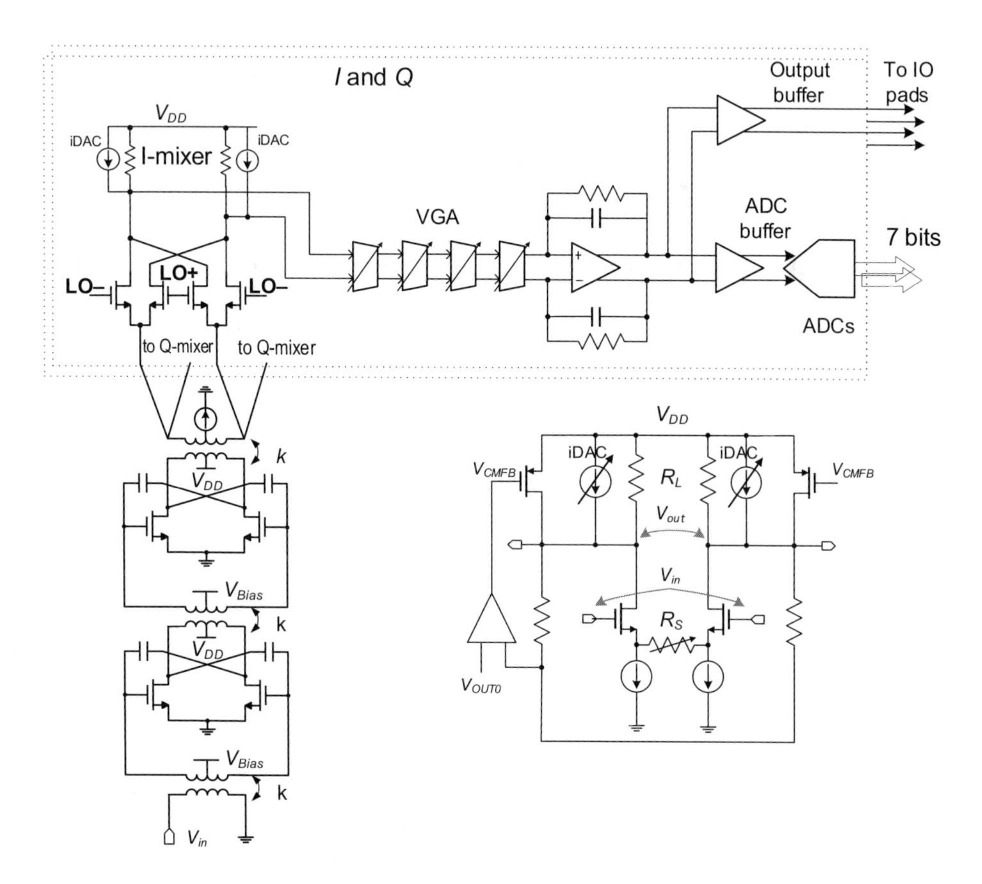

Figure 7.10 Receiver simplified schematic with details of the VGA stages [13]. (©2017 IEEE. Reprinted, with permission, from 2017 *IEEE Journal of Solid-State Circuits*.)

and for interstage matching. Input devices are optimized for simultaneous noise and impedance matching. Variable gain is implemented in the second stage by enabling or disabling parallel transistors as proposed in [33], which is not shown for simplicity in Figure 7.10.

The VGA is realized as a cascade of three programmable transconductance stages with resistive load, as shown in the simplified schematic of Figure 7.10, followed by a fourth one loaded by a transimpedance amplifier. Each transconductor stage has a programmable a gain from 3 to 10 dB in eight steps. It consists of a degenerated differential pair with programmable degeneration (Figure 7.10) in eight steps. With sufficient g_m in the transistors, the transconductance $G_m \approx 1/R_S = G_S$ with G_S being the degeneration conductance. The differential voltage gain at the load resistance R_L is approximately $-2R_L \cdot G_S$, and therefore controlled by 3-bit binary weighted resistor R_S. Practically, the gain is affected by the transistor g_m and the output conductance of the current source, limiting the maximum achievable gain to $\sim$10 dB. A PMOS current source in parallel with each load resistor R_L branches off part of the bias current and controls the output DC by means of a common mode feedback loop. Further, iDACs in parallel with the load are used for offset compensation, just as in the mixer.

The fully differential Miller-compensated opamp used in the transimpedance amplifier shown in Figure 7.10 has more than 10 GHz GBW product, which is possible thanks to the very high f_T of the 28 nm CMOS technology. A feedback resistor of 500 Ω and a parallel programmable capacitance (3 bits, 55 fF unit) fix the overall receiver bandwidth to 1 GHz for optimum SNR at the ADC input as described in Section 7.5.1. An opamp-based topology is chosen to provide a voltage swing at its output larger than the full-scale input of the ADC (0 dBm).

The transimpedance output goes to two different buffers. A unity-gain buffer is used for test purposes to carry the signal off chip on 100 Ω differential lines while the other is used to drive the ADC.

Analog-to-Digital Converters

The receiver baseband outputs I and Q are digitized by two 7-bit ADCs, realized as eight-lane time-interleaved pipelined hybrid ADCs [34]. In each of the $1.975\,\text{GHz}/8 \approx 250$ MSps lane, a 3-bit SAR is followed by a residue amplifier merged with a comparator and a second stage implemented as a 3-bit comparator-based asynchronous binary search. A fully dynamic structure is used for comparators and a residue amplifier resulting in very low power consumption (2.6 mW from 0.9 V) at 2 Gsps while achieving 6-bit effective resolution on a full-scale sinusoidal input. Data coming from the eight lanes are multiplexed on a single 7-bit bus that connects the ADC to the digital core. Programmable delays on the data lines are used to align the data with the sampling clock in the digital core avoiding metastability. In order not to affect the effective resolution, the clock jitter should be <400 fs, setting a spec for the integrated phase noise of the PLL to –49 dBc (SSB) at the sampling frequency of 1.975 GHz corresponding to –17 dBc at 79 GHz (SSB).

7.8 Experimental Results

The system described here is integrated in 28 nm CMOS technology (see Figure 7.11). Two bare dies are flip-chip mounted on a module with the antennas on the back side (Figure 7.12). The module itself is flipped on a larger PCB to connect supplies, a 25 MHz crystal reference oscillator, a connector for digital input/output (IO), controls, and other debug IOs (Figure 7.12).

7.8.1 Module and Antenna Design

The antenna module of Figure 7.12 is designed for short-range indoor sensing application. To cover an entire room, antennas are designed with a 120° × 90° beam coverage over the bandwidth from 76 to 81 GHz. A 45° linearly polarized antenna element is chosen to minimize mutual interference between modules facing each other. The antenna module is realized in a 12-layer *Panasonic Megtron 6* stack. Eight stacked patch elements (see [35]) realize a square ×4 virtual array for two chips, working as a 2 × 2 when only one chip is enabled. The elements are designed with a shielding cage around the radiators formed by metal rings and micro electrostatic discharge (ESD) via arrays, as depicted in Figures 7.13 and 7.14. This shielding reduces substrate mode excitation, which is favorable for the antenna gain and efficiency, and for the isolation between neighboring elements. The antenna module has a size of 20 × 40 mm. Transmitters and receivers antennas have a minimum spacing of 2.5 mm to target

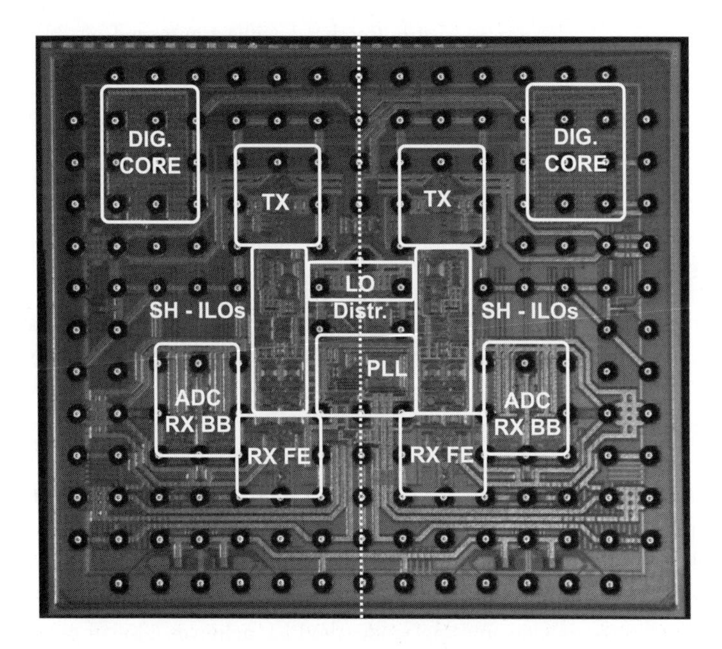

Figure 7.11 Micrograph of the die with size 3 mm × 2.63 mm [13]. (©2017 IEEE. Reprinted, with permission, from 2017 *IEEE Journal of Solid-State Circuits*.)

Figure 7.12 Antenna module die side (back), antenna side (front), and evaluation board with mounted antenna module [13]. (©2017 IEEE. Reprinted, with permission, from 2017 *IEEE Journal of Solid-State Circuits.*)

>40 dB isolation between any RX and any TX and limit the spillover at the inputs of the receivers.

7.8.2 Circuit-Level Measurements

Analog and mm-wave circuits have been characterized on the modified module with GSG probe pads instead of antennas. Attenuation from chip to probe pads is not deembedded. With a nominal supply (excluding RX output buffer at 1.8 V), power breakdown between the blocks is shown in Table 7.2.

The PLL and VCO have been characterized at the TX output (with no modulation). The tuning range goes from 78 to 87 GHz, corresponding to 1.8 GHz for the 15.8 GHz VCO. The phase noise of –92 and –116 dBc/Hz at 1 and 10 MHz offset respectively is measured at 79 GHz, corresponding to −106 and −130 dBc/Hz at the VCO output at 15.8 GHz. Figure 7.15 shows the PLL phase noise in charge pump and subsampling mode. When operating in subsampling mode, the PLL achieves a lower phase noise due to the reduction of charge pump and divider contribution, as reported in [32]. The SSB integrated phase noise is −19 dBc with a PLL bandwidth of 800 kHz.

Table 7.2 Power consumption per IC block.

Function	# per chip	Power (mW)
PLL, LO distrib, ILOs	1	284
RX	2	136
TX	2	252
Digital + CK distribution	2	340
Total	–	1,012

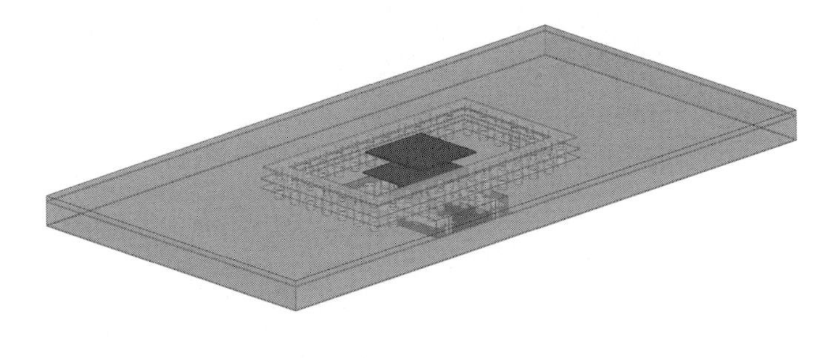

Figure 7.13 Details of the stacked patch antenna element element and shielding cage (HFSS rendering) [13]. (©2017 IEEE. Reprinted, with permission, from 2017 *IEEE Journal of Solid-State Circuits.*)

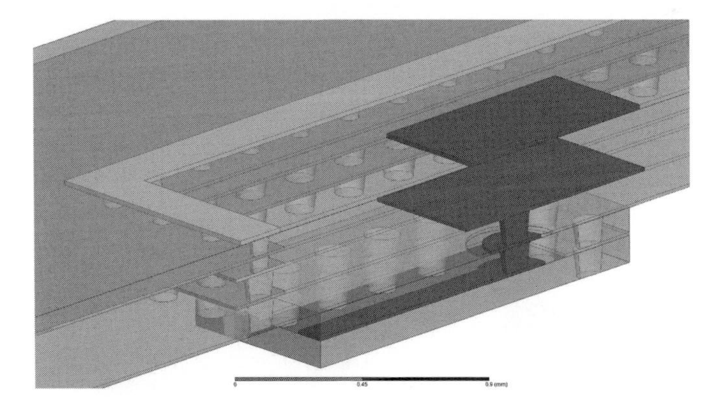

Figure 7.14 Details of the stacked patch antenna element and shielding cage (HFSS rendering): close up, cut at symmetry plane [13]. (©2017 IEEE. Reprinted, with permission, from 2017 *IEEE Journal of Solid-State Circuits.*)

The TX output power is measured to be 10 dBm without modulation, 9.5 dBm with BPSK modulation and 8.5 dBm with BPSK modulation and sidelobe suppression introduced in [22,23]. The TX output spectrum measured at 79 GHz is shown in Figure 7.16.

The RX conversion gain and noise figure are shown in Figure 7.17. Gain and noise figure have been measured either with a signal source at the input and a calibrated noise

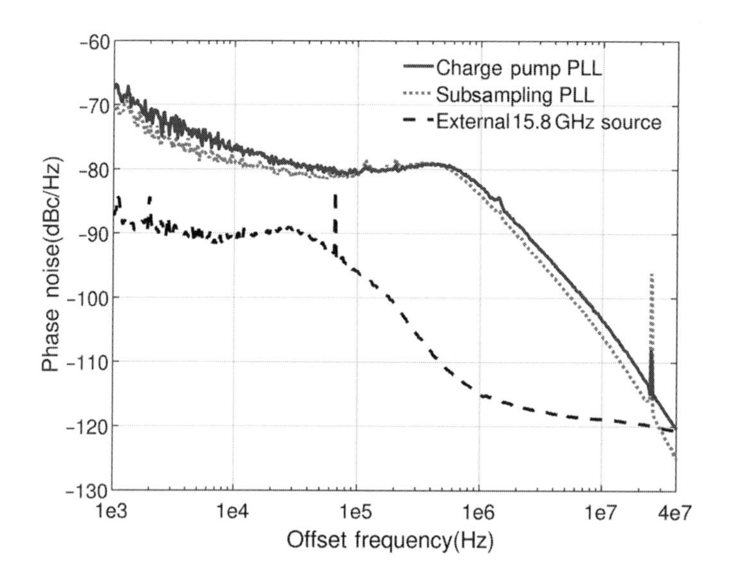

Figure 7.15 PLL phase noise in charge pump and subsampling mode compared with the output phase noise when the SH-ILO is locked on a lab source at 15.8 GHz [13]. (©2017 IEEE. Reprinted, with permission, from 2017 *IEEE Journal of Solid-State Circuits.*)

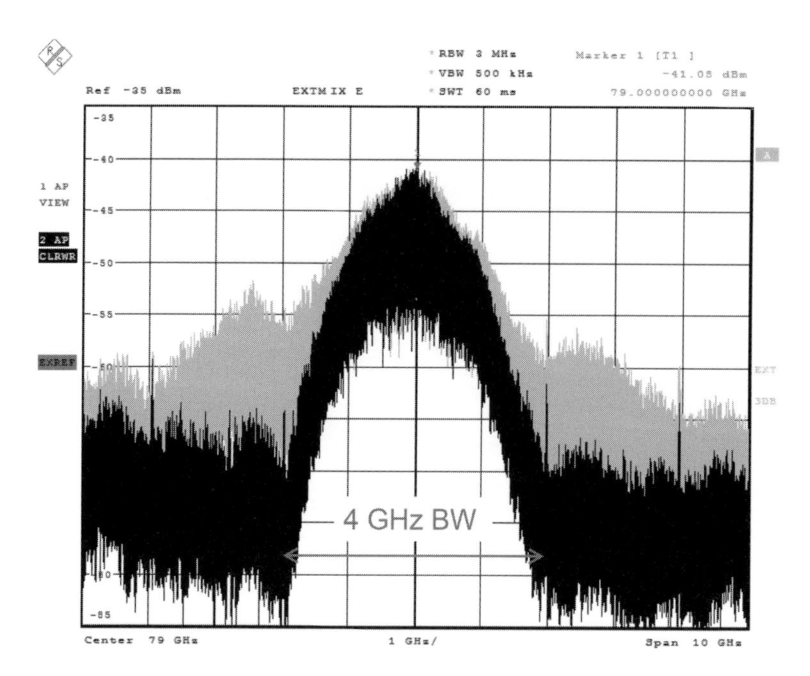

Figure 7.16 TX output spectrum – BPSK modulation (gray) and sidelobe suppression circuitry activated (black) measured at 79 GHz carrier and 1.975 Gbps sequence. The main sidelobe occupies ≈4 GHz RF bandwidth. Spurs are probably due to clock coupling [13]. (©2017 IEEE. Reprinted, with permission, from 2017 *IEEE Journal of Solid-State Circuits.*)

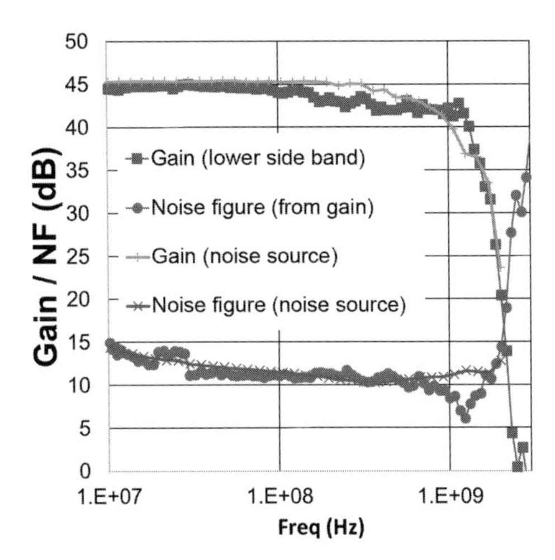

Figure 7.17 Receiver gain and noise figure, direct measurements and with Y-factor method (noise source) [14]. (©2016 IEEE. Reprinted, with permission, from *2016 IEEE Asian Solid-State Circuits Conference [A-SSCC]*.)

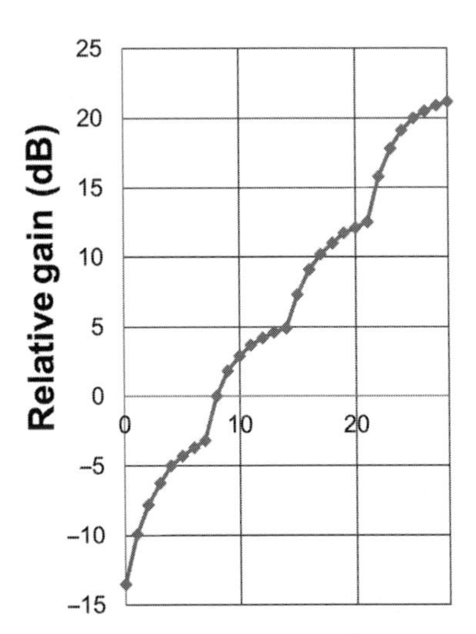

Figure 7.18 VGA gain programmability [14]. (©2016 IEEE. Reprinted, with permission, from *2016 IEEE Asian Solid-State Circuits Conference [A-SSCC]*.)

source *NOISE COM 5112* and a *R&S FSU* (Y-factor method) at a middle setting of the VGA gain. Both methods yield $NF < 12$ dB between 100 MHz and 1 GHz. The VGA gain can be tuned from −14 to +21 dB compared to the middle setting as shown in Figure 7.18. A 1 dB output compression point higher than 2 dBm is measured at the analog buffer output.

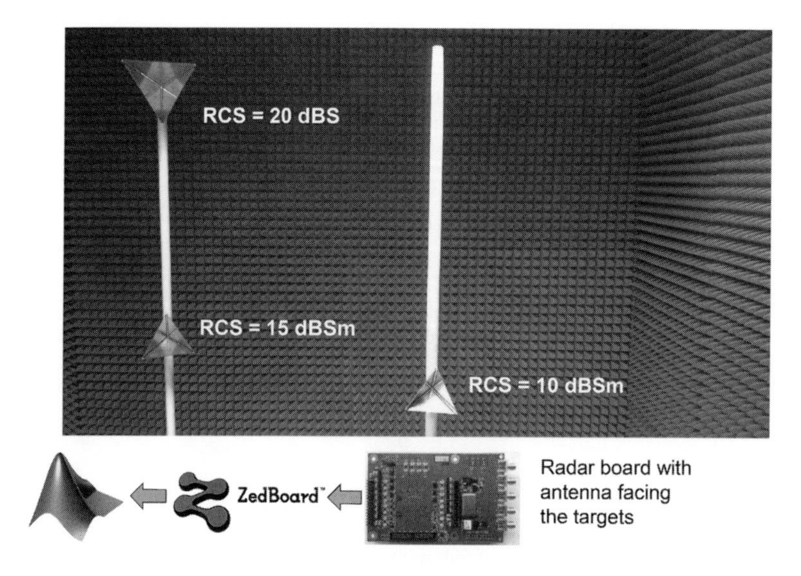

Figure 7.19 Radar measurement setup. The radar module is placed in front of the anechoic chamber with corner reflectors as targets. Data are captured by a *ZedBoard* and transferred to a PC running Matlab for processing [13]. (©2017 IEEE. Reprinted, with permission, from 2017 *IEEE Journal of Solid-State Circuits*.)

7.8.3 Radar System Measurements

To demonstrate the operation of the system as a radar, an anechoic chamber where corner reflectors can be moved to simulate targets on the front of the antennas is set up (Figure 7.19). The depth of the chamber is only 2.2 m, limiting the observable range depth to 29 range gates (2.2 m/7.5 cm). The digital data stream produced by the chip (four lines at 80 Mbps) is captured by a *ZedBoard* featuring a *Xilinx Zynq-7000* field programmable gate array (FPGA). Data acquired from the chip are packed on the FPGA board and transferred with a 1 Gbps Ethernet link to a PC running Matlab for post-processing and display.

SISO: 1 One Antenna Path

In Figures 7.20 and 7.21, the range and Doppler profile for one target (15 dBm^2) at 1.35 m and moving at 1 m/s toward the radar are shown. For these measurements, an almost perfect sequence (APS) is used with parameters $L_C = 504$, $M = 933$, $N = 64$, and a Blackman window for FFT. A high value of M is used to increase the dwell time and allow the target to move at lower speed in the relatively limited measurement space. The finite image rejection ratio of the chip (IRR $\approx$ 26 dB) is calibrated after measurement and before FFT. The range–Doppler mesh is shown in Figure 7.22.

In Figure 7.22, the noise floor, where not limited by range or Doppler sidelobes, is found to be around -100 dB, being 0 dB the spillover to which the plot is normalized.

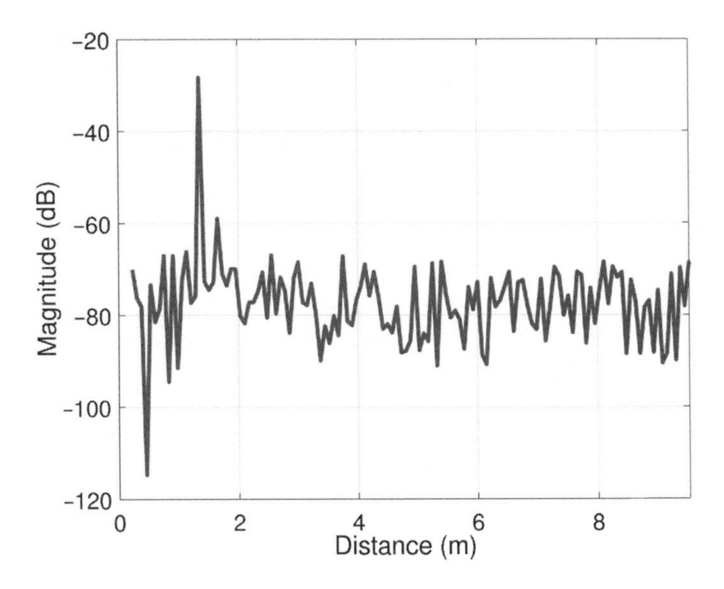

Figure 7.20 Range profile (correlations) at target speed for SISO processing. Points in the zero- and first-range bins (TX-to-RX spillover appears as a target at zero distance and zero speed) are removed for clarity [13]. (©2017 IEEE. Reprinted, with permission, from 2017 *IEEE Journal of Solid-State Circuits.*)

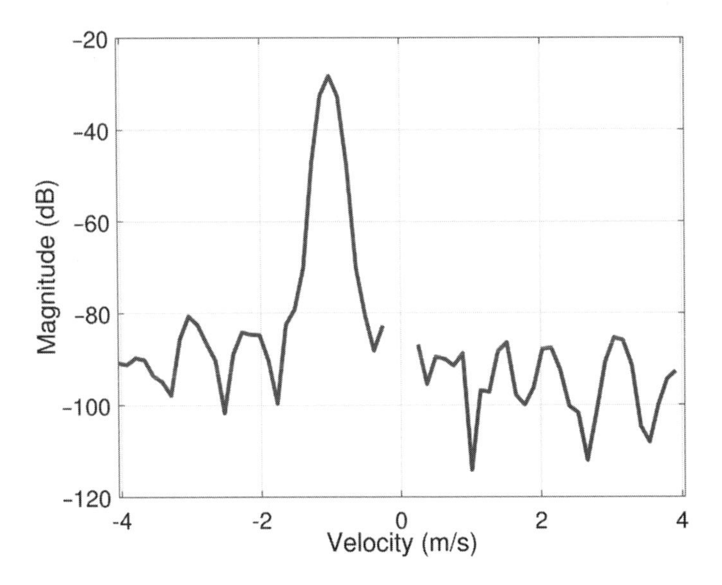

Figure 7.21 Doppler profile at target range for SISO processing. Points in the zero and first Doppler bins (TX-to-RX spillover appears as a target at zero distance and zero speed) are removed for clarity [13]. (©2017 IEEE. Reprinted, with permission, from 2017 *IEEE Journal of Solid-State Circuits.*)

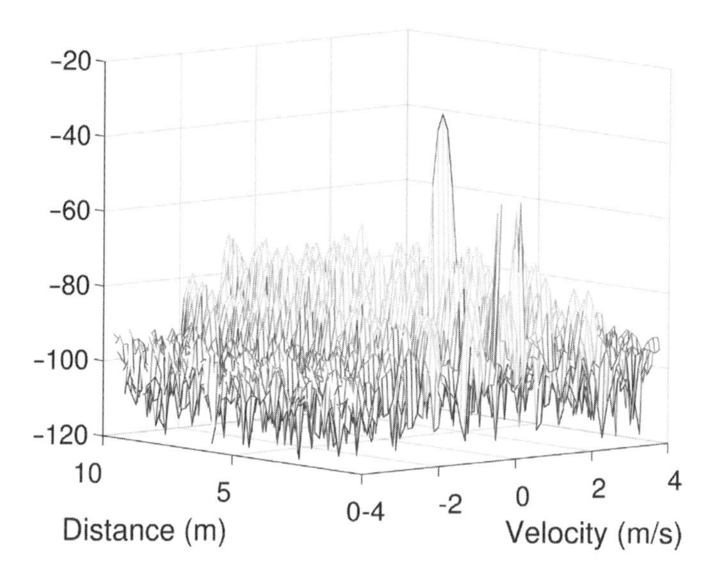

Figure 7.22 Range Doppler mesh for a target at 1.35 m moving at 1 m/s toward the radar. Image rejection canceled in postprocessing before FFT. Points in the zero and first range and Doppler bins (TX-to-RX spillover appears as a target at zero distance and zero speed) are removed for clarity [13]. (©2017 IEEE. Reprinted, with permission, from 2017 *IEEE Journal of Solid-State Circuits*.)

It is useful to compare the measured noise floor with the design target to assess the radar performances. With the parameters L_C = 504, M= 933, N = 64, and the other parameters previously used in (7.10), the input referred noise after processing gain can be written as $k_B \cdot T \cdot B_N \cdot NF_{RX}/G_p \approx -144\,\text{dBm}$, corresponding to $-109\,\text{dB}$ below the spillover level with 45 dB TX-to-RX isolation estimated for our prototype. Degradation of the measured SNR compared to the calculations is due to ADC quantization noise, differential nonlinearity (DNL), and interleaving mismatches, together with RX noise figure at low gain higher than expected. Receiver BW is also higher than expected, resulting in more noise folding at ADC input. Noise coming from the transmitter that has been neglected may also impact receiver noise floor via the spillover path. The 15 dBm^2 target at 1.35 m has a lower SNR ($\sim$73 dB) compared to what (7.10) would suggest ($\approx$ 88 dB); this is because the corner reflector behaves as a target of the nominal cross section only in the far field, which happens for distances larger than the Fraunhofer distance of $2D^2/\lambda \approx 5.2\,\text{m}$ with 10 cm reflector size (as defined in [36]). Below the Fraunhofer distance, its cross section is much smaller, explaining the SNR difference.

2 × 2 MIMO: Two Antenna Paths

When the two antenna paths on the same chip are activated, two TX and two RX can be operated in 2 × 2 MIMO mode with Hadamard outer codes, as proposed in [26]. Thanks to the orthogonal alignment of the antennas, azimuth and elevation of

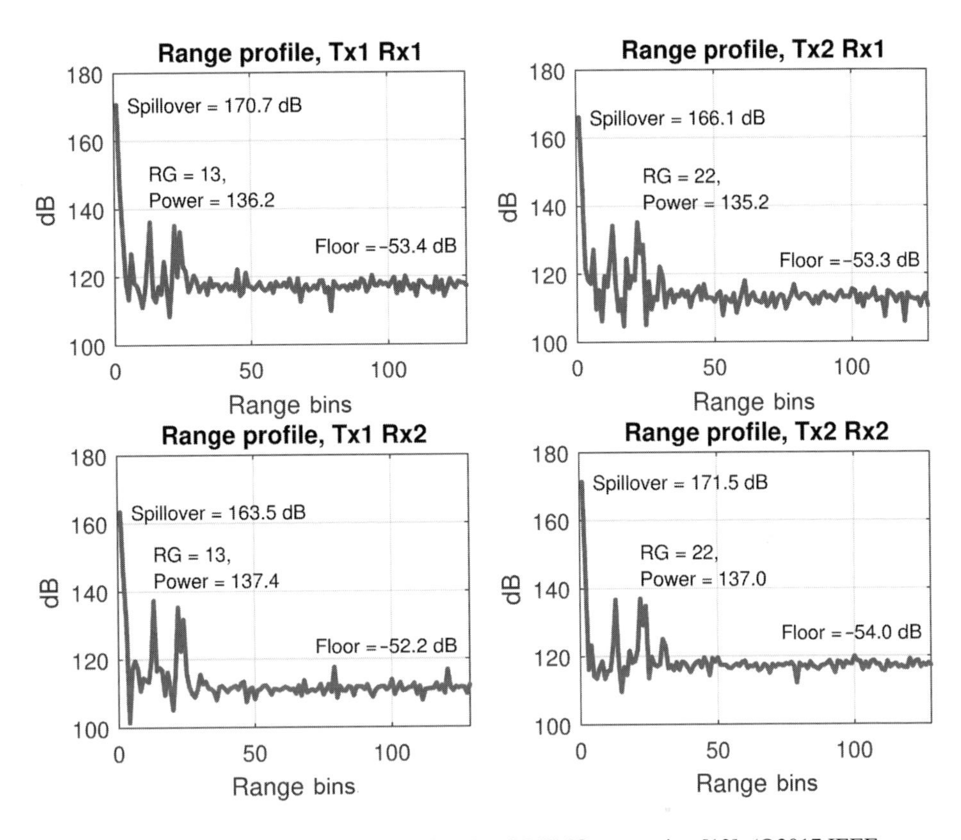

Figure 7.23 Range profiles (correlations) after 2 × 2 MIMO processing [13]. (©2017 IEEE. Reprinted, with permission, from 2017 *IEEE Journal of Solid-State Circuits*.)

targets can be computed on a 2 × 2 virtual antenna array. Figure 7.23 shows range profiles (128 range gates) in 2 × 2 MIMO mode with three targets positioned at 1, 1.52, and 1.65 m away, corresponding to the position of corner reflectors of Figure 7.19. An m-sequence is used with parameters $L_C = 511$, $M = 232$, and $N = 64$. The autocorrelation property of the m-sequence limits the dynamic range of the range profiles to $-20\log_{10}(L_C) \approx -54$ dB, and no ghost targets are found in the first 128 range gates thanks to the zero cross-correlation given by the Hadamard outer coding. The angle of arrival can be extracted from MIMO processing and is also shown in Figure 7.24. The 2D-MUSIC superresolution algorithm, proposed in [37], boosts the angular resolution up to 5°.

7.8.4 Conclusions and State-of-the-Art Comparison

This chapter demonstrates a fully integrated 2 × 2 MIMO radar SoC in 28 nm CMOS. It occupies 8 mm^2 and consumes 1 W for a range depth of 10 m with a range resolution of 7.5 cm. Angular accuracy of 5° in both azimuth and elevation is achieved in 2 × 2 MIMO

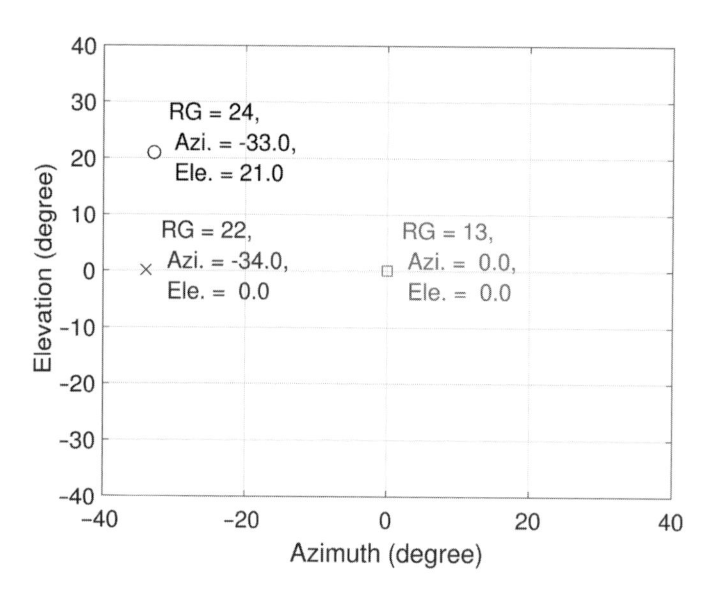

Figure 7.24 Angle of arrival after 2×2 MIMO processing [13]. (©2017 IEEE. Reprinted, with permission, from 2017 *IEEE Journal of Solid-State Circuits*.)

mode. With respect to previously published works found in the references, either FMCW or PMCW, this work achieves the highest level of integration in terms of number of TX and RX, frequency synthesis, and signal processing capability.

References

[1] M. van Schijndel-de Nooij, A. Schieben, N. Ford, and M. McDonald, "Definition of necessary vehicle and infrastructure systems for automated driving," European Commission, Information Society and Media DG, Components and Systems Directorate – ICT for Transport, Study Rep. SMART 2010/0064, Tech. Rep., June 2011.

[2] J. Wenger, "Automotive radar – Status and perspectives," in *IEEE Compound Semiconductor Integrated Circuit Symposium, 2005 (CSIC '05)*, IEEE, October 2005, pp. 21–25.

[3] J. Hasch, E. Topak, R. Schnabel, T. Zwick, R. Weigel, and C. Waldschmidt, "Millimeter-wave technology for automotive radar sensors in the 77 GHz frequency band," *IEEE Transactions on Microwave Theory and Techniques*, vol. 60, no. 3, pp. 845–860, March 2012.

[4] I. Nasr, R. Jungmaier, A. Baheti, et al., "A highly integrated 60 GHz 6-channel transceiver with antenna in package for smart sensing and short-range communications," *IEEE Journal of Solid-State Circuits*, vol. 51, no. 9, pp. 2066–2076, September 2016.

[5] H. C. Kuo, C. C. Lin, C. H. Yu, et al., "A fully integrated 60-GHz CMOS direct-conversion Doppler radar RF sensor with clutter canceller for single-antenna noncontact human vital-signs detection," *IEEE Transactions on Microwave Theory and Techniques*, vol. 64, no. 4, pp. 1018–1028, April 2016.

[6] E. J. Barlow, "Doppler radar," *Proceedings of the IRE*, vol. 37, no. 4, pp. 340–355, April 1949.

[7] B. P. Ginsburg, S. M. Ramaswamy, V. Rentala, E. Seok, S. Sankaran, and B. Haroun, "A 160 GHz pulsed radar transceiver in 65 nm CMOS," *IEEE Journal of Solid-State Circuits*, vol. 49, no. 4, pp. 984–995, April 2014.

[8] H. J. Ng, R. Feger, and A. Stelzer, "A fully-integrated 77-GHz UWB pseudo-random noise radar transceiver with a programmable sequence generator in SiGe technology," *IEEE Transactions on Circuits and Systems I: Regular Papers*, vol. 61, no. 8, pp. 2444–2455, August 2014.

[9] H. Jia, L. Kuang, W. Zhu, et al., "A 77 GHz frequency doubling two-path phased-array FMCW transceiver for automotive radar," *IEEE Journal of Solid-State Circuits*, vol. 51, no. 10, pp. 2299–2311, October 2016.

[10] S. Trotta, M. Wintermantel, J. Dixon, et al., "An RCP packaged transceiver chipset for automotive LRR and SRR systems in SiGe BiCMOS technology," *IEEE Transactions on Microwave Theory and Techniques*, vol. 60, no. 3, pp. 778–794, March 2012.

[11] J. Oh, J. Jang, C. Y. Kim, and S. Hong, "A W-band 4-GHz bandwidth phase-modulated pulse compression radar transmitter in 65-nm CMOS," *IEEE Transactions on Microwave Theory and Techniques*, vol. 63, no. 8, pp. 2609–2618, August 2015.

[12] J. Jang, J. Oh, C. Y. Kim, and S. Hong, "A 79-GHz adaptive-gain and low-noise UWB radar receiver front-end in 65-nm CMOS," *IEEE Transactions on Microwave Theory and Techniques*, vol. 64, no. 3, pp. 859–867, March 2016.

[13] D. Guermandi, Q. Shi, A. Dewilde, et al., "A 79-GHz 2×2 MIMO PMCW radar SoC in 28-nm CMOS," *IEEE Journal of Solid-State Circuits*, vol. 52, no. 10, pp. 2613–2626, October 2017.

[14] D. Guermandi, Q. Shi, A. Dewilde, et al., "A 79 GHz 2x2 MIMO PMCW radar SoC in 28 nm CMOS," in *2016 IEEE Asian Solid-State Circuits Conference (A-SSCC)*, IEEE, November 2016, pp. 105–108.

[15] A. G. Stove, "Linear FMCW radar techniques," *IEE Proceedings F – Radar and Signal Processing*, vol. 139, no. 5, pp. 343–350, October 1992.

[16] V. Winkler, "Range Doppler detection for automotive FMCW radars," in *2007 European Microwave Conference*, IEEE, October 2007, pp. 1445–1448.

[17] A. Melzer, F. Starzer, H. Jager, and M. Huemer, "Real-time mitigation of short-range leakage in automotive FMCW radar transceivers," *IEEE Transactions on Circuits and Systems II: Express Briefs*, vol. PP, no. 99, pp. 1–1, 2016.

[18] K. J. Kelley and C. L. Weber, "Principles of spread spectrum radar," in *MILCOM 1985 – IEEE Military Communications Conference*, vol. 3, October 1985, pp. 586–590.

[19] S. W. Golomb, "Shift-register sequences and spread-spectrum communications," in *IEEE Third International Symposium on Spread Spectrum Techniques and Applications, 1994 (ISSSTA '94)*, IEEE , vol. 1, IEEE, July 1994, pp. 14–15.

[20] D. Guermandi, Q. Shi, A. Medra, et al., "A 79GHz binary phase-modulated continuous-wave radar transceiver with TX-to-RX spillover cancellation in 28nm CMOS," in *2015 IEEE International Solid-State Circuits Conference (ISSCC) Digest of Technical Papers*, February 2015, pp. 355–356.

[21] W. V. Thillo, V. Giannini, D. Guermandi, S. Brebels, and A. Bourdoux, "Impact of ADC clipping and quantization on phase-modulated 79 GHz CMOS radar," in *2014 11th European Radar Conference*, October 2014, pp. 285–288.

[22] V. Giannini, D. Guermandi, Q. Shi, et al., "A 79 GHz phase-modulated 4 GHz-BW CW radar transmitter in 28 nm CMOS," *IEEE Journal of Solid-State Circuits*, vol. 49, no. 12, pp. 2925–2937, December 2014.

[23] V. Giannini, D. Guermandi, Q. Shi, et al., "A 79GHz phase-modulated 4GHz-BW CW radar TX in 28nm CMOS," in *2014 IEEE International Solid-State Circuits Conference Digest of Technical Papers (ISSCC)*, IEEE, February 2014, pp. 250–251.

[24] N. Yamada, Y. Tanaka, and K. Nishikawa, "Radar cross section for pedestrian in 76GHz band," in *2005 European Microwave Conference*, vol. 2, IEEE, October 2005, pp. 4.

[25] E. Fishler, A. Haimovich, R. Blum, D. Chizhik, L. Cimini, and R. Valenzuela, "MIMO radar: An idea whose time has come," in *Proceedings of the 2004 IEEE Radar Conference (IEEE Cat. No.04CH37509)*, IEEE, April 2004, pp. 71–78.

[26] A. Bourdoux, U. Ahmad, D. Guermandi, S. Brebels, A. Dewilde, and W. V. Thillo, "PMCW waveform and MIMO technique for a 79 GHz CMOS automotive radar," in *2016 IEEE Radar Conference (RadarConf)*, May 2016, pp. 1–5.

[27] P. P. Vaidyanathan and P. Pal, "MIMO radar, SIMO radar, and IFIR radar: A comparison," in *2009 Conference Record of the 43rd Asilomar Conference on Signals, Systems and Computers*, November 2009, pp. 160–167.

[28] D. A. Bell, "Walsh functions and Hadamard matrixes," *Electronics Letters*, vol. 2, no. 9, pp. 340–341, September 1966.

[29] R. Wu, J. Pang, Y. Seo, et al., "An LO-buffer-less 60-GHz CMOS transmitter with oscillator pulling mitigation," in *2016 IEEE Asian Solid-State Circuits Conference (A-SSCC)*, IEEE, November 2016, pp. 109–112.

[30] Q. Shi, D. Guermandi, V. Giannini, and P. Wambacq, "A 5th subharmonic, inverter-based injection locked oscillator with 72-83GHz locking range," in *2014 IEEE Radio Frequency Integrated Circuits Symposium*, June 2014, pp. 185–188.

[31] E. Hegazi, H. Sjoland, and A. Abidi, "A filtering technique to lower oscillator phase noise," in *2001 IEEE International Solid-State Circuits Conference. Digest of Technical Papers. ISSCC (Cat. No.01CH37177)*, February 2001, pp. 364–365.

[32] X. Gao, E. A. M. Klumperink, M. Bohsali, and B. Nauta, "A low noise sub-sampling PLL in which divider noise is eliminated and PD/CP noise is not multiplied by N^2," *IEEE Journal of Solid-State Circuits*, vol. 44, no. 12, pp. 3253–3263, December 2009.

[33] A. Medra, V. Giannini, D. Guermandi, and P. Wambacq, "A 79GHz variable gain low-noise amplifier and power amplifier in 28nm CMOS operating up to 125^oC," in *ESSCIRC 2014 – 40th European Solid State Circuits Conference (ESSCIRC)*, IEEE, September 2014, pp. 183–186.

[34] A. Spagnolo, B. Verbruggen, S. D'Amico, and P. Wambacq, "A 6.2mW 7b 3.5GS/s time interleaved 2-stage pipelined ADC in 40nm CMOS," in *ESSCIRC 2014 – 40th European Solid State Circuits Conference (ESSCIRC)*, September 2014, pp. 75–78.

[35] S. D. Targonski, R. B. Waterhouse, and D. M. Pozar, "Design of wide-band aperture-stacked patch microstrip antennas," *IEEE Transactions on Antennas and Propagation*, vol. 46, no. 9, pp. 1245–1251, Sepember 1998.

[36] C. A. Balanis, *Antenna Theory: Analysis and Design*. Wiley-Interscience, 2005.

[37] R. Schmidt, "Multiple emitter location and signal parameter estimation," *IEEE Transactions on Antennas and Propagation*, vol. 34, no. 3, pp. 276–280, March 1986.

8 CMOS Transceiver Design for Ultra-High-Speed Millimeter-Wave Wireless Communications

Kenichi Okada and Rui Wu

8.1 Introduction

In the near future, we will enter a society in which "anything that benefits from a connection will be connected." It is so called the networked society. This society will have two prominent characteristics: the exponential growth in the traffic volume and massive growth in connected devices, as shown in Figure 8.1. Those characteristics raise the requirements for future wireless communications, which include (a) the capability of supporting data traffic explosion (e.g., 1,000 × capacity/km^2); (b) the increase of quality of experience (QoE) for various applications (e.g., 10–100 × data rates, reduced latency, and terminal battery saving); (c) massive device connectivity (e.g., 100× connected devices); (d) intelligent network with low cost and high robustness (e.g., suitable for diverse environment).

So what are the solutions for those requirements? With the evolution of communication technologies, wireline communications can achieve hundreds of gigabits per second (Gb/s) data rate nowadays [1–4]. While the data rate of wireless communications using low-frequency bands, such as the 2.4 GHz band and 5 GHz band, seems to be saturated around hundreds of megabits per second (Mb/s) [5–7], as depicted in Figure 8.2. One of the most important reasons is the crowded frequency band at low frequencies (frequencies below 10 GHz), which limits the possible bandwidth for the multi-Gb/s wireless communication, as illustrated in Figure 8.3. Fortunately, millimeter-wave (mm-wave) bands, which have many fewer standards and wider available frequency bands, is one of the most promising candidates for high-data-rate (multi-Gb/s) wireless communications. Especially, it is predicted that the 60 GHz band as the beginning of a trend of escalating carrier frequencies has the potential of delivering unprecedented data rates [8–11], which allows uncompressed high-definition media transfers, sensing and radar applications, and virtually instantaneous access to massive libraries of information. Moreover, advances in complementary metal-oxide-semiconductor (CMOS) processes allow for the low-cost, low-power implementation of the 60 GHz transceiver system on a chip (SoC) [12–15]. Today, we are at the dawn of a new age of massively broadband devices fabricated in CMOS processes, which operate around mm-wave bands.

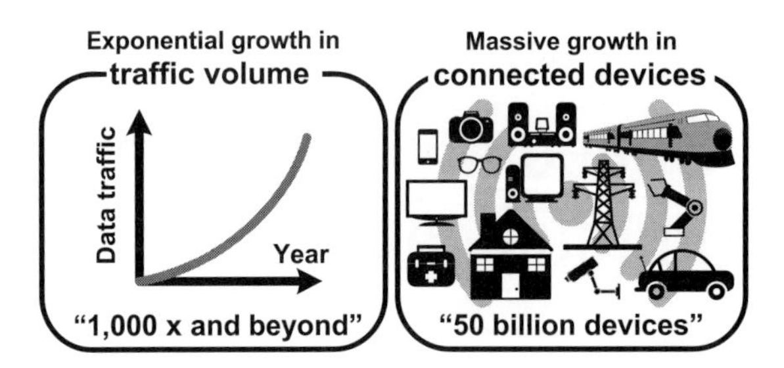

Figure 8.1 The networked society.

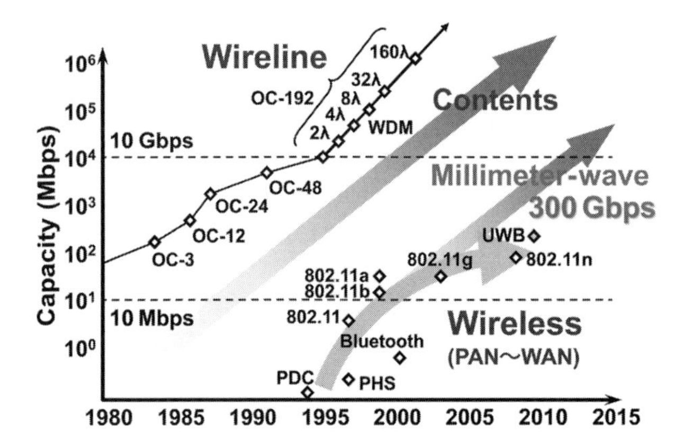

Figure 8.2 The conceptual trend of communication data rate.

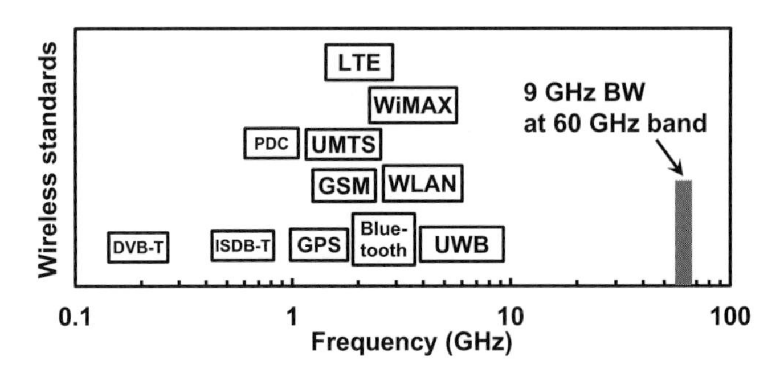

Figure 8.3 The wireless standards at different frequency bands.

8.2 60 GHz CMOS Transceiver Architecture

8.2.1 Challenges and Design Considerations

Considering the implementation of the 60 GHz CMOS transceivers for the short-range wireless communication, there are several key requirements, such as high-data-rate capability, low cost (and hence small circuit size), high reliability, very low power consumption, and high integrability. To achieve those requirements, new challenges are encountered consequently.

It is well known that the parasitics have a large effect on the device performance at 60 GHz frequencies. Unfortunately, foundries do not yet provide process design kits (PDKs) supporting 60 GHz frequencies. Therefore, device modeling for 60 GHz circuit design is indispensable. To accurately model the devices, a test elementary group (TEG) of passive and active devices including pads and interconnects has to be implemented before circuit design [16,17]. Deembedding is needed to correctly obtain the performance of the device removing the effects of extra test fixtures [18].

The maximum available gain (MAG) of the transistor is inversely proportional to the operation frequency. More stages and power consumption are demanded to reach a certain gain value at 60 GHz than at low frequencies. Customized transistor layout and new circuit design techniques are necessary to improve the gain of the 60 GHz amplifier with low power consumption [19].

The quality factor of circuit components (especially varactors) drops enormously at the 60 GHz band, which leads to the difficulties of direct generation of a low phase noise 60 GHz local oscillator (LO) signal. Therefore, with traditional design techniques, it becomes difficult to implement the required low phase noise LO with low power consumption. Different approaches are needed to overcome this issue.

It is known that many wireless standards are under discussion to satisfy the unprecedented capacity requirement. For example, the IEEE802.11ay standard is targeting over 100-Gb/s data rate by using the 60 GHz band. Unfortunately, the channel bandwidth of 2.16 GHz for the 60 GHz band is not wide enough to realize such a high data rate. A channel-bonding capability as well as high-order modulation support is strongly demanded to boost the data rate. For instance, a four-channel bonding in 64-QAM can achieve 42.24 Gb/s data rate (7.04 GS/s $\times$ 6 b/s).

To realize a four-channel bonding operation with 64-QAM, there are several issues to be considered, such as wideband gain characteristics, wide dynamic range, low LO phase noise, fine and wideband I/Q mismatch calibration, and small LO leakage. Direct-conversion architecture is widely used for the 60 GHz CMOS transceivers due to its low power consumption and wideband characteristic [14,20]. The direct-conversion 60 GHz transceivers reported in [21] and [22] achieve 4 Gb/s in QPSK and 7 Gb/s in 16-QAM, respectively. However, the low-pass nature of the baseband amplifiers limits the full use of the 60 GHz band with reasonable power consumption. Furthermore, an injection-locking technique is employed for the 60 GHz transceivers [23,24], which realizes low phase noise with a wide frequency tuning range in 60 GHz quadrature LO synthesis. Nevertheless, a method for the fine and wideband I/Q mismatch calibration is still desired to accomplish the four-channel bonding operation with 64-QAM. In addition, an 8-bit 14.08 GS/s ADC is normally required to support 42.24 Gb/s in 64-QAM, which is usually realized by a massive time-interleaved ADC, and needs unreasonably large power consumption as summarized in Figure 8.4 [25]. In this condition, a baseband signal bandwidth of 3.52 GHz is assumed. The Nyquist rate is 7.04 GS/s. Considering an oversampling ratio of 2, which can ease the implementation of the baseband filtering [26], the sampling rate of the ADC is 14.08 GS/s. With such a high sampling rate, the signal-to-noise-and-distortion ratio (SNDR) of an 8-bit ADC may be less than

Table 8.1 Link budget for short-range high-speed communication.

Parameter	Value	
	one-channel	four-channel
Carrier frequency (GHz)	61.56	
Distance (m)	0.2	0.1
Bandwidth (GHz)	1.76	7.04
TX output power (dBm)	3	
TX/RX antenna gain (dBi)	2	
LOS loss (dB)	−54.2	−48.2
Implementation loss (dB)	−3	
Received level (dBm)	−50.3	−44.3
Thermal noise (dBm)	−81.4	−75.4
NF (dB)	6	
Received SNR (dB)	25.1	
Modulation	64-QAM	
Required SNR (dB)	22.5	

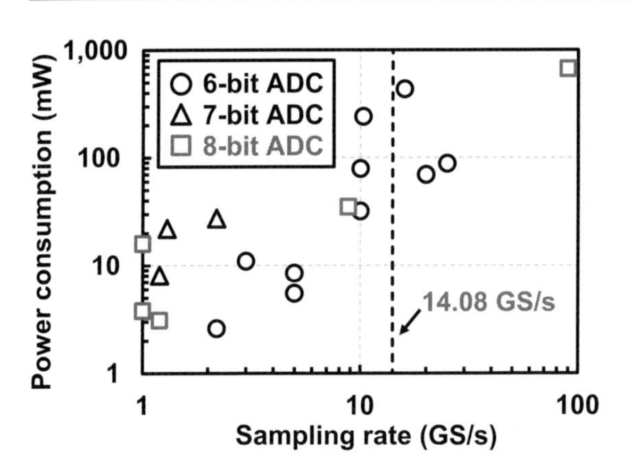

Figure 8.4 Summarized ADC power consumption versus sampling rate [28]. (©2017 IEEE. Reprinted, with permission, from *IEEE Journal of Solid-State Circuits.*)

36 dB [27], which leaves around 10 dB margin to tolerate the impairments previously mentioned.

Table 8.1 shows a link budget example of the 60 GHz transceiver, which includes the communication of one-channel (1.76 GHz) and four-channel bonding (7.04 GHz) in 64-QAM. The output power of the transmitter is 3 dBm. The target communication distances are 0.2 and 0.1 m, respectively. The transmitter and receiver antennas have a gain of 2 dBi. A noise figure of 6 dB is assumed for the receiver. It can be seen that the received signal-to-noise ratio (SNR) has about 3 dB margin from the required SNR of 22.5 dB.

However, there are many other factors, especially for the 64-QAM and four-channel bonding case, that will degrade the quality of the received signal. Firstly, the wideband gain characteristics are required for both transmitter and receiver. At the radio frequency (RF) side, a wide-and-flat gain roughly from 57 to 66 GHz is needed. On the other hand, a 5 GHz bandwidth is necessary at baseband side, which is very power consuming because of the low-pass nature of baseband amplifiers. Figure 8.5 shows Matlab simulation results of error vector magnitude (EVM) in 16-QAM considering the influence of the transceiver (TRX) gain flatness. The symbol rate is 7.04 GS/s on single-carrier mode. The frequency response of a fourth-order finite impulse response (FIR) filter is assumed for modeling the gain flatness of the TRX. A recursive least square (RLS) linear equalizer is used for demonstrating the influence of the gain flatness and equalization. It is observed that the maximum gain variation of the unequalized TRX within the 7.04 GHz bandwidth is about ±1.4 dB, which corresponds to an EVM of −16.5 dB. When the number of the equalizer taps increases, the maximum gain

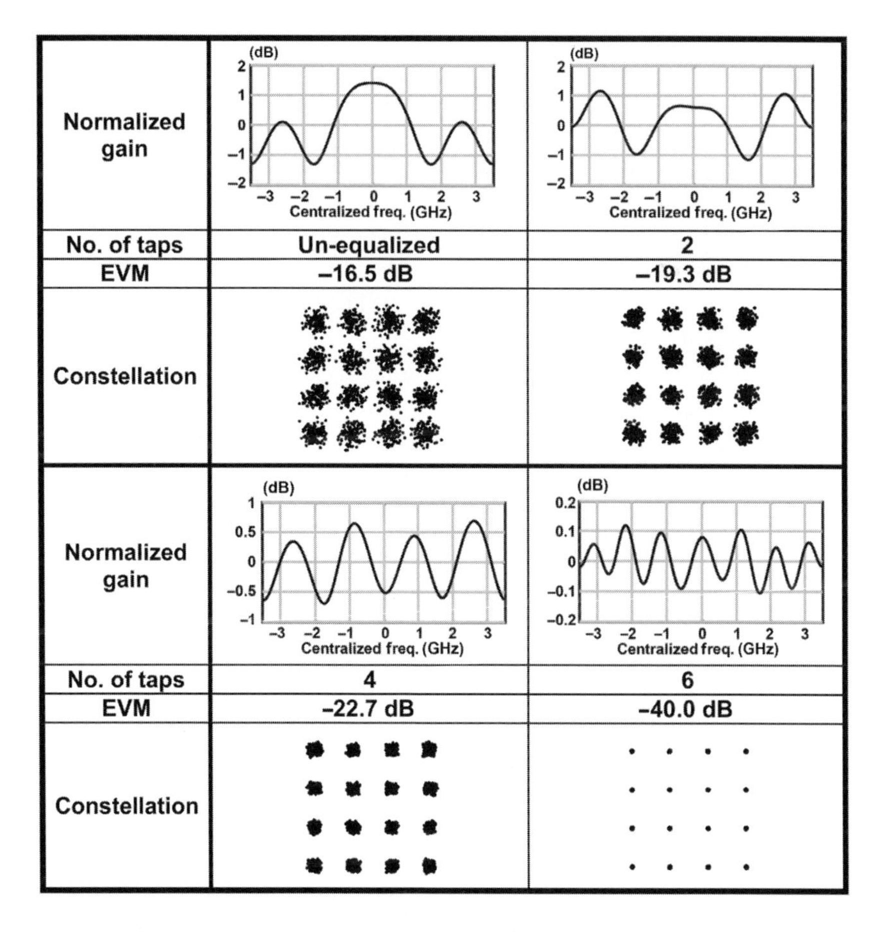

Figure 8.5 Simulated EVM in 16-QAM with 7.04-GS/s symbol rate at different equalization conditions [28]. (©2017 IEEE. Reprinted, with permission, from *IEEE Journal of Solid-State Circuits*.)

variation is reduced, and the EVM performance is improved. A gain variation of less than ±0.7 dB (four-tap equalizer) over the frequency band of interest is needed to achieve 16-QAM and four-channel bonding considering about 6 dB margin from the required EVM of −16.5 dB. As for 64-QAM and four-channel bonding, a larger number of taps with smaller gain variation (such as six taps, ±0.1 dB gain variation) may be required considering over 10 dB margin from the required EVM of −22.5 dB. In practical implementation, the actual required number of taps and the corresponding gain variations should be estimated by both the value and shape of the TRX gain characteristics.

Furthermore, the wide dynamic range has to be maintained considering linearity and noise with the 5 GHz baseband bandwidth. In addition, the phase noise of the LO, image rejection, and LO leakage are also critical [26,29]. All those impairments can disturb the signal constellation and degrade the system EVM. The relationship between the TX/RX EVM and impairments can be expressed as ([30,31])

$$EVM_{tot} \approx \sqrt{\frac{1}{SNDR^2} + \varphi_{RMS}^2 + EVM_{IMRR}^2 + EVM_{LOFT}^2 + EVM_{GF}^2} \tag{8.1}$$

where $SNDR$ is the signal-to-noise-and-distortion ratio, which considers the effects of the front-end noise and nonlinearity such as AM-AM and AM-PM distortion [32]. φ_{RMS} is the integrated double-sideband (DSB) phase noise of a carrier. EVM_{IMRR}, EVM_{LOFT}, and EVM_{GF} represent the degraded EVM due to I/Q mismatch, LO leakage, and gain flatness, respectively. It should be indicated that (8.1) is a simplified expression that emphasizes several dominant effects for the 60 GHz transceivers. In actual transceivers, many other effects could influence the EVM performance, such as the correlated effects between AM-AM/AM-PM distortion and phase noise, the resolution and clock jitter of the ADC/DAC [33,34], and so on. Thorough system analysis and simulations should be conducted to identify the major performance limiters.

As shown in (8.1), the I/Q mismatch has significant influence on the EVM performance of the 60 GHz transceiver, which is normally evaluated by the image rejection ratio (IMRR). The IMRR can be expressed as the function of gain and phase imbalance by

$$|IMRR| = \left| \frac{10^{\frac{\Delta A}{10}} + 2 \cdot 10^{\frac{\Delta A}{20}} \cos \Delta\theta + 1}{10^{\frac{\Delta A}{10}} - 2 \cdot 10^{\frac{\Delta A}{20}} \cos \Delta\theta + 1} \right| \tag{8.2}$$

where ΔA is the gain error in decibel and $\Delta\theta$ is the phase error in degree. For 64-QAM communication, $|IMRR| \geq 35$ dBc, should be satisfied in consideration of other impairment degradations. It is corresponding to less than 0.3 dB gain mismatch and less than 2 degrees phase mismatch, as illustrated in Figure 8.6. Therefore, fine calibrations of I/Q gain and phase errors are desired to obtain the high IMRR. Moreover, the IMRR of more than 35 dBc needs to be maintained over wide bandwidth for the channel-bonding cases, which further increases the design difficulty.

As we know, the nonidealities in transceivers, such as phase noise of TX's and RX's LOs and nonlinearities of PA, degrade the packet error rate (PER), especially for high-order modulation schemes, e.g., 8-PSK, 16-QAM, and 64-QAM. For a 60 GHz

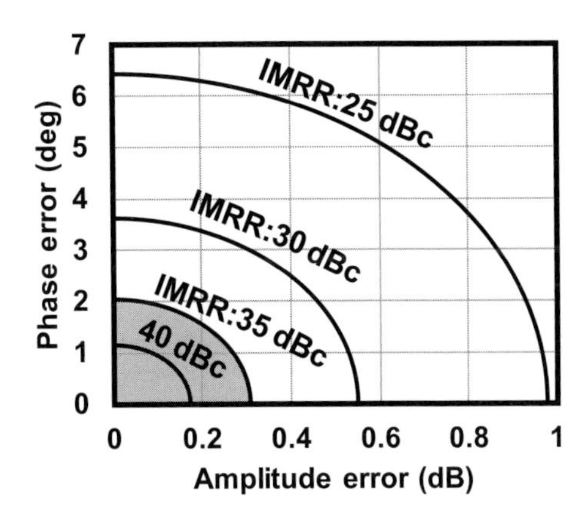

Figure 8.6 Calculated IMRR versus gain and phase error [28]. (©2017 IEEE. Reprinted, with permission, from *IEEE Journal of Solid-State Circuits*.)

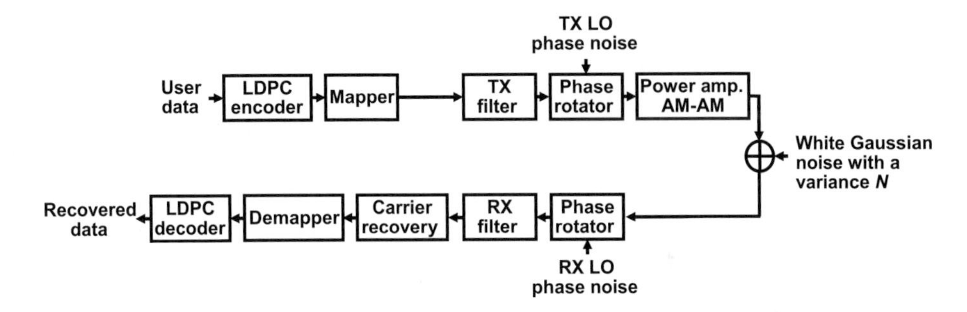

Figure 8.7 Block diagram for evaluating the effects of phase noise and nonlinearity [24]. (©2011 IEEE. Reprinted, with permission, from *IEEE Journal of Solid-State Circuits*.)

wireless system, RF impairment models for the phase noise and nonlinearities have been reported [35], and the bit error rate (BER) is simulated with the effect of RF impairments. However, a carrier phase noise can be recovered by a digital phase-locked loop (PLL) in baseband signal processing in an actual 60 GHz wireless system [36], and an evaluation of PER using a moderately short packet is more practical than that of BER to clarify the effect of phase noise when neither pilot word nor differential encoding is employed. In this section, the PER is simulated with a carrier recovery process to estimate more practical requirements of phase noise.

Figure 8.7 shows a block diagram for evaluating the effects of phase noise and nonlinearity. Single-carrier modulation schemes using QPSK, 8-PSK, and 16-QAM with a low-density parity-check (LDPC) error correction code are simulated based on Figure 8.7. The LDPC code is a (1440, 1344) code in the IEEE 802.15.3c standard [37, 38], where (n, k) represents a code word length of n in bits and a source word length of k in bits. The symbol rate is set to 1,760 Msymbols/s and a root raised-cosine roll-off filter with a roll-off factor of 0.25 is used at each Tx and Rx side. The phase noise is generated by using a one-pole one-zero model [35] with power spectral densities of -55.8 dBc/Hz at the carrier frequency and -133 dBc/Hz at the floor level. The measured AM-AM nonlinearity of the PA is used for the simulation, and the operating point of the PA is set to a 4 dB back-off from the saturated output power. A carrier recovery is applied in baseband and time domain to track the phase of received symbols with a low-frequency variation, which is performed by a digital PLL with one-symbol delay. The packet length is 2,052 bytes and no pilot word is inserted. The maximum number of iterations for LDPC decoding is 8.

Figure 8.8 shows ΔCNR, which is the increment of a required carrier-to-noise ratio (CNR) by the RF impairments to achieve a PER of 10^{-3} after LDPC decoding, as a function of phase noise at 1 MHz offset frequency. The ΔCNR can be calculated as

$$\Delta CNR = CNR_1 - CNR_0, \tag{8.3}$$

where CNR_0 and CNR_1 are required CNRs without and with the RF impairments, respectively, to achieve a PER of 10^{-3} after LDPC decoding. When the phase noise level is low enough compared to the white noise, only the nonlinearity degrades the PER performance. CNR_0 for QPSK, 8-PSK, and 16-QAM are 9.2, 14.1, and 15.9 dB, respectively. In Figure 8.8, at a low phase-noise region, it can be seen that ΔCNR for QPSK is only 0.1 dB but for 16-QAM 1.6 dB. The large ΔCNR for 16-QAM is mainly due to the nonconstant amplitude mapping. For QPSK, the degradation is still negligible when the phase noise is -84 dBc/Hz. For 8-PSK and 16-QAM, however, the required CNR increases rapidly as the phase noise increases, because the carrier recovery process cannot catch up with relatively large phase variation and the constellation is often rotated by multiples of $\pi/4$ for 8-PSK or $\pi/2$ for 16-QAM. According to the simulated result, it can be said that a phase noise should be lower than -90 dBc/Hz at 1 MHz offset frequency for 16-QAM to avoid the serious PER degradation for the 60 GHz wireless system.

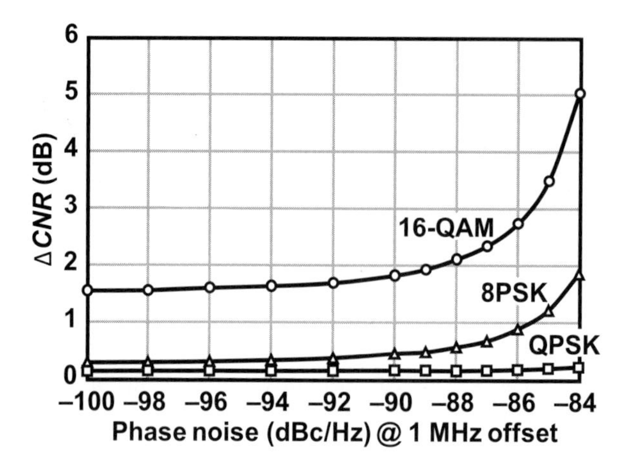

Figure 8.8 Increment of a required *CNR* by RF impairments (ΔCNR) to achieve a PER of 10^{-3} [24]. (©2011 IEEE. Reprinted, with permission, from *IEEE Journal of Solid-State Circuits.*)

To support 64-QAM, a phase noise of −96dBc/Hz at 1 MHz offset is required considering a 400 kHz bandwidth for the carrier tracking [30,39]. The carrier tracking is a common technique in baseband, which can suppress the phase noise effects within the tracking bandwidth. For the single-carrier (SC) mode, because the tracking bandwidth can be wider than the loop bandwidth of the PLL, the in-band phase noise is not a critical issue. Therefore, the out-of-band phase noise requirement is discussed as mentioned before. However, in Orthogonal Frequency Division Multiplexing (OFDM) cases, the tracking bandwidth is generally smaller than the loop bandwidth. Both in-band and out-of-band phase noise are dominant issues, which leads to a more stringent phase noise requirement. It has been demonstrated that the required phase noise performance can be achieved by using 60-GHz quadrature injection-locked oscillators (QILOs) and a 20 GHz PLL [24].

8.2.2 Direct-Conversion Transceiver Architecture

Figure 8.9 shows the block diagram of the 60 GHz one-stream front-end. A direct-conversion architecture is employed for both transmitter and receiver because of wide-bandwidth and low-power capability [9]. The transmitter adopts the mixer-first topology for achieving wideband gain characteristics with low power consumption.

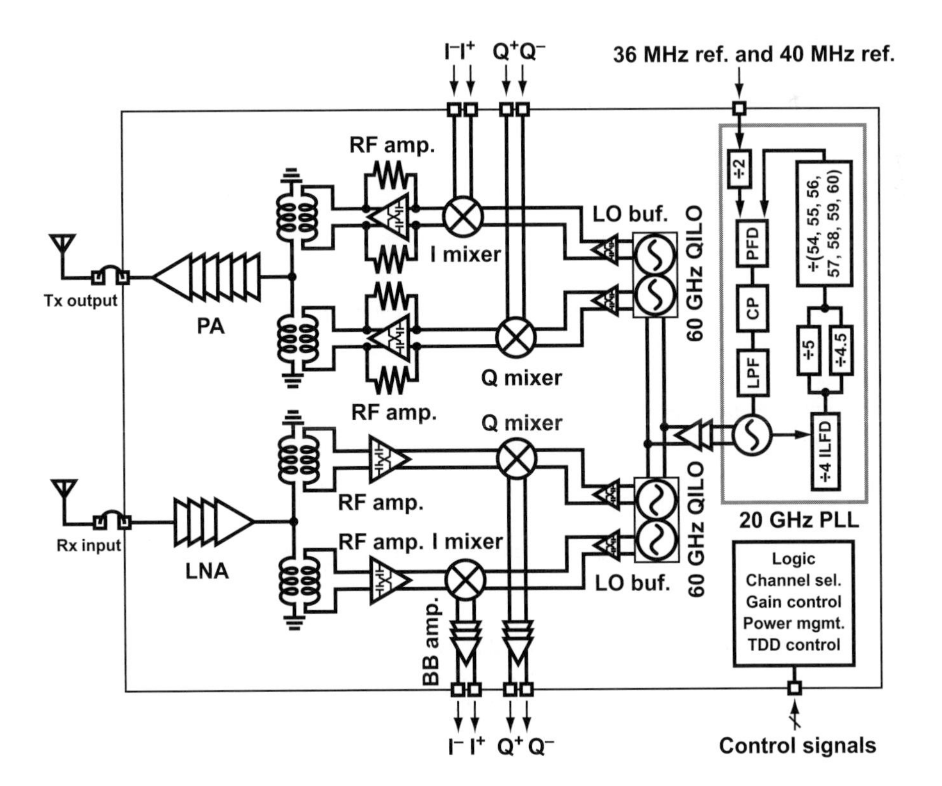

Figure 8.9 Block diagram of the 60 GHz direct-conversion TRX (one-stream). [28]. (©2017 IEEE. Reprinted, with permission, from *IEEE Journal of Solid-State Circuits.*)

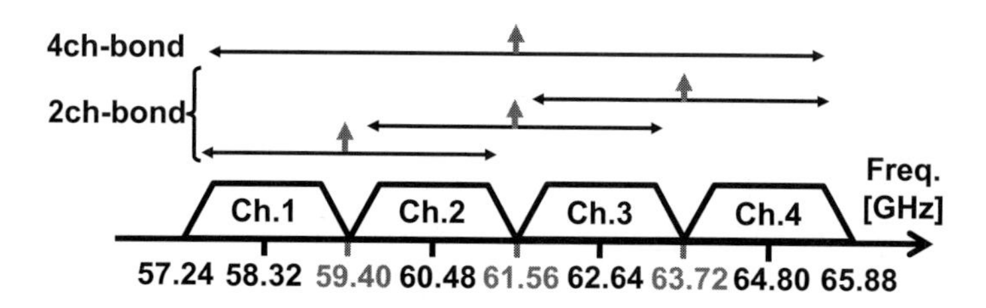

Figure 8.10 Channels defined by the IEEE802.11ad/WiGig standard. [28]. (©2017 IEEE. Reprinted, with permission, from *IEEE Journal of Solid-State Circuits.*)

It consists of a six-stage power amplifier (PA), differential preamplifiers, I/Q double-balanced passive mixers, and a QILO. To realize the wideband and linear characteristics, the receiver employs an open-loop baseband amplifier based on the flipped voltage follower (FVF) and a current-bleeding mixer. The receiver is composed of a four-stage low-noise amplifier (LNA), differential RF amplifiers, I/Q current-bleeding mixers, a QILO, and baseband amplifiers. The PA and LNA have a single-ended configuration due to the single-ended antenna, and the other parts have fully differential configurations. Each QILO is placed close to the transmitter and receiver mixers to avoid I/Q mismatch. If only a QILO is used for both transmitter and receiver, the 60 GHz differential LO distribution causes large insertion loss and I/Q phase and amplitude mismatch.

The LO consists of the 60 GHz QILO and a 20 GHz PLL. The 60 GHz QILO works as a frequency tripler with the integrated 20 GHz PLL. It can generate seven carrier frequencies with a 36/40 MHz reference, 58.32 GHz (channel 1), 60.48 GHz (channel 2), 62.64 GHz (channel 3), and 64.80 GHz (channel 4) defined in IEEE802.11ad/WiGig, 59.40 GHz (channels 1 and 2), 61.56 GHz (channels 1 and 2 or 1 through 4), and 63.72 GHz (channel 3 and 4) aiming for channel bonding in IEEE802.11ay, as illustrated in Figure 8.10. Injection-locked oscillators can be driven by a 1/N-frequency incident signal, which can be 30, 20, 15, 12, 10 GHz, etc., for 60 GHz injection-locked oscillators. There are trade-offs between locking range and phase noise, and between phase noise and frequency range. Basically, a too high frequency, e.g., 30 GHz, is not preferable for obtaining a good phase-noise performance. On the other hand, a larger divide ratio results in a poor locking range. In this work, a frequency of 20 GHz is employed as the incident signal frequency. The frequency of 20 GHz is very good for obtaining a wide frequency tuning range and good phase-noise performance. As mentioned earlier, the phase noise of the injection-locked oscillator is determined by that of the incident signal. Thus, a good phase-noise performance can also be obtained for a 60 GHz QILO.

8.3 Circuit Implementation of Key Building Blocks

8.3.1 Local Synthesizer

Figure 8.11 shows a block diagram of the proposed 60 GHz quadrature local synthesizer. The local synthesizer consists of a 60 GHz quadrature injection-locked oscillator and a

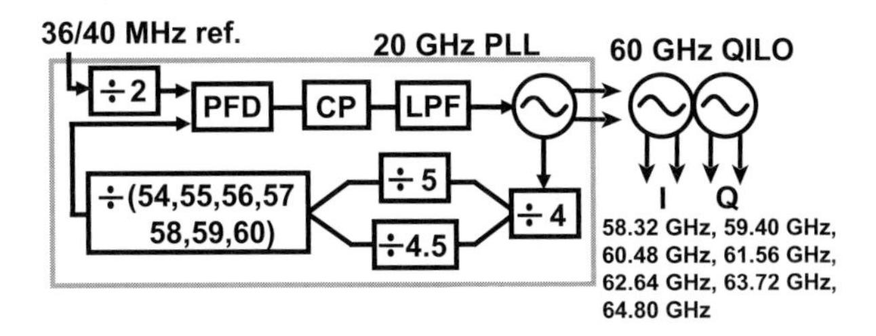

Figure 8.11 Block diagram of the 60 GHz quadrature local synthesizer.

20 GHz PLL as explained in the previous section. The QILO works as a frequency tripler with a 20 GHz injection-lock input, and it can generate quadrature outputs at 60 GHz. Basically, the phase noise of the QILO is determined by that of the 20 GHz PLL, so we can focus on frequency coverage and I/Q phase balance in the QILO design.

The 20 GHz PLL has an integer-*N* configuration [40,41]. The PLL has a two-stage divide-by-4 current-mode logic (CML) divider, a divide-by-5 static divider, and a programmable divider, /54, /55, /56, /57, /58, /59, and /60, to generate 19.44, 19.80, 20.16, 20.52, 20.88, 21.24, and 21.60 GHz with a 36 MHz reference, respectively. If a 40 MHz reference is applied, the static divider of divide-by-4.5 (instead of divide-by-5) should be used. These frequencies are exactly one-third of the required carrier frequency determined by the IEEE802.11ad standard [42] and channel bonding cases, which are used as incident signals to generate 58.32, 59.40, 60.48, 61.56, 62.64, 63.72, and 64.80 GHz, respectively. An LC-voltage-controlled oscillator (VCO) in the 20 GHz PLL has a tail-feedback to improve the phase noise [40,43]. The 20 GHz incident signal is injected through two differential buffer amplifiers to the 60 GHz QILO. The loop bandwidth is chosen to reduce the integrated phase noise in consideration of carrier recovery, which is about 76 kHz. There are two complex poles at 7.7 kHz, two real poles at 1.1 and 10.6 MHz, respectively, and one zero at 26.5 kHz.

The 60 GHz QILO design is shown in Figure 8.12, which consists of two LC oscillators. The QILO in this work employs a single-injection topology due to the layout simplicity, while the QILO in [40] uses a dual-injection technique. The free-running frequency can be adjusted by a switched capacitor array shown in Figure 8.12b. Millimeter-wave LC oscillators usually have a trade-off between phase noise and frequency tuning range because of a low Q factor of varactors and switched capacitors. On the other hand, the phase noise of the QILO is determined by the 20 GHz PLL, so the frequency range of the QILO can be widened by a larger capacitor variation.

One of the most important design considerations for the QILO is the I/Q mismatch. I/Q oscillators are connected to each other through tail transistors, and the I/Q phase balance is basically maintained by this I/Q cross-coupling. Theoretically, the operation of I/Q cross-coupling is not frequency dependent, so the phase balance can be robustly maintained across a wide frequency range.

The 20 GHz incident signal is injected through tail transistors of only the I oscillator. A mismatch between the intrinsic free-running frequencies of the I/Q oscillators degrades the I/Q phase balance, so tail transistors of the Q oscillator have dummy

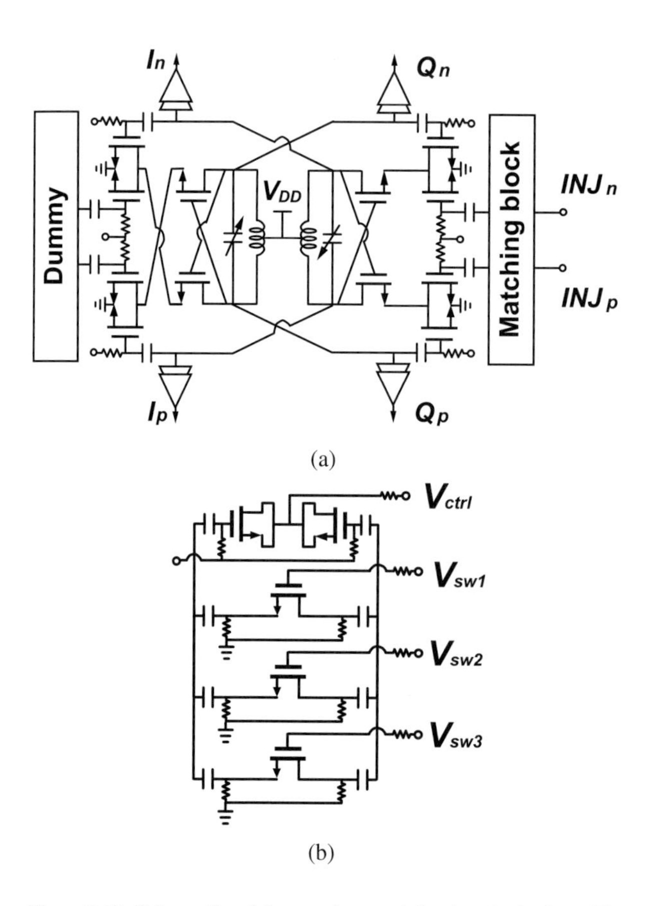

Figure 8.12 Schematic of the quadrature injection-locked oscillator [24]. (a) Quadrature injection-locked oscillator and (b) capacitor part. (©2011 IEEE. Reprinted, with permission, from *IEEE Journal of Solid-State Circuits*.)

matching blocks even though tail transistors are less sensitive to oscillation frequency [44]. As in previous work, the QILO in [45] has a polyphase filter to generate a 20 GHz quadrature signal, which is injected into both the I and Q oscillators. This is hereinafter referred to as I-Q injection while the topology in Figure 8.12 is referred to as I injection. I-Q injection contributes to widening the locking range. However, the polyphase filter generally has a phase mismatch, which becomes three times larger at 60 GHz. In addition, the phase mismatch at 60 GHz cannot be compensated by I/Q cross-coupling and is a critical problem for direct-conversion transceivers.

A QILO shown in Figure 8.12 is simulated to evaluate the locking range. As a comparison, a QILO using I-Q injection is also simulated [45], with a one-stage polyphase filter to generate an I-Q injection signal instead of the dummy matching block. The polyphase filter is designed for a center frequency of 20.16 GHz, which is one-third of 60.48 GHz (channel 2). Figure 8.13 shows simulated locking ranges for I and I-Q injections, which are 0.9 and 0.8 GHz, respectively. The polyphase filter for the I-Q injection has a 0.7 dB insertion loss, which degrades the locking range, so the QILO

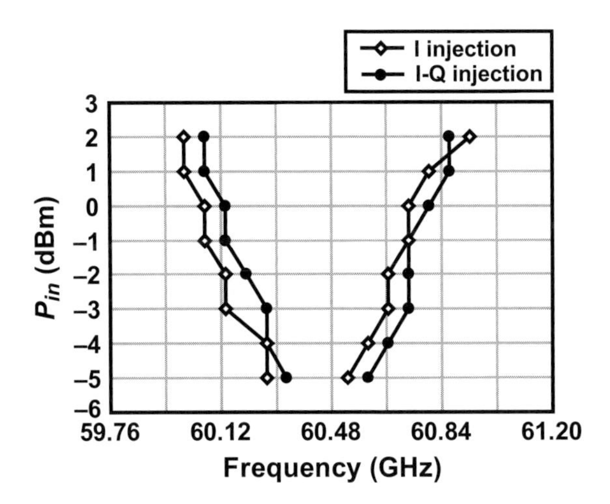

Figure 8.13 Simulated locking range [24]. (©2011 IEEE. Reprinted, with permission, from *IEEE Journal of Solid-State Circuits*.)

using I-Q injection has a slightly narrower locking range. In addition, quadrature injection signal paths tend to have a larger phase mismatch. Thus, a single-side injection (I injection) topology is employed in this work.

8.3.2 Transmitter

Starting from the TX topology consideration, Figure 8.14a shows a conventional design of the 60 GHz direct-conversion transmitter. The 5 GHz baseband bandwidth requires large power consumption for this topology because of low-pass nature of baseband amplifiers. For example, each baseband amplifier consumes 11 mW while achieving less than 3 GHz bandwidth in [12]. Therefore, in this work, the wide 5 GHz baseband bandwidth for both gain and input impedance is maintained by the proposed mixer-first topology. The detailed circuit implementation will be explained later in this section. Basically, this topology contributes to improving baseband gain characteristics and reducing power consumption. In addition, as shown in Figure 8.14b, by increasing RF gain by adding more stages for the PA and decreasing baseband gain by removing baseband amplifiers, the linearity requirement for the mixer block is relaxed. For instance, a 15 dB increase in the PA gain leads to a 15 dB reduction of the required output third-order intercept point (OIP3) for the mixer and RF amplifier group. Consequently, the number of LO buffer stages can be reduced. The LO buffers have to amplify four paths of 60 GHz signals in case of differential topology, which are very power-hungry. Even through a 13 mW increase in RF path is considered, the reduction of the LO buffers contributes to saving 45 mW power consumption in total. As a result, the mixer-first topology achieves wider bandwidth and reduces power consumption.

It is well-known that in software-defined radios the mixer-first receiver can be used to realize tunable RF characteristics with high linearity [46,47]. The baseband low-pass characteristics can be up-converted by a passive mixer. Thus, the RF bandpass

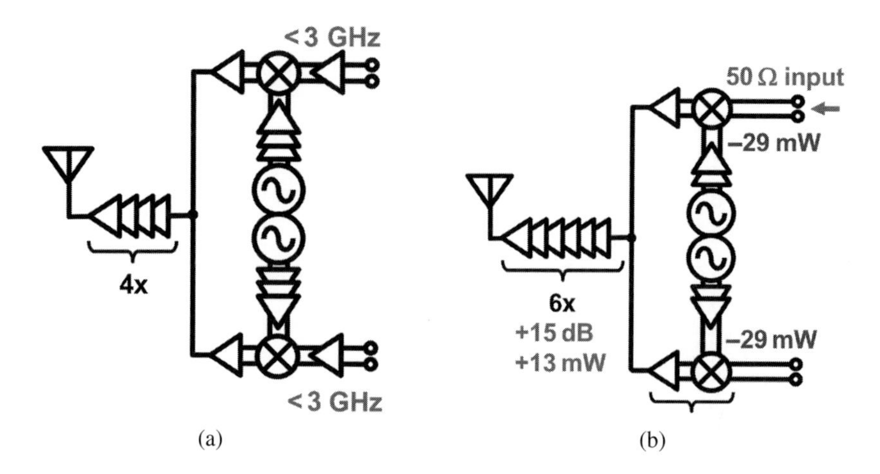

Figure 8.14 Block diagram of (a) a conventional 60 GHz direct-conversion transmitter [12] (©2012 IEEE. Reprinted, with permission, from *IEEE ISSCC Digest of Technical Papers*); (b) the proposed mixer-first transmitter [28]. (©2017 IEEE. Reprinted, with permission, from *IEEE Journal of Solid-State Circuits*.)

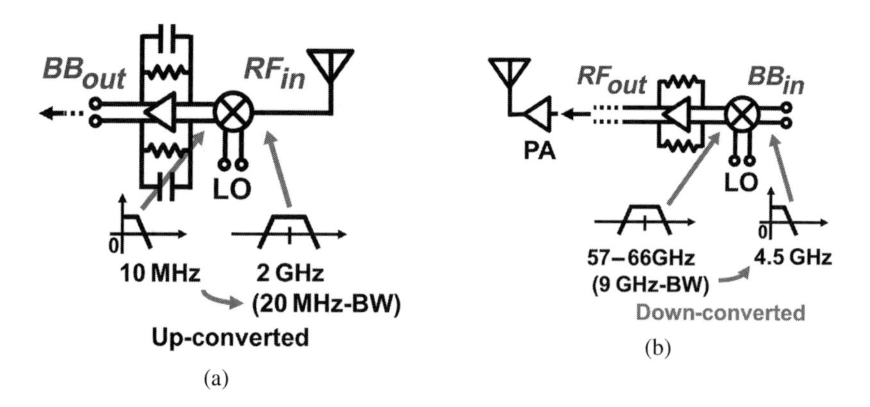

Figure 8.15 Simplified block diagram of (a) a mixer-first receiver and (b) the proposed mixer-first transmitter [28]. (©2017 IEEE. Reprinted, with permission, from *IEEE Journal of Solid-State Circuits*.)

characteristics are controlled by baseband circuitry, as shown in Figure 8.15a. In this work, the mixer-first topology is applied to a millimeter-wave transmitter for realizing wide baseband characteristics. A wide-and-flat gain characteristic such as 57–66 GHz is realized at the RF side, which is down-converted by a passive mixer. Hence, a 4.5 GHz wide-and-flat gain at the baseband side can be obtained, as depicted in Figure 8.15b. In addition, the 50 Ω input impedance is also maintained by the mixer-first topology.

Figure 8.16 shows the detailed circuit schematic of the mixer-first transmitter (I or Q path). It consists a double-balanced passive mixer and a differential resistive-feedback

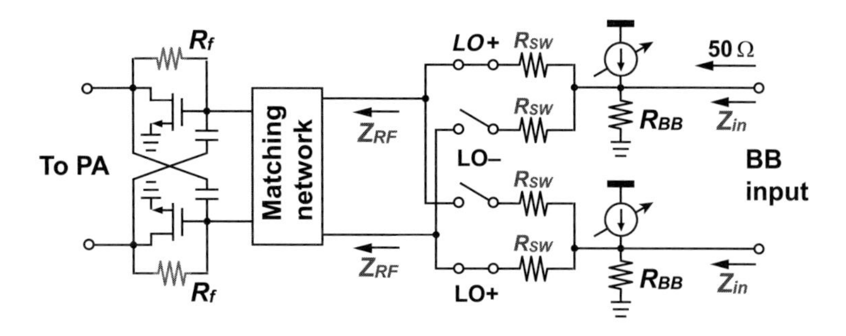

Figure 8.16 Circuit schematic of the mixer-first transmitter [28]. (©2017 IEEE. Reprinted, with permission, from *IEEE Journal of Solid-State Circuits.*)

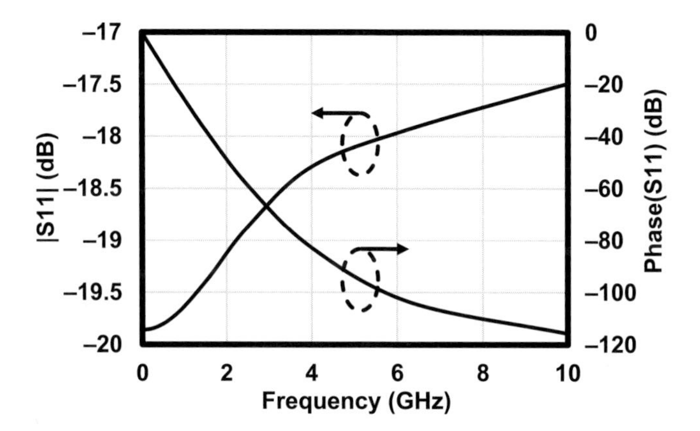

Figure 8.17 Simulated S11 seen at the baseband input port of the TX Mixer [28]. (©2017 IEEE. Reprinted, with permission, from *IEEE Journal of Solid-State Circuits.*)

RF amplifier with a matching network. The baseband input impedance (Z_{in}) can be roughly calculated as [48]

$$Z_{in}(\omega_{BB}) = R_{BB} /\!/ \left\{ R_{SW} + \frac{4}{\pi^2} \left[Z_{RF}(\omega_{BB} + \omega_{LO}) + Z_{RF}(\omega_{BB} - \omega_{LO}) \right] \right\} \tag{8.4}$$

where ω_{BB} and ω_{LO} represent the baseband frequency and LO frequency, respectively. Z_{RF} is the input impedance of the RF amplifier including the matching network. R_{SW} is the on-resistance of the switch. To achieve a wide-and-flat impedance characteristic at RF side, a resistive-feedback topology is applied to a differential amplifier with a capacitive-cross coupling neutralization. The matching block is used to compensate the imaginary part of Z_{RF}. The shunt resistors R_{BB} are also used for maintaining 50 Ω input impedance over the four channels. Figure 8.17 shows the simulated S11 seen at the baseband input port of the TX mixer. The magnitude of S11 is lower than −17 dB over 10 GHz bandwidth, which demonstrates the wideband characteristic. This will ease the implementation of the matching network between the baseband input of the mixer

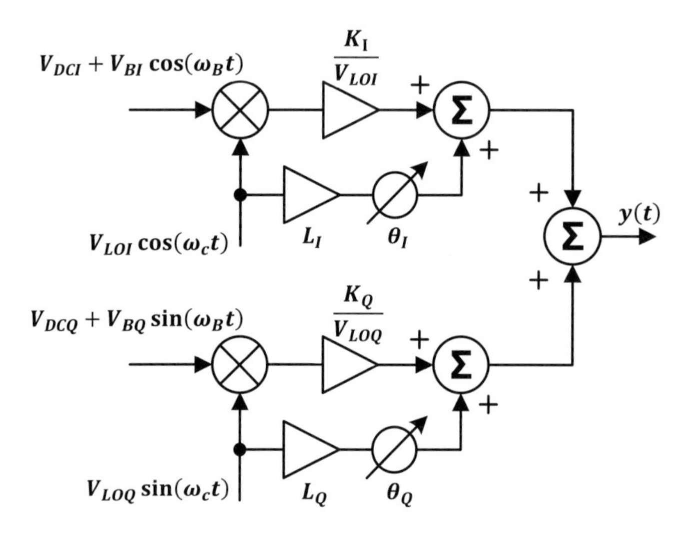

Figure 8.18 Analysis model of the TX quadrature mixer LO leakage [28]. (©2017 IEEE. Reprinted, with permission, from *IEEE Journal of Solid-State Circuits.*)

and the output of the low-pass filter (LPT) LPF in baseband circuitry, since the output of the LPF generally has a 100 Ω (differential) interface.

The current sources at the baseband input are used for LO leakage calibration. It is known that the LO leakage of the quadrature mixer can be minimized by adding the DC offset voltage at the baseband input [49,50]. Its principle is explained using the analysis model shown in Figure 8.18. V_{DCI} and V_{DCQ} are DC offset voltages applied at the baseband input of I path and Q path, respectively. The baseband signal has a frequency of ω_B and an amplitude of V_B. The frequency of the carrier signal is ω_c with an amplitude of V_{LO}. The LO leakage for each mixer is modeled as a scaled (L_I and L_Q) and phase-shifted (θ_I and θ_Q) signal of each LO input. K represents the conversion gain of the mixer. Therefore, the combined output signal is expressed as

$$y(t) = y_{leak}(t) + y_{sig}(t), \tag{8.5}$$

where $y_{leak}(t)$ is the LO leak at the combined output.

$$y_{leak}(t) = A\cos(\omega_c t + \alpha) + B\sin(\omega_c t + \beta) \tag{8.6}$$

$$A = \sqrt{K_I^2 V_{DCI}^2 + V_{LOI}^2 L_I^2 + 2K_I V_{DCI} V_{LOI} L_I \cos\theta_I} \tag{8.7}$$

$$\alpha = \arctan\left(\frac{V_{LOI} L_I \sin\theta_I}{K_I V_{DCI} + V_{LOI} L_I \cos\theta_I}\right) \tag{8.8}$$

$$B = \sqrt{K_Q^2 V_{DCQ}^2 + V_{LOQ}^2 L_Q^2 + 2K_Q V_{DCQ} V_{LOQ} L_Q \cos\theta_Q} \tag{8.9}$$

$$\beta = \arctan\left(\frac{V_{LOQ} L_Q \sin\theta_Q}{K_Q V_{DCQ} + V_{LOQ} L_Q \cos\theta_Q}\right) \tag{8.10}$$

and $y_{sig}(t)$ contains the desired signal and image signal (if $K_I V_{BI} \neq K_Q V_{BQ}$).

$$y_{sig}(t) = \frac{K_I V_{BI} + K_Q V_{BQ}}{2} \cos[(\omega_c - \omega_B)t] + \frac{K_I V_{BI} - K_Q V_{BQ}}{2} \cos[(\omega_c + \omega_B)t]. \tag{8.11}$$

Hence, the relative LO leakage is calculated using the following equation.

$$\eta|_{dBc} = 20 \log \frac{2C}{K_I V_{BI} + K_Q V_{BQ}} \tag{8.12}$$

where

$$C = \sqrt{A^2 + B^2 - 2AB \sin(\alpha - \beta)}. \tag{8.13}$$

The required DC offset voltages and voltage tuning resolution ($V_{DC,\,res}$) can be estimated by solving (8.12). In CMOS implementation, the leakage phase shift is very close to zero ($\theta_I \approx \theta_I \approx 0°$). Then (8.12) has a simplified formation:

$$\eta|_{dBc} = 10 \log \frac{4K_I^2 \left(V_{DCI} + \frac{V_{LOI} L_I}{K_I}\right)^2 + 4K_Q^2 \left(V_{DCQ} + \frac{V_{LOQ} L_Q}{K_Q}\right)^2}{(K_I V_{BI} + K_Q V_{BQ})^2}. \tag{8.14}$$

It is interesting to know that (8.14) is an ellipse equation in the V_{DCI}–V_{DCQ} plane with the semimajor axis of a and semiminor axis of b (or vice versa). The center point of the ellipse is ($V_{DCI,opt}$, $V_{DCQ,opt}$), which corresponds to $\eta = 0$.

$$a = 10^{\frac{\eta}{20}} \frac{(K_I V_{BI} + K_Q V_{BQ})}{2K_I} \tag{8.15}$$

$$b = 10^{\frac{\eta}{20}} \frac{(K_I V_{BI} + K_Q V_{BQ})}{2K_Q} \tag{8.16}$$

$$V_{DCI,opt} = -\frac{V_{LOI} L_I}{K_I} \tag{8.17}$$

$$V_{DCQ,opt} = -\frac{V_{LOQ} L_Q}{K_Q} \tag{8.18}$$

Therefore, $V_{DC,res}$ is determined by the minimum values among a, b, $|V_{DCI,opt}|$, and $|V_{DCQ,opt}|$.

$$V_{DC,res} = \min\left(a, b, |V_{DCI,opt}|, |V_{DCQ,opt}|\right). \tag{8.19}$$

Consequently, the optimum offset current ($I_{DCx,opt}$) and required current resolution ($I_{DC,res}$) are expressed as

$$I_{DCx,opt} = \frac{V_{DCx,opt}}{Z_{in}(\omega_{BB})|_{\omega_{BB}=0,x}} \tag{8.20}$$

$$I_{DC,res} = \frac{V_{DC,res}}{Z_{in}(\omega_{BB})|_{\omega_{BB}=0,x}} \tag{8.21}$$

where the subscript x is I or Q. To obtain some quantitative values, the following design parameter is used. The conversion gain of the I/Q mixer is -10 dB. The baseband input power of each path is 0 dBm. The LO amplitude is 1.2 V for I and Q path. L_I and L_Q are both −30 dB. The target η is −50 dBc. In this condition, $V_{DC,res}$ is dominated by a (=b), which is about 1.4 mV. The corresponding $I_{DC,res}$ is around 24 μA with the baseband input resistance of 59 Ω. In practical implementation, $I_{DCx,opt}$ and $I_{DC,res}$ fluctuate under process, voltage, and temperature variation. It should be taken into consideration such that recalibration schemes and enough design margin for the offset current tuning range and resolution are prepared.

Figure 8.19 shows the circuit schematic of the six-stage power amplifier. A transmission-line (TL)–based design is employed to achieve a reliable simulation and flexible layout. The loss constant of the TL is 0.8 dB/mm. A metal-insulator-metal transmission line (MIM TL) with the MIM capacitors shunt-connected alongside the TL is used for the decoupling of the power supplies. A series DC-cut capacitor is sometimes used as negative reactance in a matching block. However, capacitance is very sensitive to the process variation [51], so a large capacitance is used to reduce the influence of process variation. In this design, 300 fF capacitors are used, which are 8.8j Ω at 60 GHz.

A common-source topology is chosen for each gain stage due to its high linearity. The transistors in the PA have a finger width of 2 μm for gain optimization [52] and have an asymmetric drain-source structure [53]. The distance between the gate and drain contacts is kept long to reduce a drain-gate capacitance, and the distance between the gate and source contacts is kept short to reduce gate-source diffusion resistance. The asymmetric layout structure contributes a 0.5-dB improvement in the maximum available gain [53]. The transistor size is gradually increased from the input stage to the output stage of the PA with the consideration of power consumption and linearity. The multistage gain peaking technique [24] is adopted for realizing wide and flat gain characteristics. Figure 8.20 shows the simulated gain of the power amplifier at different temperatures and supply voltages. The maximum gain difference within the 8.64 GHz bandwidth is kept around 1.3 dB over temperature and voltage variation. However, the absolute gain of the PA varies obviously, which necessitates the use of compensation

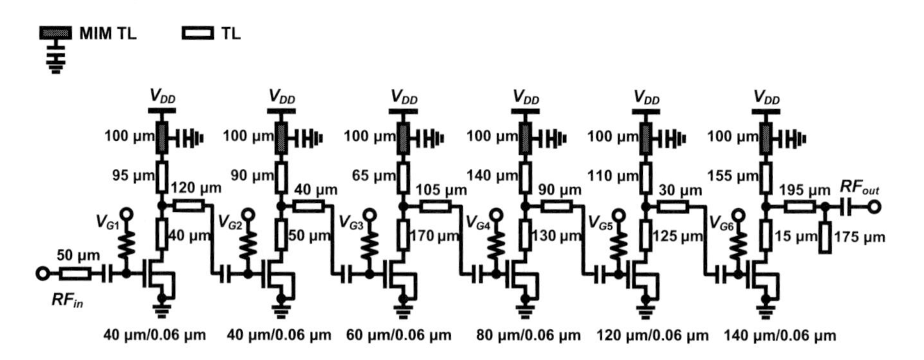

Figure 8.19 Schematic of the 60 GHz power amplifier [28]. (©2017 IEEE. Reprinted, with permission, from *IEEE Journal of Solid-State Circuits*.)

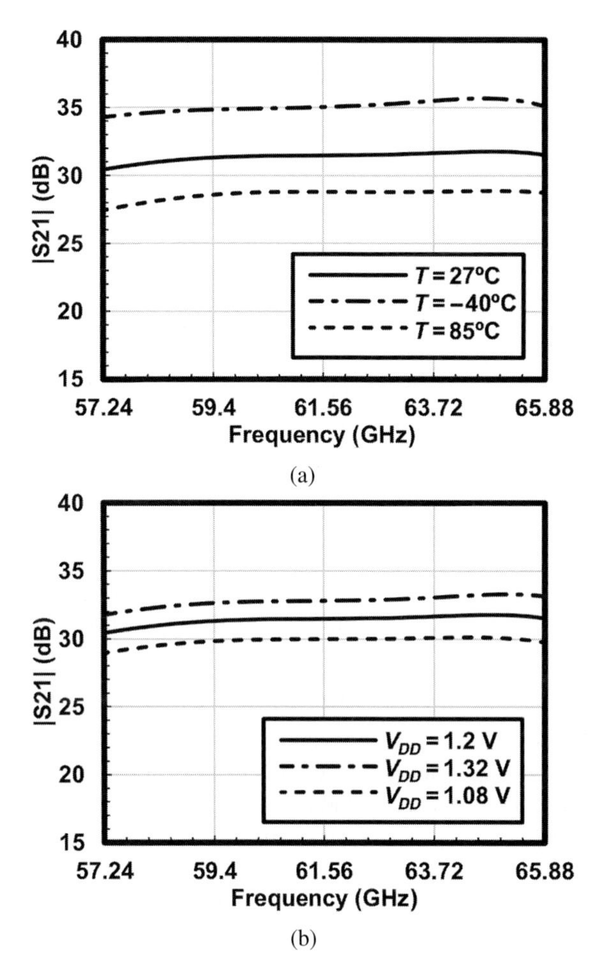

Figure 8.20 Simulated gain of the PA versus frequency (a) at different temperatures with 1.2 V supply and (b) at the temperature of 27°C with different supply voltages [28]. (©2017 IEEE. Reprinted, with permission, from *IEEE Journal of Solid-State Circuits.*)

and calibration techniques, such as constant-g_m biasing, low-ripple voltage regulators, gain calibration loop, and so on, in practical implementation.

8.3.3 Receiver

For mm-wave receiver design, the noise figure, linearity, gain flatness, and frequency-dependent I/Q mismatch should be considered. Especially for the four-channel bonding condition, the noise floor becomes at least 98 dB higher from −174 dBm/Hz. The peak SNDR of the RX also suffers from the nonlinearity of the receiver chain. In addition, it is difficult to use a close-loop baseband amplifier for improving linearity and gain flatness due to the wide bandwidth. Thus, in this work an open-loop baseband amplifier based on the FVF [54] is employed to maintain both gain flatness and linearity with reasonable power consumption.

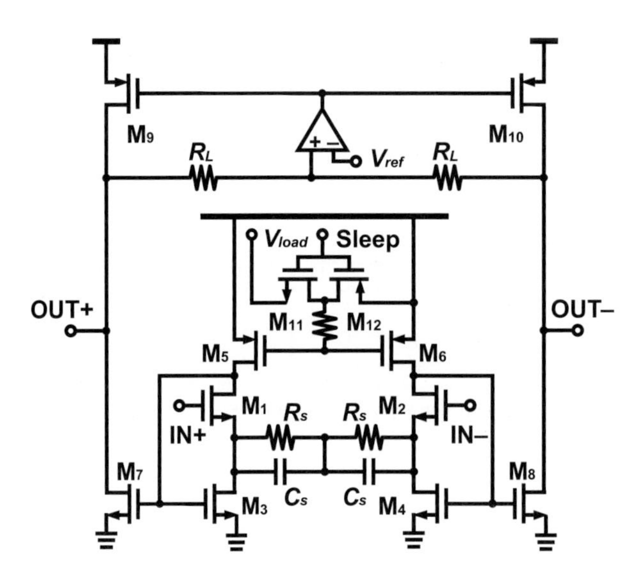

Figure 8.21 Circuit schematic of one unit cell for the first and second stage of the baseband amplifier [28]. (©2017 IEEE. Reprinted, with permission, from *IEEE Journal of Solid-State Circuits.*)

Figure 8.21 shows the detailed circuit schematic of one unit cell used for the first and second stage of the baseband amplifier. The unit cell consists two amplifier stages. The first stage is a modified FVF amplifier with resistors R_s and capacitors C_s connected between the source terminals of M1 and M2. The voltage gain of the first stage can be expressed as

$$A_{V,1st}(\omega) \approx \frac{1}{g_{m3} \cdot \left[r_{ds3} /\!/ R_s /\!/ (1/j\omega C_s)\right]}. \tag{8.22}$$

The second stage is an active-load common-source amplifier with common-mode feedback loop. The voltage gain of the second stage is

$$A_{V,2nd}(\omega) \approx g_{m7} \cdot \left[r_{ds7} /\!/ r_{ds9} /\!/ R_L /\!/ (1/j\omega C_L)\right], \tag{8.23}$$

where C_L represents the capacitor between the output node of the second stage and ground. Assume that $r_{ds3} \gg R_s$, $r_{ds7} \gg R_L$, and $r_{ds9} \gg R_L$, the voltage gain of the unit cell is

$$A_{V,BB}(\omega) = A_{V,1st}(\omega) \cdot A_{V,2nd}(\omega) \approx \frac{g_{m7} \cdot \left[R_L /\!/ (1/j\omega C_L)\right]}{g_{m3} \cdot \left[R_s /\!/ (1/j\omega C_s)\right]} \tag{8.24}$$

where g_{m3} and g_{m7} are the transconductance of M3 and M7, respectively. Basically, the voltage gain is determined by the g_m ratio and resistance ratio. The nonlinearity of the amplifier is reduced because of the similar bias conditions for M3 (M4) and M7 (M8). The capacitor C_s is used for a gain peaking at high frequencies. So, a flat-and-linear characteristic can be realized by an open-loop amplifier. Considering that M3 (M4) and

M7 (M8) have the same channel length and bias conditions, (8.24) can be simplified as the following equation at low frequency:

$$A_{V,BB} \approx N\frac{R_L}{R_s}, \quad N = \frac{W_{M7}}{W_{M3}} = \frac{W_{M8}}{W_{M4}} \tag{8.25}$$

Figure 8.22 shows the simulated voltage gain of the baseband amplifier versus frequency. A −3-dB bandwidth of 4.6 GHz is achieved at 27°C with 1.2 V supply voltage. As will be discussed later in Section 8.4, a mismatch in cut-off frequencies of baseband amplifiers can cause frequency-dependent I/Q mismatch. This FVF-based amplifier can relax the influence of the mismatch due to its high cutoff frequency. However, the variation of the supply voltage and temperature causes the change of the

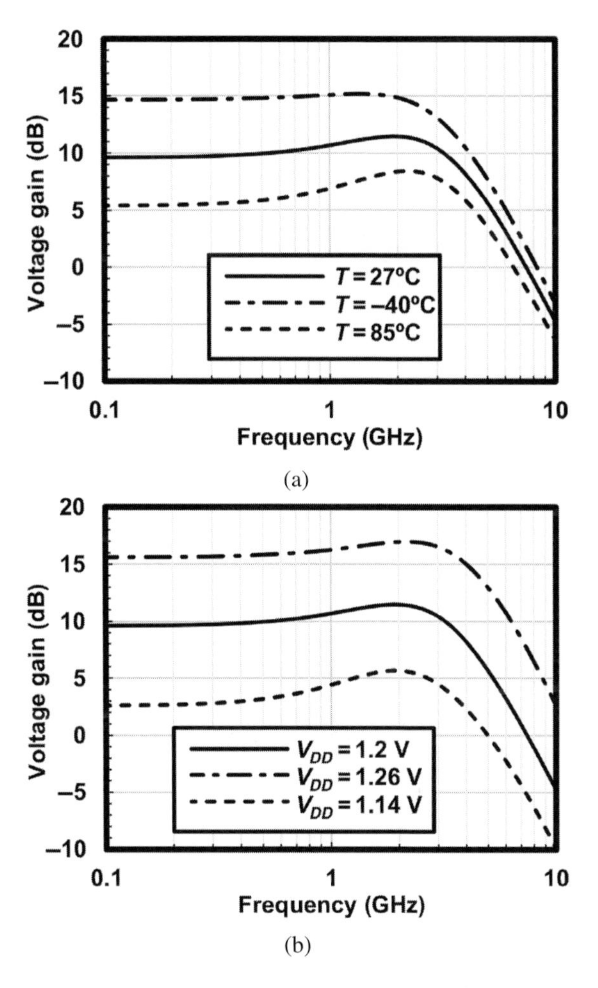

Figure 8.22 Simulated voltage gain of the baseband amplifier versus frequency (a) at different temperatures with 1.2 V supply and (b) at the temperature of 27°C with different supply voltages [28]. (©2017 IEEE. Reprinted, with permission, from *IEEE Journal of Solid-State Circuits.*)

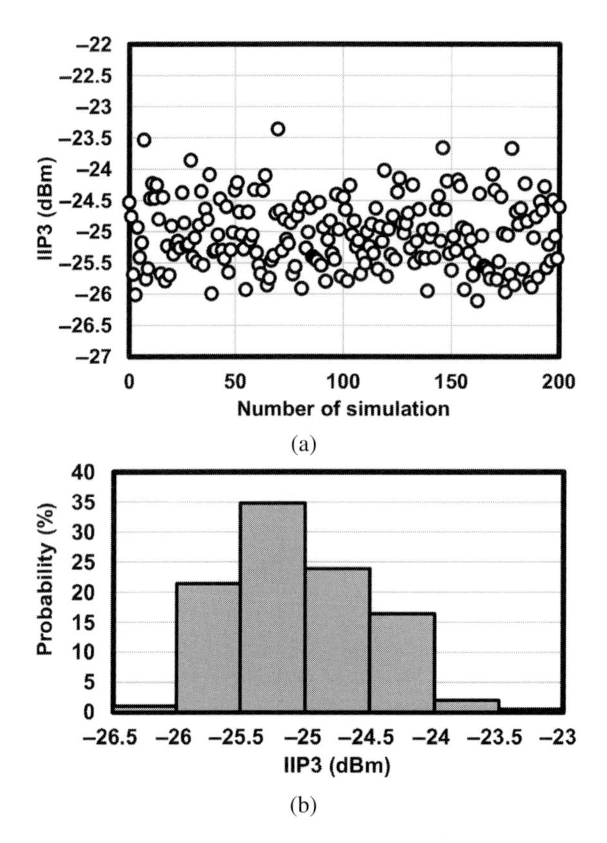

Figure 8.23 Monte Carlo simulation results of the baseband amplifier: (a) simulated IIP3 versus number of simulation and (b) calculated probability histogram of IIP3 [28]. (©2017 IEEE. Reprinted, with permission, from *IEEE Journal of Solid-State Circuits*.)

gain flatness and −3-dB bandwidth as depicted in Figure 8.22. It is mainly because the transistors deviate from the designed operation condition with constant-voltage biases when the temperature and supply voltage change. In practical uses, it is better to employ temperature-insensitive bias techniques and on-chip voltage regulators to mitigate those effects. Figure 8.23 shows the Monte Carlo simulation results of the baseband amplifier, which estimates the influence of the random mismatch between transistors M3 (M4) and M7 (M8) on the linearity of the baseband amplifier. A $\pm 10\%$ random mismatch among the transistors M3, M4, M7, and M8 of the second FVF stage is applied in the simulation. The simulated IIP3 varies between –23 and −26.5 dBm considering the random mismatch, while the simulated IIP3 is around −25 dBm without the mismatch.

Figure 8.24 shows the circuit schematic of the double-balanced down-conversion mixer. The input stage has a capacitive cross-coupling for higher gain. To reduce the required LO power, a known technique called current-bleeding [55] is applied, which contributes to reducing power consumption of the LO buffers. The transmission

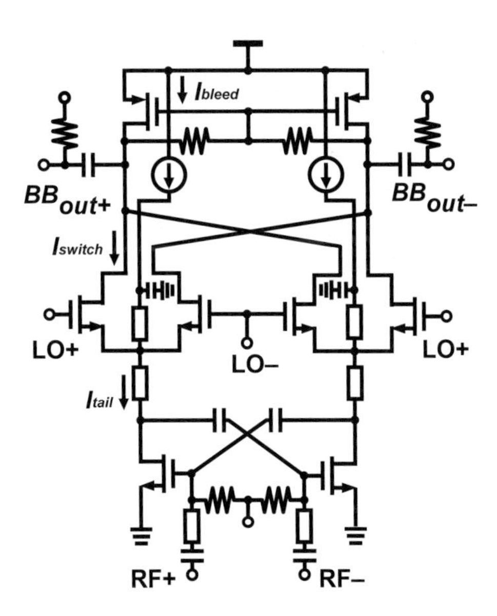

Figure 8.24 Circuit schematic of the current-bleeding down-conversion mixer [28]. (©2017 IEEE. Reprinted, with permission, from *IEEE Journal of Solid-State Circuits*.)

lines are placed between the neutralized mixer input stage and the switching stage to maximize the power transfer. In the current-bleeding paths, the series transmission lines and shunt capacitors are used to form a high impedance at the RF frequency. The simulated IIP3 and NF of the mixer are -7.5 dBm and 17 dB, respectively. The estimated (averaged) IIP2 from the Monte Carlo simulation is about 12.4 dBm. In this condition, the estimated SNDR at the output of the mixer is around 31 dB for a typical input power of -30 dBm, which has enough margin (about 9 dB) for the 64-QAM communication.

Figure 8.25a shows a circuit schematic of the LNA, which has a common-source common-source (CS-CS) topology to improve the noise figure [53]. The first and second stages use a 1 μm finger width for noise optimization [52]. Actually, the second stage still has a large noise contribution at millimeter frequency due to low gain such as 6 dB of the first stage, therefore the second stage also employs the common-source configuration to improve the noise figure rather than a common-gate configuration, i.e., a cascode configuration. The third and fourth stages have a 2 μm finger width for gain optimization [52]. The input matching block of the LNA has a shunt-grounded structure for electrostatic discharge (ESD) protection. The matching block is designed with low-impedance MIM TLs, 50 Ω transmission lines, and parallel-line transformers. The gain can be varied by adjusting the gate-bias voltages.

The LNA output is separated into I and Q paths through a parallel-line transformer, and each is connected to a down-conversion mixer. In this design, the parallel-line transformer is used for single-to-differential conversion. Since a transformer balun generally causes an imbalance in differential signals, a differential amplifier that has a high common-mode rejection ratio (CMRR) is used to reduce the imbalance as explained in the following equations:

$$\frac{A_{\text{in}+}}{A_{\text{in}-}} = \frac{A(1+k)e^{j(\omega t+\theta)}}{-Ae^{j\omega t}} = -(1+k)e^{j\theta} \tag{8.26}$$

$$\frac{A_{\text{out}+}}{A_{\text{out}-}} = -\frac{\alpha\{(1+k)e^{j\theta}+1\}+\{(1+k)e^{j\theta}-1\}}{\alpha\{(1+k)e^{j\theta}+1\}-\{(1+k)e^{j\theta}-1\}} \tag{8.27}$$

$$\simeq -\left(1+\frac{k}{\alpha}\right)e^{j\frac{\theta}{\alpha}} \tag{8.28}$$

where k and θ are gain and phase mismatches in input differential signal, and α is CMRR. Output differential signal ($A_{\text{out}+}$, $A_{\text{out}-}$) has less mismatch than input differential signal ($A_{\text{in}+}$, $A_{\text{in}-}$) due to the common-mode rejection. For example, when CMRR is 10 dB, gain and phase mismatches in a balun can be improved from 1.0 dB and 5° to 0.10 dB and 0.49°, respectively. Thus, a high CMRR amplifier is required before a double-balanced mixer to improve the dynamic range of receiver, e.g., IM2.

The common differential amplifier using a tail transistor cannot maintain a high CMRR at millimeter-wave frequency because of the parasitic capacitance at the drain of the tail transistor. Thus, the common-mode rejection (CMR) is realized in the matching blocks of differential amplifiers, as shown in Figure 8.25b. The shunt part in the matching block works as a short stub for a differential-mode signal because it forms a virtual ground. For a common-mode signal, it works as an open stub. However, the stub

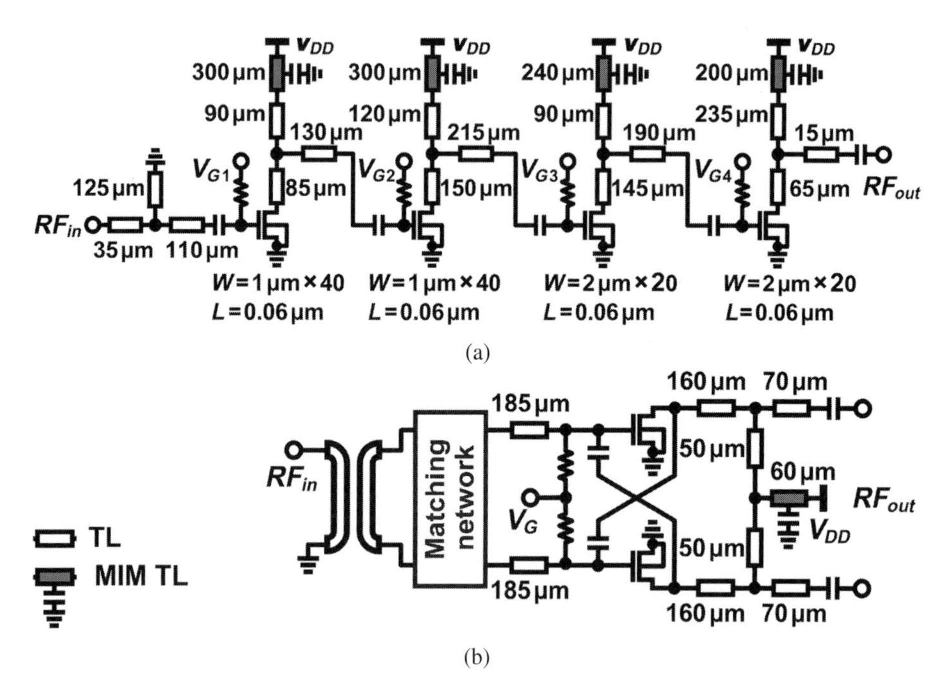

Figure 8.25 Circuit schematic of (a) the 60-GHz LNA and (b) the RF amplifier [28]. (©2017 IEEE. Reprinted, with permission, from *IEEE Journal of Solid-State Circuits*.)

parasitics reduce the impedance at the virtual ground node even for a common-mode signal, so in this work a short transmission line is inserted at the virtual ground node for obtaining a larger open-stub impedance. This technique is also used for LO buffers.

8.3.4 Calibration Techniques

It is known that the I/Q mismatch has significant influence on the EVM performance of the 60-GHz transceiver. Unfortunately, as depicted in Figure 8.26, conventional tunable RF/analog amplifiers cannot realize individual and fine-tuning for the gain and phase mismatch. Even though digital-baseband techniques can achieve fine and separate calibration of the gain and phase error, they require extra high-speed baseband circuitry with large power consumption and area penalty. A mm-wave phase shifter using an injection-locked oscillator is reported for a phased-array receiver [56], and this technique can be applied to a QILO. A fine I/Q phase calibration is realized by adjusting free-running frequency of the QILO, and a 10-bit DAC is used for a DC-domain fine-tuning of the control voltage of the QILO varactor [29,30]. The schematic of the 60 GHz QILO implemented in this work is shown in Figure 8.12. To satisfy the phase tuning requirement, the free-running frequency of the QILO is designed to cover from 58 to 66 GHz, which is wider than the required carrier frequency range (58.32–64.8 GHz). The frequency tuning is implemented by a 3-bit switched capacitor bank and a varactor controlled by the 10-bit DAC.

Figure 8.27 illustrates the I/Q mismatch-calibration technique used in this work, which can realize the fine phase calibration and compensate the I/Q gain and phase errors separately. The variable-gain RF amplifiers are used for the I/Q gain calibration. A 10-bit DAC is used to tune the gate bias voltage of the RF amplifier for gain adjusting. The QILO is used as a very-fine phase shifter for the fine I/Q phase calibration. Because the proposed phase tuning has negligible influence on the gain characteristic, individual tuning for the gain and phase mismatch can be realized. The LO leakage is minimized by adjusting the DC level at the baseband input of the TX through the current sources and shunt resistors. All of the transceiver can be calibrated by the following order: TX LO leakage, TX I/Q gain mismatch, TX I/Q phase mismatch, RX I/Q gain mismatch, and RX I/Q phase mismatch.

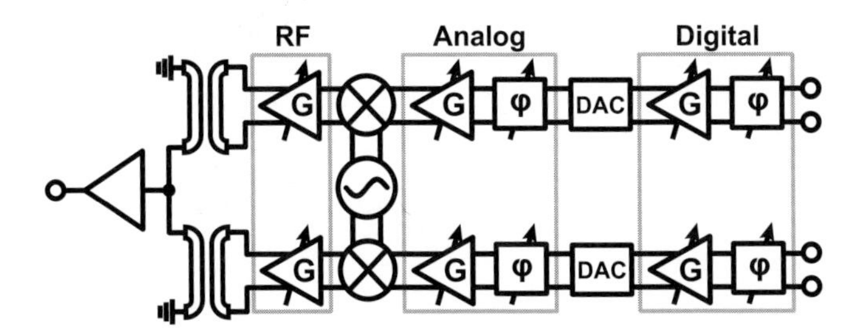

Figure 8.26 Conventional I/Q mismatch calibration method [28]. (©2017 IEEE. Reprinted, with permission, from *IEEE Journal of Solid-State Circuits*.)

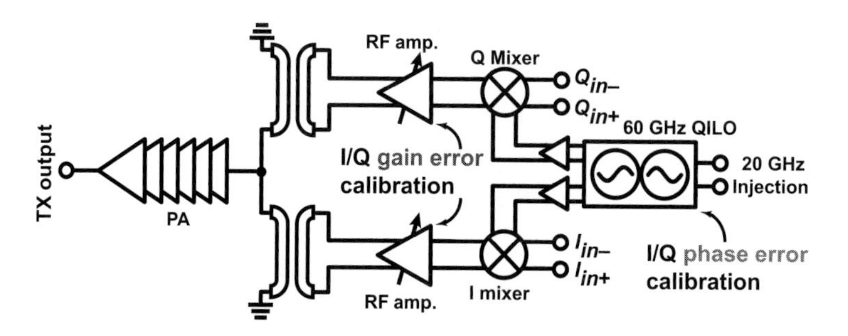

Figure 8.27 Proposed I/Q mismatch calibration method [28]. (©2017 IEEE. Reprinted, with permission, from *IEEE Journal of Solid-State Circuits.*)

It is worth knowing that the performance of the calibrated (I/Q mismatch and LO leakage) transceiver may be degraded over corners. Figure 8.28 shows the simulated gain difference of the I/Q RF amplifiers, which are used for the I/Q gain mismatch calibration. The I path amplifier is biased at 0.6 V. The Q path amplifier is biased at 0.55 V. This imitates the bias-tuning operation during the calibration. It is shown that the gain difference between I and Q path is not kept the same over temperature and supply voltage variations. In the simulation, the varied gain difference can be over 0.4 dB, which causes an IMRR of worse than 33 dBc. Figure 8.29 shows the simulated phase difference between I output and Q output of the QILO at different temperatures. The phase difference is optimized to 90° around the temperature of 27°C, while in the worst case, the phase difference is shifted to 95°, which corresponds to an IMRR of 27 dB. Therefore, a PVT-variation tolerant design [10] and automatic detection and calibration technique [57] are necessary for maintaining the system performance.

By utilizing the preceding techniques, the implemented transceiver can realize four-channel bonding with 16-QAM, which will be shown in Section 8.4. However, it is still difficult to realize four-channel bonding with 64-QAM achieving 42.24-Gb/s data rate. One of the dominant issues is the frequency-dependent I/Q mismatch, as depicted in Figure 8.30. For 64-QAM four-channel-bonding applications, an IMRR of more than 35 dBc should be satisfied over 4.32 GHz bandwidth in consideration of other impairment degradations. Unfortunately, the IMRR will be degraded along with the frequency even using the proposed calibration technique.

To gain more detailed insight of this issue, (8.2) can be extended to a frequency-dependent formation. Note that the analysis and discussion presented in the following are mainly for the single-carrier mode. Since the I/Q mismatch can be compensated for individual subcarriers in OFDM mode, the frequency-dependent IMRR has fewer effects on the system EVM.

$$|IMRR\,(\omega)| = \left|\frac{P_{sig}(\omega)}{P_{img}(\omega)}\right| = \left|\frac{\gamma^2(\omega) + 2\gamma(\omega)\cos[\Delta\phi(\omega)] + 1}{\gamma^2(\omega) - 2\gamma(\omega)\cos[\Delta\phi(\omega)] + 1}\right| \tag{8.29}$$

where

$$\gamma(\omega) = \frac{\alpha(\omega)}{\beta(\omega)} \tag{8.30}$$

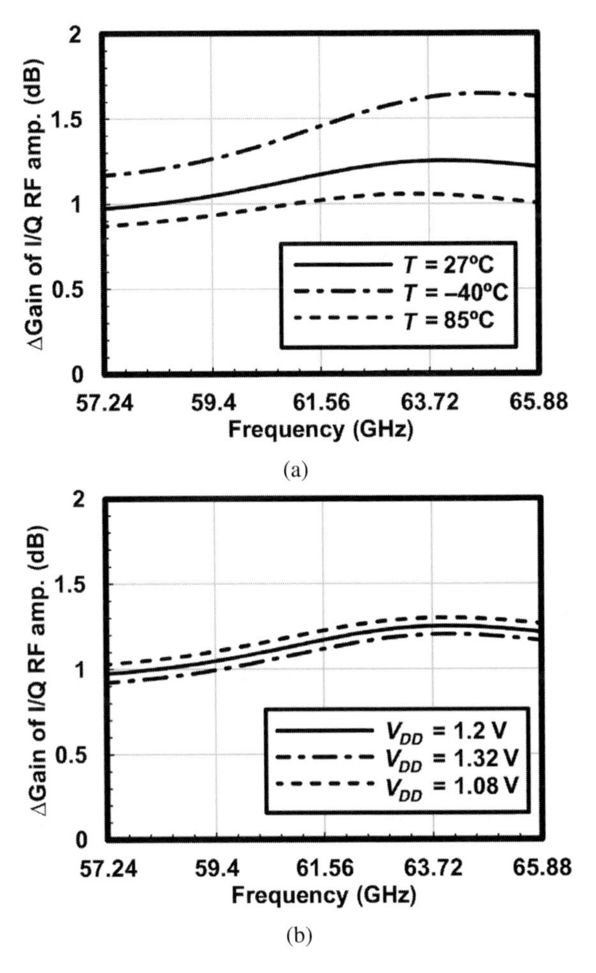

Figure 8.28 Simulated gain difference of the I/Q RF amplifier when the gate bias of the I and Q amplifier is 0.6 and 0.55 V, respectively: (a) at different temperatures with 1.2 V supply (b) at the temperature of 27°C with different supply voltages [28]. (©2017 IEEE. Reprinted, with permission, from *IEEE Journal of Solid-State Circuits*.)

$$P_{sig}(\omega) = \frac{1}{8}\left\{\alpha^2(\omega) + 2\alpha(\omega)\beta(\omega)\cos[\Delta\phi(\omega)] + \beta^2(\omega)\right\} \tag{8.31}$$

$$P_{img}(\omega) = \frac{1}{8}\left\{\alpha^2(\omega) - 2\alpha(\omega)\beta(\omega)\cos[\Delta\phi(\omega)] + \beta^2(\omega)\right\} \tag{8.32}$$

$\alpha(\omega)$ is the gain of the in-phase path and $\beta(\omega)$ is the gain of the quadrature-phase path. $\Delta\phi(\omega)$ is the phase difference between the in-phase path and quadrature-phase path. $P_{sig}(\omega)$ represents the desired signal power. $P_{img}(\omega)$ denotes the unwanted image power.

Therefore, an integrated EVM ($EVM_{IMRR,int}$) induced by the frequency-dependent I/Q mismatch can be evaluated using the following equation:

$$EVM_{IMRR,int} = \frac{1}{SNR_{IMRR,int}} = \frac{\int P_{img}(\omega)\,d\omega}{\int P_{sig}(\omega)\,d\omega} \tag{8.33}$$

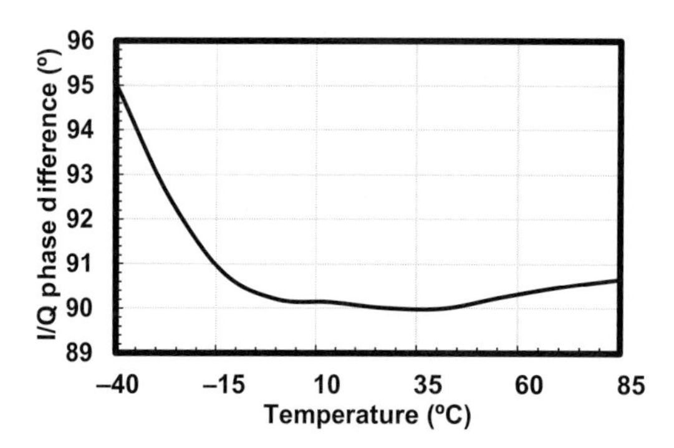

Figure 8.29 Simulated phase difference between I output and Q output of the QILO at different temperatures [28]. (©2017 IEEE. Reprinted, with permission, from *IEEE Journal of Solid-State Circuits.*)

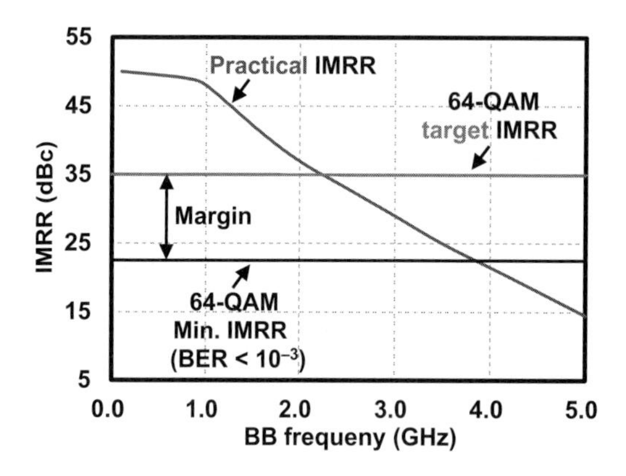

Figure 8.30 Wideband I/Q mismatch challenge [28]. (©2017 IEEE. Reprinted, with permission, from *IEEE Journal of Solid-State Circuits.*)

$SNR_{IMRR,int}$ is the ratio of the integrated desired signal power to the integrated image power over the frequency band of interest. It should be pointed out that the integrated EVM is calculated assuming the signal tones are not correlated. This assumption gives a handy expression for a quantitative but rough estimation of the frequency-dependent IMRR issues. More accurate results should be obtained by the comprehensive system simulation and measurement.

The main cause of the frequency-dependent I/Q mismatch is the mismatch in cut-off frequencies, especially at baseband paths. Figure 8.31 shows an example how the baseband mismatch and proposed calibration technique influence the IMRR and EVM of the system. It is assumed that the I path and Q path both exhibit a third-order low-pass characteristic at baseband side, which is a bandpass response referred to RF frequencies. The −3-dB bandwidth of I path is 4.1 GHz, while that of Q path is 3.9 GHz. This models the cut-off frequency mismatch that generally occurs in baseband paths. The carrier frequency is 61.56 GHz. In this example, the uncalibrated I/Q phase mismatch includes a constant phase error of 3 degree and phase errors caused by the cut-off

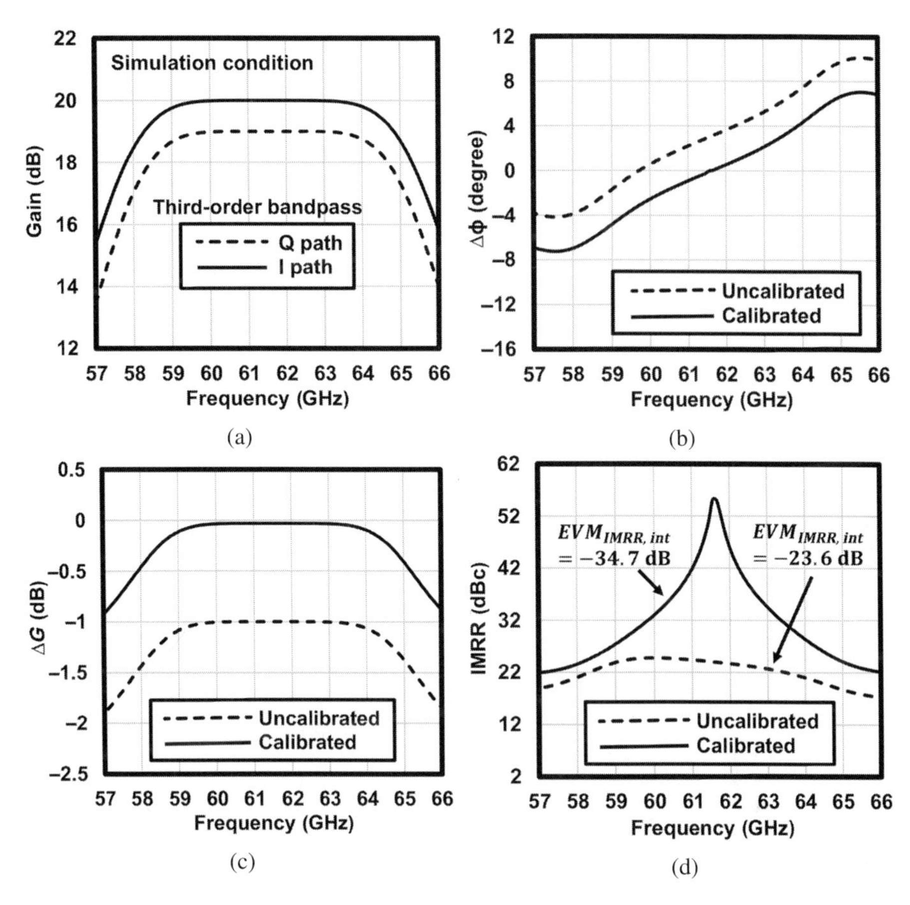

Figure 8.31 An example of frequency-dependent I/Q mismatch dominated by baseband parts: (a) uncalibrated gain characteristic of I path and Q path versus frequency; (b) phase difference between I path and Q path versus frequency; (c) gain difference between I path and Q path versus frequency; and (d) calculated IMRR versus frequency and $EVM_{IMRR,int}$ within 4.32 GHz bandwidth [28]. (©2017 IEEE. Reprinted, with permission, from *IEEE Journal of Solid-State Circuits*.)

frequency mismatch shown in Figure 8.31b. The uncalibrated I/Q gain mismatch is also composed of a constant gain error of 1 dB and gain errors due to the cut-off frequency mismatch depicted in Figure 8.31c. It can be observed in Figure 8.31d that the IMRR is less than 25 dBc over the 4.32 GHz frequency band of interest (tow-channel bonding) without calibration. The corresponding $EVM_{IMRR,int}$ is about −23.6 dB, which is far from the target value of −35 dB. For demonstration simplicity, the following assumptions are made for the calibration applied in Figures 8.31 and 8.32: (a) the phase error and gain error calibration have no mutual influence on each other and (b) the gain tuning value and phase tuning value are constant over the frequency band of interest. The calibration is performed at 61.66 GHz, where a peak IMRR is observed in Figure 8.31d. The corresponding $EVM_{IMRR,int}$ is −34.7 dB over the 4.32 GHz band. It should also be pointed out that the calibrated $EVM_{IMRR,int}$ is degraded to −28.1 dB if the bandwidth is extended to 8.64 GHz (four-channel bonding). This gives a quantitative demonstration of the critical frequency-dependent I/Q mismatch issue for four-channel bonding.

Similarly, the influence of the cut-off frequency mismatch at RF is modeled and exemplified in Figure 8.32. Both I path and Q path show an third-order bandpass characteristic with the −3-dB bandwidth of 7 GHz at RF paths. The higher-side and lower-side cut-off frequency of the Q path are 400 MHz lower than those of the I path. The uncalibrated I/Q phase mismatch consists a constant phase error of 3 degree and phase errors caused by the cut-off frequency mismatch shown in Figure 8.32b. The uncalibrated I/Q gain mismatch is composed of a constant gain error of 1 dB and gain errors due to the cutoff frequency mismatch depicted in Figure 8.32c. An obvious improvement of IMRR and $EVM_{IMRR,int}$ (within 4.32 GHz bandwidth) can be observed when the calibration is conducted at 61.66 GHz. It is also shown that the frequency-dependent

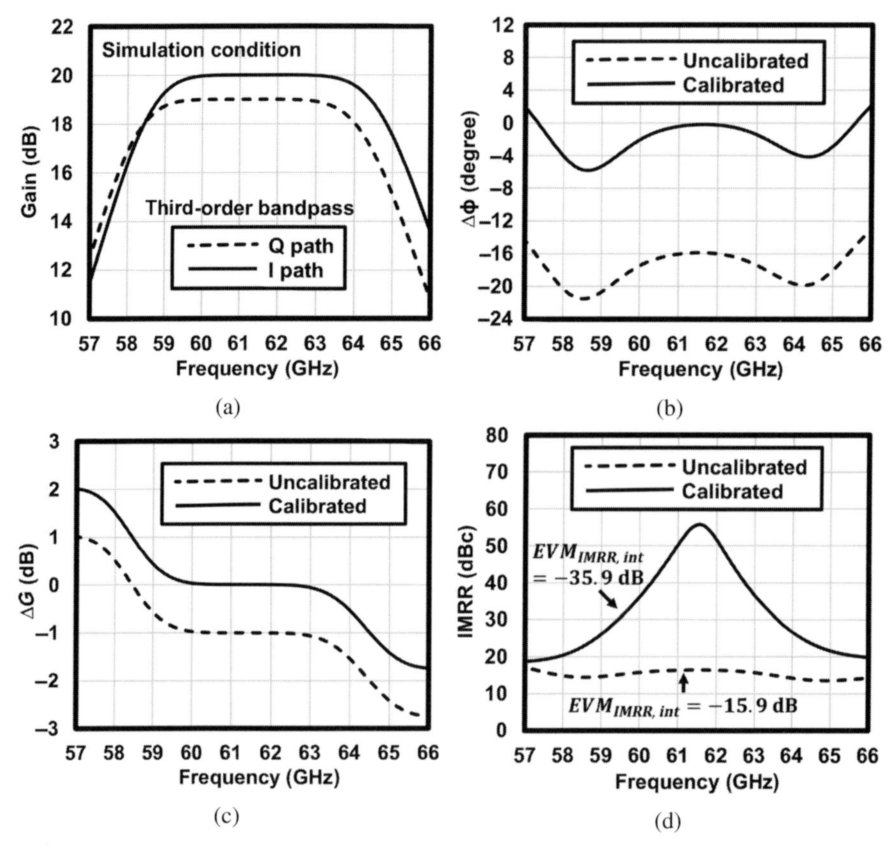

Figure 8.32 An example of frequency-dependent I/Q mismatch dominated by RF parts: (a) un-calibrated gain characteristic of I path and Q path versus frequency; (b) phase difference between I path and Q path versus frequency; (c) gain difference between I path and Q path versus frequency; (d) calculated IMRR versus frequency and $EVM_{IMRR,int}$ within 4.32 GHz bandwidth [28]. (©2017 IEEE. Reprinted, with permission, from *IEEE Journal of Solid-State Circuits.*)

I/Q mismatch limits the achievable IMRR at the frequency away from the calibration point. Furthermore, 8-bit 14.08 GS/s ADCs are normally required to support the four-channel bonding in 64-QAM, which is usually realized by massive time-interleaved ADCs. The ADC may consume several hundreds milliwatts power, as summarized in Figure 8.4 [25].

To cope with the wideband I/Q mismatch issue and the ADC requirement, the frequency-interleaved (FI) architecture can be utilized for the 60 GHz transceiver. Figure 8.33 shows the proposed FI front-end design. The transceiver is composed of two direct-conversion FI transceivers. Each FI transceiver consists of an individual FI transmitter, FI receiver, and local oscillator. A control-logic block is integrated to manage the operation of both FI transceivers. The two FI transceivers operate simultaneously within different frequency bands. One of the TRX is working in the low band (LB, 57.24–61.56 GHz), while the other one is working in the high band (HB; 61.56–65.88 GHz). Therefore, the required bandwidth of individual FI transceiver for $EVM_{IMRR,int} < -35$ dBc is reduced from 8.64 to 4.32 GHz. The corresponding sampling rate of the ADC is also reduced to 7.04 GS/s. The architecture of each FI transceiver

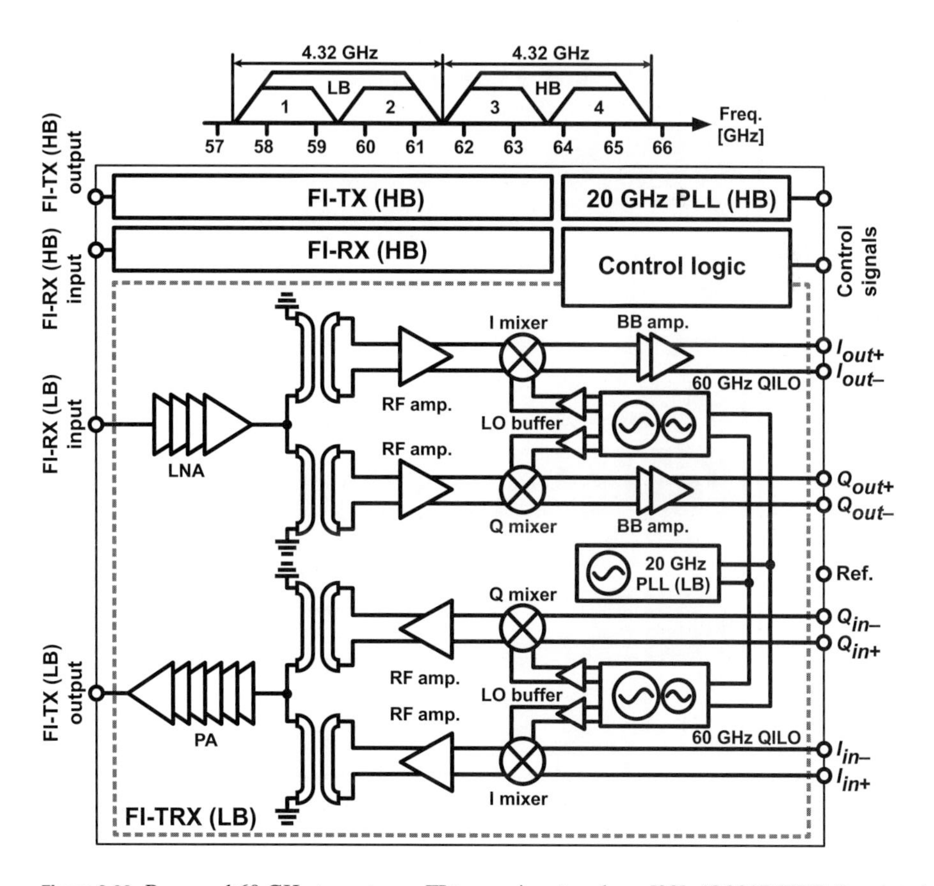

Figure 8.33 Proposed 60 GHz two-stream FI transceiver topology [28]. (©2017 IEEE. Reprinted, with permission, from *IEEE Journal of Solid-State Circuits*.)

is similar to that of the one-stream transceiver, except the asymmetric QILO used for enhancing the locking range [58], the PA with higher P_{1dB}, the injection-locked frequency dividers (ILFDs) with wider locking range, and the baseband amplifier with improved linearity.

8.4 Measurement Results of Transceiver Chips

Figures 8.34 and 8.35 show the die micrographs of the 60 GHz one-stream transceiver and two-stream FI transceiver, respectively. Both RF chips are fabricated in the standard 65 nm CMOS technology. The core area of each transceiver chip is 3.85 and 7.18 mm^2,

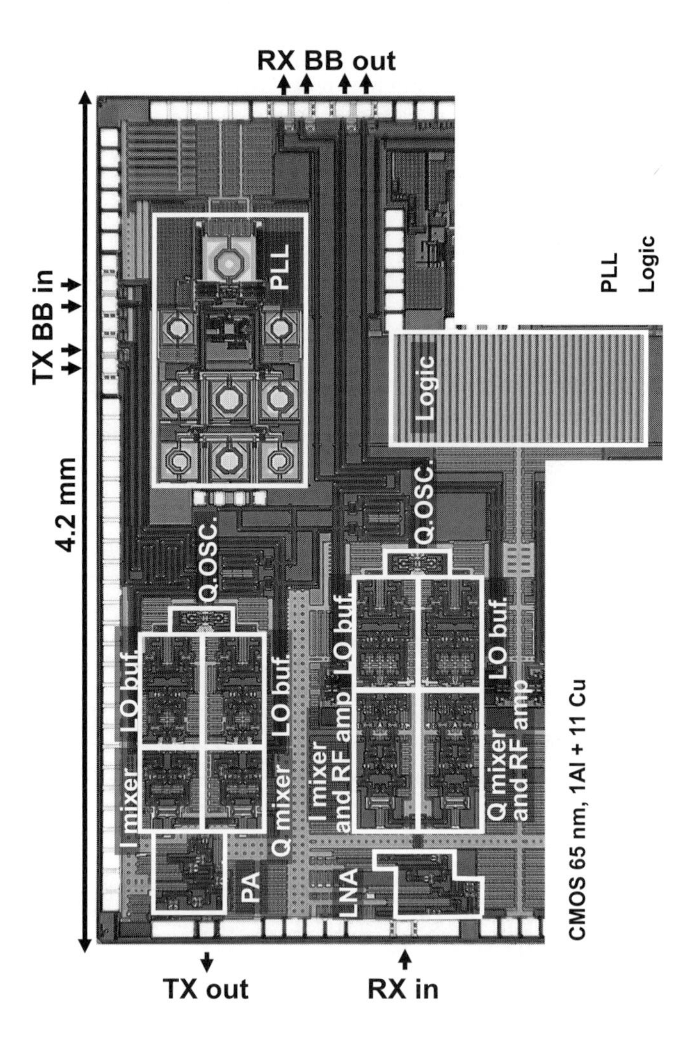

Figure 8.34 Die micrograph of the 60 GHz one-stream transceiver [28]. (©2017 IEEE. Reprinted, with permission, from *IEEE Journal of Solid-State Circuits*.)

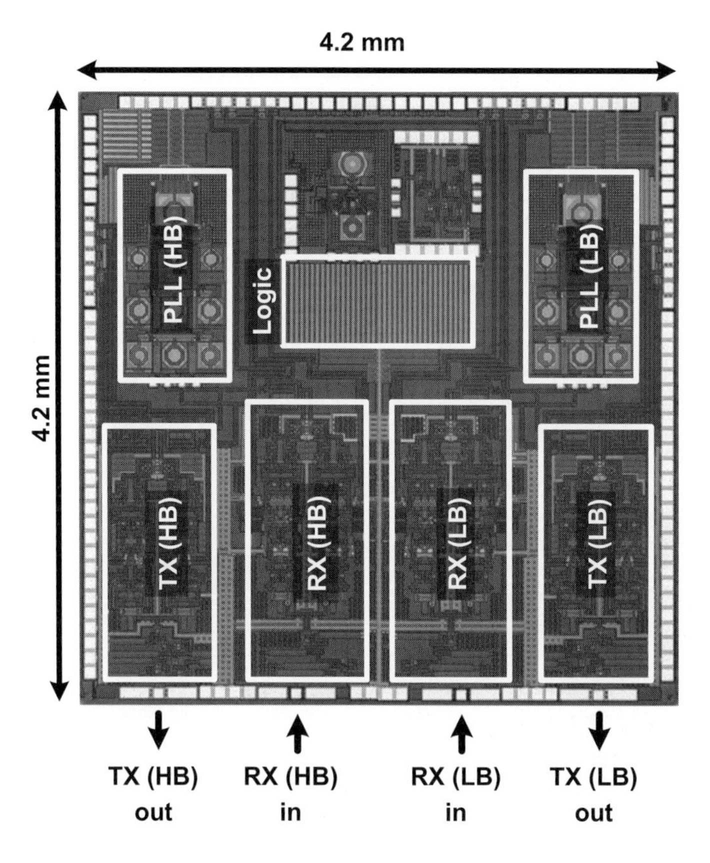

Figure 8.35 Die micrograph of the 60 GHz two-stream FI TRX [28]. (©2017 IEEE. Reprinted, with permission, from *IEEE Journal of Solid-State Circuits*.)

respectively. Tables 8.2 and 8.3 summarize the power consumption and area breakdown of the transceivers.

Figure 8.36 shows the measured conversion gain of the transmitter. The LO frequency f_{LO} is 61.56 GHz, which is the center frequency of the four channels. The wide-and-flat gain characteristic is implemented by the proposed mixer-first transmitter. Figure 8.37 shows the measured conversion gain of the receiver. The LO frequency f_{LO} is 61.56 GHz. Because the output nodes of the mixers have DC-cut capacitors, the measured lower-side cutoff frequency is 0.27 MHz. The upper side is more than 4 GHz. A very wide-and-flat gain characteristic can be observed.

Figure 8.38 shows the measured characteristics of the one-stream transceiver. Both TX and RX cover four channels. The TX conversion gain is about 15 dB, excluding the printed circuit board (PCB) loss. The saturated output power is 10.3 dBm at the center frequency of 61.56 GHz. The output power is measured for both a stand-alone PA and a transceiver chip implemented on a PCB. The PCB loss is estimated from the difference between these saturated output powers. The LO leakage is less than −47 dBc as shown in Figure 8.39. The image rejection ratio of more than 41 dB is achieved at

Table 8.2 Power consumption breakdown summary.

		Power consumption (mW)	
		One-stream	FI (LB)
20 GHz PLL		64	66
TX	PA	115	113
	RF amplifier +Mixer	16	16
	QILO	19	40
	LO buffer	37	37
Total in TX mode		251	272
RX	LNA	41	29
	RF amplifier	19	31
	Mixer	23	14
	BB amplifier	30	11
	QILO	15	35
	LO buffer	28	30
Total in RX mode		220	216

Table 8.3 Core area breakdown summary.

	Core area (mm^2)			
	TX	RX	PLL	Logic
One-stream TRX	1.03	1.25	0.90	0.67
FI TRX (LB and HB)	2.03	2.56	1.93	0.66

the 0.5 GHz offset after the I/Q calibration. The PA consumes 115 mW, and the two differential amplifiers and mixers consume 16 mW. The RX conversion gain is more than 20 dB, excluding the PCB loss. The SNDR at the center frequency of 61.56 GHz is estimated from the measured IM3 and noise figure of the RX PCB. A peak SNDR is 30.3 dB excluding the PCB loss. The power consumptions of LNA, two differential amplifiers, two mixers, and two BB amplifiers are 41, 19, 23, and 30 mW, respectively.

The phase noise measured at the TX output is −96.5 dBc/Hz at 1 MHz offset from the center frequency of 61.56 GHz, as shown in Figure 8.40. The measured free-running frequency of the QILO covers from 58 to 66 GHz. The 20 GHz PLL consumes 64 mW. The QILOs for TX and RX consume 19 and 15 mW, and I/Q LO buffers consume 37 and 28 mW, respectively.

Figure 8.41 shows the measured TX EVM as a function of the averaged output power. A single-carrier 16-QAM is applied with 7.04 Gb/s data rate in channel 3. The EVM approaches the optimum value of −28.4 dB around 5.4 dBm output power. It also can

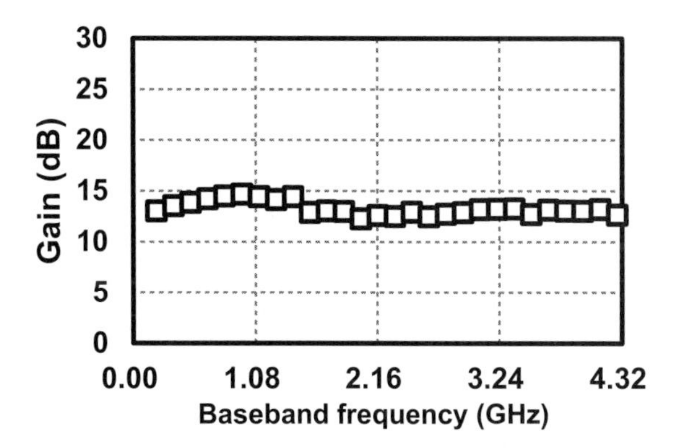

Figure 8.36 Measured conversion gain of the transmitter (lower sideband) versus baseband frequency with $f_{LO} = 61.56$ GHz [28]. (©2017 IEEE. Reprinted, with permission, from *IEEE Journal of Solid-State Circuits.*)

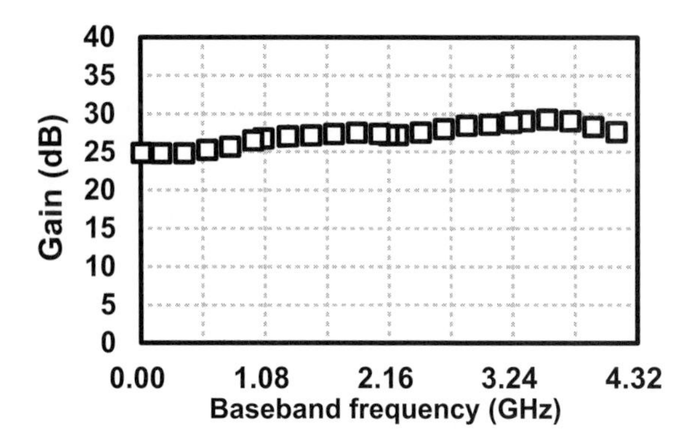

Figure 8.37 Measured conversion gain of the receiver (lower sideband) versus baseband frequency with $f_{LO} = 61.56$ GHz [28]. (©2017 IEEE. Reprinted, with permission, from *IEEE Journal of Solid-State Circuits.*)

be observed that the TX EVM is lower than −21 dB (16-QAM requirement) for a wide output power range.

The measurement setup for TX-to-RX performance of the one-stream TRX is shown in Figure 8.42a and b. Two PCBs with 14 dBi horn antennas are used. One is for TX mode and the other is for RX mode with on-board 36 MHz temperature-compensated crystal oscillators (TCXOs). The modulated I/Q signals are generated by an arbitrary waveform generator (Tektronix AWG70002A) with symbol rates of 1.76 GS/s (for one channel) and 7.04 GS/s (for a four-bonded channel), and a roll-off factor of 25%. The TX output spectrum is measured with a spectrum analyzer and a down-conversion mixer. An oscilloscope (Tektronix DSA73304D) with an adaptive RX equalizer is used to evaluate the constellation and EVM as illustrated in Figure 8.42c. To estimate the required number of equalizer taps for the implemented transceiver, Matlab simulations similar to Figure 8.5 are performed. The TRX gain characteristic is modeled by

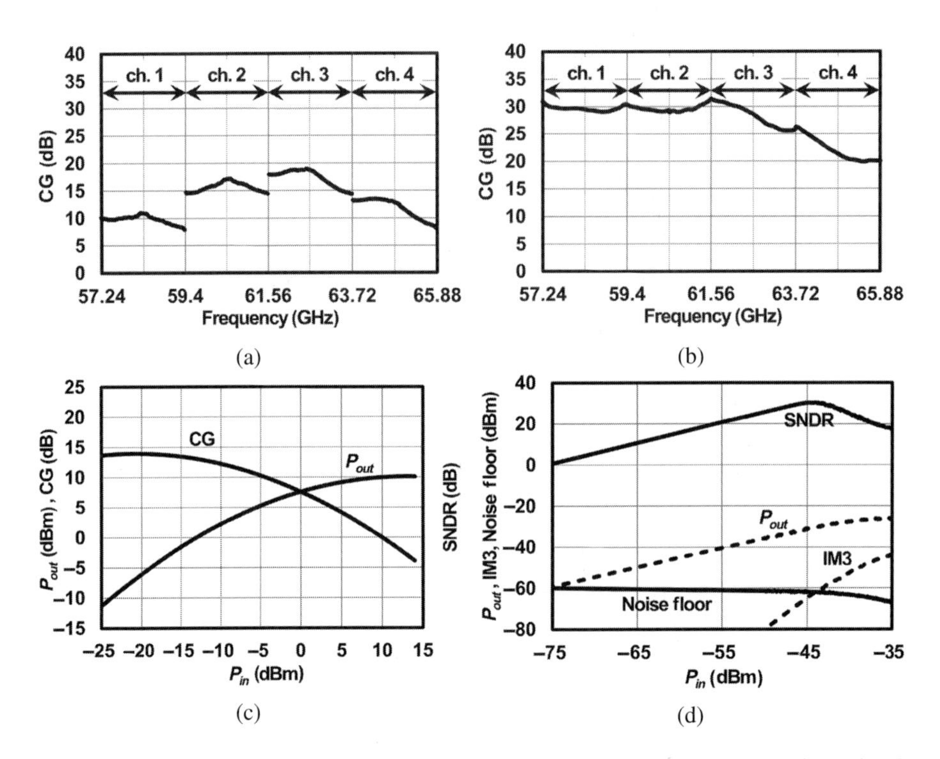

Figure 8.38 Measured characteristics of the one-stream TRX front-end: (a) conversion gain of TX; (b) conversion gain of RX; (c) output power of TX; and (d) output power, IM3, noise floor, and SNDR of RX for one channel [28]. (©2017 IEEE. Reprinted, with permission, from *IEEE Journal of Solid-State Circuits.*)

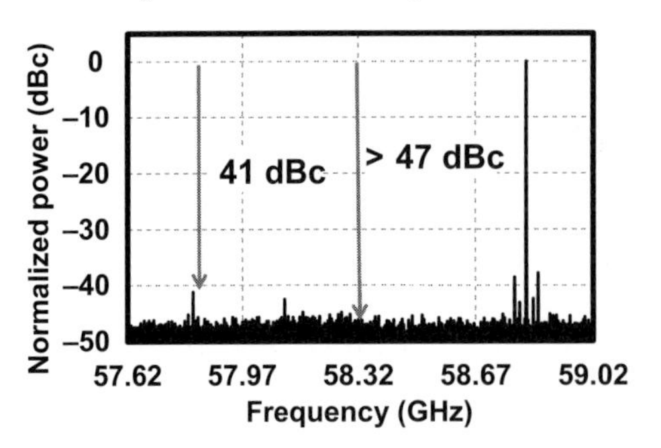

Figure 8.39 Measured spectrum of the TX output with 0.5 GHz baseband input signal [28]. (©2017 IEEE. Reprinted, with permission, from *IEEE Journal of Solid-State Circuits.*)

the frequency response of a fourth-order FIR filter with about 10 dB variation within 7.04 GHz bandwidth. It is observed that the gain flatness of this assumption is worse than the measured TX/RX gain characteristics in Figures 8.36 and 8.37. An RLS linear equalizer is used in the receiver for the estimation. The symbol rate is 7.04 GS/s in 64-QAM on SC mode. It shows that with 10 taps the equalizer can successfully suppress the gain variation between −0.4 and 0.2 dB, which leads to an EVM of −34.9 dB. Furthermore, from literature [59], it is known that a 15-tap 28 Gb/s equalizer can be

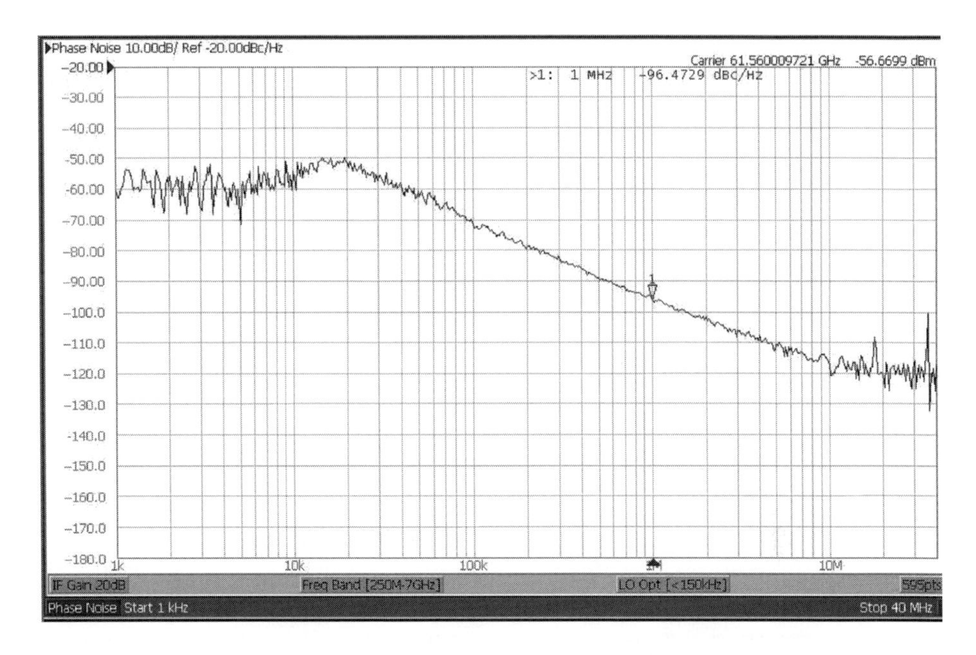

Figure 8.40 Measured phase noise at the TX output with $f_{LO} = 61.56\,\text{GHz}$ [28]. (©2017 IEEE. Reprinted, with permission, from *IEEE Journal of Solid-State Circuits.*)

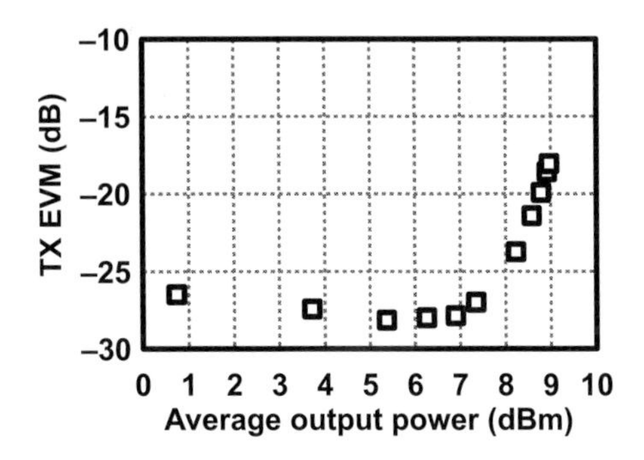

Figure 8.41 Measured TX EVM with 16-QAM in channel 3 [28]. (©2017 IEEE. Reprinted, with permission, from *IEEE Journal of Solid-State Circuits.*)

implemented in CMOS processes with reasonable power consumption and area. Therefore, the equalizer setting in the measurement equipment could be a realistic model for the baseband circuitry, which will not influence the observations and conclusions made in this chapter.

In this measurement, a maximum distance is defined by a TX-to-RX EVM of $-9.8\,\text{dB}$ for QPSK, $-16.5\,\text{dB}$ for 16-QAM, and $-22.5\,\text{dB}$ for 64-QAM for a theoretical bit error rate (BER) of 10^{-3}. Figures 8.43–8.45 show the measured constellation with spectrum, EVM, and maximum communication distance. The measured TX-to-RX EVM ($= -\text{SNR}$) in 64-QAM is less than $-23.9\,\text{dB}$ for every channel with a data rate of 10.56 Gb/s, and $-26.3\,\text{dB}$ is achieved at channel 4. By using the four-bonded channel,

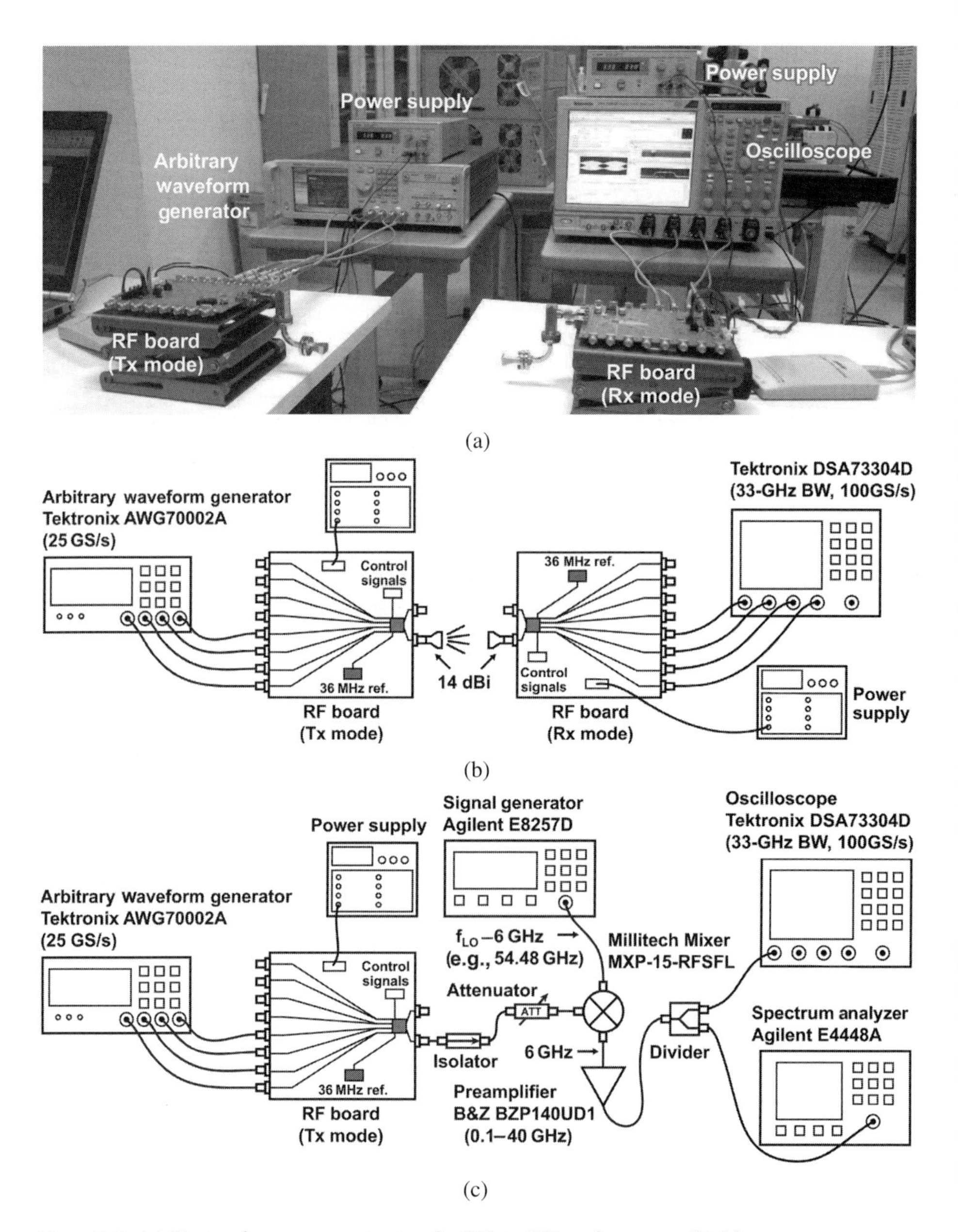

Figure 8.42 (a) Photo of measurement setup for TX-to-RX performance. (b) Measurement setup for TX-to-RX performance. (c) Measurement setup for TX performance (one-stream TRX) [28]. (©2017 IEEE. Reprinted, with permission, from *IEEE Journal of Solid-State Circuits*.)

14.08 Gb/s in QPSK and 28.16 Gb/s in 16-QAM have been achieved within a BER of 10^{-3}. The maximum communication distances with 14 dBi horn antennas are 2.4, 2.0, 2.6, and 0.9 m in QPSK; 0.7, 0.6, 0.6, and 0.4 m in 16-QAM; and 0.08, 0.08, 0.13, and

Channel/ carrier freq.	Ch. 1 58.32 GHz	Ch. 2 60.48 GHz	Ch. 3 62.64 GHz	Ch. 4 64.80 GHz
Modulation	64-QAM			
Data rate	10.56 Gb/s	10.56 Gb/s	10.56 Gb/s	10.56 Gb/s
Constellation				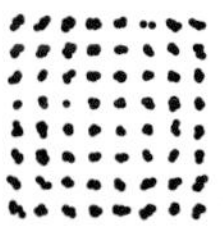
Spectrum	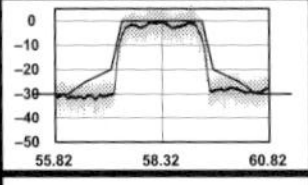	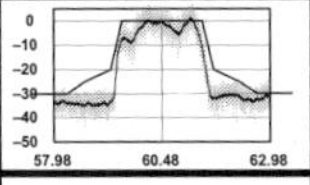	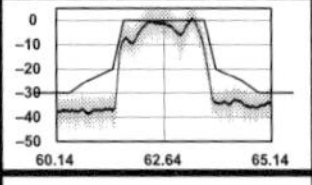	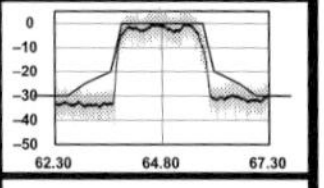
TX output power	−0.2 dBm	0.1d Bm	1.8 dBm	0.1 dBm
TX EVM	−27.1 dB	−27.5 dB	−28.0 dB	−28.8 dB
TX-to-RX EVM	−24.6 dB	−23.9 dB	−24.4 dB	−26.3 dB
Distance	0.08 m	0.08 m	0.13m	0.06 m

Figure 8.43 Measured TX-to-RX performance in 64-QAM [28]. (©2017 IEEE. Reprinted, with permission, from *IEEE Journal of Solid-State Circuits.*)

Channel/ carrier freq.	Ch. 1 58.32 GHz	Ch. 2 60.48 GHz	Ch. 3 62.64 GHz	Ch. 4 64.80 GHz	Ch. 1–4 channel bond
Modulation	16-QAM				
Data rate	7.04 Gb/s	7.04Gb/s	7.04 Gb/s	7.04 Gb/s	28.16 Gb/s
Constellation	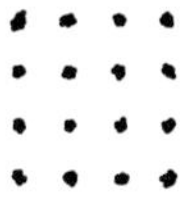	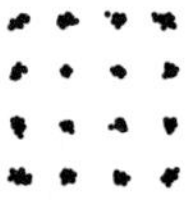			
Spectrum	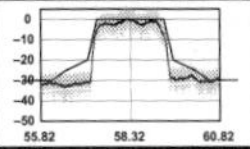	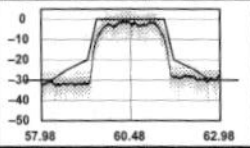	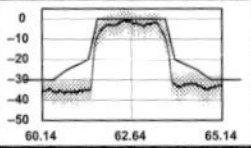	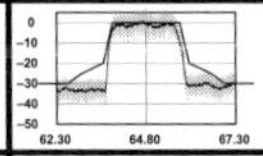	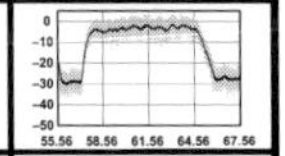
TX output power	6.3 dBm	6.5 dBm	5.4 dBm	5.8 dBm	3.2 dBm
TX EVM	−27.8 dB	−27.6 dB	−28.4 dB	−28.8 dB	−20.0 dB
TX-to-RX EVM	−24.6 dB	−24.1 dB	−24.6 dB	−27.0 dB	−17.2 dB
Distance	0.7 m	0.6 m	0.8 m	0.4 m	0.07 m

Figure 8.44 Measured TX-to-RX performance in 16-QAM [28]. (©2017 IEEE. Reprinted, with permission, from *IEEE Journal of Solid-State Circuits.*)

Channel/ carrier freq.	Ch. 1 58.32 GHz	Ch. 2 60.48 GHz	Ch. 3 62.64 GHz	Ch. 4 64.80 GHz	Ch. 1–4 channel bond
Modulation	QPSK				
Data rate	3.52 Gb/s	3.52 Gb/s	3.52 Gb/s	3.52 Gb/s	14.08 Gb/s
Constellation	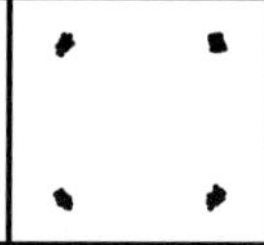		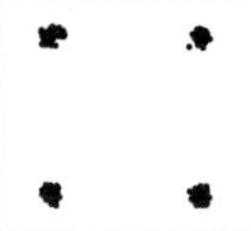		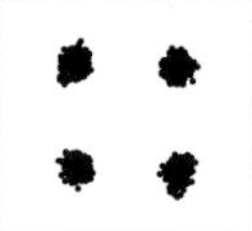
Spectrum	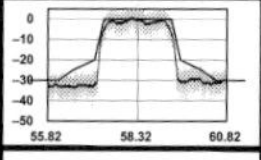	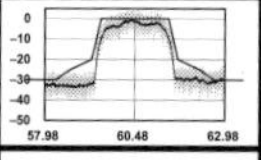	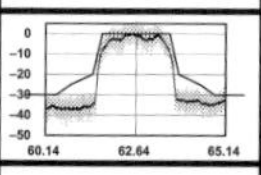	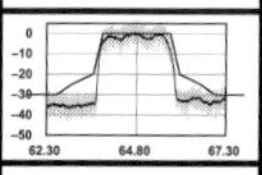	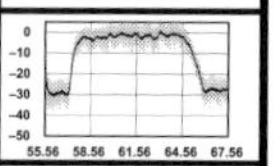
TX output power	6.3 dBm	7.3 dBm	8.7 dBm	6.9 dBm	3.2 dBm
TX EVM	–28.1 dB	–27.7 dB	–29.0 dB	–29.7 dB	–20.1 dB
TX-to-RX EVM	–25.3 dB	–24.5 dB	–24.5 dB	–26.6 dB	–17.9 dB
Distance	2.4 m	2.0 m	2.6 m	0.9 m	0.3 m

Figure 8.45 Measured TX-to-RX performance in QPSK [28]. (©2017 IEEE. Reprinted, with permission, from *IEEE Journal of Solid-State Circuits*.)

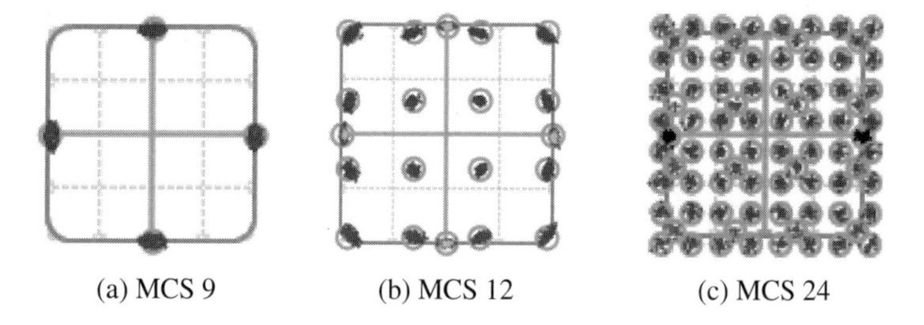

(a) MCS 9 (b) MCS 12 (c) MCS 24

Figure 8.46 Measured performance using IEEE802.11ad/WiGig packets in channel 3.

0.06 mW in 64-QAM for channels 1–4, respectively. All the spectrum meets the mask requirement defined in the IEEE802.11ad/WiGig standard.

Recalling (8.1), it would be interesting to have a breakdown of the measured TRX EVM and discuss the limiting factor. The measurement results for the four-channel bonding in 16-QAM are chosen as an example, since most of the required data for the breakdown have already been obtained. It is known that the measured TX-to-RX EVM and TX EVM are –17.2 and −20.0 dB, respectively. The estimated RX EVM is around −17.9 dB. Obviously, the impairments in the receiver limit the transceiver performance. Similarly to Figure 8.38d, the calculated maximum RX SNDR for four-channel bonding is about 26.6 dB. The measurement equipment shows the gain and phase imbalance of the received signal are 0.74 dB and 0.26°, respectively. It corresponds to an IMRR of 27.4 dB. The integrated phase noise of the carrier is −28.7 dB considering a 400 kHz carrier tracking bandwidth. Because the influence of the LO feedthrough is trivial for the receiver, it is omitted in the analysis. Therefore, the unknown impairment in (8.1),

Table 8.4 Summary of measured performance using IEEE802.11ad/WiGig packets in channel 3.

MCS	Modulation		Data rate (Mb/s)	TX EVM (dB)	
				Spec.	Meas.
9	QPSK	SC	2,502.5	−15	−27.1
12	16-QAM	SC	4,620.00	−21	−27.0
24	64-QAM	OFDM	6,756.75	−26	−26.5

EVM_{GF}, is estimated to be −18.0 dB, which indicates that the gain flatness of the receiver finally limits the transceiver performance.

The measurement result by using IEEE802.11ad/WiGig packets in channel 3 is demonstrated in Figure 8.46. For modulation and coding schemes (MCS) 9 and 12, the TX EVM are −27.1 and −27.0 dB, respectively. For OFDM MCS 24, a TX EVM is −26.5 dB, and the TX-to-RX cascaded EVM is −21.3 dB. See Table 8.4.

To demonstrate the frequency-dependent I/Q mismatch, the measured IMRR of the FI transmitter (LB) versus the baseband frequency is shown in Figure 8.47. The IMRR is calibrate with 100 MHz baseband input. The optimum IMRR at the calibration frequency is 55 dBc, which corresponds to a phase error of 0.2 degree or a gain error of 0.03 dB. A wideband IMRR characteristic ($\geq$35 dBc for around 2 GHz bandwidth) is also observed. However, the measured IMRR degrades prominently with the increasing of the baseband frequency, which indicates the degradation of EVM performance for direct four-channel bonding. The measured EVM of the FI-TX (LB) shows that the optimum EVM for two-channel bonding (channels 3 and 4) in 64-QAM is −27.4 dB, while it is only −21.5 dB for four-channel bonding, as depicted in Figure 8.48.

Figure 8.49 shows the measured constellation and performance summary of the FI transceiver. Two PCBs are used in the measurement, as illustrated in Figure 8.50. One is for TX mode, and the other is for RX mode. All the FI-transmitters on the PCB in TX mode are operating during the measurement, while the FI-receivers are working on the PCB in RX mode. The modulated baseband signals are generated by two AWGs (Keysight M8190A and M8195A) with a symbol rate of 3.52 GS/s (for two-channel bonding) and a roll-off factor of 25%. Two oscilloscopes (Keysight DSO90904 and DSO91304A) are used to evaluate the constellation and EVM performance. Horn antennas with 14 dBi gain are utilized in the TX-to-RX EVM measurement. The horizontal distance is 3 cm. The measured TX-to-RX EVM in 64-QAM is −24.1 dB for LB and −23.0 dB for HB, which satisfy the required EVM of less than −22.5 dB for a BER of 10^{-3}. The data rates for both FI-transceiver pairs are 21.12 Gb/s, which demonstrates a total data rate of 42.24 Gb/s by the FI four-channel bonding. The spectrum for each TX channel is measured using an external down-conversion mixer and the oscilloscope.

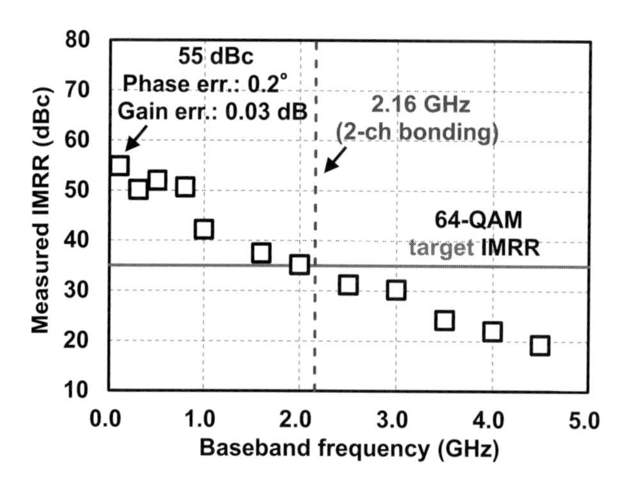

Figure 8.47 Measured FI-TX (LB) IMRR versus baseband frequency [28]. (©2017 IEEE. Reprinted, with permission, from *IEEE Journal of Solid-State Circuits*.)

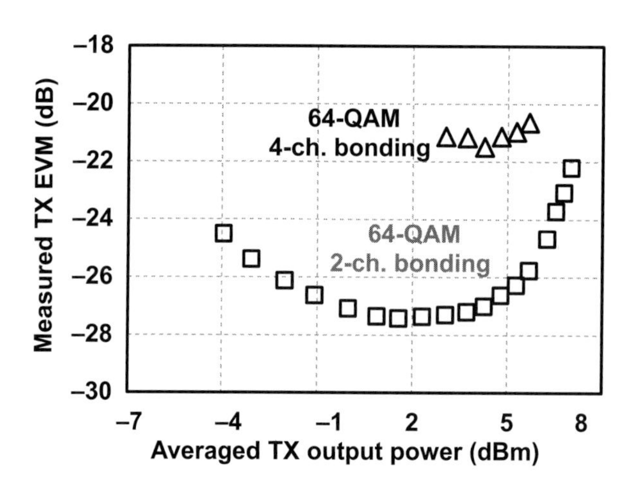

Figure 8.48 Measured FI-TX (LB) EVM versus output power [28]. (©2017 IEEE. Reprinted, with permission, from *IEEE Journal of Solid-State Circuits*.)

The measured RF front-end characteristics of the FI transceiver are shown in Figure 8.51. Both FI-TX and FI-RX realize two-channel bonding. The conversion gain of the TX is around 10 dB excluding the PCB loss. The saturated output power is 10.4 dBm at the center frequency of 63.72 GHz. The output power at 1 dB compression point is 6.1 dBm. The RX conversion gain is over 24 dB for low band and about 20 dB for high band, excluding PCB loss. The estimated SNDR at the center frequency of 63.72 GHz for two-channel bonding is calculated from the measured output power, IM3, and noise figure of the RX PCB. The peak SNDR is 32.9 dB. The estimated sensitivity of the FI-RX is −66 and −49.3 dBm for one-channel QPSK and two-channel bonding

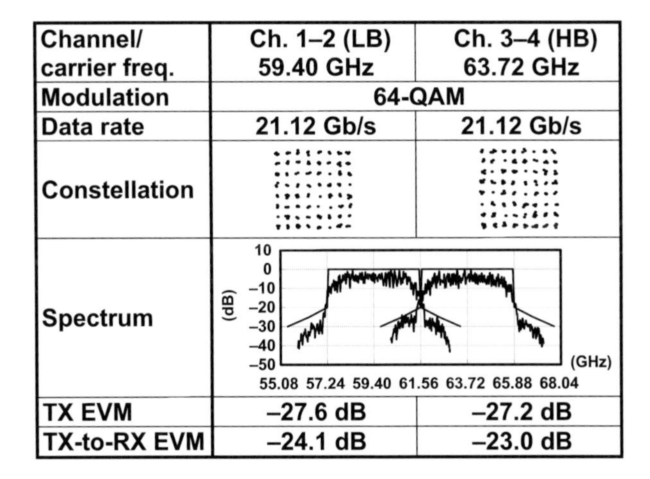

Channel/ carrier freq.	Ch. 1–2 (LB) 59.40 GHz	Ch. 3–4 (HB) 63.72 GHz
Modulation	64-QAM	
Data rate	21.12 Gb/s	21.12 Gb/s
Constellation		
Spectrum	(dB) 10 0 –10 –20 –30 –40 –50 (GHz) 55.08 57.24 59.40 61.56 63.72 65.88 68.04	
TX EVM	–27.6 dB	–27.2 dB
TX-to-RX EVM	–24.1 dB	–23.0 dB

Figure 8.49 Measured constellation and performance summary of the FI-TRX [28]. (©2017 IEEE. Reprinted, with permission, from *IEEE Journal of Solid-State Circuits.*)

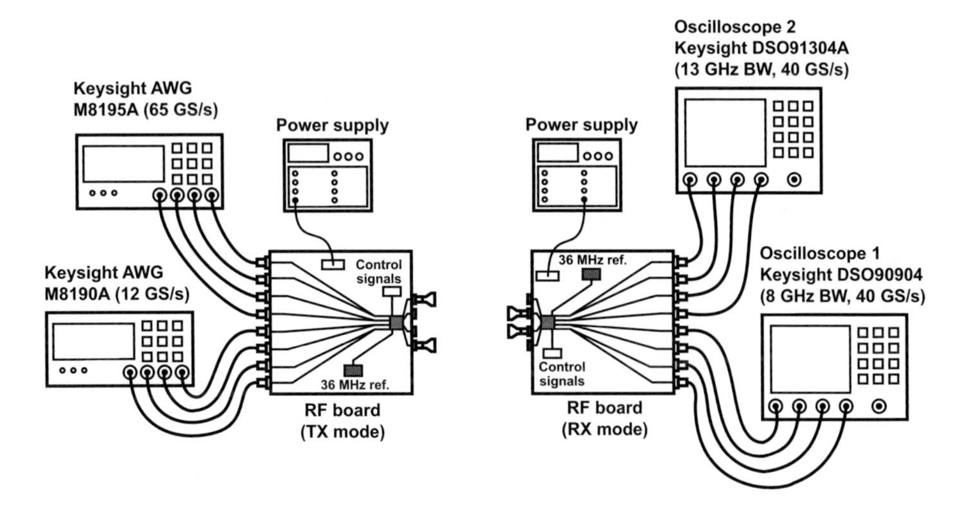

Figure 8.50 Measurement setup for TX-to-RX performance of the FI TRX [28]. (©2017 IEEE. Reprinted, with permission, from *IEEE Journal of Solid-State Circuits.*)

64-QAM, respectively. Table 8.5 summarizes the RF front-end performance of both the one-stream transceiver and two-stream FI transceiver.

Table 8.6 shows a performance comparison of the state-of-the-art 60 GHz CMOS transceivers. This work realizes the first-reported four-channel bonding in 16-QAM and 64-QAM, which achieves the data rate of 28.16 and 42.24 Gb/s, respectively. The front-end also covers all of the four channels and achieves full data rates for QPSK, 16-QAM, and 64-QAM with the best EVM.

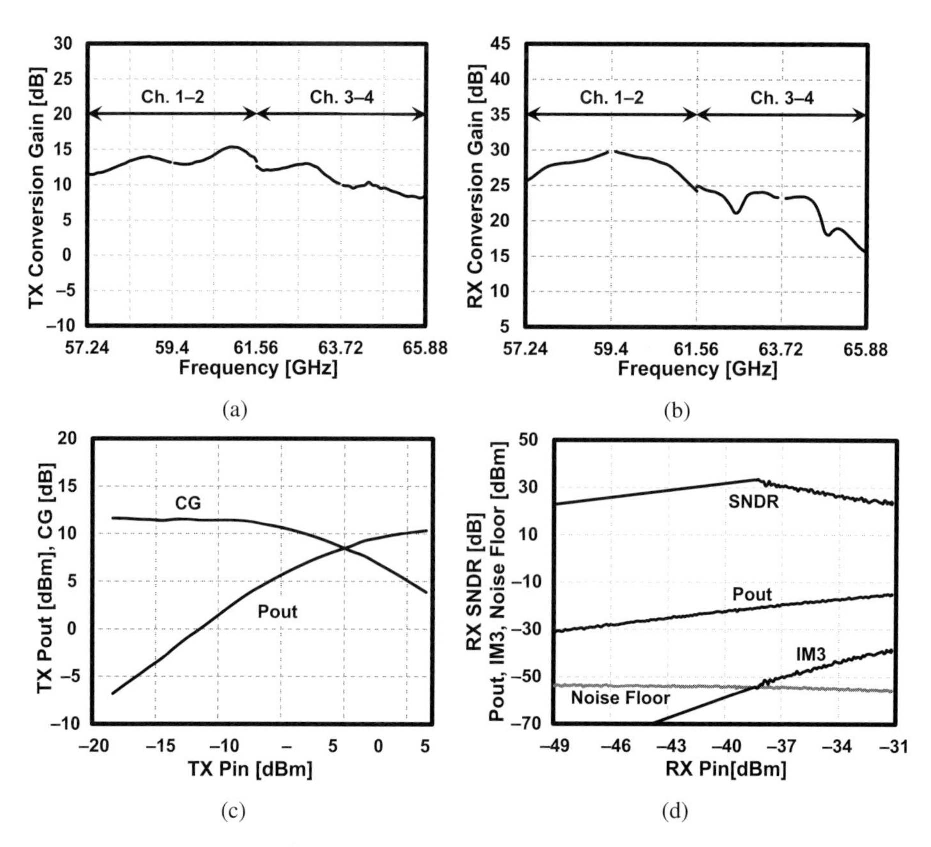

Figure 8.51 Measured characteristics of the frequency-interleaved TRX front-end: (a) conversion gain of TX; (b) conversion gain of RX; (c) output power of TX; and (d) output power, IM3, noise floor, and SNDR of RX for channel 3 and 4 bonding [28]. (©2017 IEEE. Reprinted, with permission, from *IEEE Journal of Solid-State Circuits*.)

8.5 Conclusion

A study of 60 GHz CMOS transceivers for ultra-high-speed wireless communication is introduced in this chapter. The requirements of wireless transceivers in future radio access network are discussed, which show that the 60 GHz CMOS transceiver would be one of the most promising candidates with some significant challenges. The challenges, such as CMOS gain/noise trade-off, modeling/measurement issues, SNDR, gain flatness, wideband image rejection, and LO phase noise, are discussed in detail. The system architecture and key block designs of the 60 GHz TRXs are elaborated. They include techniques such as active/passive device modeling and characterization, injection-locked local synthesizer, mixer-first transmitter design, flipped-voltage-follower baseband amplifier, I/Q mismatch and LO leakage calibration, and so on. The measurement results of some TRX chips/boards and discussions are shown to demonstrate the techniques.

Table 8.5 RF front-end performance summary.

		One-stream	Two-stream FI
TX			
	Conversion gain	13.5 dB	11.6 dB
	P_{sat}	10.3 dBm (at 61.56 GHz)	10.4 dBm (at 63.72 GHz)
	P_{1dB}	2.6 dBm (at 61.56 GHz)	6.1 dBm (at 63.72 GHz)
RX			
	Conversion gain	16 dB	18 dB
	NF	5.7 dB	5.7 dB
	IIP3	−30 dBm	−21 dBm
	Sensitivity	−66 dBm[a] −53 dBm[b]	−66 dBm[a] −52 dBm[b]
	Maximum SNDR (1-ch)	30.3 dB	35.3 dB

[a] For 1-ch QPSK, SNDR=9.8 dB.
[b] For 1-ch 64-QAM, SNDR=22.5 dB.

The implemented one-stream TRX achieves a 28.16-Gb/s data rate with an EVM of −17 dB by using a four-bonded channel in 16-QAM. A best TX-to-RX EVM of −26 dB in 64-QAM is achieved within the channel defined by IEEE802.11ad/WiGig standard. This performance is supported by flat gain characteristics, fine calibration of I/Q mismatch and LO leakage, and low phase noise. Moreover, with the help of the frequency-interleaved architecture, four-channel bonding in 64-QAM is viable for the 60 GHz transceiver, which realizes 42.24 Gb/s with reasonable power consumption.

Acknowledgment

The authors thank the contributors at the Tokyo Institute of Technology to the works presented in this chapter: Kota Matsushita, Keigo Bunsen, Rui Murakami, Dr. Ahmed Musa, Ryo Minami, Takahiro Sato, Hiroki Asada, Naoki Takayama, Shogo Ito, Dr. Win Chaivipas, Tatsuya Yamaguchi, Yasuaki Takeuchi, Hiroyuki Yamagishi, Makoto Noda, Yuuki Tsukui, Seitaro Kawai, Yuuki Seo, Shinji Sato, Kento Kimura, Satoshi Kondo, Tomohiro Ueno, Nurul Fajri, Shoutarou Maki, Noriaki Nagashima, Korkut Kaan Tokgoz, Dr. Teerachot Siriburanon, Bangan Liu, Yun Wang, Jian Pang, Dr. Ning Li, Professor Masaya Miyahara, and Professor Akira Matsuzawa

The authors thank Dr. Hirose, Dr. Suzuki, Dr. Sato, and Dr. Kawano of Fujitsu Laboratories, Ltd.

Table 8.6 Performance comparison of 60 GHz CMOS RF front-ends.

Ref.	Data rate (modulation)	P_{out} (dBm) /antenna path	TX-to-RX EVM (dB)	Integration level	Core (mm^2)	Power consumption
This work (Tokyo Tech)	28.16 Gb/s[a] (16-QAM)	8.5 @TX EVM = −21 dB	−27 (1-ch.) −17 (4-ch.)	65 nm, direct conversion, TX, RX, LO (**1-stream**)	3.9	TX:251 mW RX:220 mW
	42.24 Gb/s[a] (64-QAM)	7 @TX EVM = −22 dB	−24 (LB[b]) −23 (HB[b])	65 nm, direct conversion, 2TX, 2RX, 2LO (**2-stream FI**)	7.2	TX:544 mW RX:432 mW
[14] Panasonic	2.5 Gb/s (QPSK)	2 @TX EVM = −22 dB[d]	N/A	90 nm, direct conversion, TX, RX, LO	13.5[c]	TX:347 mW RX:274 mW
[20] NEC	2.6 Gb/s (QPSK)	5.2 @TX P_{1dB}	N/A	90 nm, direct conversion, TX, RX	3.4	TX:133 mW RX:206 mW
[21] UCB	4 Gb/s (QPSK)	N/A	N/A	90 nm, direct conversion, TX, RX, LO, BB	6.9[c]	TX:170 mW RX:138 mW
[22] IMEC	7 Gb/s (16-QAM)	7.3[e] @TX-to-RX EVM = −15 dB	−15 (1-ch.)	40 nm, direct conversion, TX, RX, LO	11.5[c]	TX:584 mW RX:397 mW 4 × 4 array
[23] IMEC	7 Gb/s (16-QAM)	6[e] @TX EVM = −23 dB	−20 (1-ch.)	28 nm, direct conversion, TX, RX, LO	7.9[c]	TX:670 mW RX:431 mW 4 × 4 array
[15] Broadcom	4.6 Gb/s (16-QAM)	-4[e,f] @TX EVM = −23 dB[d]	−20[d] (1-ch.)	40 nm, heterodyne, TX, RX, LO	26.3[c]	TX:1,190 mW RX:960 mW 16 × 16 array
[8] SiBEAM	7.14 Gb/s (16-QAM)	−2[e] @TX EVM = −19 dB	−19 (1-ch.)	65 nm, heterodyne, TX, RX, LO	149.9[c]	TX:1,820 mW RX:1,250 mW 32 × 32 array
[13] Toshiba	2.6 Gb/s (QPSK)	2.6[f] @TX P_{1dB}	N/A	65 nm, heterodyne, TX, RX, LO	2.9	TX:160 mW RX:233 mW
[60] CEA-LETI	3.9 Gb/s (16-QAM)	5 @TX P_{1dB}	−20 (1-ch.)	65 nm, heterodyne, TX, RX, LO	9.2[c]	TX:357 mW RX:454 mW

[a] Four-channel bonding.
[b] 2-ch.
[c] Chip area.
[d] Measured with integrated ADC/DAC.
[e] Estimated from literature.
[f] Measured with integrated TX/RX switch.

References

[1] M.-S. Chen, Y.-N. Shih, C.-L. Lin, H.-W. Hung, and J. Lee, "A fully-integrated 40-Gb/s transceiver in 65-nm CMOS technology," *IEEE Journal of Solid-State Circuits*, vol. 47, no. 3, pp. 627–640, March 2012.

[2] M. Harwood, S. Nielsen, A. Szczepanek, et al., "A 225 mW 28 Gb/s SerDes in 40nm CMOS with 13 dB of analog equalization for 100 GBASE-LR4 and optical transport lane 4.4 applications," in *IEEE ISSCC Digest of Technical Papers*, IEEE, 2012, pp. 326–327.

[3] R. Navid, E.-H. Chen, M. Hossain, et al., "A 40 Gb/s serial link transceiver in 28 nm CMOS technology," *IEEE Journal of Solid-State Circuits*, vol. 50, no. 4, pp. 1–14, April 2014.

[4] P.-C. Chiang, H.-W. Hung, H.-Y. Chu, G.-S. Chen, and J. Lee, "60 Gb/s NRZ and PAM4 transmitters for 400 GbE in 65 nm CMOS," in *IEEE ISSCC Digest of Technical Papers*, IEEE, 2014, pp. 42–43.

[5] "802.11n-2009 – IEEE standard for information technology: Local and metropolitan area networks–Specific requirements: Part 11: Wireless LAN medium access control (MAC)and physical layer (PHY) specifications amendment 5: Enhancements for higher throughput," IEEE Std 802.11n-2009, November 2009.

[6] Y. Zheng, K.-W. Wong, M. A. Asaru, et al., "A 0.18 μm CMOS dual-band UWB transceiver," in *IEEE ISSCC Digest of Technical Papers*, IEEE, 2007, pp. 114–115.

[7] H. Hedayati and K. Entesari, "A 90-nm CMOS UWB impulse radio transmitter with 30-dB in-band notch at IEEE 802.11a system," *IEEE Transactions on Microwave Theory and Techniques*, vol. 61, no. 12, pp. 4220–4232, December 2013.

[8] S. Emami, R. F. Wiser, E. Ali, et al., "A 60GHz CMOS phased-array transceiver pair for multi-Gb/s wireless communications," in *IEEE ISSCC Digest of Technical Papers*, IEEE, 2011, pp. 164–165.

[9] K. Okada, K. Matsushita, K. Bunsen, et al., "A 60GHz 16QAM/8PSK/QPSK/BPSK direct-conversion transceiver for IEEE 802.15.3c," in *IEEE ISSCC Digest of Technical Papers*, 2011, pp. 160–161.

[10] T. Tsukizawa, A. Yoshimoto, H. Komori, et al., "A PVT-variation tolerant fully integrated 60 GHz transceiver for IEEE 802.11ad," in *Symposium on VLSI Circuits Digest of Technical Papers*, 2014, pp. 123–124.

[11] K. Okada, R. Minami, Y. Tsukui, et al., "A 64-QAM 60GHz CMOS transceiver with 4-channel bonding," in *IEEE ISSCC Digest of Technical Papers*, 2014, pp. 346–347.

[12] K. Okada, K. Kondou, M. Miyahara, et al., "A full 4-channel 6.3 Gb/s 60 GHz direct-conversion transceiver with low-power analog and digital baseband circuitry," in *IEEE ISSCC Digest of Technical Papers*, IEEE, 2012, pp. 218–219.

[13] T. Mitomo, Y. Tsutsumi, H. Hoshino, et al. "A 2 Gb/s-throughput CMOS transceiver chipset with in-package antenna for 60 GHz short-range wireless communication," in *IEEE ISSCC Digest of Technical Papers*, IEEE, 2012, pp. 266–267.

[14] T. Tsukizawa, N. Shirakata, T. Morita, et al., "A fully integrated 60GHz CMOS transceiver chipset based on WiGig/IEEE802.11ad with built-in self calibration for mobile applications," in *IEEE ISSCC Digest of Technical Papers*, IEEE, 2013, pp. 230–231.

[15] M. Boers, I. Vassiliou, S. Sarkar, et al., "A 16TX/16RX 60GHz 802.11ad chipset with single coaxial interface and polarization diversity," in *IEEE ISSCC Digest of Technical Papers*, IEEE, 2014, pp. 344–345.

[16] K. K. Tokgoz, K. Lim, K. Okada, and A. Matsuzawa, "Shunt characterization technique of decoupling transmission line for millimeter-wave CMOS amplifier design," in *Proceedings of IEEE APMC*, IEEE, 2014, pp. 274–276.

[17] K. K. Tokgoz, N. Fajri, Y. Seo, S. Kawai, K. Okada, and A. Matsuzawa, "A characterization method of on-chip Tee-junction for millimeter-wave CMOS circuit design," in *Proceedings of IEEE SSDM*, IEEE, 2014.

[18] N. Li, K. Matsushita, N. Takayama, S. Ito, K. Okada, and A. Matsuzawa, "Millimeter-wave amplifiers design by employing multi-line de-embedding technique," *IEICE Transactions on Electronics*, vol. E93-A, no. 2, pp. 431–439, February 2010.

[19] N. Li, K. Bunsen, N. Takayama, et al., "A 24 dB gain 51-68 GHz common source low noise amplifier using asymmetric-layout transistors," *IEICE Transactions on Electronics*, vol. E95-A, no. 2, pp. 498–505, February 2012.

[20] M. Tanomura, Y. Hamada, S. Kishimoto, et al., "TX and RX front-ends for 60 GHz band in 90 nm standard bulk CMOS," in *IEEE ISSCC Digest of Technical Papers*, IEEE, 2008, pp. 558–559.

[21] C. Marcu, D. Chowdhury, C. Thakkar, et al., "A 90nm CMOS low-power 60GHz transceiver with integrated baseband circuitry," in *IEEE ISSCC Digest of Technical Papers*, IEEE, 2009, pp. 314–315.

[22] V. Vidojkovic, V. Szortyka, K. Khalaf, et al., "A low-power radio chipset in 40nm LP CMOS with beamforming for 60GHz high-data-rate wireless communication," in *IEEE ISSCC Digest of Technical Papers*, IEEE, 2013, pp. 236–237.

[23] G. Mangraviti, K. Khalaf, Q. Shi, et al., "A 4-antenna-path beamforming transceiver for 60GHz multi-Gb/s communication in 28nm CMOS," in *IEEE ISSCC Digest of Technical Papers*, IEEE, 2016, pp. 246–247.

[24] K. Okada, N. Li, K. Matsushita, et al., "A 60-GHz 16QAM/8PSK/QPSK/BPSK direct-conversion transceiver for IEEE802.15.3c," *IEEE Journal of Solid-State Circuits*, vol. 46, no. 12, pp. 2988–3004, December 2011.

[25] B. Murmann, "ADC performance survey 1997–2016," July 2016. [Online]. Available: http://web.stanford.edu/~murmann/adcsurvey.html.

[26] K. Okada, K. Kondou, M. Miyahara, et al., "Full four-channel 6.3-Gb/s 60-GHz CMOS transceiver with low-power analog and digital baseband circuitry," *IEEE Journal of Solid-State Circuits*, vol. 48, no. 1, pp. 46–65, January 2013.

[27] L. Kull, T. Toifl, M. Schmatz, et al., "A 90GS/s 8b 667mW 64× interleaved SAR ADC in 32nm digital SOI CMOS," in *IEEE ISSCC Digest of Technical Papers*, IEEE, 2014, pp. 378–379.

[28] R. Wu, R. Minami, Y. Tsukui, et al., "64-QAM 60-GHz CMOS transceivers for IEEE 802.11ad/ay," *IEEE Journal of Solid-State Circuits*, vol. 52, no. 11, pp. 2871–2891, November 2017.

[29] S. Kawai, R. Minami, Y. Tsukui, et al., "A digitally-calibrated 20-Gb/s 60-GHz direct-conversion transceiver in 65-nm CMOS," in *IEEE RFIC Symposium Digest Papers*, IEEE, 2013, pp. 137–140.

[30] T. Siriburanon, S. Kondo, M. Katsuragi, et al., "A low-power low-noise mm-wave sub-sampling PLL using dual-step-mixing ILFD and tail-coupling quadrature injection-locked oscillator for IEEE 802.11ad," *IEEE Journal of Solid-State Circuits*, vol. 51, no. 5, pp. 1246–1260, May 2016.

[31] D. Zhao and P. Reynaert, "A 40nm CMOS E-band transmitter with compact and symmetrical layout floor-plans," *IEEE Journal of Solid-State Circuits*, vol. 50, no. 11, pp. 2560–2571, November 2015.

[32] J. H. Kim, J. H. Jeong, S. M. Kim, C. S. Park, and K. C. Lee, "Prediction of error vector magnitude using AM/AM, AM/PM distortion of RF power amplifier for high order modulation OFDM system," in *Proceedings of IEEE IMS*, IEEE, 2005, pp. 2027–2030.

[33] S. Kundu, E. Alpman, J. H.-L. Lu, et al., "A 1.2 V 2.64 GS/s 8 bit 39 mW skew-tolerant time-interleaved SAR ADC in 40 nm digital LP CMOS for 60 GHz WLAN," *IEEE Transactions on Circuits Systand Systems I*, vol. 62, no. 8, pp. 1929–1939, August 2015.
[34] Y. Tan, J. Duster, C.-T. Fu, et al., "A 2.4GHz WLAN transceiver with fully-integrated highly-linear 1.8V 28.4dBm PA, 34dBm T/R switch, 240MS/s DAC, 320MS/s ADC, and DPLL in 32nm SoC CMOS," in *Symposium on VLSI Circuits Digest of Technical Papers*, IEEE, 2012, pp. 76–77.
[35] C.-S. Choi, Y. Shoji, H. Harada, et al., "RF impairment models for 60GHz-band SYS/PHY simulation," IEEE P802.15-06-0477-01-003c, November 2006.
[36] K. Kondou and M. Noda, "A new parallel algorithm for full-digital phase-locked loop for high-throughput carrier and timing recovery systems," in *Proceedings of IEEE ICECS*, IEEE, December 2010, pp. 1163–1166.
[37] "IEEE standard for information technology – Telecommunications and information exchange between systems – Local and metropolitan area networks – Specific requirements. Part 15.3: Wireless medium access control (MAC) and physical layer (PHY) specifications for high rate wireless personal area networks (WPANs) amendment 2: Millimeter-wavebased alternative physical layer extension," IEEE Std 802.15.3c-2009, October 2009.
[38] H. Yamagishi and M. Noda, "High throughput hardware architecture for (1440, 1344) low-density parity-check code utilizing quasi-cyclic structure," in *Proceedings of IEEE Turbo Coding*, IEEE, September 2008, pp. 78–83.
[39] K. Okada, "60GHz WiGig frequency synthesizer using injection locked oscillator," in *IEEE RFIC Symposium Digest of Papers*, 2014, pp. 109–134, invited.
[40] A. Musa, R. Murakami, T. Sato, W. Chaivipas, K. Okada, and A. Matsuzawa, "A 58-63.6GHz quadrature PLL frequency synthesizer in 65nm CMOS," in *Proceedings of IEEE A-SSCC*, November 2010, pp. 189–192.
[41] A. Musa, R. Murakami, T. Sato, W. Chaivipas, K. Okada, and A. Matsuzawa, "A 58-63.6 GHz quadrature PLL frequency synthesizer using dual-injection technique," in *ASP-DAC*, January 2011, pp. 101–102.
[42] "802.11ad-2012 – IEEE standard for information technology – Telecommunications and information exchange between systems – Local and metropolitan area networks-Specific requirements – Part 11: Wireless LAN medium access control (MAC) and physical layer (PHY) specifications amendment 3: Enhancements for very high throughput in the 60 GHz band," IEEE Std 802.11ad-2012, December 2012.
[43] E. A. M. Klumperink, S. L. J. Gierkink, A. P. van der Wel, and B. Nauta, "Reducing MOSFET 1/f noise and power consumption by switched biasing," *IEEE Journal of Solid-State Circuits*, vol. 35, no. 7, pp. 994–1001, July 2000.
[44] L. Fanori, A. Liscidini, and R. Castello, "3.3GHz DCO with a frequency resolution of 150Hz for all-digital PLL," in *IEEE ISSCC Digest of Technical Papers*, IEEE, February 2010, pp. 48–49.
[45] W. Chan and J. Long, "A 56-65GHz injection-locked frequency tripler with quadrature outputs in 90-nm CMOS," *IEEE Journal of Solid-State Circuits*, vol. 43, no. 12, pp. 2739–2746, December 2008.
[46] C. Andrews and A. C. Molnar, "A passive-mixer-first receiver with baseband-controlled RF impedance matching, <6dB NF, and >27dBm wideband IIP3," in *IEEE ISSCC Digest of Technical Papers*, IEEE, 2010, pp. 46–47.
[47] M. C. M. Soer, E. A. M. Klumperink, Z. Ru, F. E. van Vliet, and B. Nauta, "A 0.2-to-2.0GHz 65nm CMOS receiver without LNA achieving >11dBm IIP3 and <6.5 dB NF," in *IEEE ISSCC Digest of Technical Papers*, IEEE, 2009, pp. 222–223.

[48] C. Andrews and A. C. Molnar, "Implications of passive mixer transparency for impedance matching and noise figure in passive mixer-first receivers," *IEEE Transactions on Circuits and Systems I*, vol. 57, no. 12, pp. 3092–3103, December 2010.
[49] E. Nash, "Correcting imperfections in IQ modulators to improve RF signal fidelity," Analog Devices: Application Note AN-1039, 2009. [Online]. Available: www.analog.com/media/en/technical-documentation/application-notes/AN-1039.pdf.
[50] J. G. Baldwin and D. F. Dubbert, "Quadrature mixer LO leakage suppression through quadrature DC bias," *Sandia National Laboratories Report*, May 2002. [Online]. Available: http://prod.sandia.gov/techlib/access-control.cgi/2002/021316.pdf.
[51] T. Suzuki, Y. Kawano, M. Sato, T. Hirose, and K. Joshin, "60 and 77GHz power amplifiers in standard 90nm CMOS," in *IEEE ISSCC Digest of Technical Papers*, IEEE, February 2008, pp. 562–563.
[52] T. Yao, M. Q. Gordon, K. K. W. Tang, et al., "Algorithmic design of CMOS LNAs and PAs for 60-GHz radio," *IEEE Journal of Solid-State Circuits*, vol. 42, no. 5, pp. 1044–1057, May 2007.
[53] N. Li, K. Bunsen, N. Takayama, et al., "A 24dB gain 51-68GHz CMOS low noise amplifier using asymmetric-layout transistors," in *Proceedings of ESSCIRC*, IEEE, September 2010, pp. 342–345.
[54] R. G. Carvajal, J. Ramírez-Angulo, A. J. López-Martín, et al., "The flipped voltage follower: A useful cell for low-voltage low-power circuit design," *IEEE Transactions on Circuits and Systems I*, vol. 52, no. 7, pp. 1276–1291, July 2005.
[55] W. L. Chan and J. R. Long, "A 60-GHz band 2×2 phased-array transmitter in 65-nm CMOS," *IEEE Journal of Solid-State Circuits*, vol. 45, no. 12, pp. 2682–2695, December 2010.
[56] L. Wu, A. Li, and H. C. Luong, "A 4-path 42.8-to-49.5 GHz LO generation with automatic phase tuning for 60 GHz phased-array receivers," *IEEE Journal of Solid-State Circuits*, vol. 48, no. 10, pp. 2309–2322, October 2013.
[57] J. Pang, S. Maki, S. Kawai, et al., "A 128-QAM 60GHz CMOS transceiver for IEEE802.11ay with calibration of LO feedthrough and I/Q imbalance," in *IEEE ISSCC Digest of Technical Papers*, IEEE, 2017, pp. 424–425.
[58] R. Wu, S. Kawai, Y. Seo, et al., "A 42Gb/s 60GHz CMOS transceiver for IEEE 802.11ay," in *IEEE ISSCC Digest of Technical Papers*, IEEE, 2016, pp. 248–249.
[59] J. F. Bulzacchelli, C. Menolfi, T. J. Beukema, et al., "A 28-Gb/s 4-tap FFE/15-tap DFE serial link transceiver in 32-nm SOI CMOS technology," *IEEE Journal of Solid-State Circuits*, vol. 47, no. 12, pp. 3232–3248, December 2012.
[60] A. Siligaris, O. Richard, B. Martineau, et al., "A 65nm CMOS fully integrated transceiver module for 60GHz wireless HD applications," in *IEEE ISSCC Digest of Technical Papers*, IEEE, 2011, pp. 162–163.

9 Phased Arrays for 5G Millimeter-Wave Communications

Bodhisatwa Sadhu and Leonard Rexberg

9.1 The Role of mm-Wave in 5G Communications

A phased array is an array of antennas that can be used to create a narrow beam of radio signals that can be directed toward a specific point in space exclusively through electronic control. Most of today's 2G/3G/4G cellular communications use base station antennas with broad fixed beams like that of a street lamp, to communicate between the user and the base station. As a result, there are some inefficiencies from the energy that is wasted in unwanted directions. The future 5G radio system is expected to use advanced beamforming and beam tracking to overcome these inefficiencies. Phased array antennas will be used to create narrow "searchlight" beams that will stay focused on each user even as they move.

Phased array antennas have been developed since the early twentieth century and include developments by a number of Nobel Prize winners. Silicon-based phased array antenna solutions began to be developed only in the last decade, starting with the publication of [1]. This was closely followed by a foray into mm-wave transmitters and receiver ICs on silicon [2]. Since then, significant research on silicon-based mm-wave phased arrays has resulted in improved performance, increased complexity, and experiments at submillimeter wave and THz frequencies. Standards developed for 60 GHz communication [3,4] led to increased research and commercial developments of integrated circuits (ICs) on silicon for mm-wave. These systems are targeting indoor, high-bandwidth, networking applications [5–14], and have helped mm-wave communications, and phased array beamforming acquire the required maturity for commercial deployment. However, unlike mobile networks, these 60 GHz systems were limited to serving a few users (<5) at short distances (<20 m).

The first 5G standard released in December 2017 [15,16] supports mm-wave communications, with 2018 seeing the first product deployments. Operators are preparing their networks for the next wave of mobile communications, including fixed wireless access and improved mobile broadband as well as new use cases, such as mission critical Internet-of-Things (IoT). The 5G standard will bring important additions to mm-wave communications, which is expected to mature into a cellular technology including all current functionality associated with cellular systems, such as mobility, advanced scheduling, and quality of service. The predicted and continued increase in data consumption in mobile networks will require advanced technology additions, as well as more spectrum, to meet the expected service quality. Millimeter-wave bands will be part of this additional new spectrum, supporting future radio access networks. On the device side, it is expected that future mobile devices will include mm-wave technologies.

Phased arrays can be designed at any frequency. However, the size of the antenna array depends on the frequency. While a phased array at 2G/3G/4G frequencies will be at least a few feet in size, the smaller wavelength at mm-wave allows us to create a far more compact solution (a few inches).

This chapter introduces phased array beamforming, beam-steering, and sidelobe-level control, and describes an exemplary silicon-based mm-wave Phased Array Antenna Module (PAAM) prototype. Section 9.2 introduces beamforming as a spatial Fourier transform operation, and derives the radiated field of a continuous source as well as a discrete array of antenna elements. Beam shaping and beam steering using amplitude and phase control of each element are described. Section 9.3 discusses the several desirable features of mm-wave beamforming using silicon-based phased arrays. Section 9.4 describes an exemplary mm-wave phased array IC suitable for 5G communications. Measurement results detailing the performance of the packaged phased array IC with in-package antennas are reported.

9.2 Introduction to Beamforming

9.2.1 Beamforming as a Fourier Transform

Radiation from a continuous source can be expressed as a Fourier transform from the physical source domain to the angular space domain. Without loss in generality, we can express the radiation from a continuous line source (Figure 9.1) as (9.1).

$$P(k_z) = \int_{-\infty}^{\infty} S(z) \cdot e^{jk_z \cdot z} dz \tag{9.1}$$

As shown in the preceding equation, there exists a Fourier transform pair relating the far-field pattern $P(k_z)$ to the source distribution $S(z)$ by a parameter, k_z. $P(k_z)$ is a dimensionless scalar unit proportional to either the electric field or the magnetic field in the far field.

Since the source extension is limited over only a section that is L wide, we can modify the integration limits to only $-L/2$ to $+L/2$ instead of infinity, and rewrite the radiated field (9.1) as

$$P(k_z) = \int_{-L/2}^{L/2} S(z) \cdot e^{jk_z \cdot z} dz. \tag{9.2}$$

Until now, we have used k_z to merely denote a parameter to link the line source to the far-field radiation pattern through the use of the Fourier transform. However, there is also a physical interpretation of k_z that links the Fourier transform parameter directly to

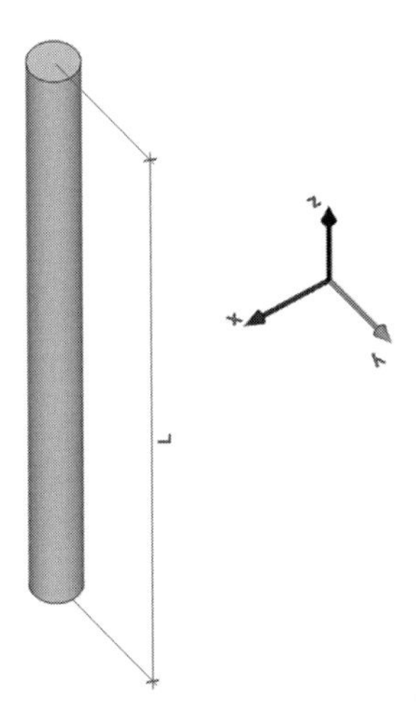

Figure 9.1 1D line source.

the angular domain: as the projection of the general free space wave vector $k_0 \cdot \hat{r}$ onto the z-axis. In other words, we may consider the wave vector as directed in the $\hat{r}$-direction, with components along Cartesian coordinates such that

$$k_0 \cdot \hat{r} = k_x \cdot \hat{x} + k_y \cdot \hat{y} + k_z \cdot \hat{z} \tag{9.3}$$

where

$$\hat{r} = \cos\phi \sin\theta \cdot \hat{x} + \sin\phi \sin\theta \cdot \hat{y} + \cos\theta \cdot \hat{z} \tag{9.4}$$

such that the Cartesian components (k_x, k_y, k_z) of the wave vector are expressed as in (9.5)–(9.7), where θ and ϕ are defined as in Figure 9.2:

$$k_x = k_0 \cdot \cos\phi \sin\theta \tag{9.5}$$

$$k_y = k_0 \cdot \sin\phi \sin\theta \tag{9.6}$$

$$k_z = k_0 \cdot \cos\theta \tag{9.7}$$

where

$$k_0 = \frac{2\pi f}{c} \tag{9.8}$$

Here, the wave number k_0 depends on the properties of the medium in which it propagates, c is the speed of light in the given medium, and f is the carrier signal frequency. $S(z)$ denotes the line source distribution as a function of the position z, and may be an electrical current or express radiation from an aperture. In the former case, we say that the source is of electric art (electrical current), and in the latter case, we say it is of the magnetic art (electric field in an aperture). Sources can be a combination of the two as well.

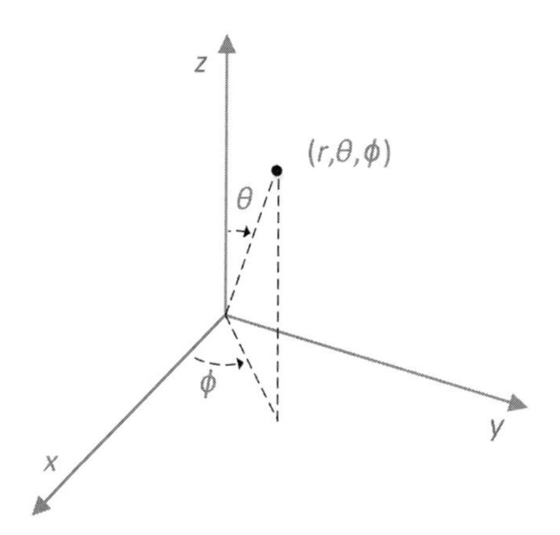

Figure 9.2 Spherical coordinate system.

Note also that

$$k_0^2 = k_x^2 + k_y^2 + k_z^2. \tag{9.9}$$

In the preceding expression for the radiated field, (9.2), we have compressed the angular dependence into the wave number in the z-direction k_z, which is the projection of the free space wave number k_0 onto the z-axis. The radiated field is therefore usually expressed as a function of the angular parameter θ as discussed in the next section.

Note that in (9.2) for the radiated field, we have omitted any decay of the propagating wave, time dependence in the form of harmonic oscillations, or any medium-related property constants such as permittivity or permeability. Moreover, note that this expression is unitless and it merely expresses a relative behavior of the radiated field in different directions.

Continuous Source of Radiation

As a simple example, we may say that the line source is a constant current over the z-range $-L/2$ to $+L/2$. Then, if the magnitude of the line source is S_0, (9.2) yields a radiation pattern showing a typical $\left(\frac{\sin x}{x}\right)$ pattern, as

$$P(k_z) = 2S_0 \frac{\sin\left(\frac{k_z L}{2}\right)}{\frac{k_z L}{2}}. \tag{9.10}$$

Using (9.5), the same expression may be written as

$$P(\theta) = 2S_0 \frac{\sin\left(\frac{k_0 L \cdot \cos\theta}{2}\right)}{\frac{k_0 L \cdot \cos\theta}{2}}. \tag{9.11}$$

This equation describes a beam pattern with a sinc-shaped cross-section whose maximum is at $\theta = 90°$, or perpendicular to the line source. The beam pattern radiates out in all ϕ as a disc-shaped beam.

Distributed Source: 1D Array Antenna

Going a step further, we consider a sampled radiation source instead of a continuous source. In this case, the source consists of a group or array of antenna elements placed at regular intervals d_z along the z-axis. We may now expand the line source representation into a finite series of element sources as in (9.12), where each element has a physical extension L, which is smaller than the interelement spacing d_z. The source expansion may be written as

$$S(z) = \sum_{n=1}^{N} A_n \cdot S_n(z). \tag{9.12}$$

The continuous line source is depicted in Figure 9.3a, and a first step corresponding discretized approximation to the line source is depicted in Figure 9.3b. Each subsection $S_n(z)$ of the approximation, as outlined in (9.12), may be individually different, but as a further simplification, it is convenient to make them all equal in shape as described in Figure 9.3c and also described by (9.13). This step will eventually let us split the radiation pattern into essentially two parts: one element factor and one array factor part as described later in (9.17).

$$S_n(z) = F(z - nd_z). \tag{9.13}$$

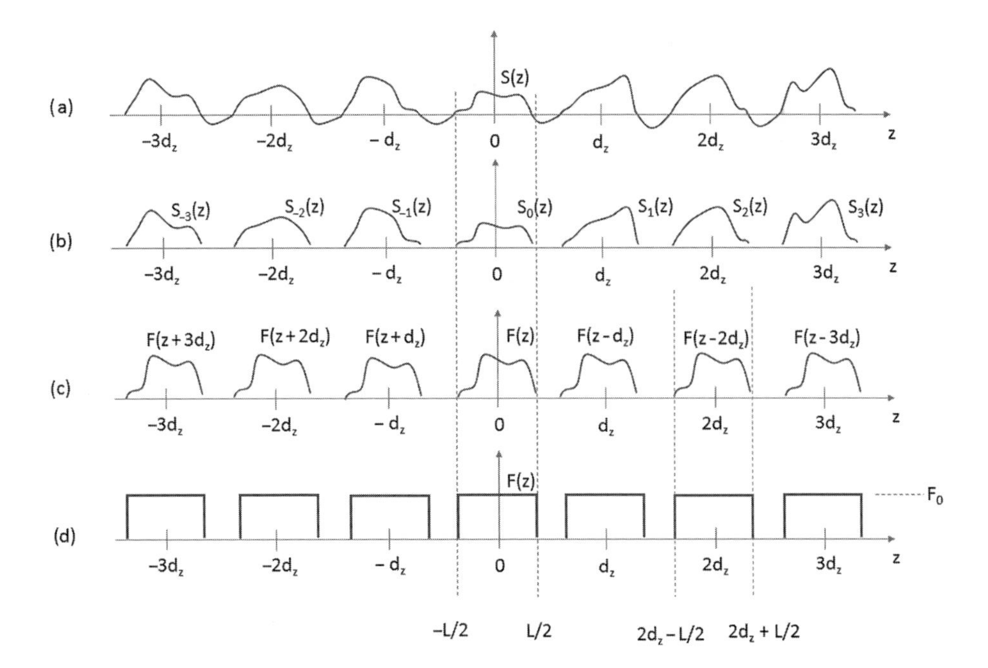

Figure 9.3 (a) Continuous line source, (b) discretized line source, (c) approximation of line source by repetition of base function, (d) base function approximated by a constant over the element definition $-L/2$ to $+L/2$.

As a further simplification, the basis function representing the source may be approximated by a simple constant function over its defined range as in (9.14) and as depicted in Figure 9.3d.

$$F(z) = F_0 \qquad \text{for } -L/2 < z < L/2 \tag{9.14}$$

$$P(k_z) = \int_{-L/2}^{L/2} \sum_{n=1}^{N} A_n \cdot F(z - nd_z) \cdot e^{jk_z z} dz \tag{9.15}$$

Here, we have introduced the ability to have individual signal amplitudes, A_n, applied to each element. Reversing the order of summation and integration in (9.15) gives a result that expresses just a multiplication of an element factor and a summation of complex harmonics according to the following:

$$P(k_z) = \int_{-L_0/2}^{L_0/2} F(z)e^{jk_z z} dz \cdot \sum_{n=1}^{N} A_n e^{jk_z n d_z} \tag{9.16}$$

$$P(k_z) = E(k_z) \cdot AF(k_z), \tag{9.17}$$

where we have separated the radiation pattern into its element factor (E)

$$E(k_z) = \int_{-L_0/2}^{L_0/2} F(z)e^{jk_z z} dz \tag{9.18}$$

and its array factor (AF)

$$AF(k_z) = \sum_{n=1}^{N} A_n e^{jk_z n \cdot d_z} \tag{9.19}$$

The element factor, E, represents the field of each element positioned at the origin, while the array factor, AF, represents the interference pattern created by the array of elements and depends on the number of elements, their relative spacing, magnitudes, phases and geometric arrangement. The AF does not depend on the directionality of the individual antennas.

This is a significant result, as it allows us to compute the AF assuming isotropic antennas, and later multiply the resulting AF with the individual element radiation patterns to calculate the overall radiation pattern of the real source.

And finally, if the source element is constant over its limits L_0, as in (9.14) and Figure 9.3d, we may again use the $\left(\frac{\sin x}{x}\right)$ formulation for the element factor to arrive at

$$P(k_z) = \underbrace{2\frac{\sin\left(\frac{k_z L_0}{2}\right)}{\frac{k_z L_0}{2}}}_{\text{E}} \underbrace{\sum_{n=1}^{N} A_n e^{jk_z n d_z}}_{\text{AF}} \tag{9.20}$$

This final equation expresses the radiated field from a group of individual antennas placed at regular intervals of distance d_z.

9.2.2 Beam Shaping and Beam Steering

The ability to excite each element with different amplitudes offers a degree of freedom that allows us to influence the radiation pattern. This ability, for example, can be used to obtain other sidelobe levels than prescribed by the simple $\left(\frac{\sin x}{x}\right)$ pattern that will always place its first sidelobe at -13.7 dB below the main lobe. This ability to control sidelobe levels by applying different amplitude levels to the different elements is called amplitude tapering of an array antenna.

The parallels between amplitude tapering in phased arrays and digital filter design provide insight. For example, in filter design, each tap in an finite impulse response (FIR) filter may be given a specific amplitude weight in a technique called windowing. In signal theory, windowing is commonly employed to reduce sidelobes in the signal spectrum when emulating an infinite duration signal using a time-limited one. In antenna array design, a similar approach may be used to lower the sidelobes that are pointing at other (unwanted) directions in space. The most commonly used tapering windows in phased array design are the Chebyshev and Taylor windows. Amplitude control can also be used to steer nulls toward a given direction to spatially filter out interferers.

In addition to amplitude tapering by changing the magnitude of A_n, one can also change the phase of each A_n and replace it with the following

$$A_n \rightarrow A_n e^{j\alpha_n} \tag{9.21}$$

This results in a radiation pattern given by

$$P(k_z) = 2\frac{\sin\left(k_z\frac{L_0}{2}\right)}{k_z\frac{L_0}{2}}\sum_{n=1}^{N} A_n e^{j\alpha_n} e^{jk_z n d_z} \tag{9.22}$$

If specifically we choose to pick the phases such that

$$\alpha_n = -k_0 \cdot nd_z \cdot \cos\theta_0 \tag{9.23}$$

then (9.21) expresses the required complex weighting of each antenna element so as to obtain a beam that is directed toward θ_0. For omni-directional antenna elements, the beam direction is independent of the magnitude weights of the individual elements and depends only on the phases of the individual elements. The resulting general expression for the radiation pattern in a 1D phased array with uniform phase shifts is given by

$$P(\theta) = 2\frac{\sin\left(k_0\frac{L_0}{2}\cos\theta\right)}{k_0\frac{L_0}{2}\cos\theta}\sum_{n=1}^{N} A_n \cdot e^{jk_0 n d_z[\cos\theta - \cos\theta_0]}. \tag{9.24}$$

This last equation shows that by choosing appropriate phases in (9.21), the whole radiation pattern may be steered toward a direction θ_0, independently from the sidelobe level. Moreover, by controlling the amplitude $|A_n|$, the sidelobe level can be controlled, independently from the steering angle. This orthogonal control of beam direction and beam shape is critical in phased array antennas. As a result, it is vital that these two parameters be kept independent such that the beam angle and beam shape can be controlled orthogonal to each other.

9.2.3 2D Antenna Array

As a final step, now consider the general case of a 2D distribution of antenna elements (one element is shown in Figure 9.4), thus providing the array antenna (Figure 9.5) the ability to steer not only in one direction, but also in the second angular direction, ϕ. This result can be easily achieved as to be given the freedom of steering the array antenna not only in one direction, but also in the second angular direction, ϕ. This result can be easily achieved by multiplying the 1D linear array antenna expression with a second one, so as to obtain a 2D radiation pattern.

It is important to choose the correct angular (θ, ϕ)-dependence to employ. It is noted that the z-direction only involves one angle, θ while in the x-direction both angular dependences are present θ and ϕ. This dependence is explicitly given by (9.5), so the general radiation pattern may be found as

$$P(\theta, \phi) = 4 \frac{\sin\left(k_0 \frac{L_0}{2} \cos\theta\right)}{k_0 \frac{L_0}{2} \cos\theta} \cdot \frac{\sin\left(k_0 \frac{W_0}{2} \cos\phi \cos\theta\right)}{k_0 \frac{W_0}{2} \cos\phi \cos\theta} \cdot \sum_{n=1}^{N} \sum_{m=1}^{M} \alpha \cdot \beta \cdot \gamma \tag{9.25}$$

with

$$\alpha = | A_{n,m} | \tag{9.26}$$

$$\beta = e^{jk_0 n d_z [\cos\theta - \cos\theta_0]} \tag{9.27}$$

$$\gamma = e^{jk_0 m d_x [\cos\phi \cos\theta - \cos\phi_0 \cos\theta_0]}. \tag{9.28}$$

It is often convenient to express the radiation pattern as being in a separable form in that it may be expressed as a clear product of two summations, one in the x-direction and one in the z-direction. In that case, the excitation amplitude matrix $A_{n,m}$ can also be split into an x-part and a z-part, and can be described by (9.29), which is a multiplication of two amplitude vectors. In this simple example, one vector is only three elements long while the second one is four elements long. This gives the excitation for a 3×4 antenna array, which has a separable source distribution.

$$A = \begin{bmatrix} v_1 \\ v_2 \\ v_3 \end{bmatrix} \cdot \begin{bmatrix} w_1 & w_2 & w_3 & w_4 \end{bmatrix} = \begin{bmatrix} A_{1,1} & A_{1,2} & A_{1,3} & A_{1,4} \\ A_{2,1} & A_{2,2} & A_{2,3} & A_{2,4} \\ A_{3,1} & A_{3,2} & A_{3,3} & A_{3,4} \end{bmatrix} \tag{9.29}$$

$$A_{n,m} = v_n \cdot w_m. \tag{9.30}$$

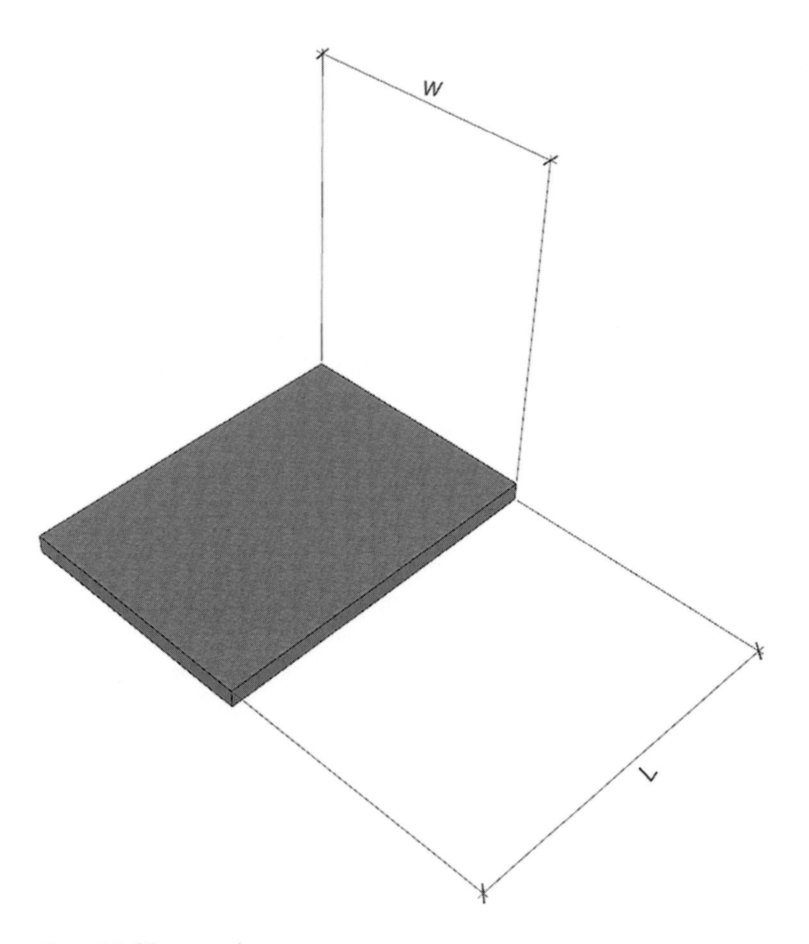

Figure 9.4 2D rectangle source.

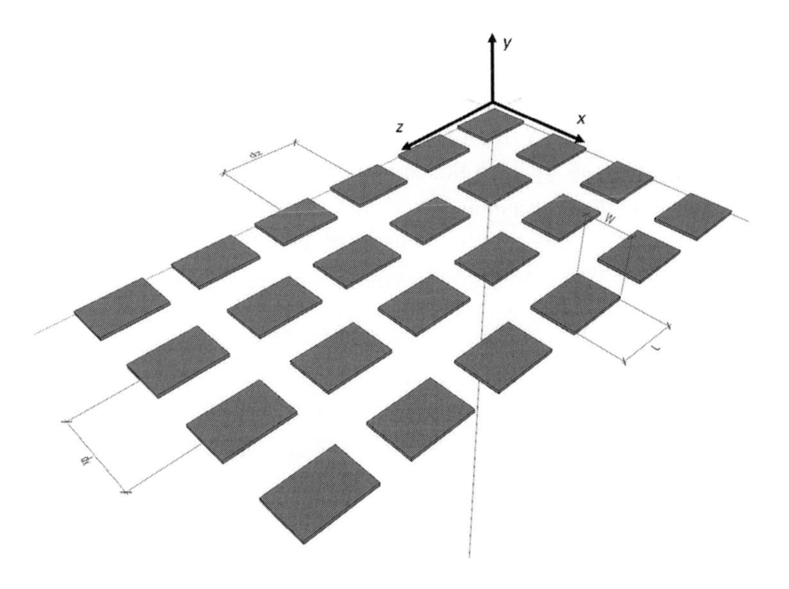

Figure 9.5 2D array antenna.

9.3 Desired Features of Millimeter-Wave Phased Arrays

9.3.1 Accurate Beam Control

Accurate beam control *while maintaining a low sidelobe level* is achieved using a high resolution and accurate phase shifter. For phased arrays where a uniform phase gradient is applied (i.e., each element is progressively phase-shifted by a fixed amount relative to its adjacent elements), the phase resolution directly translates to beam steering resolution. For example, as shown in row 2 in Table 9.1, when a uniform phase gradient is applied across the phased array, the beam steering resolution is 1.4° at broadside for a phase shifter resolution of 5°, and an antenna element spacing of 0.55λ. In contrast, the beam steering resolution is only 6.5° for a phase shifter resolution of 22.5°. However, for large arrays, it is possible to apply a nonuniform phase gradient and use an averaging effect to interpolate to a finer beam steering resolution, as shown in row 3 of Table 9.1. This effect of beam interpolation using phase averaging is shown in the simulations results in Figure 9.6a–9.6c for an 8×8 square array with $\lambda/2$ element spacing. Figure 9.6a shows a family of simulated beam patterns for attempted beam pointing at an arbitrary steering angle. Each beam pattern in Figure 9.6a corresponds to a fixed phase shifter resolution; the phase shifter resolution is swept to create the family of beam patterns shown. As seen in the figure, while fairly accurate beam pointing is achieved in an arbitrary direction even with low-resolution phase shifters (e.g., 22.5° resolution), the sidelobes are severely compromised when compared to beam patterns created using high-resolution phase shifters (e.g., 5° resolution). Figure 9.6b and 9.6c show the resultant sidelobe suppression as the beam is pointed across ±45°. In general, as shown in Figure 9.6b and 9.6c, limited phase resolution produces a periodic degradation in the sidelobe suppression versus steering angle. While attempting ~20 dB sidelobe suppression using a Taylor window tapering function, and with ideal phase resolution, a uniform sidelobe suppression of ~19.5 dB is obtained across a ±45° scanning range. However, for a coarse phase resolution of 22.5°, the sidelobe suppression at some angles is degraded by as much as 5 dB. Moreover, a recent study has shown that a coarse (2-bit) resolution can also result in antenna array gain reduction of up to 3 dB for nonboresight directions [17]. In this context, a phase resolution of 5° is an attractive target as it results in beam pointing with <1° beam steering resolution with sidelobe suppression levels within ~1 dB of an ideal phase shifter when attempting 20 dB sidelobe suppression.

Table 9.1 Simulated resolution of the phase shifter and beam steering with 0.55λ element spacing.

Phase shifter resolution	0°	5°	22.5°
Beam steering resolution (uniform phase)	0°	1.4°	6.5°
Beam steering resolution (nonuniform phase)	<1°	<1°	<1°
Sidelobe suppression (attempting 20 dB)	19.5 dB	>18.5 dB	<18 dB

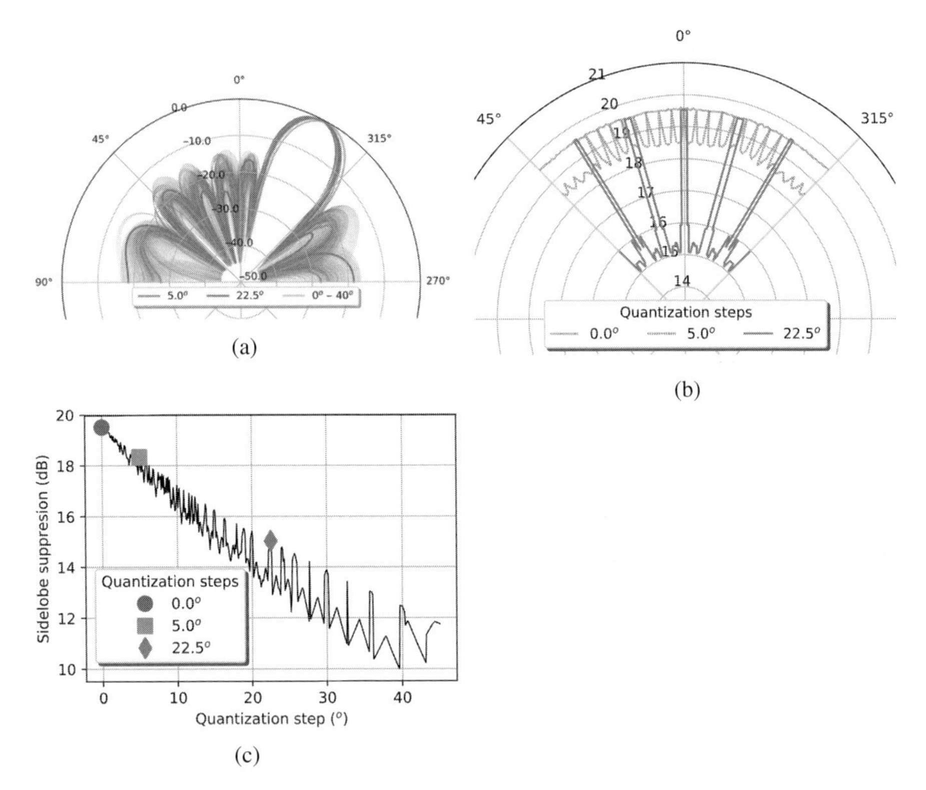

Figure 9.6 Simulation of the effect of phase shift quantization in an 8 × 8 phased array showing, (a) beam pointing at an arbitrary direction (30°) for different phase shifter resolutions showing a large impact on sidelobe suppression, (b) sidelobe suppression vs. beam steering angle for three different phase shifter resolutions while attempting 20 dB Taylor window tapering, and (c) the worst case sidelobe suppression vs. quantization steps while attempting ∼20 dB Taylor window tapering [18]. (©2017 IEEE. Reprinted, with permission, from *IEEE Journal of Solid States Circuits*.)

9.3.2 Architecture Scalability

The ability to achieve beamforming provides a unique opportunity for spatial multiplexing, thereby reusing the same bandwidth in multiple spatially exclusive directions. Each beam can be formed using a unit phased array. Moreover, by using scalability techniques [13,19], unit phased arrays can be combined to create larger apertures, resulting in narrower beams and higher directivity in transmit (TX) and receive (RX) modes. In this context, a system-level consideration relates to the optimal way to divide the total number of elements among multiple modular ICs. An approach using multiple small ICs with few elements per IC can be used to target a reduction in the interconnect length from the ICs to the antennas by enabling the small ICs to be placed close to the antennas. However, from a system perspective, the small IC approach suffers from drawbacks:

1. System verification simulations are complicated and need to include many ICs and significant IC-to-IC connectivity on the package.

Table 9.2 Summary of the trade-off between different IC modularity options.

Modular IC approach	Many small ICs with low integration	Few highly integrated large ICs
Interconnect length (IC FEs $\leftrightarrow$ in-package antennas)	Small interconnect length when aggregate IC area is much lower than aggregated antenna area	Small interconnect length when aggregate IC area is similar to aggregate antenna area
mm-wave performance verification of complete phased array	More challenging: significant package routing and many ICs in phased array module make verification simulations challenging	Less challenging: Less package/board-level routing and fewer ICs makes phased array module verification easier
Digital control	Challenging: Digital infrastructure is repeated per IC, control is not centralized	Superior: Centralized digital control
Total IC area per complete phased array	Potentially larger aggregate IC area due to repeating digital infrastructure	Potentially smaller aggregate IC area due to fewer digital infrastructure instantiations
Package assembly	More challenging to assemble many small ICs, increasing number of assembly steps	Less challenging to assemble fewer ICs requiring fewer assembly steps
Overall yield	Trade-off: Higher yield for each individual IC with smaller area but lower assembly yield due to larger number of assembly steps	Trade-off: Lower yield for each individual IC with larger area but higher assembly yield due to smaller number of assembly steps

2. Digital infrastructure needs to be repeated per IC, making digital control of the phased array cumbersome, impacting both chip area and digital control speed.
3. Package assembly involving many small ICs is potentially significantly more complex, increasing the number of assembly steps and affecting assembly yield.

The trade-offs between an approach using many small ICs and one using a few large ICs are listed in Table 9.2. In this work, we selected 4 ICs with 32 elements per IC to implement 128 element phased array elements feeding 64 dual-polarized in-package antennas. The IC scaling at the package level was implemented using a scaling approach introduced in [19,20].

9.3.3 Dual-Polarized Operation

Channel capacity can be increased by using the two available orthogonal polarizations [21] to create simultaneous beams with independent data. In the far-field region, the field vector is perpendicular to the direction of propagation; hence, the maximum number of orthogonal field vectors that can be supported is two. For example, if the direction of wave propagation is $\hat{z}$, then the field vector is polarized in the xy-plane. In general, the two transmitted field vectors may be expressed as $\hat{E}_1 = a \cdot \hat{x} + b \cdot \hat{y}$ and

$\hat{E}_2 = c \cdot \hat{x} + d \cdot \hat{y}$, where $\hat{E}_1$ and $\hat{E}_2$ are independent of each other (not in the same direction). Two independent data streams can be transmitted on the two independent field vectors using antennas that have independent polarizations. This would effectively improve the overall channel capacity by up to 2×.[1]

In the case of dual-polarized operation with independent data streams, if the receiver is able to determine the channel matrix $[a\ b;\ c\ d]$, either or both of the two data streams transmitted on the two independent field vectors can be demodulated. The channel matrix is determined by transmitting and receiving known sequences of signals in a process called channel estimation. Also, since only two orthogonal polarizations can be supported in the far field, this limits the dimension of the polarization space to two, and a maximum of two independent data streams on two independent polarizations can be transmitted. Beyond this, independence cannot be achieved, and the channel matrix becomes underdetermined and cannot be inverted even in a least mean square (LSM) sense.

Note that $\hat{E}_1$ and $\hat{E}_2$ need to be independent (not in the same direction), but need not be orthogonal to each other. This is critical since polarization orthogonality can be impossible to maintain at the RX. This is because it can be impossible to align the RX antenna to the TX, and reflections in the channel cause depolarization [22]. However, given that the more orthogonal the RX signals are, the higher the RX signal-to-interference-plus-noise ratio (SINR), it is beneficial to transmit orthogonal signals from the TX.

The support for multiple simultaneous beams not only increases the maximum number of simultaneous users that can be supported, but also opens a path to the implementation of self-backhauling [23] by using one polarization to establish a link between pico cells.

9.3.4 Small Solution Footprint

Given the advantages of supporting dual polarization operation and multiple spatial beams, the next challenge is to realize such functionality in a small form factor. Unlike current macrocell base stations, forthcoming 5G access points are expected to be deployed in multiple indoor and outdoor urban spaces, making a compact, Wi-Fi access pointlike form factor desirable. At 28 GHz, the area occupied by the phased array is primarily determined by the area of the $\lambda/2$ spaced antenna array. As a result, sharing the antennas among multiple functions has a significant impact on the overall size of the solution. For example, to support half-duplex transmit and receive in both horizontal (H) and vertical (V) polarizations, the phased array needs to support four modes of operation: simultaneous TX-H and TX-V, and simultaneous RX-H and RX-V. As shown in the illustration (Figure 9.7) and accompanying Table 9.3, this functionality can be implemented using different strategies with significantly different area implications:

[1] Note that this is different from traditional polarization diversity, where the same data were transmitted in two polarizations in order to improve the signal-to-noise ratio (SNR) and make the system more robust.

Table 9.3 Requirements for an N element dual-polarized transceiver phased array.

Parameter	Option 1	Option 2	Option 3	Option 4
Aggregate array area	4×	2×	2×	1×
Aggregate number of antennas	$4N$	$2N$	$2N$	N
Aggregate number of front-ends	$4N$	$4N$	$2N$	$2N$
Front-end type	TX or RX	TX or RX	TRX	TRX
IC complexity	Lowest	Low	High	Highest

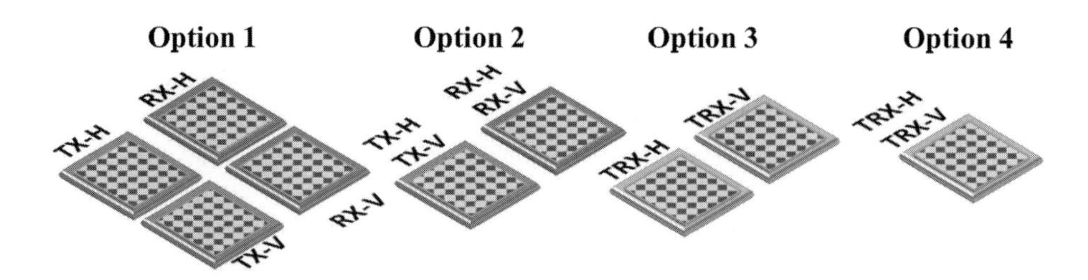

Figure 9.7 Different options for functional integration for 5G mm-wave phased arrays [18]. (©2017 IEEE. Reprinted, with permission, from *IEEE Journal of Solid States Circuits.*)

- Option 1: Four separate modules for each mode of operation: This option leads to lower complexity for each IC, but also results in the largest overall area. Each mode can be designed using a separate IC that interfaces with an antenna module, as shown in Figure 9.7 and Table 9.3. For an N element phased array, $4N$ antennas connect to $4N$ front-ends.
- Option 2: One TX module with dual-polarized antennas, and one RX module with dual-polarized antennas: This strategy requires half the area as option 1, as shown in Figure 9.7 and Table 9.3. In this case, for an N element phased array, $2N$ antennas connect to $4N$ front-ends (FEs). At the IC level, each dual-polarized antenna needs to be fed from two independent signal chains. As a result, it is useful for the two FEs feeding the two ports of a dual-polarized antenna to be colocated for ease of connectivity. This entails two independent phased arrays to be integrated on a single IC with interleaved FEs, adding to design complexity as compared to option 1. However, TX and RX operations remain on separate ICs.
- Option 3: One TRX H module, and one TRX V module: This strategy also requires half the area as option 1, as shown in Figure 9.7 and Table 9.3. However, compared to option 2, it is possible to switch between TX and RX functions for each front-end on the IC so that $2N$ antennas connect to only $2N$ FEs. The reduction in the number of FEs can achieve significant reduction in the IC area by sharing hardware between the TX and RX modes.
- Option 4: One TRX-H-V module: This strategy requires a quarter of the area occupied by option 1, and represents the most area economical solution, as shown in Figure 9.7 and Table 9.3. For an N element phased array, N antennas connect to $2N$ FEs. However, this entails a higher design complexity: each FE on the IC

can switch between TX and RX functions, while H and V FEs are interleaved on the IC to minimize connection complexity between each set of H and V front-ends and each dual-polarized antenna.

As compared to a single polarized TX RX chipset implementation in option 1, option 4 achieves a 4× reduction in area of the package. This makes option 4 a compact, practical approach for a dual polarized mm-wave 5G phased array solution.

9.3.5 Orthogonal Phase and Gain Control

The two most critical performance parameters of an antenna array are beam-steering and sidelobe-level control. From a system perspective, it is imperative to achieve orthogonal beam-steering and sidelobe suppression functions. In other words, sidelobe suppression should be maintained during beam steering, and beam-pointing direction should be maintained while achieving sidelobe suppression.

The orthogonality between sidelobe level and beam steering directly translates to the input parameter plane: amplitude control and phase control per antenna branch. As discussed in Section 9.2.2, amplitude control per element directly controls the sidelobe level, whereas phase control directly controls beam steering. One can separately adjust the sidelobe by setting the relative amplitudes of the antenna elements; if the relative phases of the antenna elements are maintained, the pointing direction is not affected. Likewise, one can independently steer the beam direction by adjusting the relative phase settings of the antenna elements; if the relative amplitude settings are maintained, the sidelobe level is not affected. Therefore, to achieve orthogonal sidelobe and beam-steering control, the phase, and amplitude control in each element must be orthogonal, as shown in Figure 9.8a.

However, in most current Si-based phased array implementations, variable gain amplifiers show a significant phase variation over gain settings, and mm-wave phase

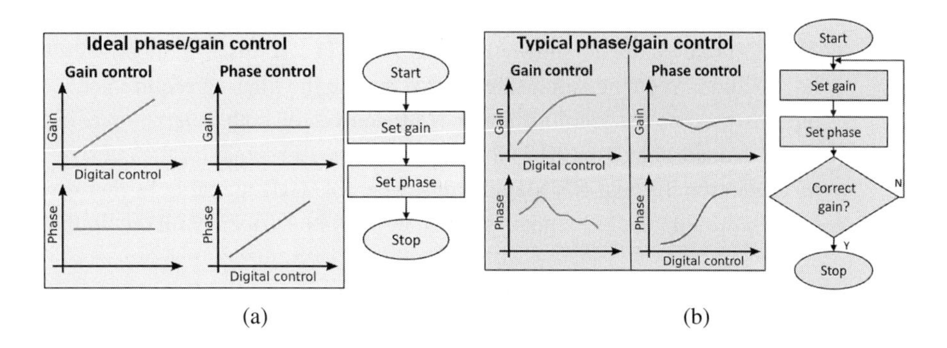

Figure 9.8 (a) Ideal phase and gain control showing a simple beam control algorithm enabled by orthogonal phase and gain control and (b) typical phase and gain control in state-of-the-art phased arrays showing a complex beam control algorithm due to nonorthogonal phase and gain control [18]. (©2017 IEEE. Reprinted, with permission, from *IEEE Journal of Solid States Circuits.*)

shifters typically show gain or loss variation of a few decibels over phase settings, as shown in Figure 9.8b. In theory, it is possible to precondition the input parameter setting by, for example, a multiple entry look-up table to provide the variable gain amplifier (VGA) control and phase shifter control settings to achieve a specific beam direction and sidelobe level. The integration of memory units or beam tables at each front-end to store precalibrated phase and amplitude settings has been demonstrated previously [6]. However, to obtain such calibration tables for both beam steering and tapering across temperature, supply, and electromagnetic (EM) environment variations implies a complex and expensive calibration process. Such calibration not only requires large amounts of data storage and complex algorithms, but also relies on accurate phase measurements over the air. Moreover, the achievable effective phase and amplitude control resolution would be defined by the performance of such calibration. As a result, it is desirable to map the functional orthogonality directly to the actuators by implementing inherently orthogonal phase and gain control functions. Circuit-level techniques to achieve such orthogonality are described in Section 9.4.

9.4 Exemplary Si-Based Millimeter-Wave Phased Array

This section describes the design of a prototype 28 GHz phased array IC for mm-wave communications [18,24]. The IC addresses the various challenges in mm-wave phased array communication outlined in Section 9.3. It features a 32-element transceiver IC with half-duplex TX and RX functions. Simultaneous dual polarization with independent data in each polarization is supported in both TX and RX modes. Each polarization uses 16 elements per IC.

9.4.1 Circuit Details

The IC architecture is shown in Figure 9.9. The IC uses a superheterodyne sliding intermediate frequency architecture [25] to ease the filtering requirements. A 5.17 GHz input is used as the local oscillator (LO) source for both polarizations in the IC. This signal is quadrupled to a 20.7 GHz to provide the LO signal for the radio frequency (RF) mixers in the TX and RX signal paths. The 20.7 GHz signal is further divided to 10.3 GHz to provide the LO signal for the intermediate frequency mixers in the TX and RX signal paths.

In each polarization, the TX intermediate frequency mixer employs a high side injection Hartley architecture for image rejection. This is followed by a low-pass filter to filter out the LO and the remnant image. The TX RF mixer employs a low side injection architecture and is followed by a high-pass filter to attenuate the LO leakage and image signals. Two TX RF mixers provide signals to two 1:8 Wilkinson-based, passive splitters as shown in Figure 9.9. Each output of the splitter interfaces with a phased array front-end element.

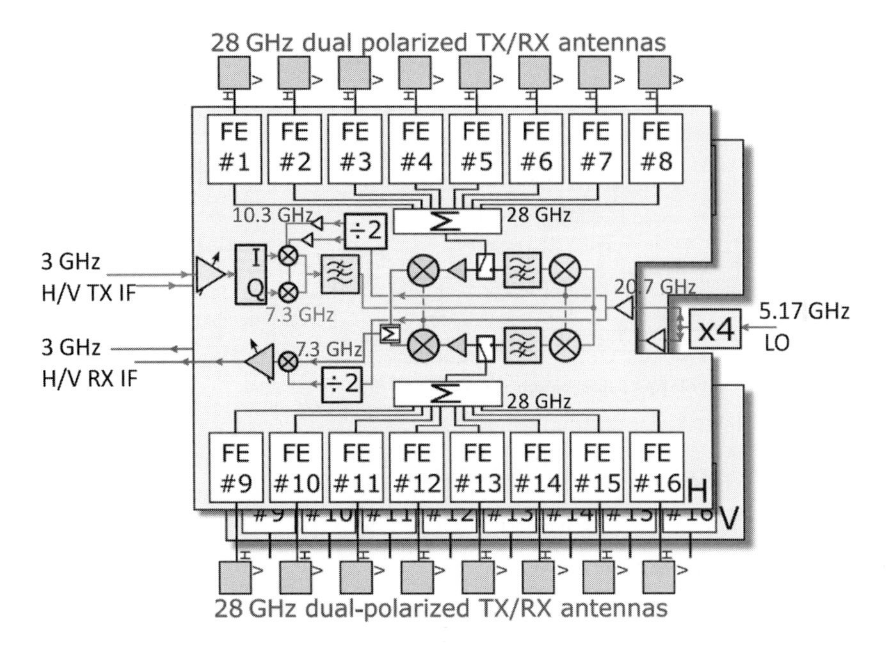

Figure 9.9 Block-level architecture of the 28 GHz phased array IC [18]. (©2017 IEEE. Reprinted, with permission, from *IEEE Journal of Solid States Circuits*.)

The RX architecture employs a similar frequency plan as the TX as shown. The passive splitters function as combiners in the RX mode. Three switches, two in the front-ends, and one quarter-wave transmission line–based switch at the RF mixer are used to switch between TX and RX modes as shown.

A detailed schematic of a phased array front-end is shown in Figure 9.10.

Loss Invariant Linear Phase Shifter

Each front-end shares a passive transmission line–based phase shifter [26] between TX and RX modes. The phase shifter comprises a series of transmission line unit cells with varying inductance and capacitance per unit cell while maintaining a constant inductance-to-capacitance ratio (constant characteristic impedance). The aggregate variable delay transmission line achieves monotonic phase control with constant phase steps, unlike reflection-type phase shifters (RTPS) [27,28]. Moreover, using a loss-control switched resistance per unit element, loss-invariant phase variation is obtained providing orthogonal phase control in the phased array. A unit-cell t-line–based phase shifter offers several advantages. Compared to RTPS designs [27,28], the proposed approach enables high phase accuracy due to unit-cell matching. Moreover, unlike an RTPS, which is limited by the C_{max}/C_{min} varactor ratio of a technology, the total phase range in a t-line phase shifter can be scaled easily by adding more unit cells, making it possible to achieve $>180^{\circ}$ phase shift. Furthermore, the phase resolution can be improved simply by reducing the unit-cell size. Additionally, the t-line approximates a broadband true-time delay and is not limited by the narrowband approximation of a phase shifter. Another key advantage of the proposed approach

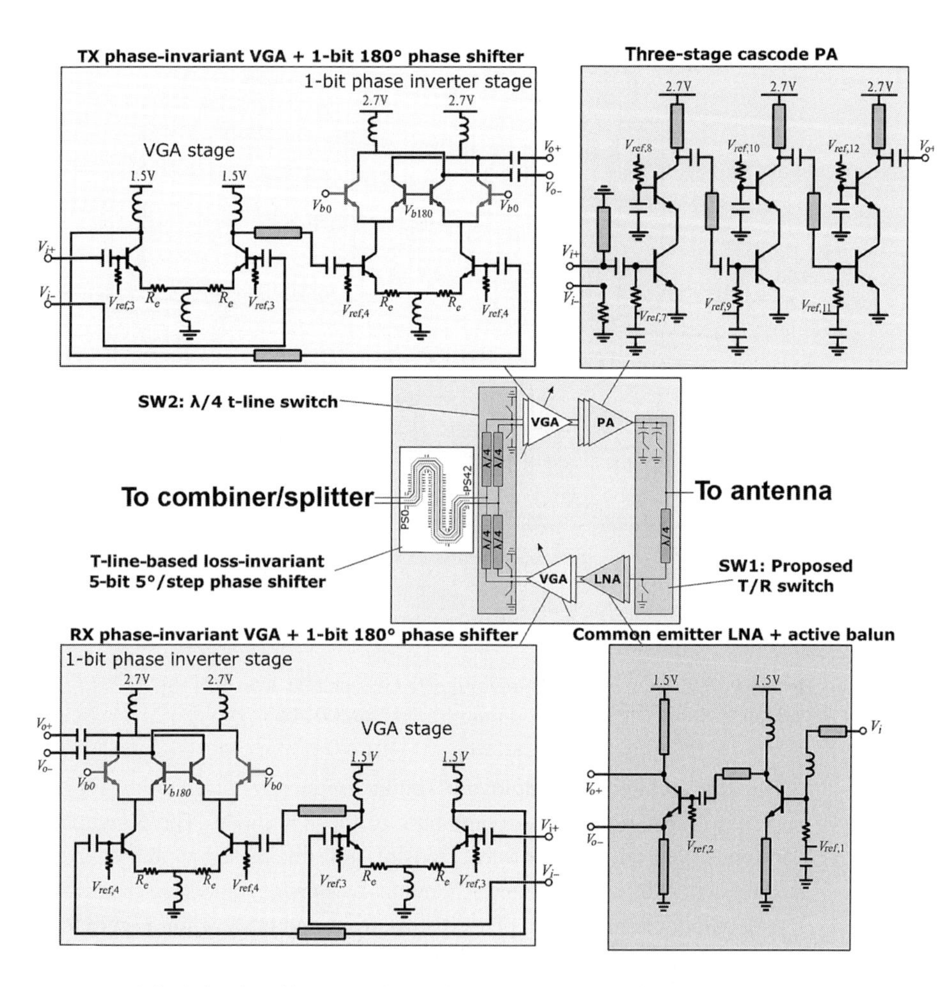

Figure 9.10 Block-level architecture of each front-end of the 28 GHz phased array IC [18]. (©2017 IEEE. Reprinted, with permission, from *IEEE Journal of Solid States Circuits*.)

is a significantly shorter phase switching time. A quarter-wave transmission line–based switch is used to share the bidirectional phase shifter between transmit and receive modes. The reader is referred to [26] for further circuit details of the standalone phase shifter, as well as detailed measurement results across temperature and frequency.

Phase Invariant Variable Gain Amplifier

In both TX and RX modes, phase invariant VGAs are used for tapering control. Amplifier gain is controlled by varying the bias current of emitter degenerated differential common emitter amplifiers. The phase dependence in this resistively degenerated common emitter topology is given by [29],

$$g_{m,eff} = \frac{g_m}{(1 + g_m R_e) + j\omega(c_{je} + c_\pi)(r_b + R_e)} \tag{9.31}$$

with a phase φ of

$$\varphi\left(g_{m,eff}\right) = \tan^{-1}\left((r_b + R_e)\omega c_{je}\frac{(1+\frac{c_\pi}{c_{je}})}{(1+g_m R_e)}\right). \tag{9.32}$$

Here, $g_{m,eff}$ is the effective transconductance of the emitter degenerated amplifier (single-ended), g_m is the transconductance of the bipolar transistor, R_e is the external emitter degeneration resistance, c_{je} is the junction capacitance, C_π is the diffusion capacitance between the base and the emitter of the bipolar transistor, r_b is the intrinsic base resistance. For simplicity, model parameters such as $R-\pi$ and c_μ are ignored in this analysis. For the argument of this equation to be constant as a function of bias current, the emitter degeneration resistance, say R_{e0}, can be calculated: $R_{e0} = \frac{c_\pi}{g_m c_{je}} = \frac{\tau_b}{c_{je}}$ (using $c_\pi = g_m \tau_b$). Consequently, for this optimized value of emitter degeneration resistance, phase invariant gain control can be obtained, achieving orthogonal gain control in the phased array.

A cascode stage with switchable transistors follows the VGAs in both TX and RX modes and provides a phase inverter function. The phase inverters, along with the 180°+ coverage of the phase shifter, provides more than 360° phase coverage per element. Both the VGA and phase inverter stages use differential inductors as loads and single-ended inductors for common mode rejection as shown in Figure 9.10. The reader is referred to [29] for further circuit details of the standalone VGA, as well as detailed measurement results across temperature and frequency.

PA, LNA, and Front-End Switch

In the TX mode, each front-end employs a three-stage cascode power amplifier (PA). The stages are internally matched to each other without using a 50 Ω interface. Transmission lines are used in the PA to achieve greater control of the electromagnetic field, in order to reduce the possibility of instability through unwanted field coupling. Capacitive coupling between stages is employed for matching as well as bias isolation. The PA is single-ended in order to provide a single-ended output for the antennas in the package.

In the RX mode, each front-end uses a single-ended inductively degenerated common emitter low-noise amplifier (LNA) followed by an active balun stage to interface with the differential signal chain in the rest of the IC.

There are 16 front-end blocks per polarization per IC, making the front-end power dissipation key to the overall power budget. In this regard, PA efficiency and T/R switch insertion loss are critical to the effective isotropic radiated power (EIRP) vs. power dissipation trade-off.

In the front-end of this phased array, a new T/R switch topology was used to minimize the insertion loss in the TX mode. In a traditional T/R switch implementation, $\lambda/4$ t-line based switches are used both at the output of the PA and input of the LNA. As a result, the signal flows through one of the $\lambda/4$ t-lines in the TX and RX modes respectively, resulting in approximately equal insertion losses in either mode. In the proposed design, shown in Figure 9.11a, the $\lambda/4$ t-line-based switch is omitted on

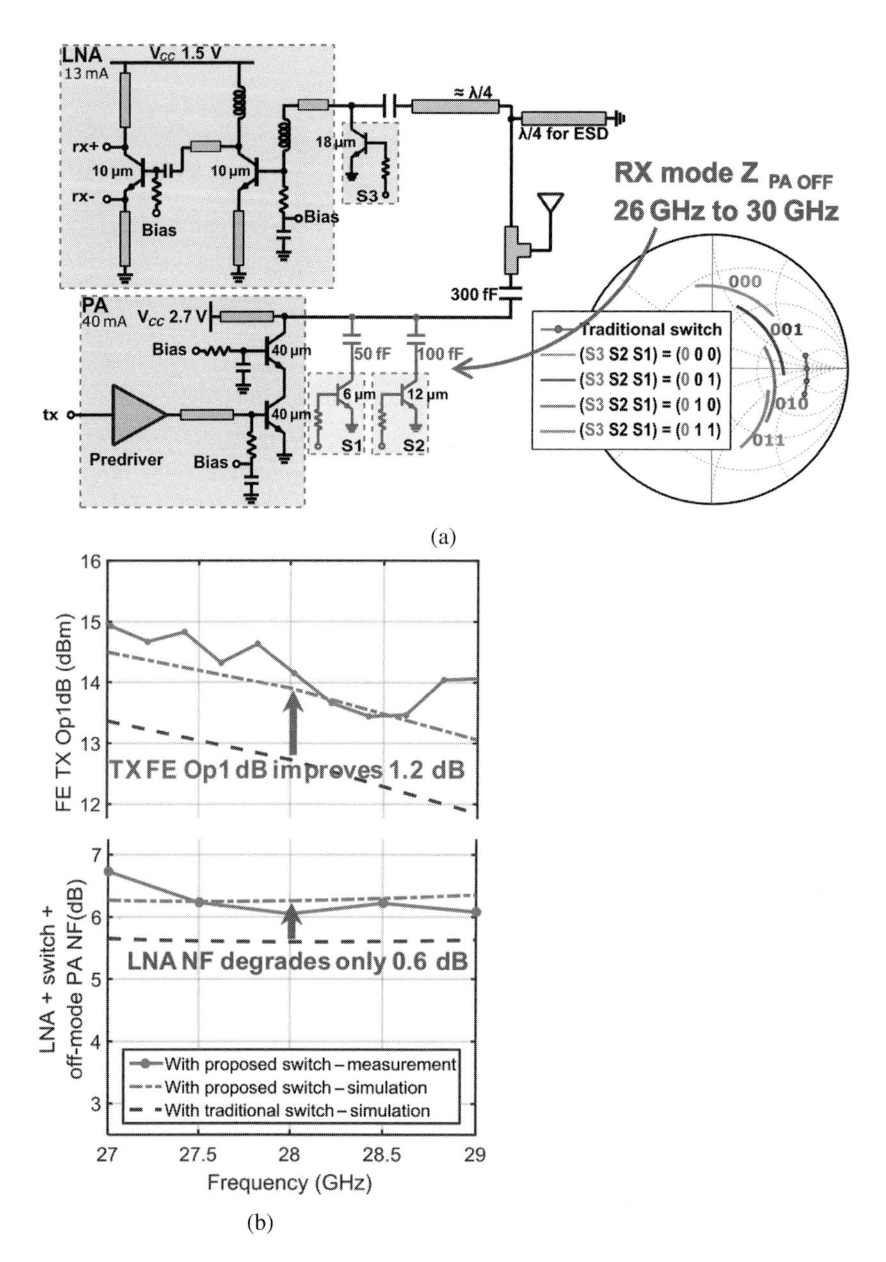

Figure 9.11 (a) Detailed schematic of the implemented TX/RX switch; (b) simulations and measurements showing performance (P_{1-dB}) of TX FE and NF of LNA + switch + PA) with the proposed TX/RX switch compared to simulations of the same with a traditional TX/RX switch across frequency [18]. (©2017 IEEE. Reprinted, with permission, from *IEEE Journal of Solid States Circuits.*)

the TX side. Consequently, the PA is connected directly to the antenna, resulting in negligible insertion loss in the TX mode. In the RX mode, it is desirable to create a high impedance at the TX input to ensure that most of the RX signal from the antenna flows into the LNA instead of the PA. In the implemented design, when the PA is turned off, its output impedance is comprised of a low conductance real part in parallel with a high susceptance inductive load. As shown in Figure 9.11a, a bank of digitally controlled switched capacitors is placed in parallel with the output of the PA and creates a suitable negative susceptance to resonate out this PA output inductance. This strategy results in an effective high real impedance in the RX mode, reducing the impact of the switch on the noise figure of the LNA.

To demonstrate the resulting output power vs. noise figure trade-off, Figure 9.11b compares the TX front-end P_{1-dB} and P_{sat}, and RX LNA + switch noise figure of the proposed approach and the traditional switch approach. The removal of the traditional PA $\lambda/4$ switch improves the P_{1-dB} and P_{sat} by 1.2 dB while incurring only a 0.6 dB penalty in the RX noise figure. This results in P_{sat} >16 dBm per signal path and PA + switch peak efficiency >20%, while still maintaining a 6 dB LNA + switch noise figure. If we translate this to power savings, the additional 1.2 dB TX loss per path of the traditional approach would have consumed 2.35 W (or 23%) more power for the same P_{sat}, compared to the chosen approach.

9.4.2 Measurement Results

The IC was implemented in the Global Foundries 130 nm SiGe BiCMOS 8HP technology with an f_T/f_{MAX} of 200/280 GHz. A die photo is shown in Figure 9.12a. The IC measures 15.8 mm by 10.5 mm. Four ICs were incorporated in an antenna in package PAAM [30], as shown in Figure 9.12b using a phased array scaling approach based on [19,20]. The 128 TRX elements in the 4 ICs interface with 64 dual-polarized antennas in the package. The antenna array also includes 36 dummy antennas to maintain coupling symmetry. All phased array over-the-air measurement results reported in this section are based on testing PAAMs such as the one shown in Figure 9.12b.

Figure 9.13 shows the measured increase in EIRP as a function of the progressive turning on of 64 elements in a single polarization. The EIRP is expected to increase as $20 \times \log(N)$ gaining 36 dB for 64 elements, close to the measured 35 dB as shown.

As discussed in Section 9.3.5, gain invariant phase control and phase invariant gain control are extremely beneficial for efficient beam-steering and tapering control in phased arrays. These functions have been implemented in the current IC, as discussed in Section 9.4.1. Figure 9.14 shows the measured gain invariant phase control for a representative element showing 360° of phase control with $<\pm 0.7$ dB gain variation.

Similarly, phase invariance of the VGA is measured indirectly using two-element over-the-air notch-forming. In this experiment, the phase of front-end 2 is swept while keeping front-end 1 at a constant phase, measuring the resulting output power at multiple front-end 2 VGA settings. With a constant loss phase shifter, a minimum is obtained at a relative phase of 180° (phase setting = 25 in this example), and the depth of the minimum varies depending on how close the relative amplitudes of front-end 1 and

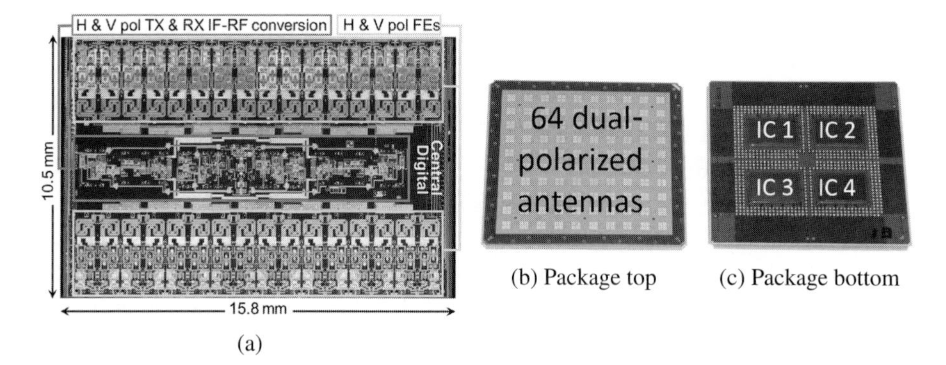

Figure 9.12 (a) Annotated die photo of the implemented IC in GF 130nm SiGe BiCMOS technology, and (b) top view of the PAAM comprising 4 ICs and 64 dual-polarized antennas and (c) its bottom view [18]. (©2017 IEEE. Reprinted, with permission, from *IEEE Journal of Solid States Circuits.*)

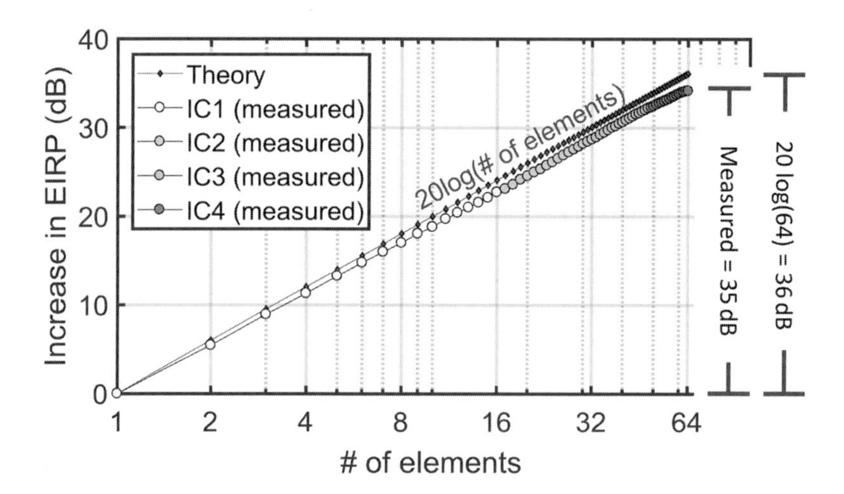

Figure 9.13 Measured increase in EIRP as a function of the number of elements turned on [18]. (©2017 IEEE. Reprinted, with permission, from *IEEE Journal of Solid States Circuits.*)

front-end 2 are; the depth is greatest when the front-end 1 and front-end 2 outputs are most equal. In this measurement, the location of the minimum is seen to remain unchanged, demonstrating the phase invariance of the VGA within half a phase shifter least significant bit (LSB) step of $\sim 2.5^\circ$ (Figure 9.15).

Beamforming and beam-steering measurements are shown in Figures 9.16–9.20. To demonstrate the accuracy of the phase and gain control, as well as the matching among elements, beamforming tests reported in this chapter were made without gain or phase calibration, i.e., all FEs are assumed to have identical gain and phase settings for broadside beams; for beam steering, mathematically computed phase shifts for each front-end were linearly translated to front-end phase settings using the average phase shifter step of 4.9°. All beam measurements have an angular measurement resolution of 1°.

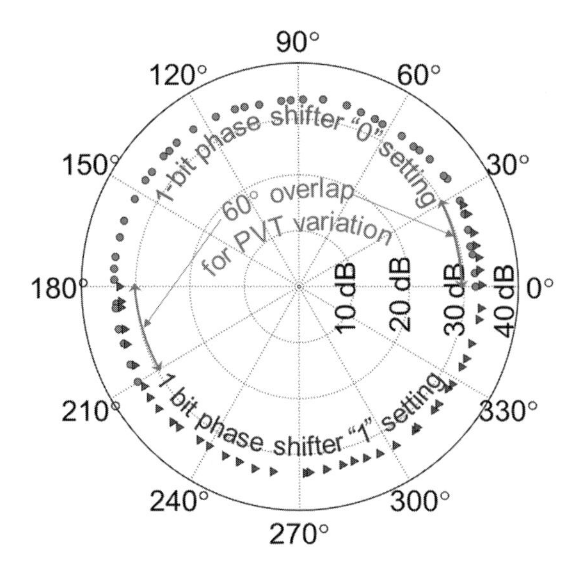

Figure 9.14 Over-the-air measurement of gain invariant phase control demonstration for the full IC showing <±0.7 dB gain variation [18]. (©2017 IEEE. Reprinted, with permission, from *IEEE Journal of Solid States Circuits*.)

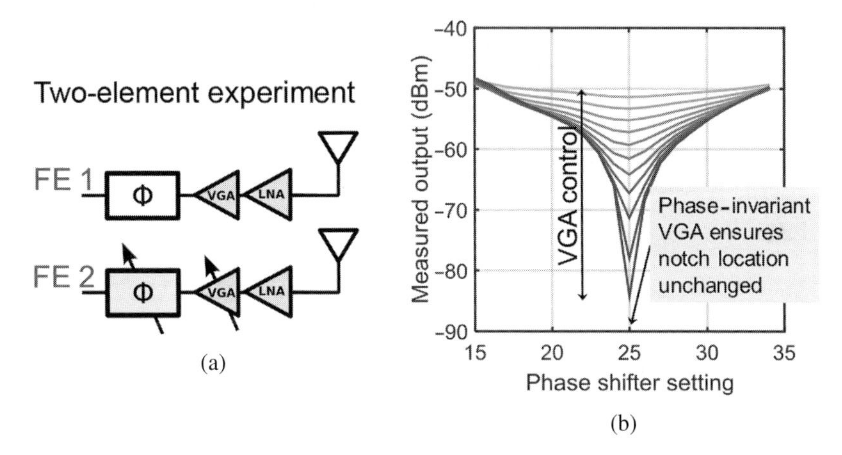

Figure 9.15 Over-the-air measurement of phase-invariant gain control demonstration using two-element notch-forming with different VGA settings on the second element (a); varying notch depths at the same phase shifter setting proves phase invariance (b).

Figure 9.16 shows uncalibrated beam steering in both TX and RX modes with a single IC enabled. Simultaneous independent beams in two polarizations are formed in both TX and RX modes, achieving maximum polarization diversity. Ideal mathematical beam patterns with the same simulation resolution as the measurement (1°) shows similar beam patterns and notch depths, demonstrating the near-ideal performance of the uncalibrated PAAM. Moreover, the similarity in notch depths between theory and measurement demonstrates that the notch depths obtained in measurement are limited by the measurement resolution, not by the hardware.

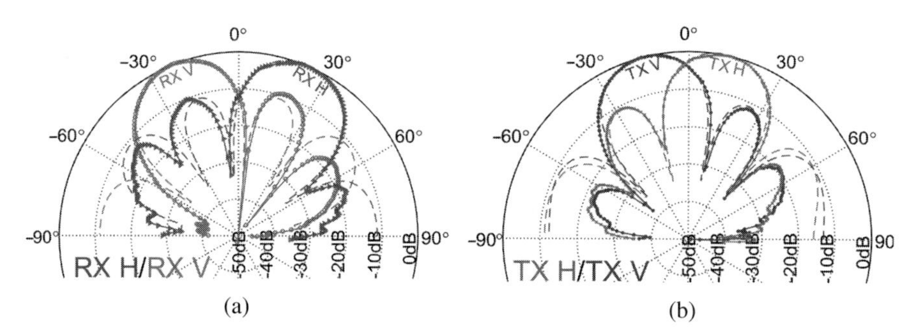

Figure 9.16 Solid lines with markers showing different measured operating modes of the IC-package module showing uncalibrated 16-element RX horizontal and vertical beams pointed at ±20° respectively (a) and TX H/V beams pointed at ±10° respectively (b) using 1 IC. Superimposed dashed lines show the ideal mathematical patterns for comparison with the same angular resolution as the measurement setup. While the beams are shown to be symmetric in this example, they can be pointed independent of each other [18]. (©2017 IEEE. Reprinted, with permission, from *IEEE Journal of Solid States Circuits.*)

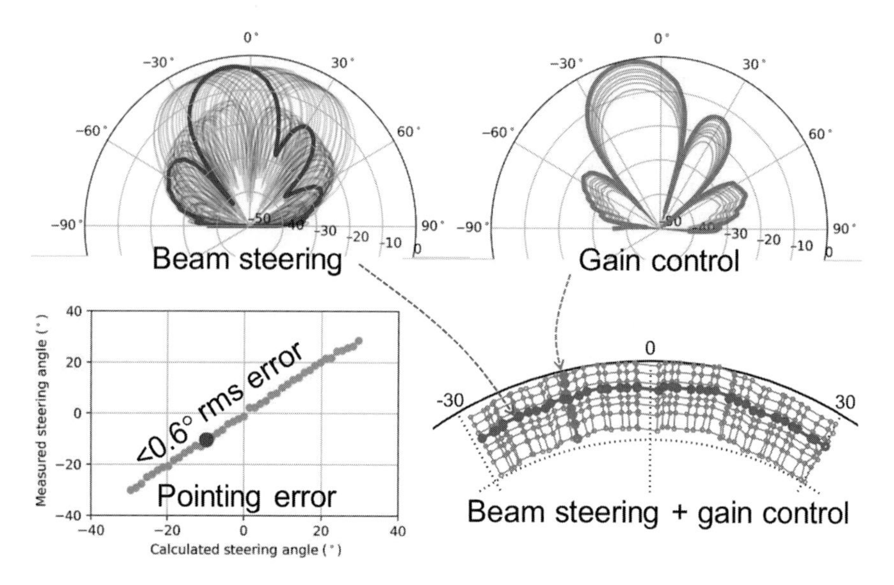

Figure 9.17 Measured one-beam steering example at a fixed VGA setting (top-left), and one gain control example at a fixed phase setting (top-right). Uncalibrated steering angle vs. calculation error (bottom-left). Beam pointing directions for uncalibrated 16-element beam steering precision between ±30° with 8 dB VGA control (bottom-right) [18]. (©2017 IEEE. Reprinted, with permission, from *IEEE Journal of Solid States Circuits.*)

Figure 9.17 shows uncalibrated beam steering and gain control using uniform phase settings in the phased array, resulting in a beam-steering resolution of 1.4°. Each data point along a given arc on the polar plot slice represents a beam-pointing direction. Phase-invariant VGA control is used to change the beam gain over a 9 dB range, represented by the data points on the radial axis. The 43 different 16-element beams along one arc are shown in the inset. Gain control on one of these beams is also shown. Without

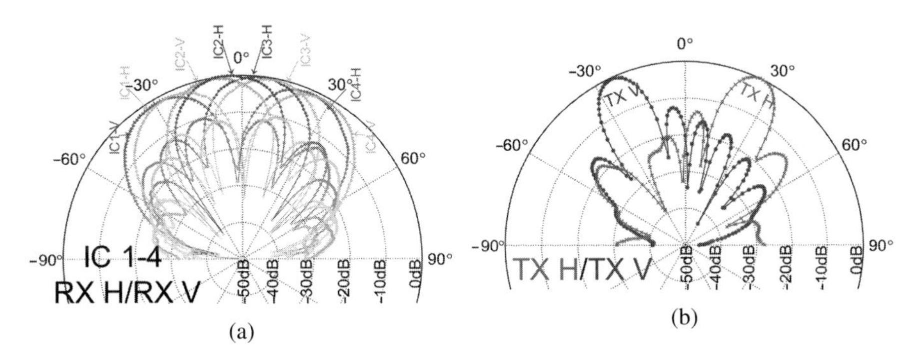

Figure 9.18 Measured (a) eight simultaneous 16-element RX beams and (b) two simultaneous 64-element TX beams using a four-IC module, without requiring calibration [18]. (©2017 IEEE. Reprinted, with permission, from *IEEE Journal of Solid States Circuits.*)

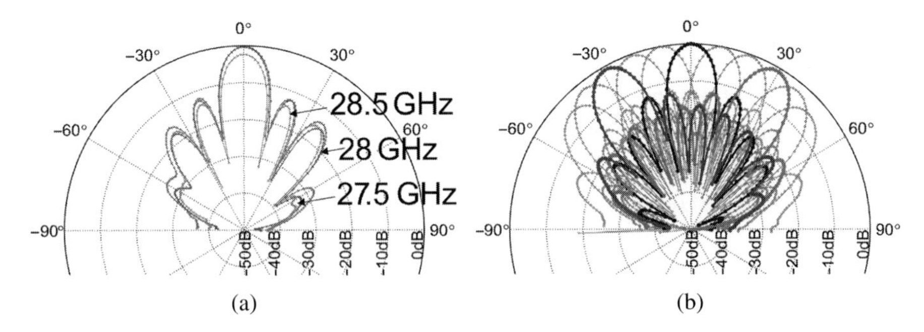

Figure 9.19 Measured 64-element TX H beam pattern measurements across frequency at 27.5, 28, and 28.5 GHz (a) and 64-element TX V beam steering measurements from -50° to 50° in steps of 10° achieving <10 dB sidelobe suppression without applying any tapering (b) [18]. (©2017 IEEE. Reprinted, with permission, from *IEEE Journal of Solid States Circuits.*)

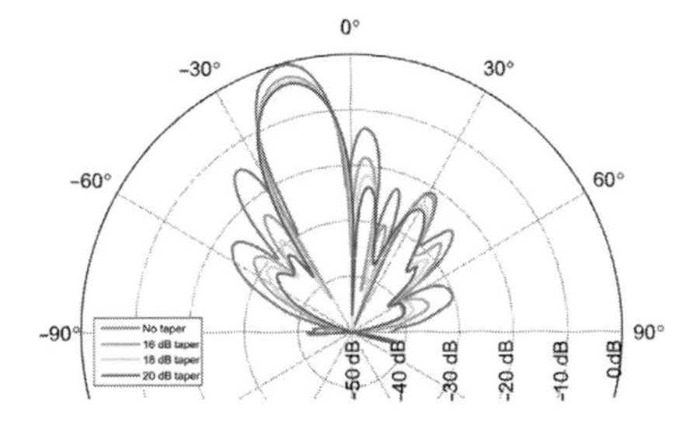

Figure 9.20 Measured tapering of the 64-element phased array without requiring calibration [18]. (©2017 IEEE. Reprinted, with permission, from *IEEE Journal of Solid States Circuits.*)

any gain or phase calibration, the error across all directions and gain settings for this measurement is only 0.6° rms.

Since each IC supports two simultaneous beams, each package housing four ICs can support eight simultaneous 16-element beams, four simultaneous 32-element beams, or two simultaneous 64 element beams. Moreover, these modes are supported in both TX and RX modes. To demonstrate some of these modes in measurement, the ICs have been configured to form eight simultaneous 16-element beams in Figure 9.18a in the RX mode, and 2 simultaneous 64-element beams in Fig. 9.18b in the TX mode. These examples comprise only two (among many) different modes of operating the $4 \times$ IC configuration by using the 128 elements to create different types of beams. Such reconfigurability can be useful during phased array calibration as well as beam finding in a possible 5G usage scenario.

Beam measurements were taken across frequency. As shown in Figure 9.19a, the beam shape is maintained across frequencies. Furthermore, in Figure 9.19b, beam steering across $\pm 50^\circ$ is shown with sidelobe suppression below 10 dB before applying tapering.

Beam tapering was performed in measurement using a Taylor window function to scale the amplitudes of the individual elements so as to achieve lower sidelobe levels. The phase invariant VGA was used for gain control; the phase invariance property ensures that the beam pointing direction is not affected by the gain control operation. Results from the tapering experiment showing varying levels of tapering are shown in Figure 9.20. Using tapering, sidelobe suppression up to 20 dB is demonstrated without significantly impacting the beam pointing direction. Again, no gain or phase calibration was applied during the tapering measurement.

Fast beam switching is expected to be critical in possible usage scenarios to reduce the data rate impact of time spent switching the beam. The PAAM was originally designed to operate in an Ericsson testbed system that is based on Orthogonal Frequency Division Multiplexing (OFDM) modulation with 75 kHz subcarrier spacing. As such, the signal modulation, scheduling, and user handling have much in common with the legacy Long Term Evolution (LTE) structure that is defined through the 3GPP body. For the legacy LTE system, which has a subcarrier spacing of 15 kHz, there is an added cyclic prefix that is of roughly 5 μs size. For the Ericsson testbed system, a five times shorter cyclic prefix was used compared to the LTE system, in the order of 1 μs due to the larger subcarrier spacing compared to 15 kHz. It is assumed that any beam switching has to be performed within that length of cyclic prefix and preferably be of an order or two faster than that. For the testbed system, the beam switching time is set to be at 1/100 of the cyclic prefix, which then amounts to 10 ns.

The measured beam switching speed is shown in Figure 9.21, and shows extremely fast <4 ns beam switching speeds.

Similarly, fast switching between transmit and receive modes is expected to be important in 5G usage scenarios. The switching time between Tx and Rx is triggered by the time domain duplex (TDD) frame structure and defined through the 3GPP body to have a maximum of 17 μs transient period length (TPL) for an LTE system having 15 kHz subcarrier spacing. In the Ericsson testbed system, the subcarrier spacing was set to 75 kHz, which led to a five times shorter TPL <3 μs .

The "transition time" is defined to safeguard TX-to-RX emissions, and the noise power from the PA has to go down to the ambient noise floor within a certain time period. 3GPP has introduced a Guard Period (GP) between downlink and uplink

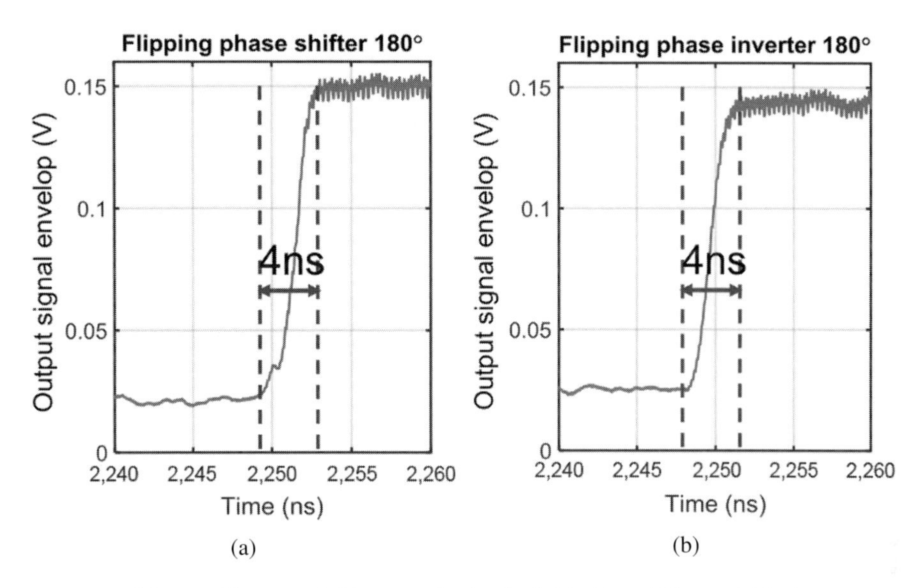

Figure 9.21 Measured beam-switching speed using both (a) phase shifter and (b) phase inverter [18]. (©2017 IEEE. Reprinted, with permission, from *IEEE Journal of Solid States Circuits.*)

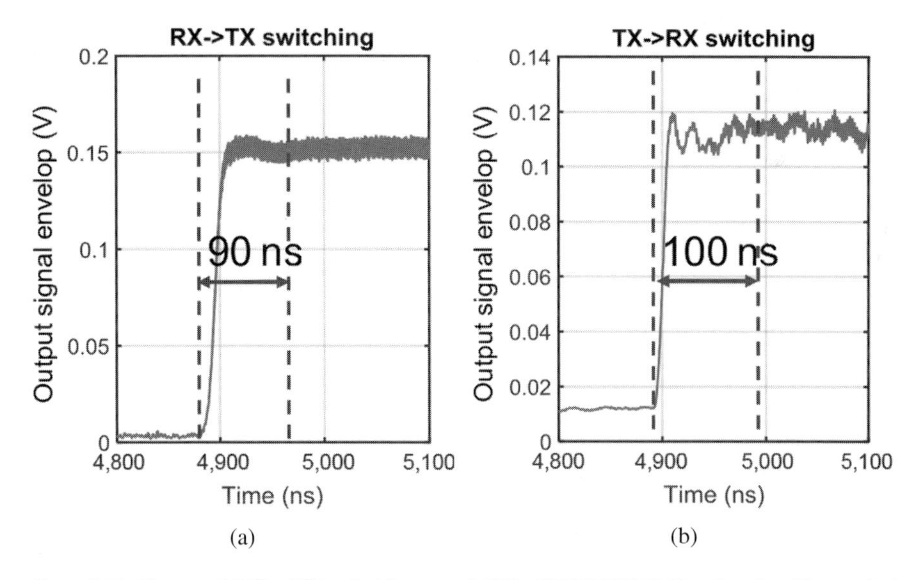

Figure 9.22 Measured RX↔TX switching speed [18]. (©2017 IEEE. Reprinted, with permission, from *IEEE Journal of Solid States Circuits.*)

subframes that is in the order of one symbol in length in order to ensure that the noise level goes down to levels that do not disturb reception in the uplink subframe. With 75 kHz subcarrier spacing, the symbol time is 13.3 μs, and one would like to use a fraction of that to reduce the noise to the ambient level. As a result, one wants to have <3 μs for the TX-to-RX switch time for a 75 kHz subcarrier testbed system.

The measured TX↔RX switching speed is shown in Figure 9.22, and shows extremely fast <100 ns TX↔RX switching speeds.

The reader is referred to [31] for additional beam and package measurements.

9.5 Conclusion

Silicon-based integrated mm-wave circuits have come a long way, from disproving skeptics in the early years of the last decade, to being touted as a key technology that will revolutionize 5G cellular communications in this decade. This chapter lays the foundations of phased array technology, explaining beamforming, beam steering, and beam shaping using a spatial Fourier transform perspective. The chapter then outlines some of the desired features of mm-wave phased arrays in the context of directional communications for 5G and other related technologies. An exemplary silicon-based phased array is then described. Measurement results are presented to demonstrate a level of performance that is able to achieve the outlined features. Based on the phased array example described in this chapter [18,24,30] and other mm-wave phased arrays developed in recent years [32–34], we believe that silicon-based integrated phased arrays are ready for widespread commercial deployment and for use in everyday consumer electronics devices.

Acknowledgment

The exemplary 28 GHz phased array antenna module described in this chapter is the outcome of a collaboration between Ericsson and IBM T. J. Watson Research Center. The authors acknowledge several individuals at IBM and Ericsson for their contributions to the phased array.

References

[1] H. Hashemi, X. Guan, and A. Hajimiri, "A fully integrated 24 GHz 8-path phased-array receiver in silicon," in *2004 IEEE International Solid-State Circuits Conference (IEEE Cat. No.04CH37519)*, IEEE, February 2004, pp. 390–534, vol. 1.

[2] S. K. Reynolds, B. A. Floyd, U. R. Pfeiffer, et al., "A silicon 60-GHz receiver and transmitter chipset for broadband communications," *IEEE Journal of Solid-State Circuits*, vol. 41, no. 12, pp. 2820–2831, December 2006.

[3] "IEEE draft standard for information technology–telecommunications and information exchange between systems local and metropolitan area networks–specific requirements part 11: Wireless LAN Medium Access Control (MAC) and Physical Layer (PHY) specifications amendment 3: Pre-association discovery," *IEEE P802.11aq/D14.0, December 2017*, International Organization Standard, pp. 1–75, January 2017.

[4] "IEEE standard for information technology – Local and metropolitan area networks – Specific requirements – Part 15.3: Amendment 2: Millimeter-wave-based alternative physical layer extension," *IEEE Std 802.15.3c-2009 (Amendment to IEEE Std 802.15.3-2003)*, pp. 1–200, October 2009.

[5] A. Natarajan, S. K. Reynolds, M. D. Tsai, et al., "A fully-integrated 16-element phased-array receiver in SiGe BiCMOS for 60-GHz communications," *IEEE Journal of Solid-State Circuits*, vol. 46, no. 5, pp. 1059–1075, May 2011.
[6] A. Valdes-Garcia, S. T. Nicolson, J. W. Lai, et al., "A fully integrated 16-element phased-array transmitter in SiGe BiCMOS for 60-GHz communications," *IEEE Journal of Solid-State Circuits*, vol. 45, no. 12, pp. 2757–2773, December 2010.
[7] M. Boers, B. Afshar, I. Vassiliou, et al., "A 16TX/16RX 60 GHz 802.11ad chipset with single coaxial interface and polarization diversity," *IEEE Journal of Solid-State Circuits*, vol. 49, no. 12, pp. 3031–3045, December 2014.
[8] E. Cohen, M. Ruberto, M. Cohen, et al., "A CMOS bidirectional 32-element phased-array transceiver at 60 GHz with LTCC antenna," *IEEE Transactions on Microwave Theory and Techniques*, vol. 61, no. 3, pp. 1359–1375, March 2013.
[9] S. Emami, R. F. Wiser, E. Ali, et al., "A 60GHz CMOS phased-array transceiver pair for multi-Gb/s wireless communications," in *2011 IEEE International Solid-State Circuits Conference*, IEEE, February 2011, pp. 164–166.
[10] K. Okada, K. Kondou, M. Miyahara, et al., "Full four-channel 6.3-Gb/s 60-GHz CMOS transceiver with low-power analog and digital baseband circuitry," *IEEE Journal of Solid-State Circuits*, vol. 48, no. 1, pp. 46–65, January 2013.
[11] A. Tomkins, A. Poon, E. Juntunen, et al., "A 60 GHz, 802.11ad/WiGig-compliant transceiver for infrastructure and mobile applications in 130 nm SiGe BiCMOS," *IEEE Journal of Solid-State Circuits*, vol. 50, no. 10, pp. 2239–2255, October 2015.
[12] N. Saito, T. Tsukizawa, N. Shirakata, et al., "A fully integrated 60-GHz CMOS transceiver chipset based on WiGig/ieee 802.11ad with built-in self calibration for mobile usage," *IEEE Journal of Solid-State Circuits*, vol. 48, no. 12, pp. 3146–3159, December 2013.
[13] S. Zihir, O. D. Gurbuz, A. Karroy, S. Raman, and G. M. Rebeiz, "A 60 GHz single-chip 256-element wafer-scale phased array with EIRP of 45 dBm using sub-reticle stitching," in *2015 IEEE Radio Frequency Integrated Circuits Symposium (RFIC)*, IEEE, May 2015, pp. 23–26.
[14] B. Sadhu, A. Valdes-Garcia, J. O. Plouchart, et al., "A 60GHz packaged switched beam 32nm CMOS TRX with broad spatial coverage, 17.1dBm peak EIRP, 6.1dB NF at < 250mW," in *2016 IEEE Radio Frequency Integrated Circuits Symposium (RFIC)*, IEEE, May 2016, pp. 342–343.
[15] 3GPP, "First 5G NR specs, 2017 rel 15, 38-series." [Online] Available: www.3gpp.org/DynaReport/38-series.htm, 2017. online.
[16] 3GPP TS 38.104 V15.0.0 (2017-12), "Base station (BS) radio transmission and reception." [Online] Available: www.3gpp.org/DynaReport/38104.htm.
[17] S. Chang, W. Hong, and J. Oh, "Quantization effects of phase shifters on 5G mmwave antenna arrays," in *2015 IEEE International Symposium on Antennas and Propagation USNC/URSI National Radio Science Meeting*, IEEE, July 2015, pp. 2119–2120.
[18] B. Sadhu, Y. Tousi, J. Hallin, et al., "A 28-GHz 32-element TRX phased-array IC with concurrent dual-polarized operation and orthogonal phase and gain control for 5G communications," *IEEE Journal of Solid-State Circuits*, vol. 52, no. 12, pp. 3373–3391, December 2017.
[19] X. Gu, A. Valdes-Garcia, A. Natarajan, B. Sadhu, D. Liu, and S. K. Reynolds, "W-band scalable phased arrays for imaging and communications," *IEEE Communications Magazine*, vol. 53, no. 4, pp. 196–204, April 2015.
[20] A. Valdes-Garcia, A. Natarajan, D. Liu, et al., "A fully-integrated dual-polarization 16-element W-band phased-array transceiver in SiGe BiCMOS," in *2013 IEEE Radio Frequency Integrated Circuits Symposium (RFIC)*, IEEE, June 2013, pp. 375–378.

[21] C. Guo, F. Liu, S. Chen, C. Feng, and Z. Zeng, "Advances on exploiting polarization in wireless communications: Channels, technologies, and applications," *IEEE Communications Surveys Tutorials*, vol. 19, no. 1, pp. 125–166, first quarter 2017.

[22] G. Valenzuela, "Depolarization of em waves by slightly rough surfaces," *IEEE Transactions on Antennas and Propagation*, vol. 15, no. 4, pp. 552–557, July 1967.

[23] R. A. Pitaval, O. Tirkkonen, R. Wichman, et al., "Full-duplex self-backhauling for small-cell 5g networks," *IEEE Wireless Communications*, vol. 22, no. 5, pp. 83–89, October 2015.

[24] B. Sadhu, Y. Tousi, J. Hallin, et al., "7.2 a 28GHz 32-element phased-array transceiver IC with concurrent dual polarized beams and 1.4 degree beam-steering resolution for 5G communication," in *2017 IEEE International Solid-State Circuits Conference (ISSCC)*, February 2017, pp. 128–129.

[25] B. Razavi, *RF Microelectronics*. Prentice-Hall, Inc., 1998.

[26] Y. Tousi and A. Valdes-Garcia, "A Ka-band digitally-controlled phase shifter with sub-degree phase precision," in *2016 IEEE Radio Frequency Integrated Circuits Symposium (RFIC)*, May 2016, pp. 356–359.

[27] A. Natarajan, A. Valdes-Garcia, B. Sadhu, S. K. Reynolds, and B. D. Parker, "*W*-band dual-polarization phased-array transceiver front-end in SiGe BiCMOS," *IEEE Transactions on Microwave Theory and Techniques*, vol. 63, no. 6, pp. 1989–2002, June 2015.

[28] M. D. Tsai and A. Natarajan, "60GHz passive and active RF-path phase shifters in silicon," in *2009 IEEE Radio Frequency Integrated Circuits Symposium*, June 2009, pp. 223–226.

[29] B. Sadhu, J. F. Bulzacchelli, and A. Valdes-Garcia, "A 28GHz SiGe BiCMOS phase invariant VGA," in *2016 IEEE Radio Frequency Integrated Circuits Symposium (RFIC)*, May 2016, pp. 150–153.

[30] X. Gu, D. Liu, C. Baks, et al., "A multilayer organic package with 64 dual-polarized antennas for 28GHz 5G communication," in *2017 IEEE MTT-S International Microwave Symposium (IMS)*, June 2017, pp. 1899–1901.

[31] A. Valdes-Garcia, B. Sadhu, X. Gu, et al., "Circuit and antenna-in-package innovations for scaled mmWave 5G phased array modules," in *IEEE Custom Integrated Circuits Conference*, IEEE, pp. 1–8 2018.

[32] H. T. Kim, B. S. Park, S. M. Oh, et al., "A 28GHz CMOS direct conversion transceiver with packaged antenna arrays for 5G cellular system," in *2017 IEEE Radio Frequency Integrated Circuits Symposium (RFIC)*, June 2017, pp. 69–72.

[33] K. Kibaroglu, M. Sayginer, and G. M. Rebeiz, "A low-cost scalable 32-element 28-GHz phased array transceiver for 5G communication links based on a 2×2 beamformer flip-chip unit cell," *IEEE Journal of Solid-State Circuits*, vol. PP, no. 99, pp. 1–15, 2018.

[34] J. D. Dunworth, A. Homayoun, B.-H. Ku, et al., "A 28GHz bulk-CMOS dual-polarization phased-array transceiver with 24 channels for 5G user and basestation equipment," in *2018 International Solid State Circuits Conference (ISSCC)*, February 2018.

10 Millimeter-Wave Frequency Synthesis Based on Frequency Multiplications

Payam Heydari

10.1 Introduction and Motivation

A critical component in W-band imaging system, automotive radars, point-to-point data communication, and 100 Gbps Ethernet is the frequency synthesizer. A number of W-band frequency synthesis techniques have been recently realized in SiGe BiCMOS [1,2] and CMOS [3,4] technologies. As the operation frequency increases, the implementation of low phase-noise oscillators with adequate tuning range and output power will become increasingly difficult. One solution is to employ the push–push voltage-controlled oscillator (VCO), where the oscillator now operates at half of the output frequency. However, the output second harmonic signal has limited power and it is single-ended. A quadrature VCO is required to generate differential output, but the phase noise may be compromised due to the quadrature structure. Another solution is to use a frequency multiplier, preceded by a lower-frequency phase-locked loop (PLL) [5,6]. Low-order frequency multipliers (e.g., frequency doubler or tripler) are more amenable to on-chip integration than higher multiplication ratios, since they rely on the nonlinearity of the transistors, and the harmonic energy decreases with frequency.

Furthermore, as the operating frequency (f_0) increases, the implementation of low phase-noise fundamental PLL will become increasingly difficult. In an integrated PLL-based frequency synthesizer, assuming VCO phase noise dominates and other blocks in the PLL contribute negligible noise, the root mean square (rms) phase noise is roughly given by [7]

$$\sigma_{\Phi}^{PLL} = \Delta f \sqrt{\frac{\pi S_{VCO}}{f_L}} \tag{10.1}$$

where S_{VCO} is the single-sideband phase noise of the free-running VCO at the offset Δf from the carrier, and f_L is the loop bandwidth. The term S_{VCO} is given by Leeson's equation and is proportional to $\left[f_0/(2V_0 Q \Delta f)\right]^2$, where V_0 and Q are the oscillation amplitude and the tank quality factor, respectively.

Several factors contribute to difficulty of achieving low phase-noise PLL at mm-wave frequencies. From Leeson, the VCO's phase noise is expected to degrade in proportion with f^2. Moreover, the tank Q begins to be dominated by capacitances, in particular, varactors as their Q degrades with increasing frequency (simulated Q of <3 at 96 GHz). Large K_{VCO} values on the order of GHz/V are commonly attained in W-band VCOs [4]. Design techniques such as segmentation with switched capacitor or varactor, which is commonly used in VCO design to reduce K_{VCO}, are not effective for VCOs whose resonant frequency is closer to f_{max}. This is because device and interconnect parasitic starts to dominate the overall capacitance, leaving little room for any form of segmentation capacitors.

Additionally, for a VCO with a large K_{VCO} inside a PLL loop, this results in a large up-conversion of noise coming from charge pump and loop filter to the PLL's center frequency. Finally, a fundamental-frequency VCO must be accompanied by a fundamental divider, which consumes considerable power if a static architecture is used; otherwise, it may limit the PLL locking range if an injection-locked divider (e.g., [4]) is used.

In comparison, a synthesizer comprising a subharmonic PLL followed by a frequency multiplier ($\times M$) [5], allows a larger tuning range and a lower phase. Though the in-band phase noise in this synthesizer is magnified by $20\log_{10}(M)$ due to frequency multiplication, this degradation would be exactly offset by the f_0^2 term in Leeson's equation, leaving an improvement in Q, output swing (V_0), and tuning range as added bonuses to overall phase noise improvement. Furthermore, practical mm-wave systems will require relatively large phased arrays [8] or multipixel focal plane arrays [9,10]. This solution readily lowers the PLL's operating frequency, thereby making local oscillator (LO) routing and distribution easier in these systems.

The system architecture of the PLL is depicted in Figure 10.1. Low-order multipliers, i.e., doubler or tripler, are more amenable to on-chip active implementation since they rely upon the nonlinearity of the transistor, where the harmonic energy decreases with

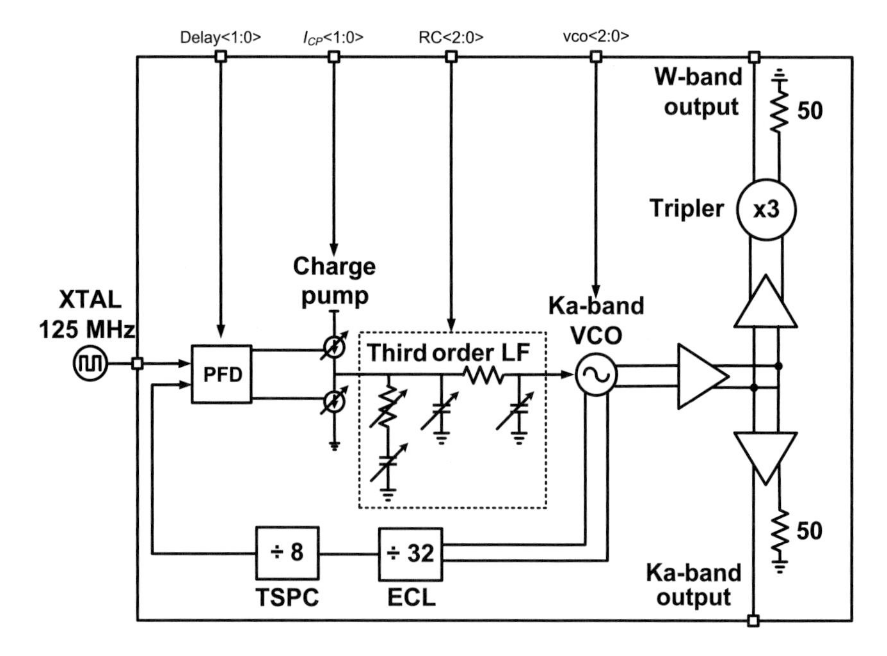

Figure 10.1 Block diagram of the W-band frequency synthesis [11]. (©2012 IEEE. Reprinted, with permission, from *IEEE Transactions on Microwave Theory and Techniques*.)

frequency. In Figure 10.1, the W-band signal is synthesized by cascading a frequency tripler after a Ka-band (30.3 through 33.8 GHz) PLL. Chip A incorporates the injection-locked frequency tripler (ILFT), whereas Chip B uses the harmonic-based frequency tripler (HBFT) after the PLL.

The Ka-band PLL is comprised of a differential Colpitts VCO, a frequency divider chain with the division ratio of 256, a phase/frequency detector (PFD), a charge pump (CP), and a third-order loop filter (LF). The divider chain utilizes emitter-coupled logic (ECL) circuits for asynchronous divide-by-32 and true single-phase-clock (TSPC) circuits for divide-by-8. Programmable PFD delay, CP current, and loop bandwidth (BW) compensate for model inaccuracy and process, voltage, and temperature (PVT) variation. To facilitate in situ characterization of the PLL, an additional ground–signal–ground (GSG) pad was used to monitor the PLL's output.

Figure 10.2 illustrates three possible LO generation and distribution schemes that can be implemented using the PLL system depicted in Figure 10.1. Shown in Figure 10.2a is a dual-conversion zero–intermediate-frequency (IF) superheterodyne 120 GHz phased-array transceiver. The 30 GHz LO signal is routed to both the RX and TX, where it drives a tripler and an I/Q generator to perform frequency conversion. Figure 10.2b shows M-pixels 94 GHz direct-conversion passive imaging [9,10]. In this case, since phase carries no information, there is no need for I/Q, and the LO is distributed to M-pixels local triplers for zero-IF down-coversion. A direct-conversion 94 GHz transceiver (TRX) for active imaging/communication is depicted in Figure 10.2c. Here, the LO is distributed to both RX and TX for both down- and up-conversion.

From the preceding discussion, the W-band frequency synthesis of this work is amenable to multichannel systems and allows stringent phase noise requirements to be easier to meet. Hence, this PLL system can serve as the LO generation and distribution of several W-band applications and even 120 GHz high-data communication for next-generation 6G applications. Next, the circuit design details of key building blocks of the PLL are described.

Two methods for implementation of a frequency multiplier are conceived. The first method is to leverage the idea of injection-locking oscillators (ILOs) and inject a signal into the ILO using a harmonic pregenerator [11,12]. The second method is based on harmonic generation and amplification, which generates the harmonics of the input signal, amplifies the desired harmonic, and filters out all the other harmonics [5–11]. Based on the preceding discussions, we discuss three different approaches for W-band frequency synthesis: (I) using a W-band fundamental PLL, as shown in Figure 10.3a; (II) using a Ka-band PLL cascaded with an ILFT; and (III) using a Ka-band PLL cascaded with an HBFT, as shown in Figure 10.3b. Here, a "fundamental PLL" means the PLL's output frequency being equal to the fundamental frequency of the VCO. Moreover, the difference between method II and method III lies in the tripler design. From another perspective, all of the three methods can be treated as scaling up a Ka-band PLL by a factor of three. In method I, this is done by directly tripling the design frequency of the VCO, whereas for the other two methods, it is achieved by cascading a frequency tripler. For ideal frequency scaling, as the frequency triples, the absolute tuning range also multiplies by three and phase noise degrades by 9.54 dB ($20 \log_{10} 3$). It is, however,

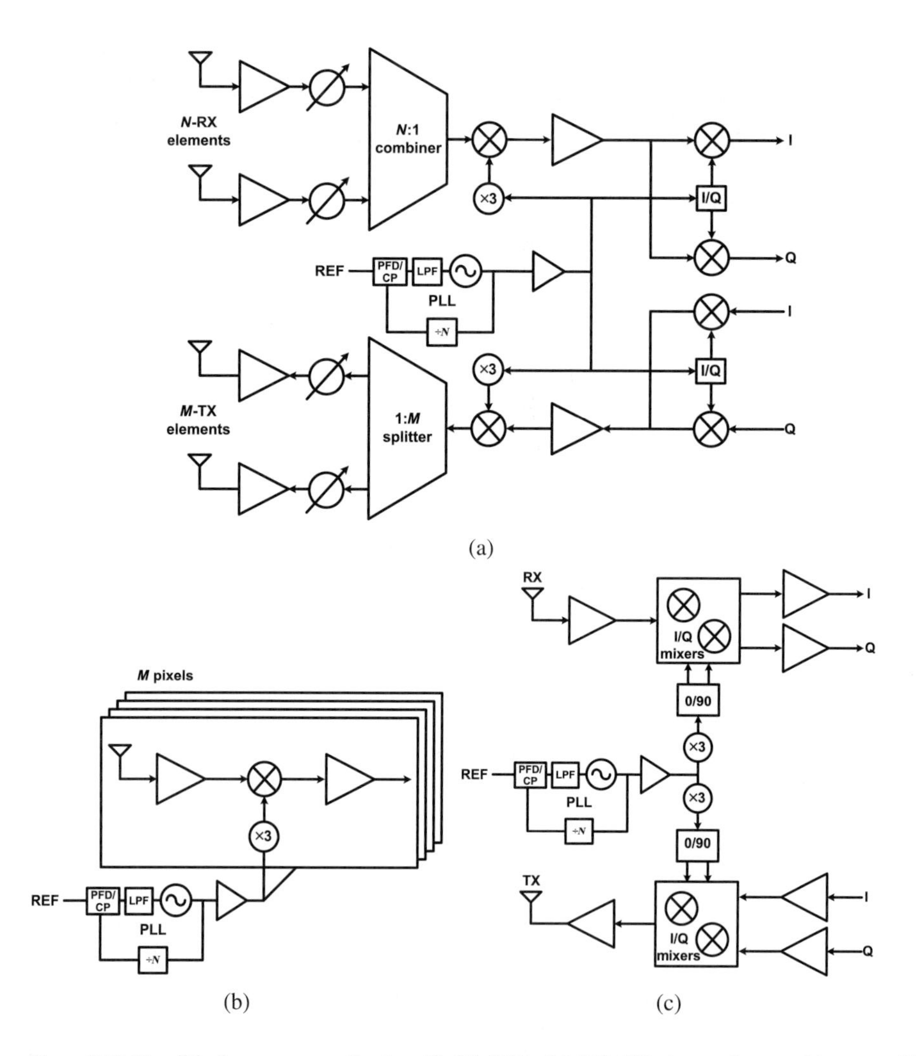

Figure 10.2 Possible frequency synthesis with this PLL: (a) 120 GHz heterodyne architecture, (b) 94 GHz direct-conversion passive imaging, and (c) 94 GHz direct-conversion active imaging/communication [11]. (©2012 IEEE. Reprinted, with permission, from *IEEE Transactions on Microwave Theory and Techniques*.)

extremely challenging to achieve the ideal scaling for method I due to the impact of parasitic capacitances and large device noise at W-band.

In order to make a fair comparison among these three methods, we apply a constraint that the circuits are designed under the same power consumption. The following discussion explains why it is a valid precondition. First, we compare the power consumption of a W-band PLL to a Ka-band PLL followed by an ILFT. By comparing Figure 10.3a and 10.3b, we notice that two building blocks are different: the W-band PLL employs a W-band VCO and an injection-locked divider-by-3 (ILFD), whereas the other system employs a Ka-band VCO and an ILFT. All the other blocks are exactly the same. Note that an ILFD has a free-running frequency at $\omega_0/3$ (within Ka-band), and it is

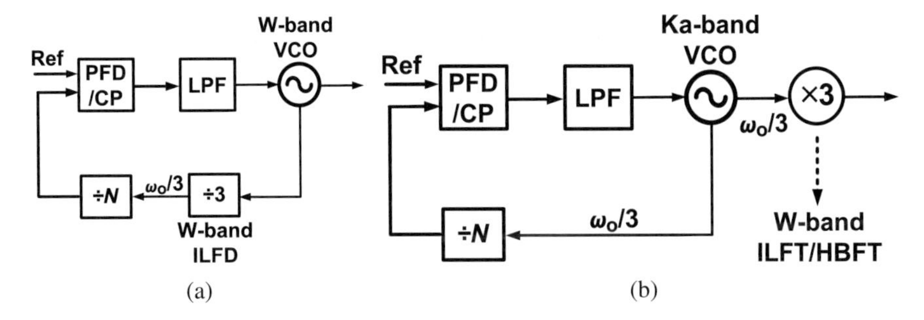

Figure 10.3 System architecture of (a) a W-band fundamental PLL and (b) a Ka-band PLL cascaded by a frequency tripler.

nothing but a Ka-band VCO when there is no signal injected. Therefore, it consumes the same amount of power as the Ka-band VCO in Figure 10.3b. For the same reason, the ILFT in Figure 10.3b has the same power consumption as the W-band VCO in Figure 10.3a. So far, it has been proved that method I and II have the same DC power consumption. As for method II and III, the two kinds of tripler are preceded by the same Ka-band PLL, and we can intentionally design these two triplers under the same power consumption.

The premise that the preceding three approaches have equal power consumption allows us to make a fair comparison among them in other aspects such as phase noise, tuning range, and output power. It is worth mentioning that in the preceding discussions, we focus on the power consumption of frequency synthesizers, which is only part of the entire LO path in an actual wireless system. The complete LO path consists of two parts: LO generation and LO distribution. Although the three LO generation techniques consumes the same DC power, the power consumed by LO distribution circuits may vary a lot when using different LO generation methods. For instance, in a TRX or a multichannel TX/RX system, where the LO signal needs to be split and distributed to several mixers, for method I, the whole LO distribution network operates at W-band frequency, and the power consumed by the LO buffers/amplifiers will be very large [1], because the T-lines exhibit larger loss and the active devices have lower available gain. The other two solutions readily lower the PLL's operation frequency, thereby making LO routing and distribution in a transceiver or a multipixel imager considerably easier [9,10].

Furthermore, the use of frequency tripler offers better phase noise than directly designing a W-band PLL. In order to compare the three methods in terms of phase noise, we first start with a Ka-band PLL. The first method is to scale up the PLL frequency by a factor of three, while the second and third methods involve cascading a frequency tripler after the Ka-band PLL. For the first method, the phase noise of a general oscillator can be expressed by Leeson's model [13,14], i.e.,

$$S_{\Delta\varphi,out} = S_{\Delta\varphi,in}\left[1+\left(\frac{\omega_0}{2Q\Delta\omega}\right)^2\right] = 2\frac{\langle|V_n|^2\rangle}{|V_0|^2}\left(\frac{\omega_0}{2Q\Delta\omega}\right)^2 \qquad (\Delta\omega \ll \omega_0), \tag{10.2}$$

where $S_{\Delta\varphi,in}$ and $S_{\Delta\varphi,out}$ denote the input and output phase noise power spectral density, respectively. ω_0 is the center frequency of the oscillator, Q is the quality factor of the tank, $\Delta\omega$ is offset frequency, V_0 is the oscillation amplitude, and $|V_n|$ represents the equivalent input referred noise voltage. Equation (10.2) states that for ideal frequency tripling of an oscillator, the phase noise will degrade by 9.54 dB ($20\log_{10} 3$). However, this ideal number is based on the assumption that other parameters such as Q factor, V_0, and $|V_n|$ do not change with frequency, which is not the case, in practice. As the operation frequency increases, especially when it goes into the high side of the mm-wave frequency range (e.g., W-band), both V_o and Q (usually limited by varactor's Q) tends to drop, while the intrinsic device noise parameter $|V_n|$ increases dramatically [15]. As a result, the phase noise degradation due to scaling up a VCO operation frequency from Ka-band to W-band would be much more than the ideal 9.54 dB. For the same reasons, it is extremely difficult to achieve good phase noise for a fundamental PLL running above 90 GHz (half of f_{max} for the BiCMOS technology in use), as corroborated in [1] and [4].

As for the second method, the phase noise penalty after an ILFT can be expressed by (10.3) [16]:

$$S_{out,ILFT}(\Delta\omega) = \frac{3^2}{1+\left(\frac{\Delta\omega}{\omega_p}\right)^1} S_{inj}(\Delta\omega) + \frac{1}{1+\left(\frac{\Delta\omega}{\omega_p}\right)^2} S_{harm}(\Delta\omega) + \frac{\left(\frac{\Delta\omega}{\omega_p}\right)^2}{\left(\frac{\Delta\omega}{\omega_p}\right)^2} S_{VCO}(\Delta\omega), \tag{10.3}$$

where $S_{out,ILFT}$, S_{inj}, S_{harm}, and S_{VCO} are the phase noise spectral density of the ILFT's output, injected signal, harmonic generator, and free-running VCO, respectively. Equation (10.3) implies that S_{VCO} has a high-pass nature, while both S_{inj} and S_{harm} exhibit a low-pass nature, and the noise contribution of S_{inj} is nine times larger than S_{harm}. Therefore, the first term dominates the output phase noise for small offset frequency compared to the corner frequency, given as [16], where $|I_o|$ and $|I_3|$ represent oscillation current and the amplitude of third-harmonic injection current, respectively. In this design, the tripler's output phase noise is determined by the injected signal (i.e., Ka-band PLL output) for frequency offset up to around 100 MHz. The preceding analysis indicates that the phase noise degradation after the ILFT can be very close to theoretical minimum value, which is also verified by several reported works [17], as well as the measurement results that will be presented in Section 10.4.2.

For the HBFT, the multiplier and amplifier chain can impart phase noise in addition to the minimum $20\log_{10} 3$ dB degradation. Using a linear phase model [18], we can express the output phase noise of the HBFT as follows:

$$S_{out,HBFT}(\Delta\omega) = 3^2 S_{in}(\Delta\omega) + S_{harm}(\Delta\omega) + S_{amp}(\Delta\omega), \tag{10.4}$$

where $S_{out,HBFT}$, S_{in}, S_{harm}, and S_{amp} are the phase noise spectral density of the HBFT's output, input signal, harmonic generator, and LO amplifier, respectively. Unlike the

ILFT, the HBFT does not resemble a PLL. Therefore, noise from the harmonic generation stage and amplification stage will increase the integrated phase noise. Typically, the amplifier's noise contribution is lower than the oscillator, implying that S_{in} dominates (10.4). Since the phase noise of an amplifier operating in linear region is a function of the input power and noise figure [18], the noise level of LO amplifier stages following the multiplier is kept low by biasing the circuit at minimum NF current density.

The preceding discussions conclude that the frequency synthesis solution incorporating frequency tripler offers better phase noise under the same power consumption comparing to fundamental PLL for frequencies around and higher than the device f_{max}.

Referring to the W-band frequency synthesizer circuit in Figure 10.1, two chips operating at 96 GHz incorporating the same Ka-band PLL and (1) an ILFT and (2) an HBFT were fabricated in 0.18 μm SiGe BiCMOS [11]. To facilitate the in situ characterization of the PLL, an additional GSG pad was used to monitor the Ka-band PLL's output. Note that two of three buffers shown in Figure 10.1 can be eliminated if the test pad for the Ka-band PLL is removed.

10.2 Design of a Silicon-Based Ka-Band PLL

The Ka-band PLL is comprised of a differential Colpitts VCO, a frequency divider chain with the division ratio of 256, a PFD, a CP, and a third-order loop filter (cf. Figure 10.1). The divider chain utilizes ECL circuits for asynchronous divide-by-32 and TSPC circuits for divide-by-8.

Programmable PFD delay (Delay<1:0>), CP current (Ic<1:0>), and loop BW (RC<2:0>) compensate for model inaccuracy and PVT variation. The PLL consumes 65 mW from 1.8 V/2.5 V supply (1.8 V for digital circuits).

Figure 10.4 shows the VCO schematic where a differential Colpitts topology with emitter degeneration is chosen for better phase noise performance [19]. The VCO employs a 3-bit digital band selection in addition to an analog varactor control. Also shown in Figure 10.4 is a simplified half-circuit equivalent model of the VCO. The admittance in parallel to the tank inductor is (assuming $g_m \ll (C_1 + C_{eff})\omega$)

$$Y_{in} = \frac{1}{R_{buf}} + \frac{1}{R_{LP}} - g_m n(1-n) j\omega \left(C_\mu + C_{buf} + \frac{C_1 C_{eff}}{C_1 + C_{eff}} \right) \tag{10.5}$$

where C_{eff} is the effective source degeneration capacitance, n is capacitive ratio $C_1/(C_1 + C_{eff})$, R_{LP} models the tank inductor's loss, and R_{buf} and C_{buf} are the equivalent shunt resistance and capacitance of the buffer. Measured from a breakout circuit, the free-running VCO exhibits a tuning range of 11.5 %, phase noise of −110 dBc/Hz at 1 MHz offset from 32 GHz carrier, and consumes 17 mW power, resulting in a figure-of-merit (*FoM*) of 184.7 dB. The *FoM* is defined as $FoM = 20\log_{10}(f/\Delta f) - 10\log_{10}(P_{DC}/1\,\text{mW}) - L\{\Delta f\}$.

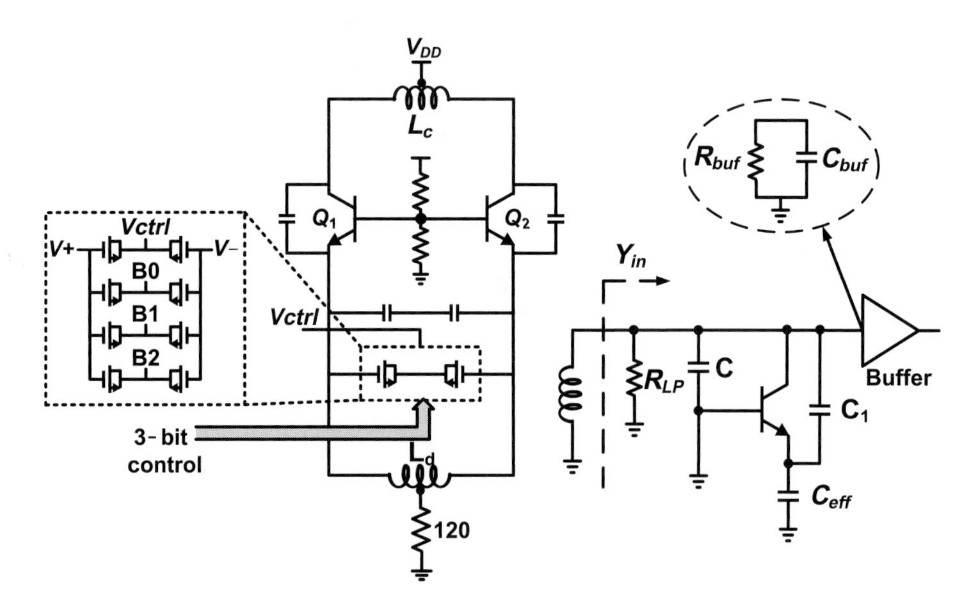

Figure 10.4 Schematic of the Ka-band differential Colpitts VCO and its half-circuit model.

10.3 Design of a W-Band ILFT

10.3.1 Harmonic Generation of HBT

Before discussing the design of frequency tripler, it is instructive to study the nonlinearity of the bipolar transistor. heterojunction bipolar transistors (HBTs) exhibit nonlinearities in two ways: (1) inherently exponential I–V relationship and (2) distortion caused by waveform clipping when the transistor is driven into saturation by a large input swing. In a frequency tripler design, the overall third harmonic strength generated by the HBT depends on input voltage swing and bias condition. The I–V equation of a bipolar transistor can be expressed as follows [20]:

$$i_c = I_S \exp\left(\frac{v_{BE}}{V_T}\right)\left(1 + \frac{v_{CE} - v_{BE}}{V_{AF}}\right) = f(v_{BE}), \tag{10.6}$$

where I_S, V_{AF}, and V_T are the transistor saturation current, the forward early voltage, and the thermal voltage ($V_T = kT/q$), respectively. If a DC bias voltage V_{BE} and an input AC signal $A\cos(\omega_o t)$ are applied, following the analysis in [21], the third harmonic can be written as follows:

$$I_3 = \frac{2A^3}{5!\,\pi}\int_{-A}^{A}\left(1 - \frac{V_{AC}^2}{A^2}\right)^{5/2} \left.\frac{d^3 f(v_{BE})}{dv_{BE}^3}\right|_{v_{BE}+V_{AC}} \frac{dV_{AC}}{A} \tag{10.7}$$

where $V_{AC} = A\cos(\omega_o t)$. The third harmonic is an average of the third-order derivative of $f(v_{BE})$ over the input voltage swing with a weighting function $(1 - V_{AC}^2/A^2)^{5/2}$, as shown by (10.7). Figure 10.5 plots the simulated collector current (I_c) and third-order derivative of I_c over V_{BE} versus V_{BE}. The simulation is done with the HICUM model

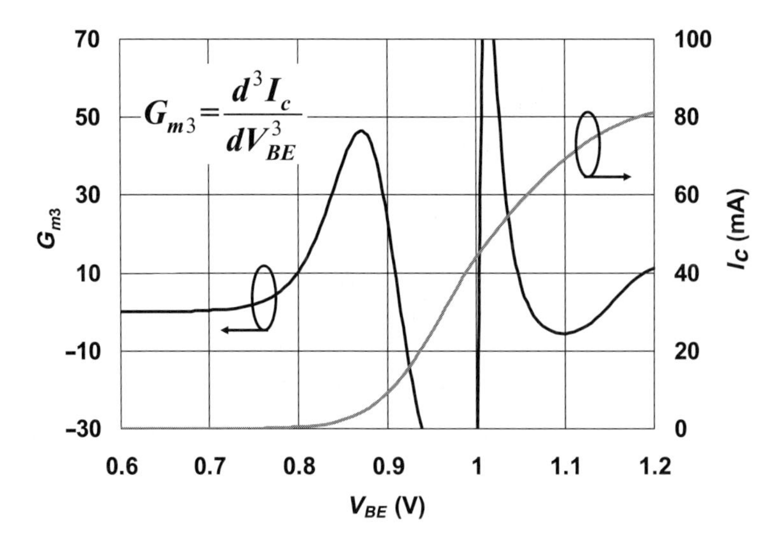

Figure 10.5 Simulated SiGe HBT I_c and its third derivative as a function of V_{BE}.

provided by the foundry. The term G_{m3} represents the small signal transconductance of the third harmonic. From the figure, we can tell that biasing the transistor at the Class-A region ($V_{BE} > 1$ V) gives the highest third harmonic transconductance. However, the corresponding DC current is huge, which leads to very poor efficiency. On the other hand, biasing the transistor in the Class-AB regime (V_{BE} close to 0.87 V) offers good third harmonic strength with a relatively small I_c.

In summary, given an input signal swing, there is an optimal bias point where the average integral in (10.7) is maximized. With the 0 dBm signal power from the Ka-band PLL in this design, simulation shows biasing the transistors slightly higher than their threshold in Class-AB regime offers the strongest third harmonic.

10.3.2 Circuit Design of the mm-Wave ILFT

The schematic of the W-band ILFT circuit is shown in Figure 10.6, which consists of two parts: (1) a pair of harmonic generating transistors $Q_3\&Q_4$ and (2) an ILO. The harmonic generator takes advantage of the nonlinearity of the HBTs and generates all the harmonics of the input fundamental frequency, which are then injected to the ILO. The ILO is based on a differential Colpitts oscillator with a free-running frequency close to three times the input fundamental frequency, and therefore exhibits a loop gain greater than unity for the third harmonic component only. The injection-locking operation is realized by feeding the third harmonic of the input signal generated by $Q_3\&Q_4$ into the emitters of ILO. Transistors $Q_3\&Q_4$ reuse part of the DC current of the tank and are biased in the Class-AB regime for maximum third harmonic generation, as explained in Section 10.3.1. The 96 GHz output signals are taken out from the collector terminals of $Q1\&Q2$ through a cascode buffer stage to minimize leakage of the 32 GHz injection signal at the output. Differential operation is achieved by connecting two

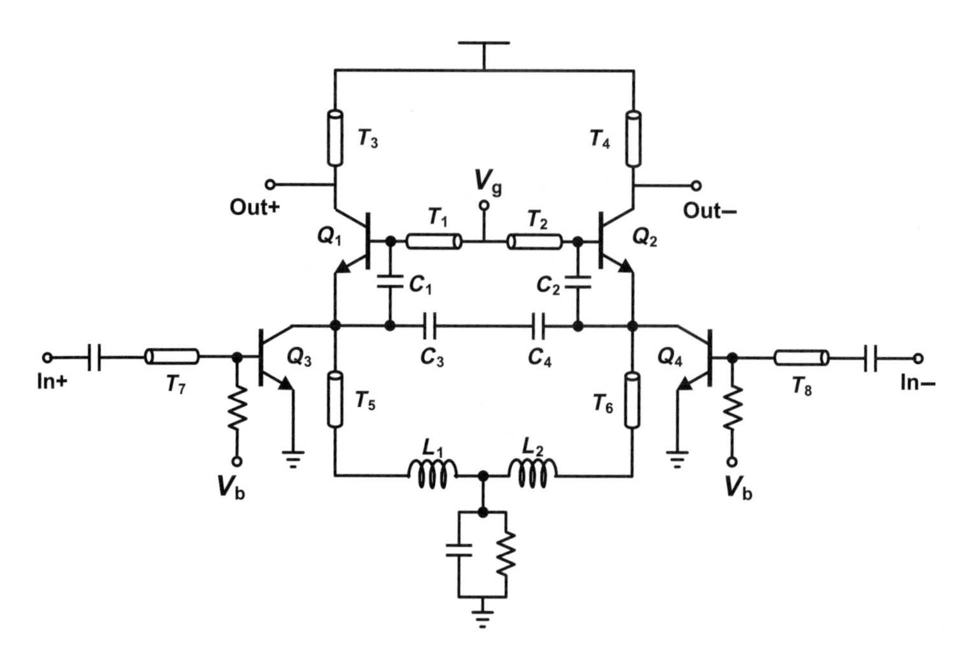

Figure 10.6 Schematic of the W-band ILFT [11]. (©2012 IEEE. Reprinted, with permission, from *IEEE Transactions on Microwave Theory and Techniques*.)

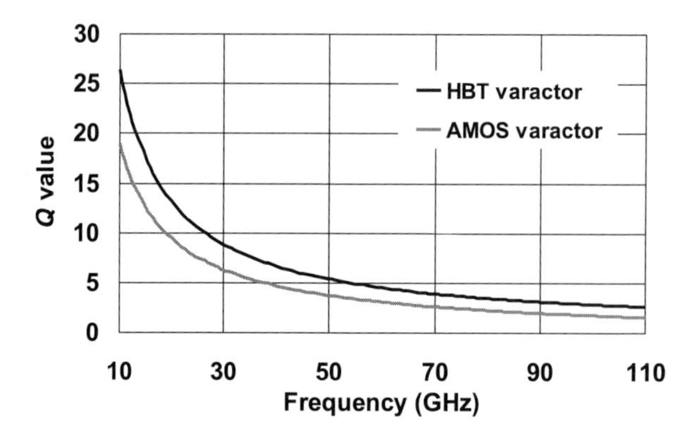

Figure 10.7 Simulated Q value of the AMOS varactor and HBT varactor.

interdigitated metal-oxide-metal (MOM) capacitors (C_3&C_4) back-to-back across the emitters of Q_1&Q_2.

Although adding varactors to the tank helps to improve the locking range by varying the self-oscillation frequency [12], no varactors are used in this design since neither accumulation-mode metal-oxide-semiconductor (MOS) varactors (AMOS) nor HBT varactors has a Q value larger than 3 at 96 GHz in this technology, as shown in Figure 10.7. However, since HBTs exhibit much stronger nonlinearity compared to MOS transistors due to its exponential I–V relationship, a wide locking range can still be achieved without varactor tuning.

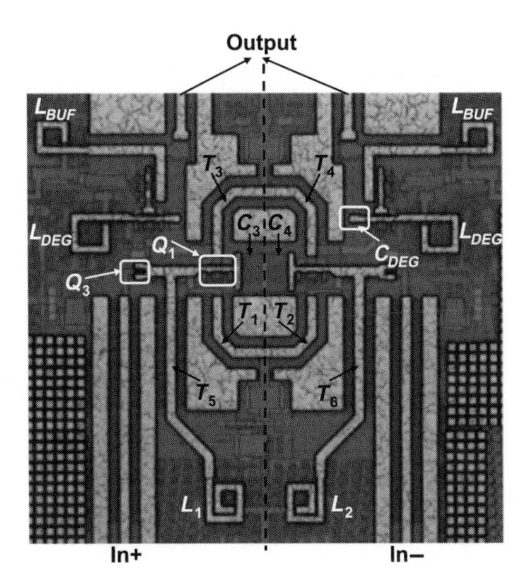

Figure 10.8 Die photo of the ILFT.

Besides the schematic design, special attention must be paid to the layout floor plan of the ILFT so as to ensure the following aspects are achieved: (1) maintaining good symmetry for common-mode noise/disturbance rejection; (2) achieving compact layout structure to save area and minimize interconnection; and (3) minimizing the mutual coupling among inductive elements connected to different transistor terminals, which may cause multimode oscillation and poor phase noise performance.

Figure 10.8 shows the die photo of the ILFT indicating its constituent circuit elements. To attain compact layout yet avoid mutual electromagnetic and substrate noise coupling, ground-shielded coplanar waveguide (GCPW) lines (T_1–T_6) are used at base, collector, and emitter to provide the tank, the load, and part of the degeneration inductances. Additional emitter degeneration inductance is realized by spiral inductor (L_1&L_2) to save area. The GCPW structure is favored, as it can realize small inductance with good modeling accuracy and adequate quality factor. The tank and load inductors (T_1–T_4) are the most critical components that are placed in close proximity, but on opposite sides, of the core transistors (Q_1&Q_2). Ground sidewalls are shared between adjacent GCPW lines, for instance, between T_1 and T_5.

One observation from simulation is that the loss introduced by the buffer to the collector nodes of Q_1&Q_2 has a noticeable impact on the tank's Q factor due to the coupling via C_μ (base-to-collector capacitance) of these transistors. The same phenomenon was also observed in [22], where the authors chose source/emitter as the output node instead. As a qualitative explanation, in the presence of C_μ, the resistive part of the buffer's input impedance, R_{buf} (mainly due to the base resistance of the buffer's input transistor), adds loss to the tank that may quench the oscillation. Moreover, the capacitive part of the buffer's input impedance, C_{buf}, increases capacitive loading of the ILFT's tank through the Miller effect, thereby causing frequency downshift. For quantitative analysis, the input impedance seen into the base of a Colpitts oscillator is derived based on a general

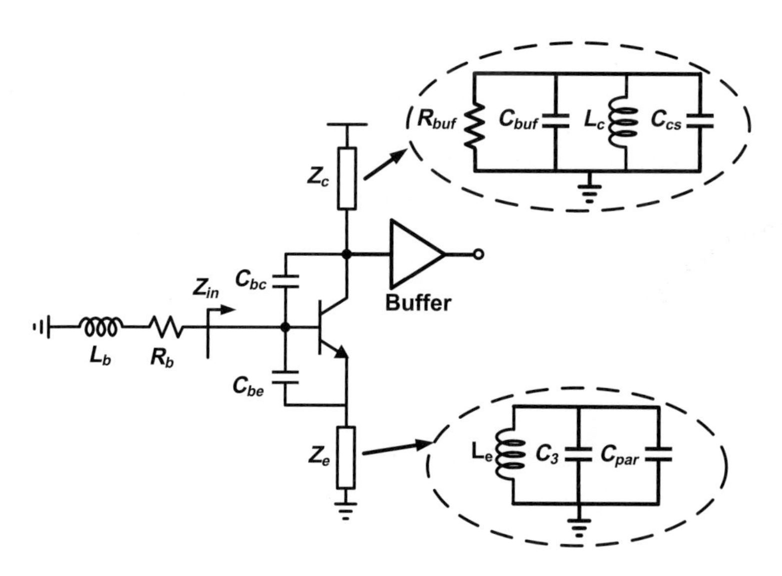

Figure 10.9 General circuit model of an emitter degenerated Colpitts oscillator.

circuit model shown in Figure 10.9. Note that the base current has been ignored for this derivation.

$$Z_{in} = \frac{(1 + sC_{bc}Z_c)(1 + g_mZ_e + sC_{be}Z_e)}{sC_{be}(1 + sC_{bc}Z_c) + sC_{bc}(1 + g_mZ_e + g_mZ_c + sC_{be}Z_e)} \tag{10.8}$$

If the effect of C_{bc} is ignored, the preceding equation is simplified to the following:

$$Z_{in} = \frac{1}{sC_{be}} + Z_e + \frac{g_mZ_e}{sC_{be}}. \tag{10.9}$$

which is commonly seen in the textbooks [23] for the input impedance of a source/emitter degenerated transistor. However, this simplification gives rise to nonnegligible errors at the W-band frequency.

In the case of a Colpitts oscillator in this design (cf. Figure 10.4), C_{bc} stands for C_μ; C_{be} equals the sum of C_1 and C_π (base-to-emitter capacitance of $Q_1\&Q_2$); Z_e is emitter degeneration impedance, which must be capacitive at the oscillation frequency to ensure the oscillation; and Z_c is the load impedance. Equations (10.10)–(10.13) interpret the general variables in (10.8) in terms of the design parameters of the ILFT (cf. Figure 10.6).

$$C_{bc} = C_\mu \tag{10.10}$$

$$C_{be} = C_\pi + C_1 \tag{10.11}$$

$$Z_e = \frac{1}{s(C_3 + C_{par})\left(1 + \frac{1}{s^2L_e(C_3+C_{par})}\right)} \tag{10.12}$$

$$Z_c = \frac{sR_{buf}L_c}{R_{buf} + sL_c + s^2R_{buf}L_c(C_{cs} + C_{buf})} \tag{10.13}$$

where L_e is the total emitter degeneration inductance from L_1 and T_5, C_{par} includes all the parasitic capacitances associated with emitter node contributed by Q_1 and Q_3, C_{cs} is the collector-to-substrate capacitance of Q_1, L_c is the load inductance from T_3, and R_{buf} and C_{buf} are the equivalent shunt resistance and capacitance of the buffer. In order to avoid the loss introduced by the buffer, capacitive degeneration is employed in the buffer design to reduce the real part of the buffer's input impedance.

The schematic of the buffer is shown in Figure 10.10a. By choosing the resonant frequency of $L_{DEG} - C_{DEG}$ lower than the operation frequency (i.e., $1/2\pi\sqrt{L_{DEG}C_{DEG}} < 96\,\text{GHz}$), the emitter degeneration impedance (L_{DEG} in parallel with C_{DEG}) becomes capacitive at 96 GHz, and the real part of buffer's input impedance is written as follows:

$$\text{Re}[Z_{in,buf}] = R_{b,Q5} - \frac{g_{m,Q5}}{\omega^2 C_{eff} C_{\pi,Q5}} \tag{10.14}$$

where $R_{b,Q5}$ is the base resistance of Q_5 (cf. Figure 10.10a) and C_{eff} is the effective emitter degeneration capacitance.

$$C_{eff} = C_{DEG}\left(1 - \frac{1}{\omega^2 L_{DEG} C_{DEG}}\right) \tag{10.15}$$

Figure 10.10b plots the simulated $\text{Re}[Z_{in,buf}]$ Re(Zin,buf) with and without capacitive emitter degeneration. The combination of negative input resistance from the buffer and avoidance of using low Q varactors improves tank's Q factor of the ILFT, and therefore, decreases the required DC current to sustain the oscillation. The tripler plus output buffer consumes 75 mW from a 2.5 V supply. This design was the first implementation of an injection-locked-based frequency multiplier in the SiGe BiCMOS process [11].

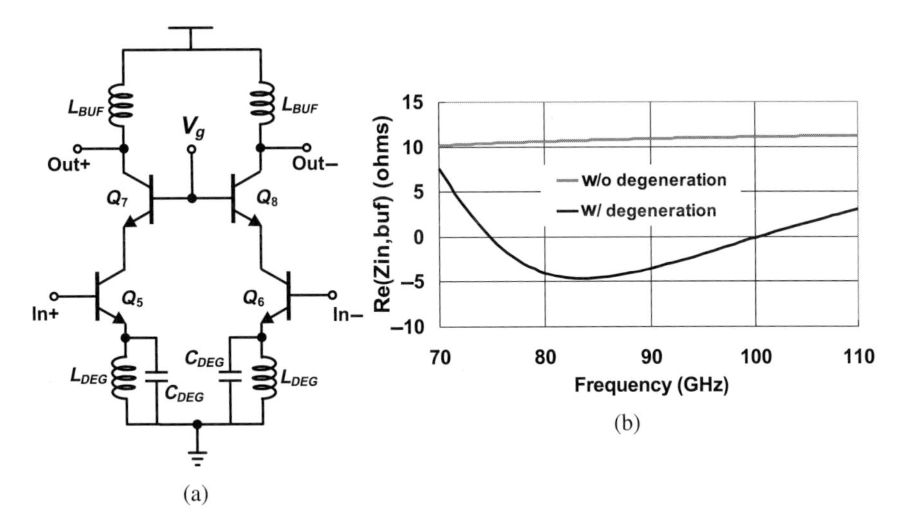

Figure 10.10 (a) Schematic of the tripler's buffer and (b) the simulated real part of the buffer's input impedance with and without capacitive degeneration.

Figure 10.11 Die photo of Ka-band PLL with ILFT [11]. (©2012 IEEE. Reprinted, with permission, from *IEEE Transactions on Microwave Theory and Techniques.*)

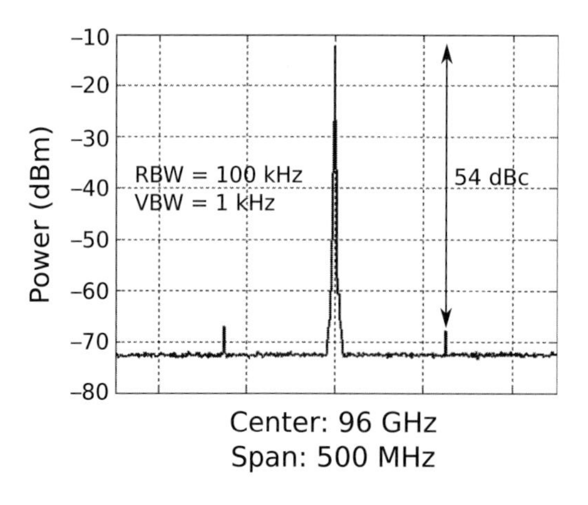

Figure 10.12 Single-ended spectrum measured at the output of the ILFT.

10.3.3 Measurement Results

The chip has been fabricated in the 0.18 μm SiGe BiCMOS process and characterized using on-wafer probing with all DC pads wirebonded to a printed circuit board (PCB). Figure 10.11 shows die micrograph of PLL plus ILFT, and the circuit occupies an area of 1.8 mm^2. The chip area can be further reduced, since two of the three buffers included only for test purpose could be removed. In addition, a separate breakout circuit has also been fabricated for ILFT to enable full characterization of the frequency tripler.

The single-ended 96 GHz output spectrum measured at the output of the ILFT is shown in Figure 10.12. Losses from measurement setups are not deembedded. The Ka-band PLL achieves a measured tuning range of 30.3–33.8 GHz and delivers an average differential output power of 0 dBm. However, the tuning range measured after the ILFT is 92.8–98.1 GHz, which is limited by the locking range of the ILFT. Figure 10.13

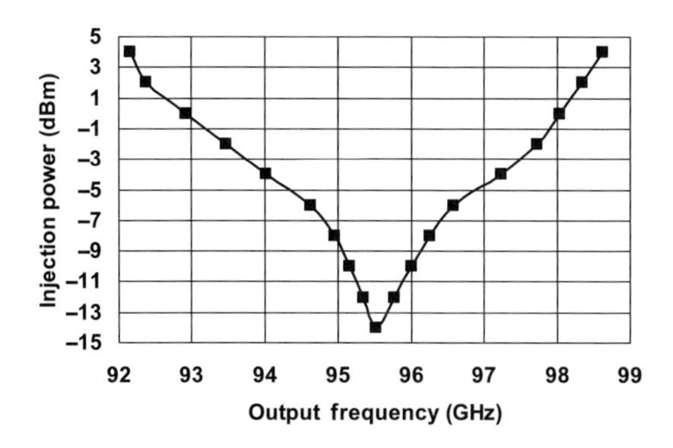

Figure 10.13 Measured input sensitivity of the standalone ILFT [11]. (©2012 IEEE. Reprinted, with permission, from *IEEE Transactions on Microwave Theory and Techniques*.)

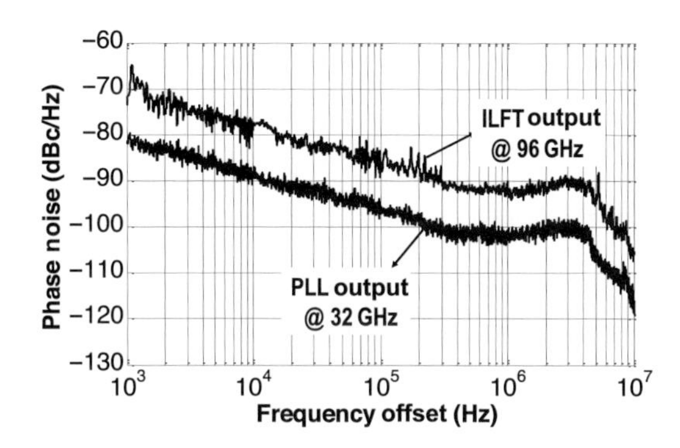

Figure 10.14 Measured phase noise at the output of Ka-band PLL and ILFT [11]. (©2012 IEEE. Reprinted, with permission, from *IEEE Transactions on Microwave Theory and Techniques*.)

plots the measured input sensitivity curve of the ILFT, from which we can see that the ILFT has a free-running frequency of 95.5 GHz, achieves an input sensitivity of −14 dBm, and exhibits a locking range of 6.5 GHz under 4 dBm injection power. Phase noise performance was measured using a 125 MHz crystal oscillator as a reference signal. Shown in Figure 10.14, the phase noise at 1 MHz offset measured at the output of the PLL and ILFT are −103 and −93 dBc/Hz, respectively. The phase noise degradation after the ILFT is 10 ± 1 dB, and this value was maintained for frequency offset ranging from 1 kHz to 10 MHz. Suppression for the first and second harmonics at the output of ILFT is observed to be better than 20 dB.

10.4 Design of a W-Band Silicon-Based HBFT

10.4.1 Circuit Design of the mm-Wave HBFT

The architecture of HBFT is shown in Figure 10.15. It consists of three stages: the harmonic generation stage, which converts a 32 GHz input signal to 96 GHz, followed by two LO amplification and filtering stages working at 96 GHz.

All three stages adopt the pseudodifferential cascode topology. Again the first stage transistor is biased at optimum third-harmonic efficiency bias voltage. In contrast, the latter two stages are biased in Class-A. In a symmetric design of a differential amplifier, the even harmonics of the collector currents are in common mode and should cancel out in the differential output voltage. The load of the amplifier is tuned to the third harmonic to maximize the gain at 96 GHz and suppress all other harmonics.

Interstage matching is achieved using metal-insulator-metal (MIM) capacitors and 50 Ω GCPW t-lines. The simulated collector impedance is $17 - j72\,\Omega$, p_1 on the Smith Chart (normalized to 100 Ω) in Figure 10.16, and is matched to 100 Ω impedance using a GCPW T-junction, p_2, and a series MIM capacitor. For the input matching, starting from the 100 Ω input, a shunt MIM capacitor moves the impedance to p_3. The matching is finalized with a different length GCPW T-junction, p_4, to $23 - j12\,\Omega$ input impedance of the transistor. Figure 10.16 illustrates the matching procedure on the Smith Chart.

For the HBFT, the multiplier and chain of amplifiers can impart phase noise in addition to the minimum $20\log_{10}(3)$ dB degradation. Using a linear phase model [3], we can express the output phase noise of HBFT as

$$S_{out,HBFT}(\Delta\omega) = 3^2 S_{PLL}(\Delta\omega) + S_{harm}(\Delta\omega) + S_{amp}(\Delta\omega). \tag{10.16}$$

Unlike the ILFT, HBFT does not resemble a first-order PLL, so noise from the harmonic generator stage and two-stage amplifier can increase the integrated phase noise. Typically, the amplifier's noise contribution is lower than the oscillator. Therefore, S_{PLL} dominates in (10.16). The simulated P_{1-dB} of the cascaded amplifier is −2 dBm. Amplifiers following the multiplier do not suffer multiplication by 3^2. Again, assuming the amplifier operates in linear region, its phase noise is a function of the input power and noise figure. Therefore, the noise level of the amplifier stages after the multiplier is kept low by biasing the first stage at minimum *NF* current density, whereas the second stage is biased at optimal f_{MAX}.

A breakout circuit of HBFT is also fabricated and measured separately. Figure 10.17 presents the frequency response of the circuit. The tripler achieves an output power of −10 dBm and a 3 dB bandwidth of 20 GHz for an input power of 0 dBm.

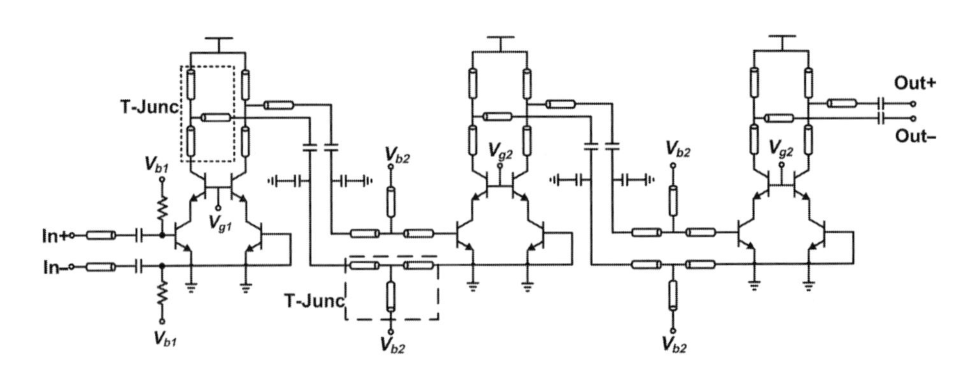

Figure 10.15 Simplified schematic of harmonic frequency tripler [11]. (©2012 IEEE. Reprinted, with permission, from *IEEE Transactions on Microwave Theory and Techniques*.)

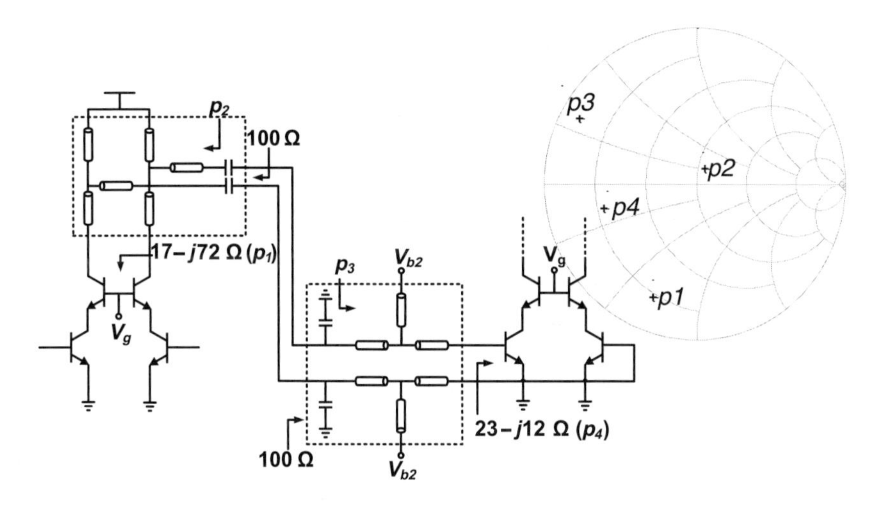

Figure 10.16 Interstage matching procedure in HBFT [11]. (©2012 IEEE. Reprinted, with permission, from *IEEE Transactions on Microwave Theory and Techniques.*)

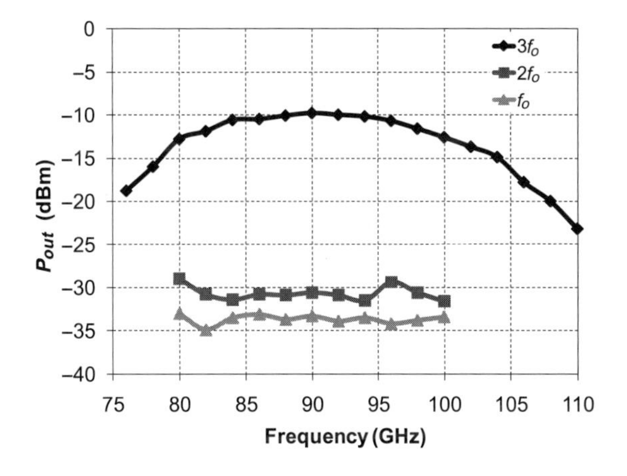

Figure 10.17 Measured breakout of HBFT output power of harmonic across frequency [11]. (©2012 IEEE. Reprinted, with permission, from *IEEE Transactions on Microwave Theory and Techniques.*)

The measurements show that the first and second harmonics are suppressed more than 20 dB compared to the desired harmonic. Again, for a differential output, the second harmonic should be suppressed further. The harmonic stage and subsequent two-stage driver amplifiers consume 5 and 70 mW, respectively, under 2.5 V supply.

10.4.2 Measurement Results

The chip has been fabricated in a 0.18 μm SiGe BiCMOS process. Figure 10.18 shows a die micrograph of a PLL plus HBFT, and the circuit occupies an area of 1.9 mm^2, which is very close to the other chip (i.e., PLL + ILFT). In addition, separate

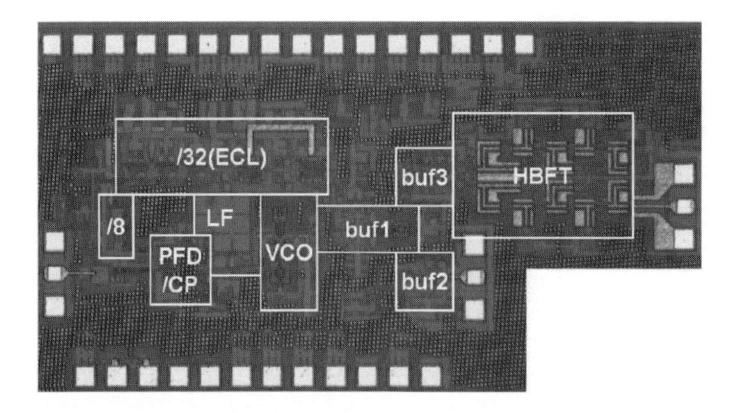

Figure 10.18 Die photo of Ka-band PLL with HBFT [11]. (©2012 IEEE. Reprinted, with permission, from *IEEE Transactions on Microwave Theory and Techniques*.)

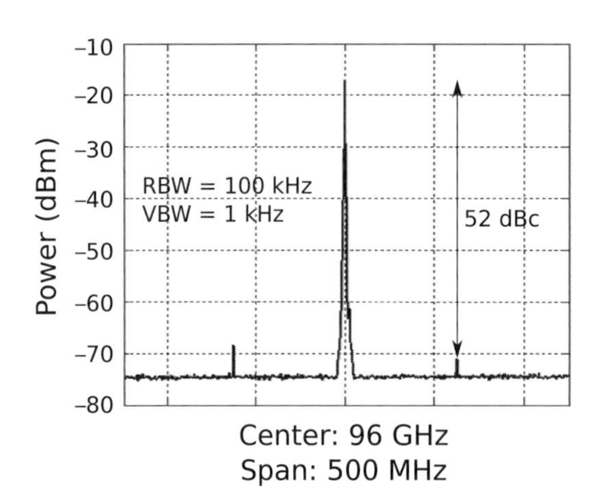

Figure 10.19 Single-ended spectrum measured at the output of HBFT.

breakout circuit has also been fabricated for HBFT to enable full characterization of the HBFT.

The single-ended 96 GHz output spectrum measured at the output of the HBFT is shown in Figure 10.19. Losses from measurement setups are not deembedded. The tuning range measured at the output of HBFT is 90.9–101.4 GHz, which is exactly three times of the PLL tuning range, as expected. Phase noise performance was measured using a 125 MHz crystal oscillator as the reference signal. Shown in Figure 10.20, the phase noise at 1 MHz offset measured at the output of the PLL and HBFT are −103 and −92 dBc/Hz, respectively. The phase noise degradation after the ILFT is 11 dB, and this value was maintained for frequency offset ranging from 1 kHz to 10 MHz. Suppression for the first and second harmonics at the output of HBFT is observed to be better than 20 dB.

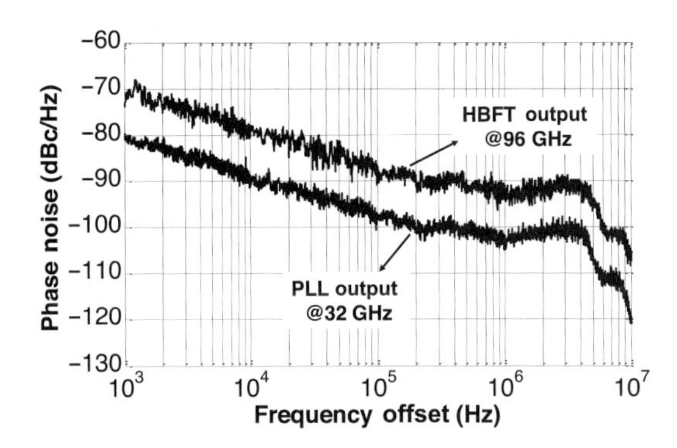

Figure 10.20 Measured phase noise at the output of Ka-band PLL and HBFT [11]. (©2012 IEEE. Reprinted, with permission, from *IEEE Transactions on Microwave Theory and Techniques.*)

10.5 Design of a Transformer-Based CMOS ILFT

10.5.1 Harmonic Generation of an MOS Transistor

Following the same procedure we used to describe the HBT tripler, we discuss the harmonic generation of an N-type metal-oxide-semiconductor (NMOS) transistor before describing the CMOS ILFT design. For an ILFT design, in order to achieve a wide locking range, it is highly desired to maximize the amplitude of the generated third harmonic signal, which is mainly determined by bias conditions and the size of the transistor. When a DC bias voltage V_{bias} and an input AC signal $V_i \cos(\omega t)$ are applied to the gate of an NMOS transistor, a certain conduction angle can be identified from the waveform of the gate-source voltage $V_{GS}(t)$, as shown in Figure 10.21.

The conduction angle ϕ of the harmonic-generating transistor (M_1) can be expressed by (10.17) in terms of threshold voltage V_{th}, gate bias voltage V_{bias}, and input amplitude V_i [24]:

$$\phi = 2\cos^{-1}\left(\frac{V_{th} - V_{bias}}{V_i}\right). \tag{10.17}$$

Figure 10.22 plots the waveform of drain current i_D of transistor M_1, which can be approximated by periodic rectified cosine function [24]. In order to analyze the harmonic component of the drain current, we perform Fourier series expansion of i_D. The Fourier coefficients of each harmonic can be expressed by (10.19). The strength of the third harmonic component under different conduction angles can be evaluated by normalizing its amplitude I_3 to that of the peak drain current I_{max} (see Figure 10.23). The results are shown by (10.20) and plotted in Figure 10.23.

$$i_D = I_0 + \sum_{n=1}^{\infty} I_n \cos(n\omega_i t) \tag{10.18}$$

$$I_n = \frac{2I_{max}\left[\sin\left(\frac{n\phi}{2}\right)\cos\left(\frac{\phi}{2}\right) - n\cos\left(\frac{n\phi}{2}\right)\sin\left(\frac{\phi}{2}\right)\right]}{\pi n\left(n^2-1\right)\left(1-\cos\frac{\phi}{2}\right)} \quad (n \geq 2) \tag{10.19}$$

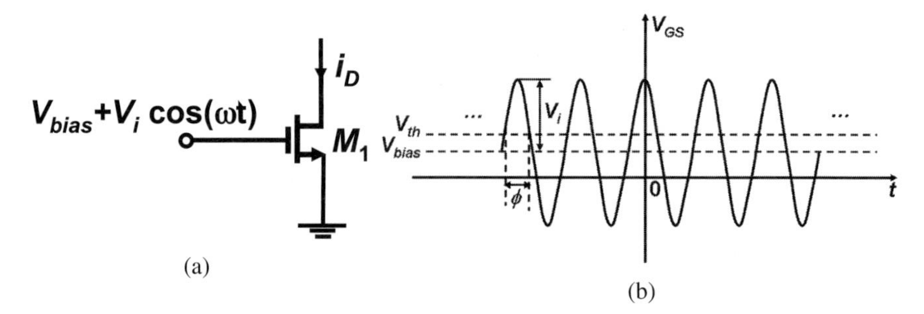

Figure 10.21 Waveform of gate-source voltage $V_{GS}(t)$ under a given bias condition (V_{bias}).

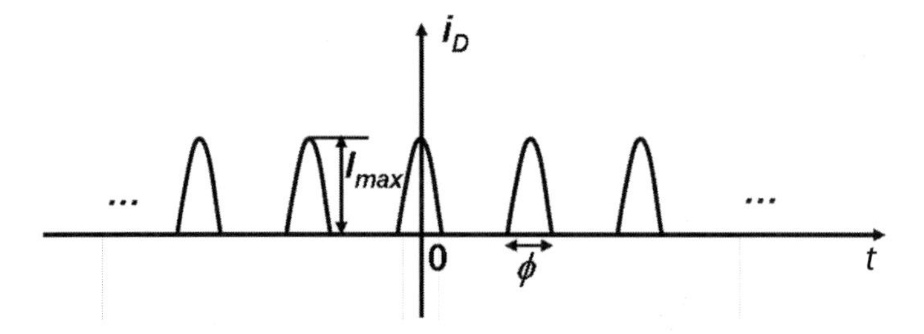

Figure 10.22 Waveform of the drain current of transistor M_1.

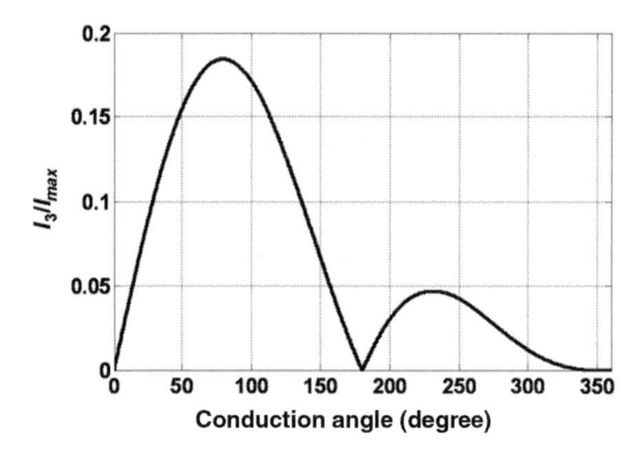

Figure 10.23 Normalized third harmonic amplitude under different conduction angles. [12]. (©2010 IEEE. Reprinted, with permission, from *IEEE MTT-S International Microwave Symposium.*)

$$\frac{I_3}{I_{max}} = \frac{\sin\left(\frac{3\phi}{2}\right)\cos\left(\frac{\phi}{2}\right) - 3\cos\left(\frac{3\phi}{2}\right)\sin\left(\frac{\phi}{2}\right)}{12\pi\left(1 - \cos\frac{\phi}{2}\right)} \tag{10.20}$$

It is clear from Figure 10.23 and (10.17) that biasing the transistor in the weak inversion region results in the strongest third harmonic. From a large signal perspective, this means that the harmonic generator should operate in the class-C regime. Although the first and second harmonics are also generated together with the desired third harmonic,

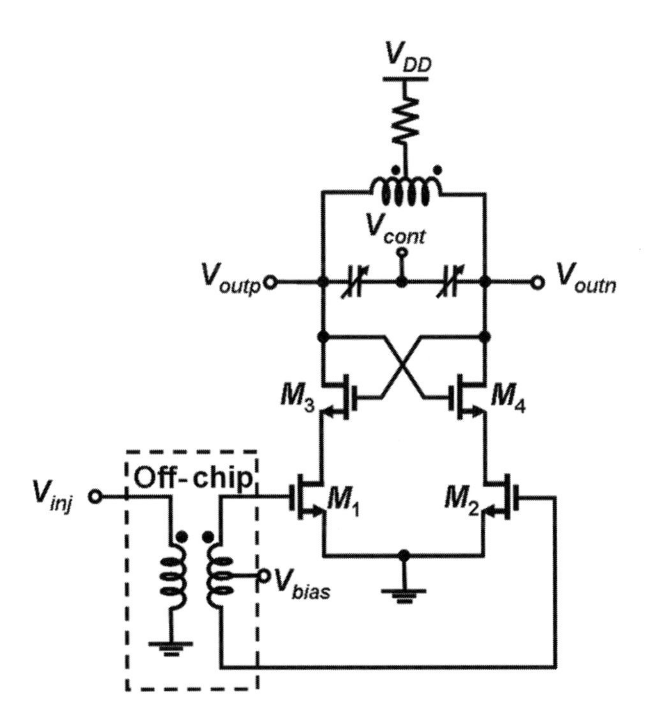

Figure 10.24 Schematic of a V-band ILFT [12]. (©2010 IEEE. Reprinted, with permission, from *IEEE MTT-S International Microwave Symposium.*)

they will be filtered out by the LC tank. Additionally, the second harmonic will be further rejected by the differential nature of the ILO.

10.5.2 Millimeter-Wave T-ILFT Structure

Figure 10.24 shows a V-band ILFT presented in [16]. An off-chip transformer is used to feed the input signal differentially to M_1–M_2 serving as harmonic generator. Because the harmonic generator shares the same current with the ILO, the gate bias of M_1 and M_2 (V_{bias} in Figure 10.24) has to be higher than V_{th} to maintain a sustainable oscillation for the ILO. In other words, instead of operating in the desired class-C region with $\phi < \pi$ (cf. Figure 10.23), the harmonic generator in Figure 10.24 has to be biased in the class-AB region with $\pi < \phi < 2\pi$.

The circuit schematic of the mm-wave T-ILFT is shown in Figure 10.25.

An on-chip transformer is employed to feed the third harmonic to the ILO and, more importantly, decouple the harmonic generator from the ILO. Several benefits can be attained from the this T-ILFT structure: (1) The harmonic-generating transistor M1 can be biased independently from the ILO to achieve optimum conduction angle (cf. Figure 10.23), while consuming negligible DC power. (2) In contrast to the circuit in Figure 10.24, the cross-coupled pair M_2–M_3 is not stacked on top of M_1, which saves voltage headroom for the ILO. As a result, the ILO can operate at lower supply voltage with less power consumption, and larger output swing can be obtained for a

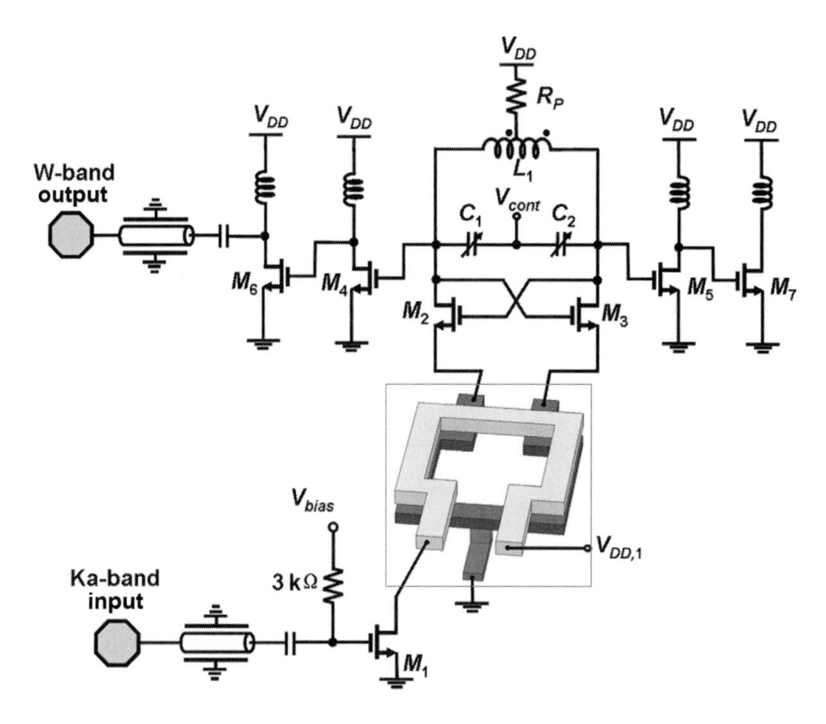

Figure 10.25 Schematic of the mm-wave T-ILFT [12]. (©2010 IEEE. Reprinted, with permission, from *IEEE MTT-S International Microwave Symposium.*)

given supply. (3) The transformer also converts the output impedance of M_1 to a much smaller value so as to reduce the source degeneration impedance of the ILO. As will be shown in the next section, the T-ILFT circuit does not exhibit severe loop gain degradation, due to the impedance transformation offered by the transformer. (4) The transformer carries out on-chip single-to-differential conversion.

A major concern about this T-ILFT is the loop gain degradation due to the source degeneration. The source degeneration impedance Z_s is derived in (10.21) using the equivalent AC model shown in Figure 10.26a, where L_P is the self-inductance of the primary coil of the transformer, L_s is the half-inductance of the secondary coil, $_o$ is the output resistance of M_1, and C_D is the total parasitic capacitance at the drain node of M_1.

$$Z_s = sL_s - \frac{(sM)^2}{\left(sL_p || \frac{1}{sC_D} || r_o\right)} \tag{10.21}$$

In order to guarantee stable oscillation startup accounting for PVT variation, the loop gain is chosen to be larger than two in this design. The simulated $|Z_s|$ and the loop gain are shown in Figure 10.26b from which we can see that $|Z_s|$ is around 9 Ω at 90 GHz compared to 60 Ω (due to r_o and C_D) without impedance transformation (i.e., more than sixfold reduction in source degeneration impedance), and the loop gain is above 2.5 for the whole frequency range.

Although Q is inversely proportional to the locking range of an ILO [25], decreasing Q is not an effective solution to improve locking range, because more DC current is

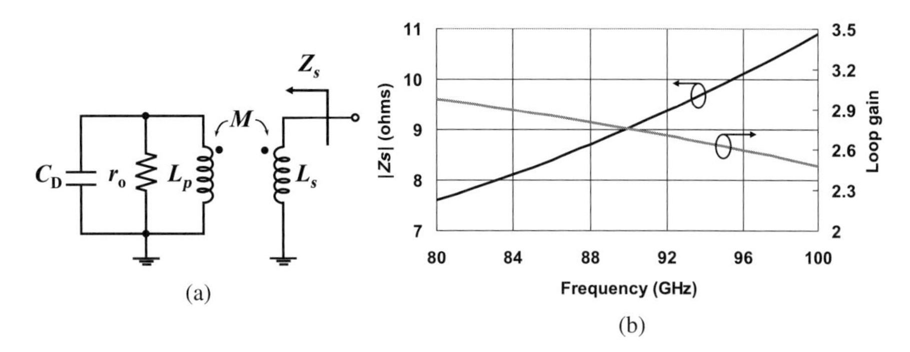

Figure 10.26 (a) Equivalent AC model illustrating the impedance transformation and (b) simulated magnitude of the source degeneration impedance and loop gain of the ILO [12]. (©2010 IEEE. Reprinted, with permission, from *IEEE MTT-S International Microwave Symposium.*)

required to maintain the loop gain, and an increase in DC current eventually keeps the locking range almost unchanged [17]. In summary, the Q of the LC tank should be maximized for better loop gain, low phase noise, and low power consumption. The DC current of the ILO is chosen as low as possible to widen locking range [25] and reduce power, but large enough to guarantee the stable oscillation startup.

The Ka-band input signal is AC-coupled to harmonic generator M_1, which is biased in weak inversion to maximize third harmonic generation. The drain voltage of M_1, $V_{DD,1}$ in Figure 10.25, is fed through the primary coil of the transformer. The value of $V_{DD,1}$ does not have any noticeable impact on third harmonic generation and is chosen to be 0.4 V to prevent oxide breakdown when a large input swing is applied. The stacked transformer is realized using the top two metal layers with 26 μm outer diameter and 4 μm metal width. The secondary coil is center-tapped to ground using lower metal layers for better port-to-port isolation. The simulated coupling coefficient is 0.71 at 90 GHz. The cross-coupled pair (M_2–M_3) provides negative resistance to compensate for the losses of the LC tank. The tank consists of a center-tapped inductor with 35 pH half-inductance, two accumulation-mode varactors C_1 and C_2, and parasitic capacitance at the drain nodes of M_2 and M_3. The one-turn octagonal center-tapped inductor L_1 is built by paralleling the top two metal layers to reduce resistive loss. Varactors C_1 and C_2 are employed to vary the natural oscillation frequency of the ILO so as to further increase the locking range of the T-ILFT. The common-mode resistor R_p is a 20 Ω polyresistor that lowers the gate voltage of C_1 and C_2, thereby making full use of the tuning capacity of the varactors to increase the tuning range of the ILO. Moreover, R_p also improves the common-mode rejection for even harmonics [16]. Two-stage common source buffers are designed to isolate the core circuits from the 50 Ω load. The buffer sizes are tapered to minimize the capacitive loading on the LC tank, while maintaining the output amplitude and driving capability. The buffer's output is AC-coupled through 100 μm ground-shielded CPW line to the GSG pad, whose capacitance is estimated to be 32 fF. The simulated total insertion loss of the two-stage buffers, the CPW, and the GSG pad driving a 50 Ω load is about 0 dB. All passive devices and structures are electromagnetically (EM) simulated using Sonnet.

10.5.3 Measurement Results

The micrograph of the T-ILFT fabricated in 65 nm standard CMOS process is shown in Figure 10.27. The chip area is 0.53 mm^2 including pads while the core circuits including buffers occupy 370 × 240 μm^2. The prototype is characterized using on-wafer probing. The W-band output signal is downconverted by an Agilent 11970W harmonic mixer and measured using an Agilent E4448A spectrum analyzer. The test setup is shown in Figure 10.28. The T-ILFT consumes 5.2 mW power and output buffers consume 14.6 mW power, all from a 0.8 V supply.

The measured single-ended output spectra under free-running and injection-locked modes are shown in Figure 10.29a and 10.29b, respectively. The T-ILFT shows an output power of −13 dBm at 93.01 GHz under free-running, and an output power

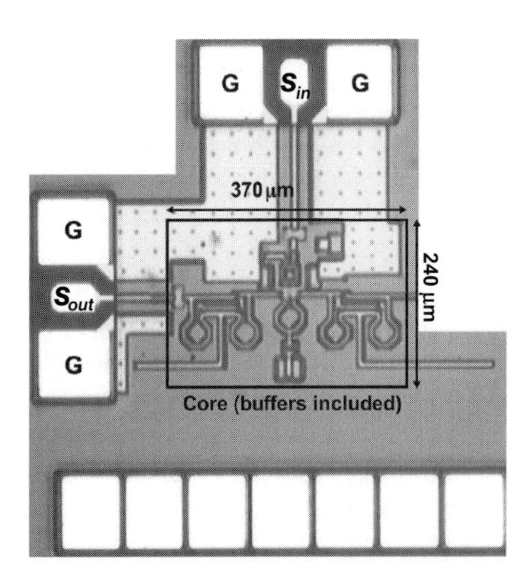

Figure 10.27 Die micrograph of the T-ILFT [12]. (©2010 IEEE. Reprinted, with permission, from *IEEE MTT-S International Microwave Symposium.*)

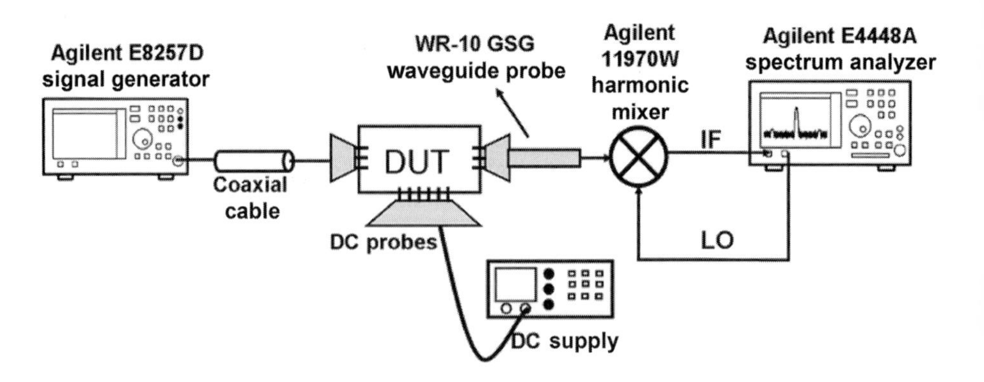

Figure 10.28 Test setup to measure the T-ILFT.

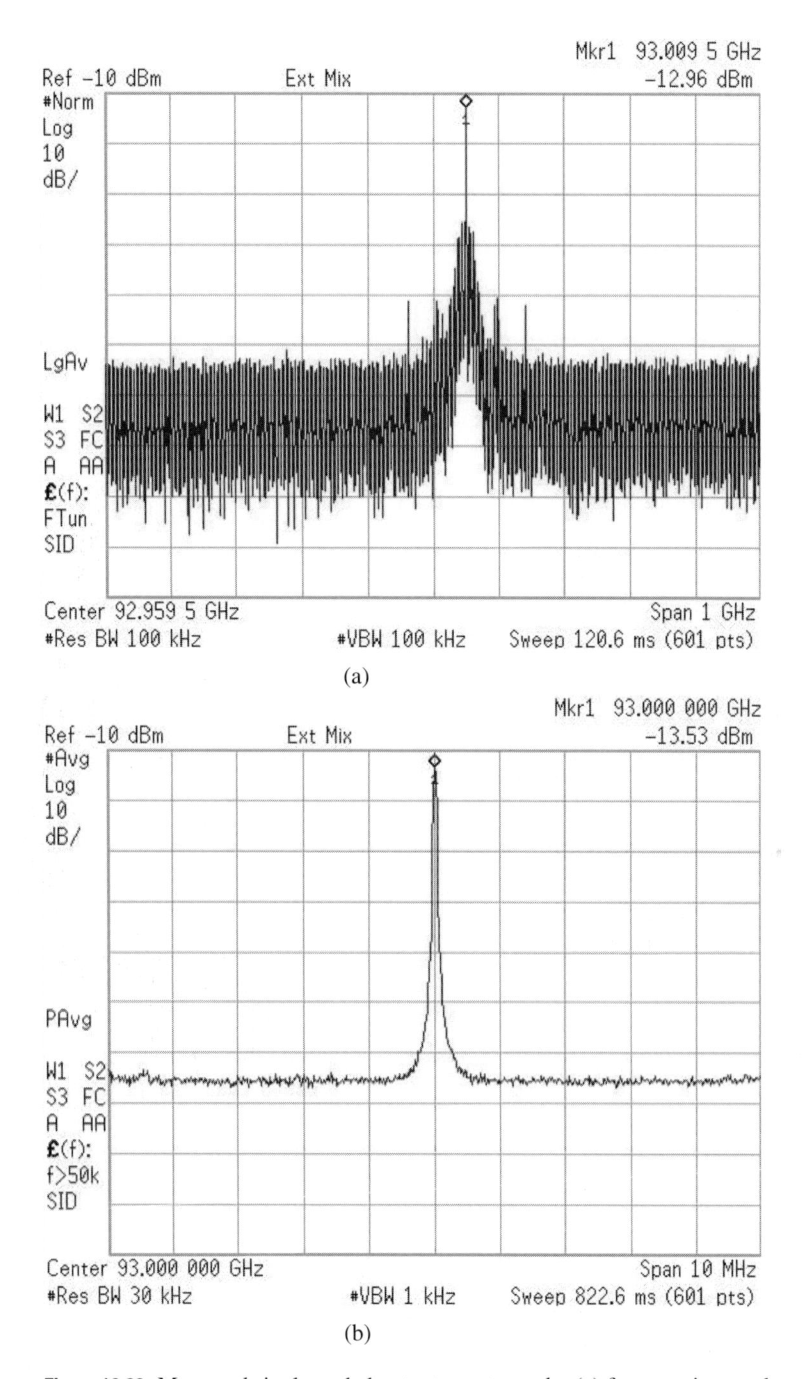

Figure 10.29 Measured single-ended output spectra under (a) free-running mode and (b) injection-locked mode.

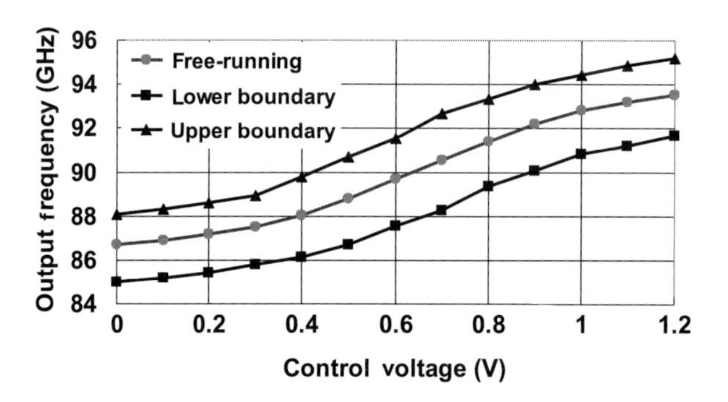

Figure 10.30 Measured tuning and locking range of the T-ILFT [12]. (©2010 IEEE. Reprinted, with permission, from *IEEE MTT-S International Microwave Symposium.*)

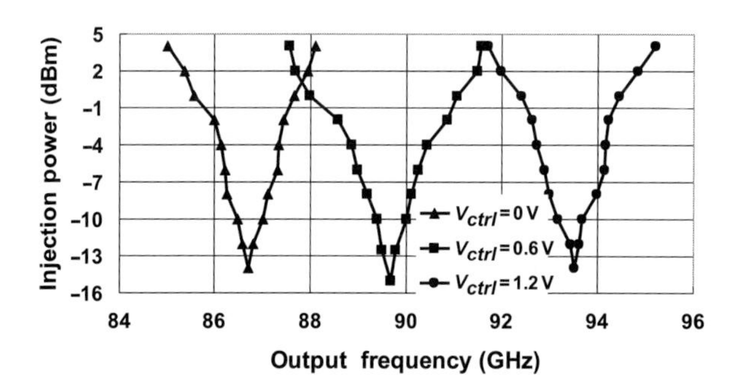

Figure 10.31 Measured input sensitivity of the T-ILFT under different control voltages [12]. (©2010 IEEE. Reprinted, with permission, from *IEEE MTT-S International Microwave Symposium.*)

of −13.5 dBm at 93 GHz when locked to a 0 dBm 31 GHz input signal. The losses from measurement setups are not deembedded.

The free-running ILO achieves a measured tuning range from 86.7 to 93.5 GHz while the T-ILFT exhibits a locking range varying from 3.1 to 4.4 GHz for different control voltages, as shown in Figure 10.30. By tuning the varactors, the T-ILFT covers a continuous locking range from 85 to 95.2 GHz. Note that the results in Figure 10.30 are measured under an input power level of 4 dBm, which is limited by the maximum output power (10 dBm) of the signal source after deembedding of loss associated with the probe and cables. In simulation, the locking range of the T-ILFT keeps increasing with the input power and saturates at an input power of 10 dBm. Figure 10.31 shows the measured input sensitivity curve of the T-ILFT under three different control voltages.

Both simulated and measured locking ranges vs. gate bias voltage of M_1 (V_{bias} in Figure 10.25) with 0 dBm input power and 0.6 V control voltage are shown in Figure 10.32. The shape of the curve is consistent with the curve of Figure 10.23, as expected, because the strength of the generated third harmonic plays a key role in determining the locking range. However, the locking range does not decrease to zero at

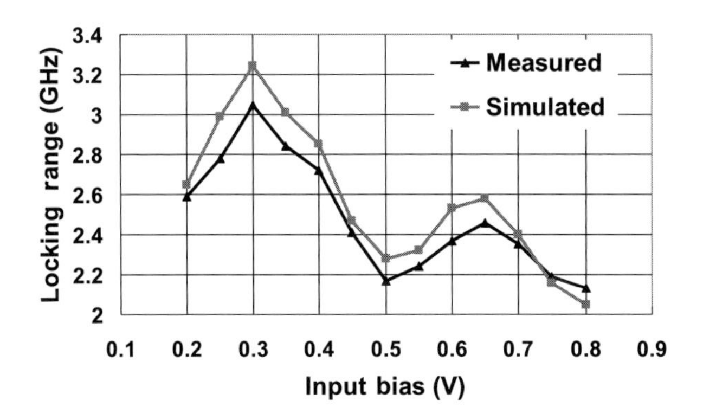

Figure 10.32 Simulated and measured locking ranges vs. gate bias of M_1 [12]. (©2010 IEEE. Reprinted, with permission, from *IEEE MTT-S International Microwave Symposium.*)

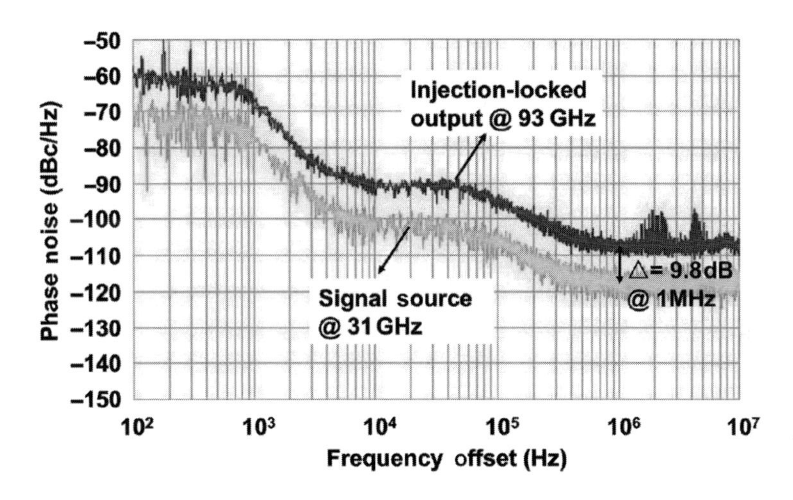

Figure 10.33 Measured phase noise of the input source and output signal under injection-locked condition [12]. (©2010 IEEE. Reprinted, with permission, from *IEEE MTT-S International Microwave Symposium.*)

the points where Figure 10.23 shows no third harmonic generation. One reason is that the ILO itself (particularly, M_2 and M_3) also exhibits nonlinearity besides M_1.

The measured phase noise performance of the output signal under injection-locked condition is plotted in Figure 10.33 together with the phase noise of the input source. The phase noise degradation from the input signal is 9.9 ± 0.3 dB for frequency offset from 100 Hz to 10 MHz, which is close to the 9.54 dB ($20 \log_{10} 3$) theoretical limit. The measured average differential output powers for the entire locking range after deembedding the loss of measurement setup is −5.5, −4.2, and −3.8 dBm with input power of −6, −1, and 4 dBm, respectively. The measured harmonic suppressions for the first and second harmonics are 32.9 and 38.5 dB, respectively.

10.6 Comparisons and Discussions

Based on the measurement results of the prototypes, we now compare two kinds of SiGe frequency triplers in terms of phase noise, tuning range, and output power. Again, the comparison is made under the same power consumption, as described in Section 10.1. The HBFT degrades the phase noise by 11 dB and triples the tuning range, i.e., it maintains the same fractional tuning range. The tuning range of the ILFT is limited by its locking range, which is highly dependent on the injection power, and the phase noise degradation is only 10 dB for the ILFT. Figure 10.34 plots P_{out} vs. P_{in} for both HBFT and ILFT measured from the tripler breakout circuits. The ILFT's P_{out} is determined by the oscillation amplitude, not P_{in}, while the HBFT's P_{out} is heavily dependent on P_{in}. The results indicate that the ILFT is more energy efficient and is capable of providing good output power, especially when input power is low.

The preceding comparison is made between two different tripler structures under the same technology (i.e., 0.18 μm SiGe BiCMOS). Another interesting comparison can be made between two injection-locking-based frequency triplers fabricated using different technologies. The SiGe ILFT and CMOS ILFT described in Section 10.3.2 and 10.5.2 are based on the same behavior model shown in Figure 10.35. The harmonic generation stage generates all the harmonics of the input signal and feeds them into

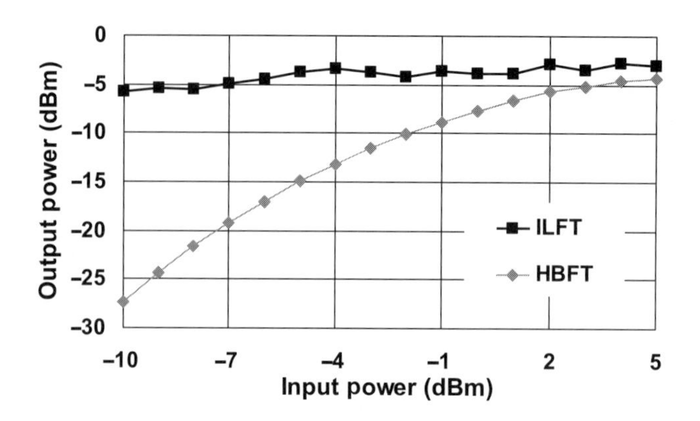

Figure 10.34 Measured P_{out} at 96 GHz vs. P_{in} at 32 GHz for ILFT and HBFT [11]. (©2012 IEEE. Reprinted, with permission, from *IEEE Transactions on Microwave Theory and Techniques.*)

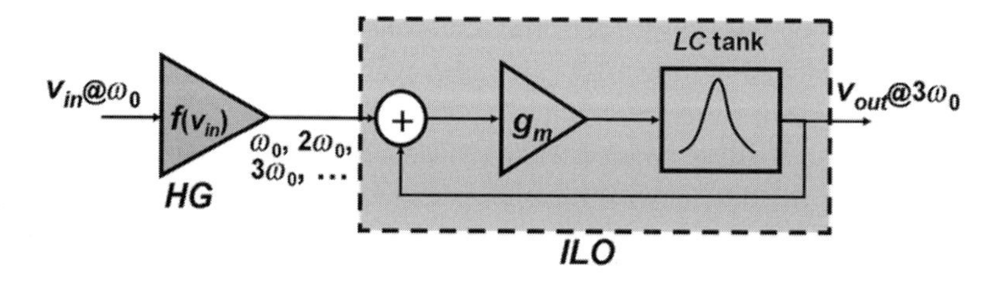

Figure 10.35 Behavior model for injection-locking-based frequency tripler.

Table 10.1 Comparison of state-of-the-art W-band frequency synthesizers.

	[1]	[3]	[4]	[2]	PLL+ ILFT	PLL+ HBFT
Center frequency (GHz)	89	74	96	96.5	96	96
Divider ratio	/16, /32, /64, /128	/1024∼ /1984	/256	/64	/256	/256[c]
Tuning range (%)	6.7, 1	10.8	1.5	7.8	5.5	10.9
Phase noise 1 MHz (dBc/Hz)	−98, −82[a]	−83	−76	−102	−93	−92
Ref. spurs (dBc)	N/A	−49	−52	< −60	−54	−52
DC power (mW)	550	65	43.7	570	140	140
Technology	0.13 μm SiGe BiCMOS	65 nm CMOS	65 nm CMOS	0.13 μm SiGe BiCMOS	0.13 μm SiGe BiCMOS	0.13 μm SiGe BiCMOS

[a] Division ratio for the Ka-band PLL.
[b] Using /16 division ratio.
[c] Using /128 division ratio.

Table 10.2 Comparison of state-of-the-art CMOS ILFTs.

System parameter	[17]	[16]	[6]	This work
Technology (nm CMOS)	90	130	180	65
Frequency (GHz)	60	60	26.5	90
Supply voltage (V)	1	1.2	1.5	0.8
Locking range (GHz)	8 (13.3%)	0.8 (1.3%)	1 (3.8%)	3.1 (3.4%)
Output power (dBm)	−27	−15	−9	−13
Phase noise degradation 1 MHz (dB)	12.2		10.5	9.8
DC power (mW)	4.8	1.9	3.0	5.2

the ILO, which is locked by the third harmonic and suppresses all the other harmonics. The different I–V relation for HBT and CMOS transistor leads to different optimum bias conditions for the harmonic generator (HG). To maximize the third harmonic, the HBT needs be biased in the Class-AB region, whereas the NMOS transistor needs to be biased in the Class-C region. As for the ILO design, the CMOS ILO is based on the cross-coupled oscillator, while the SiGe ILO is based on a differential Colpitts oscillator since the design of cross-coupled oscillator in SiGe technology is constrained by the relatively low maximum achievable oscillation frequency, which is around 77 GHz in this technology [26]. Moreover, the supply voltage and power consumption for a CMOS ILO is lower than an SiGe ILO, meaning that the CMOS oscillator can achieve higher efficiency.

Table 10.1 summarizes performance of the two SiGe chips and compares them with other state-of-the-art W-band PLLs. Table 10.2 summarizes performance of the CMOS

T-ILFT chip and compares it with other state-of-the-art injection-locking-based CMOS frequency triplers.

10.7 Conclusions

In this chapter, different methods for W-band frequency synthesis were discussed and compared in terms of phase noise, tuning range, and output power. The optimum bias condition of HBT and MOS transistors for maximum harmonic generation are also investigated as a foundation for frequency tripler design.

Two different tripler topologies (i.e., HBFT and ILFT) together with a Ka-band PLL, have been demonstrated in a 0.18 μm SiGe BiCMOS. Both chips exhibit good phase noise and harmonic suppressions and consume the same amount of power, while the main trade-off is between tuning range and output power or conversion loss.

A W-band T-ILFT has been designed and implemented in 65 nm standard CMOS technology. The use of transformer enables optimum bias for the harmonic generator by decoupling it from the ILO and also reduces the source degeneration impedance of the ILO through impedance transformation. Based on the measurement results and analytical studies, the benefits of using a frequency tripler following a Ka-band PLL for W-band frequency generation were discussed and highlighted.

References

[1] S. Shahramian, A. Hart, A. Tomkins, et al., "Design of a dual W- and D-band PLL," *IEEE Journal of Solid-State Circuits*, vol. 46, no. 5, pp. 1011–1022, May 2011.

[2] S. Kang, J. C. Chien, and A. M. Niknejad, "A 100GHz phase-locked loop in 0.13μm SiGe BiCMOS process," in *2011 IEEE Radio Frequency Integrated Circuits Symposium*, IEEE, June 2011, pp. 1–4.

[3] Z. Xu, Q. J. Gu, Y. C. Wu, et al., "An integrated frequency synthesizer for 81-86GHz satellite communications in 65nm CMOS," in *2010 IEEE Radio Frequency Integrated Circuits Symposium*, IEEE, May 2010, pp. 57–60.

[4] K. H. Tsai and S. I. Liu, "A 43.7mW 96GHz PLL in 65nm CMOS," in *2009 IEEE Int. Solid-State Circuits Conference: Digest of Technical Papers*, IEEE, February 2009, pp. 276–277,277a.

[5] S. K. Reynolds, B. A. Floyd, U. R. Pfeiffer, et al., "A silicon 60-GHz receiver and transmitter chipset for broadband communications," *IEEE Journal of Solid-State Circuits*, vol. 41, no. 12, pp. 2820–2831, December 2006.

[6] C. Y. Wu, M. C. Chen, and Y. K. Lo, "A phase-locked loop with injection-locked frequency multiplier in 0.18-*muhboxm* CMOS for V-band applications," *IEEE Transactions on Microwave Theory and Techniques*, vol. 57, no. 7, pp. 1629–1636, July 2009.

[7] W. Winkler, J. Borngraber, B. Heinemann, and F. Herzel, "A fully integrated BiCMOS PLL for 60 GHz wireless applications," in *2005 IEEE International Solid-State Circuits Conference, 2005*, IEEE, February 2005, pp. 406–407, vol. 1.

[8] B. Sadhu et al., "A 28-GHz 32-element TRX phased-array IC with concurrent dual-polarized operation and orthogonal phase and gain control for 5G communications," *IEEE J. Solid-State Circuits*, vol. 52, no. 12, pp. 3373–3391, December 2017.
[9] C. C. Wang, Z. Chen, H. C. Yao, and P. Heydari, "A fully integrated 96GHz 22 focal-plane array with on-chip antenna," in *2011 IEEE Radio Frequency Integrated Circuits Symposium*, IEEE, June 2012, pp. 1–4.
[10] Z. Chen, C. C. Wang, H. C. Yao, and P. Heydari, "A BiCMOS W-band 2×2 focal-plane array with on-chip antenna," *IEEE Journal of Solid-State Circuits*, vol. 47, no. 10, pp. 2355–2371, October 2012.
[11] C. C. Wang, Z. Chen, and P. Heydari, "W-band silicon-based frequency synthesizers using injection-locked and harmonic triplers," *IEEE Transactions on Microwave Theory and Techniques*, vol. 60, no. 5, pp. 1307–1320, May 2012.
[12] Z. Chen and P. Heydari, "An 85-95.2 GHz transformer-based injection-locked frequency tripler in 65nm CMOS," in *2010 IEEE MTT-S International Microwave Symposium*, IEEE, May 2010, pp. 776–779.
[13] D. B. Leeson, "A simple model of feedback oscillator noise spectrum," *Proceedings of the IEEE*, vol. 54, no. 2, pp. 329–330, February 1966.
[14] J. C. Nallatamby, M. Prigent, M. Camiade, and J. Obregon, "Phase noise in oscillators – Leeson formula revisited," *IEEE Transactions on Microwave Theory and Techniques*, vol. 51, no. 4, pp. 1386–1394, April 2003.
[15] S. P. Voinigescu, M. C. Maliepaard, J. L. Showell, et al., "A scalable high-frequency noise model for bipolar transistors with application to optimal transistor sizing for low-noise amplifier design," *IEEE Journal of Solid-State Circuits*, vol. 32, no. 9, pp. 1430–1439, September 1997.
[16] M. C. Chen and C. Y. Wu, "Design and analysis of CMOS subharmonic injection-locked frequency triplers," *IEEE Transactions on Microwave Theory and Techniques*, vol. 56, no. 8, pp. 1869–1878, August 2008.
[17] W. L. Chan and J. R. Long, "A 56–65 GHz injection-locked frequency tripler with quadrature outputs in 90-nm CMOS," *IEEE Journal of Solid-State Circuits*, vol. 43, no. 12, pp. 2739–2746, December 2008.
[18] X. Zhang, X. Zhou, and A. S. Daryoush, "A theoretical and experimental study of the noise behavior of subharmonically injection locked local oscillators," *IEEE Transactions on Microwave Theory and Techniques*, vol. 40, no. 5, pp. 895–902, May 1992.
[19] H. Li and H. M. Rein, "Millimeter-wave VCOs with wide tuning range and low phase noise, fully integrated in a SiGe bipolar production technology," *IEEE Journal of Solid-State Circuits*, vol. 38, no. 2, pp. 184–191, February 2003.
[20] P. Wambacq and W. Sansen, *Distortion Analysis of Analog Integrated Circuits*. Kluwer, 1998.
[21] C. P. Lee, W. Ma, and N. L. Wang, "Averaging and cancellation effect of high-order nonlinearity of a power amplifier," *IEEE Transactions on Circuits and Systems I: Regular Papers*, vol. 54, no. 12, pp. 2733–2740, December 2007.
[22] B. Heydari, M. Bohsali, E. Adabi, and A. M. Niknejad, "Millimeter-wave devices and circuit blocks up to 104 GHz in 90 nm CMOS," *IEEE Journal of Solid-State Circuits*, vol. 42, no. 12, pp. 2893–2903, December 2007.
[23] T. H. Lee, *The Design of CMOS Radio-Frequency Integrated Circuits*. Cambridge University Press, 2004.

[24] S. A. Maas, *Nonlinear Microwave and RF Circuits*. Artech House, 2003.
[25] B. Razavi, “A study of injection locking and pulling in oscillators,” *IEEE Journal of Solid-State Circuits*, vol. 39, no. 9, pp. 1415–1424, September 2004.
[26] V. Jain, B. Javid, and P. Heydari, “A BiCMOS dual-band millimeter-wave frequency synthesizer for automotive radars,” *IEEE Journal of Solid-State Circuits*, vol. 44, no. 8, pp. 2100–2113, August 2009.

11 Digitally Intensive PLL and Clock Generation

Wanghua Wu and R. Bogdan Staszewski

The incessant demand for higher integration level and lower production cost has driven mm-wave electronics to be implemented in complementary metal-oxide semiconductors (CMOS). The unique properties of submicron CMOS technologies motivate the digitization of the mm-wave systems for improved radio frequency (RF) performance. Digitally intensive phase-locked loops (PLLs) have demonstrated superb performance over the conventional charge-pump PLLs in recent literature. This chapter focuses on a digitally intensive architecture using time-domain circuitry together with calibration techniques for mm-wave frequency generation. Two major digital PLL (DPLL) architectures that are suitable for wireless applications are elaborated here with design examples. The first demonstrator is a 60 GHz all-digital PLL (ADPLL)–based FM transmitter fabricated in a 65 nm bulk CMOS process. It consists of a 60 GHz digitally controlled oscillator (DCO) and a time-to-digital converter (TDC) for fractional-N synthesis and two-point frequency modulation. It proves viability of the DPLL in the 60 GHz band and achieves wideband frequency modulation. The second DPLL architecture employs a digital-to-time converter (DTC). A DTC-assisted DPLL targeted for IEEE 802.11ac applications has demonstrated 160 fs-rms jitter and the state-of-the-art figure of merit. Although this prototype is in the RF range, the same architecture can be applied to mm-wave band by using the mm-wave DCO demonstrated in the first design example. An alternative approach is to cascade the high-performance fractional-N DPLL with a simple integer-N PLL to multiply the RF output frequency up to the mm-wave band. The decision on direct frequency generation versus frequency multiplication-based approach depends on the frequency plan of the transceiver system, the output frequency range, the phase noise requirements, and the power consumption budget. These considerations in the mm-wave clock generation are also briefly discussed in this chapter.

11.1 Introduction to Digitally Intensive PLL

The mm-wave industry has been historically dominated by high-performance technologies intended for low-volume production in communication, security, and defense applications. However, transistor scaling extends the capabilities of CMOS circuits and systems into the mm-wave range, where the integration density and cost/volume advantages demonstrated by CMOS systems on a chip (SoCs) for cellular and wireless local area networks may be applied to mm-wave applications. With the 65 nm bulk

CMOS technologies in production offering peak transit frequency (f_T) and maximum frequency of oscillation (f_{MAX}) close to 200 GHz [1], several experimental 60 GHz transceivers achieving >4 Gbps data rate over a 2 m link have been reported [2–4]. These 60 GHz prototypes employing analog transceiver architectures have demonstrated the potential to use deep-submicron CMOS for RF/baseband cointegration. More importantly, there is room for improved RF performance (lower phase noise, higher output power, and etc.), and reduced power dissipation and chip area, as the analog RF circuits cannot fully share the benefits of CMOS scaling.

To maintain reliability when scaling metal-oxide-semiconductor (MOS) devices, the supply voltage has been reduced to <1 V, while threshold voltage remains almost constant (to suppress leakage current). This reduces the available voltage headroom when transistors are intended to operate as current sources. Moreover, metal capacitors of tens of picofarads, thus occupying considerable chip area, are usually required to integrate on-chip baseband filters and loop filters in charge-pump PLLs.

On the other hand, the digital gate density doubles and the basic gate delay improves linearly with every node of CMOS technology scaling (i.e., from 90 nm to 65 nm, then to 40 nm, and so on). The fast switching characteristics of CMOS logic (rising/falling time of 20 ps in 40 nm CMOS) enable high-speed clocks and fine control of timing transitions. The high density of digital logic (1 Mgates/mm^2) and SRAM (4 Mb/mm^2) makes programmable digital functions and software flexible and inexpensive in SoC applications. Thus, the digital signal processing is more amenable to integrated circuit (IC) implementation in the future compared to the voltage-domain operation [5], which motivates the digitization of frequency synthesizer design to fit into the deep submicron CMOS paradigm.

Figure 11.1 depicts a typical analog charge-pump PLL. The phase and frequency detector (PFD) estimates the phase difference between the frequency reference (FREF) input and the divided-by-N voltage-controlled oscillator (FDIV) clock by measuring the time difference between their respective closest edges, and generates either an up or a down current pulse whose width is proportional to the time difference measured. At the loop filter (LF), this current pulse is integrated onto capacitors $C_1 + C_2$ (which is approximately equal to C_1) to generate a control voltage (V_{Tune}), which sets the average frequency of the voltage-controlled oscillator (VCO). R_1 provides instantaneous phase correction without affecting the average frequency. The combination of C_2 and R_1 forms a first-order pole to smooth the dynamic voltage ripple on V_{Tune} due to the charge-pump and PFD noise, but may make the loop unstable. The combination of C_1 and R_1 forms a zero to stabilize the loop. In a charge-pump PLL, periodic glitches arise from mismatches between the width of up and down pulses produced by the PFD as well as charge injection and clock feedthrough mismatches between p- and n-type metal-oxide-semiconductor (NMOS) devices in the charge pump. These periodic glitches modulate the VCO output frequency, giving rise to spurious tones.

For wireless applications, fractional-N PLLs are often preferred if not outright required. A fractional-N PLL can achieve arbitrarily fine time-averaged frequency-division ratio, $N_{ave} = (N + F)$, by modulation between the instantaneous integer division ratios of N and $N + 1$, where F corresponds to the fractional part of the

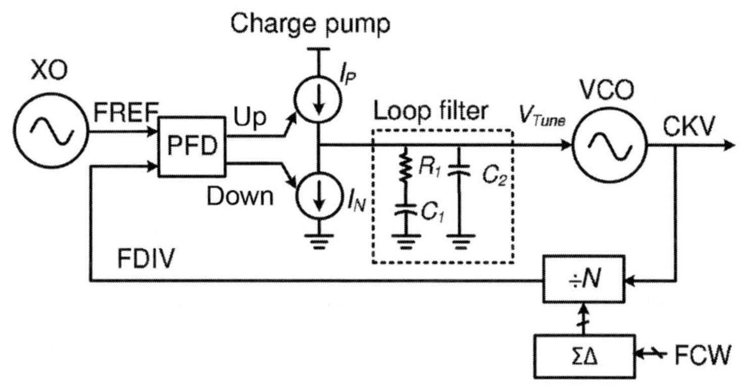

(a) Charge-pump fractional-N PLL

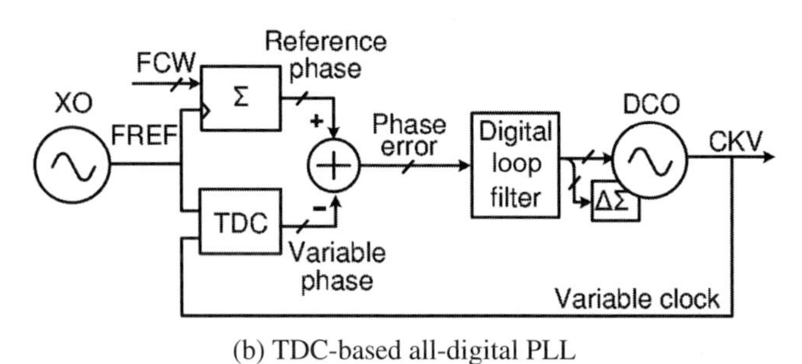

(b) TDC-based all-digital PLL

Figure 11.1 (a) Charge-pump PLL and (b) all digital PLL architectures.

frequency-division ratio. The phase detector will operate at a frequency of $f_{ref}+(F/N)\cdot f_{ref}$, and the phase error of the phase detector causes VCO fractional spurs at a multiple of the offset frequency $F\cdot f_{ref}$. One widely used method to suppress these spurs is a $\Sigma\Delta$-modulated clock divider, as shown in Figure 11.1, described by Miller and Conley [6] and Riley et al. [7], which trades the reduction in fractional spurs for the increase in the noise floor. In the fractional-N synthesizers, the output frequency can increment by fractions of the reference frequency, advantageously allowing the former to be much smaller than the latter. Compared to integer-N PLLs, this allows a wider loop bandwidth at the expense of fractional spurs, resulting in improved loop dynamics and attenuation of the oscillator-induced noise. The loop bandwidth of a fractional-N PLL is normally designed to be a few hundred hertz to sufficiently suppress the quantization noise of the $\Sigma\Delta$-modulator. Quantization noise cancellation techniques were demonstrated in several papers for charge-pump PLLs to extend the loop bandwidth without sacrificing the phase noise performance [8]. In practice, the gain of the digital-to-analog converter (DAC) in the compensation path is never perfectly matched to that of the signal path through the PFD and charge pump, so the cancellation of quantization noise is imperfect.

Although the charge-pump PLL is still the dominant architecture for mm-wave synthesizers, the standard analog PLL implementation is problematic in many applications, especially where the analog building blocks on a mostly digital chip pose design and verification challenges. The implementation cost is another concern. Because of the spur reduction requirements, the analog LF shown in Figure 11.1 usually requires large resistors and capacitors, most likely external to the IC chip, to achieve a low PLL bandwidth of several kilohertz. Realizing a monolithic capacitance on the order of a few hundred picofarads would require a prohibitively large area if implemented as a metal-oxide-metal (MOM) capacitor. Implementing it as an MOS capacitor would take less area, but it would probably be unacceptable because of its high leakage current and nonlinearity. The output impedance and the mismatch of the charge-pump currents are not improving with the CMOS scaling. Therefore, the analog-intensive PLLs do not lend themselves easily to silicon integration and lack portability from one process technology to another.

To strive for a better PLL implementation, migrating to a more digitally intensive PLL architecture has been proven possible in the past 10 years, and has begun to replace the charge-pump analog PLLs in many wireless applications [9–12]. Figure 11.1b depicts a simplified block diagram of a DPLL. A digital LF, which is compact and insensitive to transistor leakage current, replaces the analog LF in Figure 11.1a. The VCO is substituted by a DCO, and the phase error measurement is performed with the aid of a TDC converter subsystem. Thus, the aforementioned implementation difficulties associated with the charge pump and analog LF are avoided. However, the DCO and the TDC are analog circuits with digital interfaces that present new and different design challenges. More design detail of these two critical circuits will be elaborated in Sections 11.3 and 11.4.

The TDC-based DPLL operates in the digitally synchronous, fixed-point phase domain: the variable phase is obtained via a TDC, which measures and quantizes time differences between the FREF and DCO edges. The reference phase is determined by accumulating the frequency word control (FCW) on each FREF edge. Then, the sampled variable phase (sampled by FREF) is subtracted from the reference phase to obtain the digitized phase error, which is filtered by an LF and converted to a command word that tunes the DCO to the desired frequency. A $\Sigma\Delta$-modulator is often used to dither the least significant bit (LSB) of the DCO control word to obtain ultrafine frequency resolution (e.g., 100 Hz of a 5 GHz carrier).

The compact digital LF type and its coefficients can be dynamically configured during normal operation to control loop dynamics without disturbing the phase error (e.g., gear-shifting techniques [13]). Moreover, the FCW in Figure 11.1b, which is a fixed-point word to control the PLL frequency, can be changed dynamically to frequency/phase modulate the synthesizer output. The digitally intensive implementation facilitates the calibration of DCO tuning characteristic and TDC gain over process, voltage, and temperature (PVT) variations. Thus, wideband frequency modulation can be incorporated into the DPLL with less hardware overhead. In addition, the digitization of the phase error information makes many advanced calibration techniques possible. One good example is elaborated in Section 11.5 for ultralinear chirp generation for a 60 GHz frequency-modulated continuous wave (FMCW) radar transmitter. Compared to state-of-the-art FMCW generators [14–16], the digitally intensive architecture achieves wider modulation range for varying modulation slopes and better phase noise with lower power consumption. Another example, presented in [17], incorporates the pulling cancellation into the digital PLL to mitigate the pulling from power amplifier to the DCO so that a simple divide-by-2 frequency plan can be used for the Bluetooth transmitter to minimize the power consumption.

In a nutshell, digitally intensive PLL (often being named as all-digital PLL) digitizes the majority of the PLL blocks with an emphasis on using time-domain techniques to obtain scalability and a higher level of system performance. An "analog"-to-digital conversion takes place in the very early stage of the system (i.e., TDC) in order to benefit the most from the digital signal processing. Compared to the charge-pump-based analog PLLs, the digitally intensive PLLs increase the reconfigurability and testability of the PLL, harnesses the digital power of calibration to improve RF performance, reduces the design turnaround cycles by using automated digital implementation tools and flows, has lower parameter variability than with analog circuits, is easier to migrate between technology nodes, and may lead to a smaller silicon area and less power consumption by incorporating advanced calibration techniques, e.g., pulling cancellation [17], supply noise cancellation [18], spur cancellation [19], etc.

Note that "digitally intensive" doesn't mean that analog/RF design techniques are not important in the high-performance DPLLs. On the contrary, they are as crucial as before. The overall system performance is usually still dominated by the few analog building blocks. The essence of digitally intensive approach is to make the inputs/outputs (IOs) of the unavoidable RF/analog building blocks digital so that their analog nature does not propagate beyond the boundary, and thus the system can be modeled and analyzed in a digital way to make use of the advantages of digital design flow and digital signal processing power. Consequently, it requires the RF/analog designers to have knowledge of digital circuits and systems, to analyze the system from both analog and digital perspectives.

The following sections of the chapter focus on digitally intensive PLL for mm-wave frequency generation. A multirate TDC-based DPLL architecture for a mm-wave frequency synthesis and wideband frequency modulation is presented in Section 11.2. There are two critical time-domain circuits in the proposed mm-wave DPLL topology, i.e., a high-resolution mm-wave DCO and a TDC. They are elaborated in Sections 11.3 and 11.4, respectively. On-chip digital calibration techniques and experimental results of the 60 GHz multirate DPLL prototype are discussed in Section 11.5. Built-in self-test and self-characterization of the DPLL are described in Section 11.6. Section 11.7 introduces another major DPLL architecture, i.e., DTC-assisted DPLL, which is proposed to mitigate the design challenges in high-performance TDC and potentially further improve the integrated phase noise of the DPLL.

11.2 Multirate DPLL-Based Frequency Modulator Architecture

With the improved RF capability of nanoscale CMOS technology, digitally intensive frequency synthesis can now be explored in the mm-wave range, which is over 10× of the previously-proven frequency range. The DPLL architecture and implementation are not restricted to a particular application/standard, and are applicable to frequency

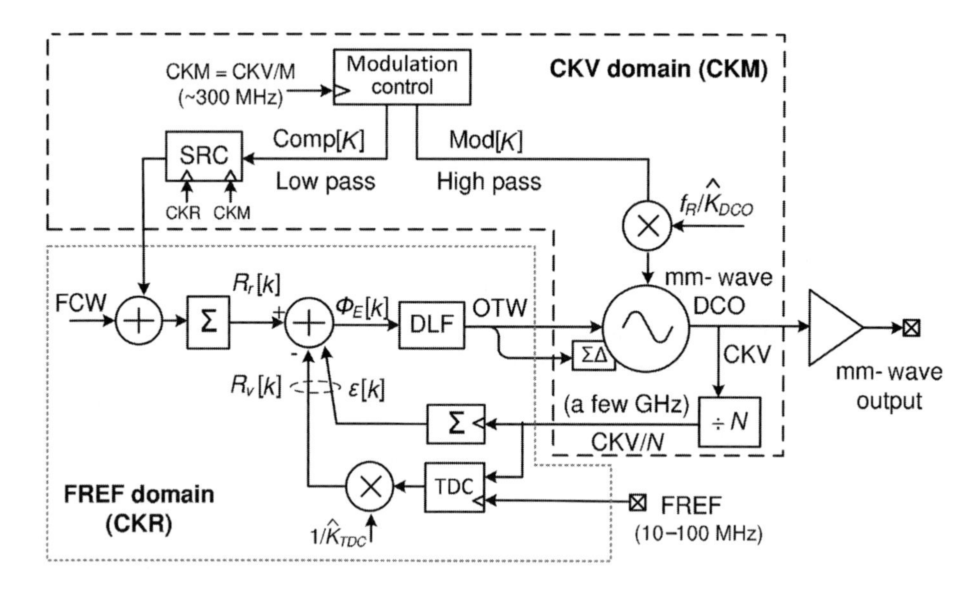

Figure 11.2 Multirate DPLL-based frequency modulator.

generation in various mm-wave systems, e.g., high data rate communication at 60 GHz, 77/79 GHz automotive radar, and imaging at 94 GHz.

The simplified block diagram of a mm-wave, DPLL-based FM modulator is depicted in Figure 11.2. The DCO operates directly in the mm-wave band (e.g., 60 GHz) and is followed by a prescaler to generate a feedback clock of several gigahertz for the TDC. The main part of DPLL operates synchronously in the phase domain at the reference clock rate (f_R). The underlying frequency stability of the system is derived from an external FREF crystal oscillator (f_R = 10–100 MHz). The FCW is defined as the desired frequency-division ratio f_V/f_R and is expressed in a fixed-point format. The variable-phase signal $R_v[k]$ is determined by counting the number of rising clock transitions of the divided DCO oscillator clock (CKV/N). The reference phase $R_r[k]$ is obtained by accumulating the FCW with every rising edge of the retimed FREF clock (CKR). The variable-phase $R_v[k]$ together with the fractional correction $\epsilon[k]$, is subtracted from the reference phase $R_r[k]$ in a synchronous arithmetic-phase detector. The $\epsilon[k]$ corrections by means of the TDC system increase the instantaneous phase resolution of the system to below the basic 2π radians of the variable phase. The digital phase error $\Phi_E[k]$ is conditioned by a reconfigurable LF. The LF can be a proportional attenuator, forming a type-I loop (i.e., only one pole due to the DCO frequency-to-phase conversion). The type-I loop generally features faster dynamics and is used for fast frequency/phase acquisition during the locking process. The LF can also be configured as proportional/integral (PI) controller to give rise to a type-II loop in order to offer better filtering of oscillator noise within the loop bandwidth. A high-order IIR filter can also be added to suppress the TDC and reference noise outside of the loop bandwidth, thus leading to improvements in the overall phase noise performance. This digital, phase-domain operation keeps the phase information in fixed-point digital numbers after conversion that cannot be corrupted further by noise. Consequently, the phase detector can be realized simply as an arithmetic subtractor that performs an exact digital operation without generating reference spurs, which is not the case in a charge-pump PLL. The dynamic range of the phase error could be made arbitrarily large by increasing the word length of the phase accumulators. This compares favorably to more conventional implementations, which typically are limited to only 2π of the comparison rate with a three-state phase/frequency detector [20].

The frequency modulation (FM) method is an exact digital two-point scheme. One data path (Mod[k] in Figure 11.2) directly modulates the DCO, while the other path (Comp[k]) compensates the frequency reference and prevents the modulating data from affecting the phase error. The former data path has a high-pass characteristic to the synthesizer output, while the latter path performs a low-pass filtering. When both paths are combined perfectly, an all-pass discrete-time transfer function is realized. The maximum data modulation rate is not limited by the PLL closed-loop bandwidth, and can be as high as one-half of the sampling rate in the modulation paths. If the modulation data paths are sampled at FREF, the maximum achievable modulation rate is limited to $f_R/2$ (e.g., 50 MHz for a 100 MHz crystal reference), which may not be sufficient for some wideband applications. To further boost the modulation rate, the direct modulation data path in Figure 11.2 can operate at a higher clock rate (CKM of e.g., ~300 MHz) obtained by a low integer division of the DCO output clock (CKV). Thus, the maximum achievable modulation data rate can be as high as one-half of CKM, which is independent of the phase detection rate of the DPLL (i.e., f_R). Sampling rate conversion (SRC) may be needed to synchronize the two modulation data paths that operate in different clock domains (i.e., CKR and CKM) in this multirate operation.

The two-point FM shown in Figure 11.2 actually operates in an open-loop fashion. The modulation data must be normalized accurately to the DCO gain (K_{DCO}, which is defined as the frequency tuning step in Hertz per LSB) in the direct modulation path (i.e., $f_R/\hat{K}_{DCO}$ in Figure 11.2, where f_R is the FREF frequency) in order to work properly. If the normalization is exact, the modulating transfer function is flat from DC to half of the sampling rate (CKM/2) in the z-domain, and has only a sinc-type response in the s-domain caused by the zero-order hold in the DCO interface. The exact K_{DCO} can be obtained from a digital calibration algorithm, which is explained in Section 11.5. In addition, there are potential timing misalignments between the two paths due to routing in the IC layout, which can be observed from the postlayout simulation results and should be compensated for in the design.

This FM capability extends the DPLL to a wideband frequency modulator, which can perform complex modulation in the polar domain in a digital transmitter. The digitally intensive architecture provides flexibility, reconfigurability, and transfer-function precision in order to meet the diverse and strict requirements imposed by mm-wave applications. A 60 GHz DPLL-based digital transmitter prototype has been implemented in 65 nm CMOS, employing the multirate frequency modulator architecture shown in Figure 11.2 [21,22]. The 60 GHz DPLL was designed for a short-range FMCW radar transmitter. The wideband frequency modulation capability and reconfigurability of the DPLL is especially attractive for this application. The following three sections will use this as a design example to elaborate on the mm-wave DCO and TDC design techniques and digital calibration algorithm required for high-performance mm-wave frequency synthesis. Before going to these advanced design techniques, it is noteworthy to explain the behavioral modeling and simulation approach for the digitally intensive PLL system, as it is quite different from the design approach for analog charge-pump PLLs.

While Simulation Program with Integrated Circuit Emphasis (SPICE)–based simulation tools are extremely useful for small RF circuits containing several components (such as an RF oscillator), their long simulation times prevent investigation of larger circuits (such as a digitally intensive PLL). Alternatively, the behavioral modeling and simulation environment based on a standard event-driven simulator (e.g., Verilog-AMS) is well suited for DPLLs with a fair amount of analog/RF circuitry. The DCO and TDC are modeled behaviorally using the same simulation engine as that used for the digital back-end, which is likely to contain thousands of gates. The main advantage of the single simulation engine at the top level is that it allows seamless integration of all hardware abstraction levels (such as behavioral, register transfer level [RTL], and gate level) in a uniform environment. This way, complex interactions and performance of the entire SoC could be validated and verified prior to tape-out. Examples include, the effect of the TDC resolution and nonlinearity on the close-in PLL phase noise and generated spurs, and the effect of the DCO frequency resolution and frequency tuning nonlinearity on PLL and transmitter performance.

11.3 High-Resolution mm-Wave DCOs

The TDC-based DPLL is now used in numerous wireless applications in the low-gigahertz frequency range [9,11,12]. However, synthesizers at mm-wave frequencies still rely on the charge-pump PLL topology, as high-resolution, wide-tuning range mm-wave DCOs were unavailable in the past. In the multirate DPLL architecture shown in Figure 11.2, the DCO converts the digital tuning word to an analog quantity of a tank resonant frequency, thus acting as a DAC. The TDC, on the other hand, translates the edge difference between the DCO and the reference clocks into a digital word, behaving as an analog-to-digital converter (ADC). These two building blocks typically dominate the out-of-band (due to DCO) and in-band (due to TDC) phase noise in a DPLL. Millimeter-wave DCOs must be capable of tuning across a wide range ($>10\%$) with fine frequency resolution (<1 MHz) and low phase noise (e.g., <-90 dBc/Hz at 1 MHz offset for a 60 GHz carrier). When used for a direct frequency modulation, linearity of the frequency tuning also becomes critical.

11.3.1 Distributed Switched Metal Capacitor Bank for mm-Wave DCOs

A conventional DCO developed for low-gigahertz oscillation is shown in Figure 11.3a, which consists of a large array of either MOS capacitors that operate in the flat region of the capacitance-voltage (C-V) curve or metal-to-metal-based switchable capacitors [23]. The total switchable capacitance consists of several subbanks with different tuning

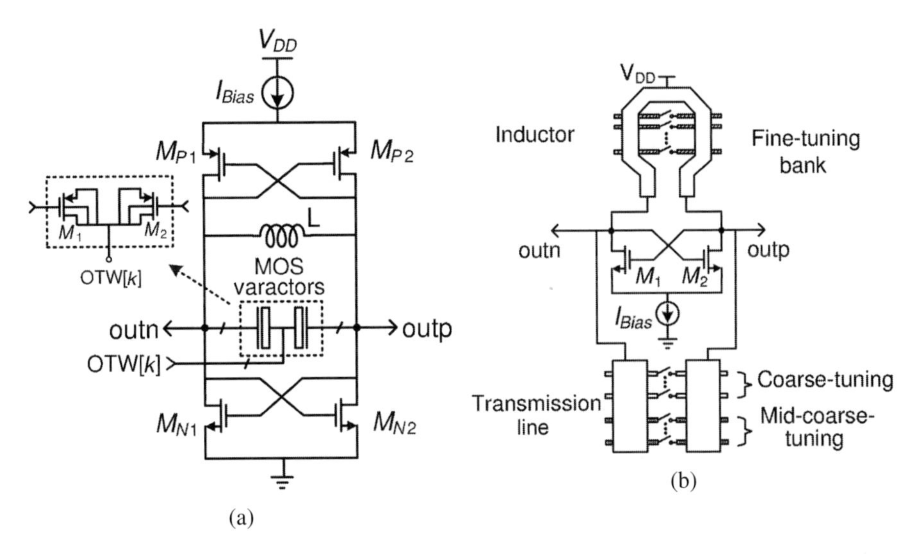

Figure 11.3 (a) Schematic of a conventional RF DCO (b) schematic of an mm-wave DCO with distributed switched metal capacitors [25]. (©2013 IEEE. Reprinted, with permission, from *IEEE Journal of Solid-State Circuits*.)

step sizes to obtain ~20% tuning range and fine-tuning steps simultaneously. The tank quality factor (Q) is dominated by the on-chip inductor Q (typically 10–20), whereas the varactor banks typically have a Q-factor over 100 when operating below 10 GHz.

The scenario is different for an LC tank operating at mm-wave frequencies. For example, the typical inductance (L_0) and capacitance (C_0) values suitable for an IC implementation of a 60 GHz LC-oscillator are 90 pH and 70 fH, respectively. The simulated Q-factor of a 50 capacitance-voltage (C-V) fF MOS capacitor in a 65 nm CMOS is as low as 5 at 60 GHz, which severely affects the phase noise of a mm-wave DCO. Parasitic capacitance from interconnections contribute a significant fixed capacitance to the DCO tank (e.g., 30 fH out of C_0 of 70 fH), which reduces the capacitive tuning ratio of the varactor (C_{max}/C_{min}) and results in a fractional tuning range smaller than 10% in practice [24]. Besides the tuning range and tank Q-factor degradations, the frequency resolution (Δf_0) obtained by digitally switching the minimum-sized MOS capacitor, which is used in low-GHz DCOs, is not sufficient for mm-wave DCOs. The finest varactor step size made possible by the fine lithography is on the order of 40 aF in nanoscale CMOS, which corresponds to 12 kHz frequency step size at the 2 GHz DCO output. When used in the aforementioned 60 GHz LC tank, the resultant frequency resolution becomes ~17 MHz for a 60 GHz carrier, which contributes significant quantization noise when integrated in an DPLL (i.e., $f_0 = 1/(2\pi\sqrt{L_0C_0})$), assuming $\Delta f_0/\Delta C_0 \approx \partial f_0/\partial C_0 = -f_0/(2C_0)$, thus $\Delta f_0 = 60\,\text{GHz}/(2 \cdot 70\,\text{fF}) \cdot (40\,\text{aF}) = 17\,\text{MHz}$). Therefore, it is very challenging to design a wideband mm-wave DCO with a tuning range above 10% and a fine-tuning step of less than 1 MHz.

The MOS-varactor-based DCO is not suitable for mm-wave DCOs. A more suitable tank topology for mm-wave DCOs is to use distributed switched-metal capacitors, as shown in Figure 11.3b, which can achieve wider tuning range and better phase noise

(PN) [25]. The distributed LC tank consists of a transmission line (TL), inductor or transformer (e.g., Figure 11.3b uses a TL and inductor), and pairs of metal shield strips located beneath the resonator and distributed along its major dimension in various metal layers. These metal strips form digital tuning banks that are distributed along the length of the resonator. Each metal shield strip pair is connected to an MOS switch driven by a digital tuning signal. Activating the switch varies the capacitive load on the resonator and introduces a distinct phase shift in the DCO loop that varies the oscillator frequency [26]. Moreover, the phase shift introduced by each metal strip pair varies with its position along the resonator, the metal layer used for implementation, and its physical dimension (width and spacing). This attribute is further exploited to form coarse- and fine-tuning banks (i.e., tuning band segmentation), and thus to optimize the tuning range and frequency resolution, simultaneously.

For coarse-tuning, a digitally controlled TL is used. Its 3D view is depicted in Figure 11.4a. The oscillation signal (e.g., at 60 GHz) runs along the TL in thick top metal to reduce losses from the conduction from the silicon substrate. Shorting metal strips beneath the differential TL via NMOS switches increases the capacitance per unit length, thus reducing the wavelength ($\lambda = 1/(f\sqrt{LC})$) of the RF signal. This increases the phase shift along the TL and reduces the tank's resonant frequency. The tuning elements are connected via the signal path of the resonator in top metal without any additional interconnecting wires, which eliminates the complex wiring scheme required for conventional varactor tuning and reduces the wiring capacitance. The design of the NMOS switch (see Figure 11.4b) involves a trade-off between the tuning range and Q-factor, which is further exacerbated by the increased operating frequency [27].

The switched-metal capacitors comprising the coarse-tuning and mid-coarse-tuning cells (see Figure 11.3b) share the same temperature coefficient, which simplifies the calibration procedure when used in a DPLL, and especially when it undergoes a direct frequency modulation. Electromagnetic (EM) simulations are required for the entire resonator structure to capture the distributed LC effects, including all metal switched-capacitor pairs and the inductor/TL/transformer as part of the DCO design procedure.

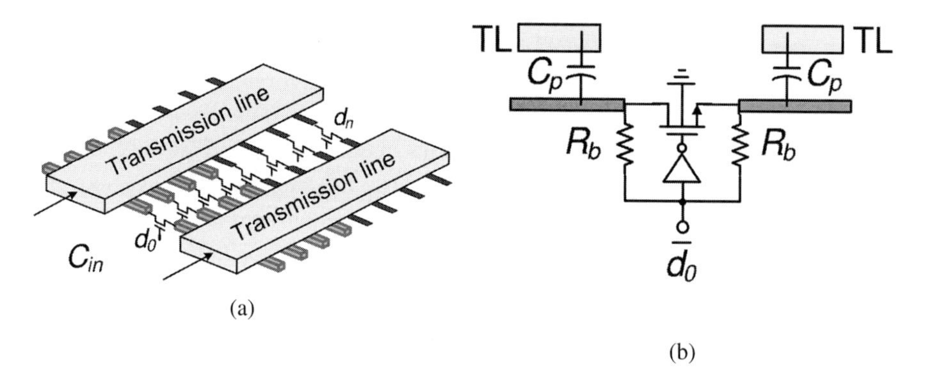

Figure 11.4 (a) Three-dimensional view of a digitally controlled TL for a coarse-tuning bank and (b) schematic of the NMOS switch [25]. (©2013 IEEE. Reprinted, with permission, from *IEEE Journal of Solid-State Circuits.*)

The tank losses and DCO tuning characteristics can only be ascertained fairly accurately via EM simulations of the entire resonator physical layout (including the unwanted capacitive coupling between adjacent unit-tuning cells). The MOS switches cannot be included in an EM simulation, but they are added in subsequent circuit simulations of the DCO to analyze the tuning linearity and the effect of switch losses.

11.3.2 Transformer-Coupled Fine-Tuning Bank

Minimum-size switched MoM capacitors employed for fine-tuning in [28] realized a frequency tuning step of 1.8 MHz for a 53 GHz DCO. However, the series parasitic capacitance of the MOS switch and the interconnections within the capacitor bank are much higher than the minimum MOM capacitance, which affects precision and matching. The intention is to generate fine-tuning steps without employing minimum-size structures so that the interconnection parasitics do not limit the frequency step size and uniformity. This can be achieved by a transformer-coupled technique, which achieves fine-tuning via magnetic coupling between the primary and a switched-metal capacitor bank placed beneath the secondary coil. Its operation principle is illustrated in Figure 11.5.

The tunable resonator consists of a transformer and a tunable load capacitor (C_L) connected to its secondary coil, as shown in Figure 11.5a. Resistor R_L models the losses of C_L. The primary coil of the transformer is connected directly to the oscillator core

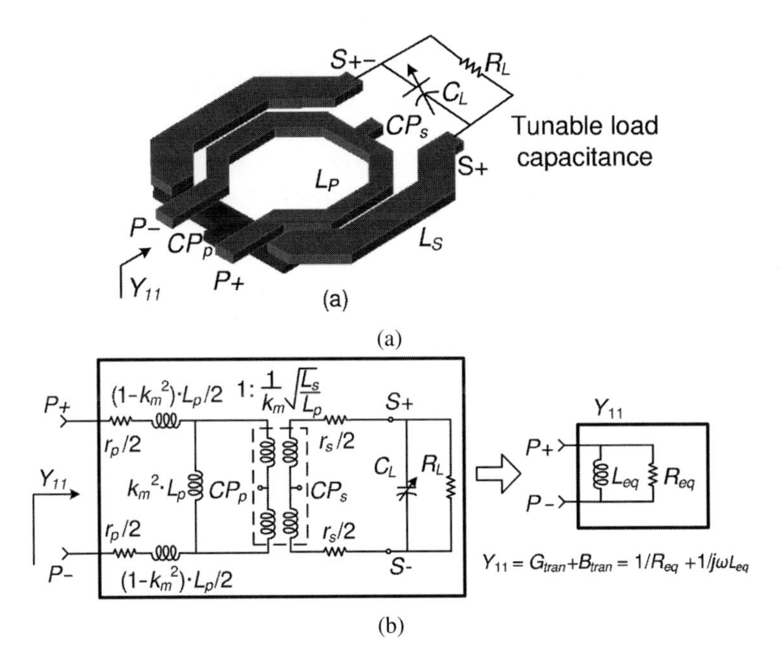

Figure 11.5 (a) Illustration on transformer-coupled fine-tuning technique and (b) a simplified lumped-circuit model for (a) [25]. (©2013 IEEE. Reprinted, with permission, from *IEEE Journal of Solid-State Circuits.*)

and a coarse switched-capacitor bank. The tunable transformer is analyzed as a one-port network for the admittance seen at the primary terminals (Y_{11}). The real part of Y_{11} (i.e., conductance G_{tran}) models the transformer losses. The imaginary part (i.e., inductive susceptance B_{tran}) in combination with the capacitive susceptance B_1 seen at the primary coils from the rest of oscillator determines the oscillation frequency, f_{osc}, such that $B_{tran} + B_1 = 0$. The coupling factor between the primary (L_p) and secondary (L_s) coils is k_m. When either C_L or R_L changes, the change in admittance is reflected back to the primary coil and varies B_{tran}, thereby altering the oscillation frequency. The susceptance seen across the primary terminals (i.e., $L_{eq} = 1/j\omega B_{tran}$) can be varied in ultrafine steps even when the discrete tuning steps in C_L are moderate.

Numerical analysis is required to determine the Y_{11} and L_{eq} accurately since parasitic capacitances are difficult to determine precisely at high frequencies, and capacitive effects are best investigated from simulating a particular case. However, some qualitative observations on the behavior of tunable transformers can be made from a simplified lumped-element circuit model, shown in Figure 11.5b.

The parasitics to the substrate are neglected when analyzing the impedance transformation from the secondary to primary. The conductor losses r_s and r_p have a negligible effect on B_{tran} and are also ignored in the following analysis. Assuming that C_L is lossless (i.e., R_L very large), the equivalent inductance determined from Im[Y_{11}] is given by (11.1).

When C_L is connected to L_s:

$$\begin{aligned} L_{eq}|_{R_L=\infty} &= L_p\left(1 + k_m^2 \frac{w^2 L_s C_L}{1 - \omega^2 L_s C_L}\right) \\ &\Longrightarrow L_p(1 + k_m^2 \omega^2 L_s C_L), \quad \text{when } w^2 L_s C_L \ll 1. \end{aligned} \tag{11.1}$$

For comparison, the terminal inductance L_{eq} seen when C_L is connected directly to the primary coil is given by (11.2).

When C_L is connected to L_p,

$$\begin{aligned} L_{eq}|_{R_L=\infty} &= \frac{L_p}{1 - \omega^2 L_s C_L} \\ &\Longrightarrow L_p(1 + \omega^2 L_p C_L), \quad \text{when } w^2 L_p C_L \ll 1. \end{aligned} \tag{11.2}$$

Placing capacitor C_L across the secondary coil results in the same L_{eq} as when a capacitor of value equal to $C_L \cdot (k_m^2 L_s/L_p)$ is connected to the primary turn. In other words, the tuning sensitivity is attenuated by a factor of $k_m^2 L_s/L_p$, which can be much smaller than unity for a weakly coupled transformer (e.g., 0.04 for $k_m = 0.2$). Therefore, fine-tuning of L_{eq} is possible using a capacitor bank with a moderate tuning step size. Furthermore, L_{eq} increases linearly with increasing C_L when the self-resonant frequency of the secondary coil ($1/\sqrt{L_s C_L}$) is much higher than the desired operating frequency, ω, (i.e., $\omega^2 L_s C_L \ll 1$). Thus, an FB with a uniform tuning step can be achieved using a unit-weighted capacitor bank for C_L.

To investigate the effect of losses, variations in the primary admittance Y_{11} and L_{eq} (at 60 GHz) across C_L and R_L are simulated and plotted in Figure 11.6. A single-

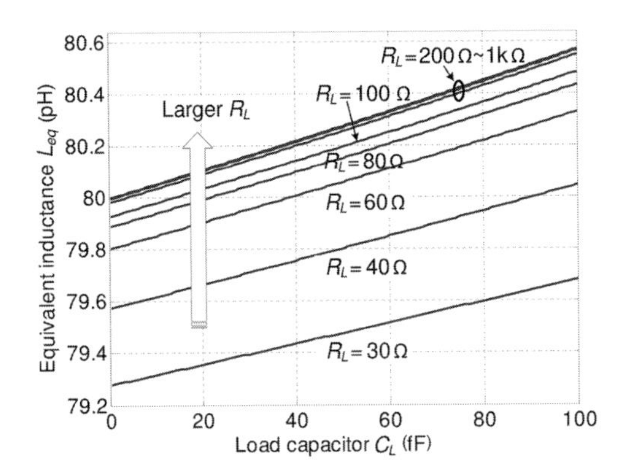

Figure 11.6 L_{eq} for variations in R_L and C_L at transformer secondary winding [25]. (©2013 IEEE. Reprinted, with permission, from *IEEE Journal of Solid-State Circuits.*)

turn transformer with $L_p = 80\,\text{pH}$ and $L_s = 60\,\text{pH}$ at 60 GHz, and k_m of 0.2 is used in the simulation. As seen in Figure 11.6, the previous results derived assuming a lossless tuning capacitance C_L are still valid when R_L is larger than 200 Ω (i.e., L_{eq} is insensitive to R_L when $R_L > 200\,\Omega$), which is satisfied easily in a practical DCO. For a C_L of 10 fF in the FB, R_L ranges between 500 Ω and 1 kΩ at 60 GHz (i.e., Q-factor of 10–20).

The losses of the tunable transformer (i.e., $\text{Re}(Y_{11}) = G_{tran}$) can be modeled by a resistor ($R_{eq} = 1/G_{tran}$) in parallel with L_{eq}. Figure 11.7 depicts the simulated R_{eq} for variations in C_L and R_L. It remains above 1 kΩ when R_L is higher than 200 Ω, indicating a negligible effect on the total tank Q-factor. On the other hand, for a small R_L (less than 100 Ω), L_{eq} depends not only on C_L but also increases rapidly with R_L, as shown in Figure 11.6. This attribute can also be employed to implement a variable inductor and a wide tuning range, but at the cost of poor tank Q-factor [29]. It is more desirable to vary C_L rather than R_L in order to achieve high DCO frequency resolution with less degradation in the tank Q-factor.

In order to obtain a linear frequency tuning characteristic, it was shown that capacitor C_L should satisfy the condition $\omega^2 L_s C_L \ll 1$. The capacitance attenuation factor can be increased either by reducing the ratio of secondary to primary inductance (L_s/L_p), or by reducing the coupling coefficient k_m. However, k_m cannot be made lower than 0.1 because the transformer bandwidth also depends upon k_m [30], and it should be wide enough to cover the entire tuning range of the oscillator (e.g., 10% of 60 GHz for the DCO design example shown next).

11.3.3 A 60 GHz DCO Design Example

A simplified schematic of the DCO used in the 60 GHz DPLL prototype is shown in Figure 11.8 with the detailed implementation of the transformer-coupled FB illustrated in Figure 11.8b. An NMOS cross-coupled pair ($M_{1,2}$) sustains the oscillation. The DCO

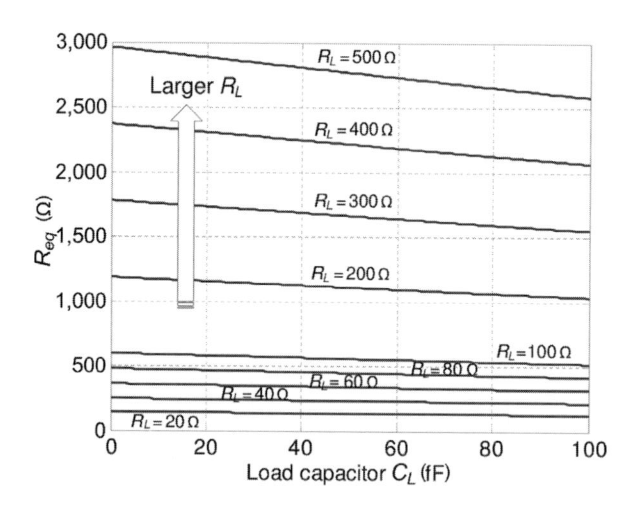

Figure 11.7 R_{eq} for variations in R_L and C_L at transformer secondary winding [25]. (©2013 IEEE. Reprinted, with permission, from *IEEE Journal of Solid-State Circuits*.)

is segmented into three banks, each with a linear tuning characteristic: coarse-tuning bank (CB), FB, and mid-coarse-tuning bank (MB), which bridges the gap in step sizes between CB and FB. The CB and MB are integrated with the TL and configurable floating metal shield to form a compact, digitally controlled frequency tuning scheme. A smaller tuning step is attained by placing metal strips on (lower) metal M6 compared to the coarse-tuning strips on metal M7. The variable capacitor load at the secondary coil is implemented as another digitally controlled differential TL with a much smaller tuning step compared to the one used for CB to form the fine-tuning bank. The weak mutual coupling factor k_m of 0.28 further attenuates its frequency tuning sensitivity by a factor greater than 10. The transformer and the TL-based tuning bank are codesigned using EM simulations to achieve the required Δf with a high Q-factor. Unwanted coupling between adjacent metal strips adds nonlinearity to the tuning curve. It is minimized by optimizing the width of the metal and the gap between adjacent strips with the aid of the EM simulator, EMX, from Integrant Software. The simulated Q-factor of the transformer-based FB is 16.5 in the 60 GHz band and varies by ± 0.03 across the tuning range.

When employed in a DPLL, the FB of the DCO serves two purposes, i.e., to track the DCO frequency drift and to apply frequency modulation. To optimize the DPLL operation in both continuous-wave (CW) and FM modes, the FB is split into two parts as shown in Figure 11.8b; FB_{Mod} at the center of the TL is dedicated for frequency modulation, and FB_{Loop}, located above and below FB_{Mod}, is used to correct DCO frequency wander in the loop at slow rates. In this way, only K_{DCO} of FB_{Mod} needs to be well matched (e.g., $<5\%$ mismatch) and applied to the $f_R/\hat{K}_{DCO}$ multiplier in the direct modulation path of Figure 11.8. The FB_{Loop} can tolerate more tuning-step mismatch (e.g., 15%) and only a rough approximation of the DCO transfer function gain, K_{DCO} (e.g., within $\sim$20 %) is required to establish an acceptable range for the

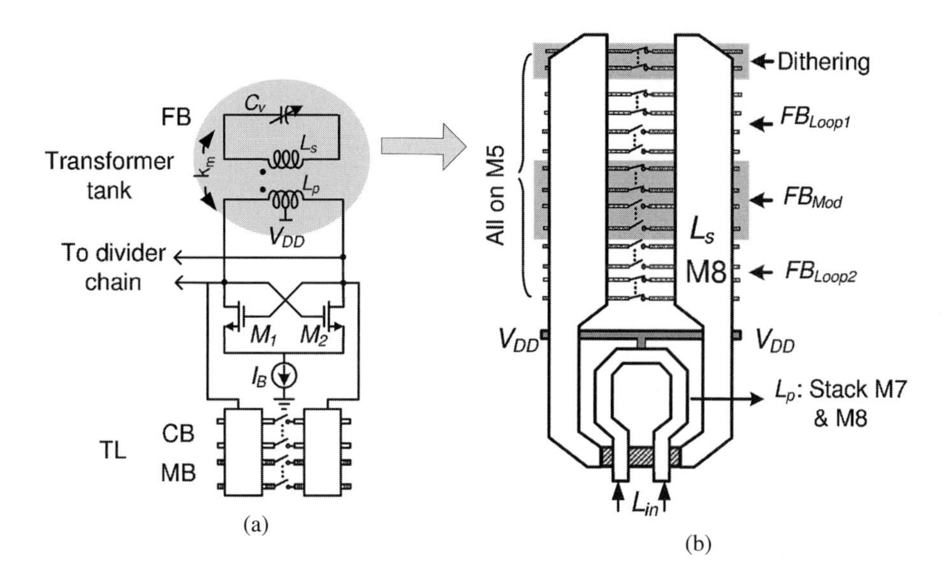

Figure 11.8 (a) Simplified schematic of the 60 GHz DCO [25]. (©2013 IEEE. Reprinted, with permission, from *IEEE Journal of Solid-State Circuits*) (b) top layout view of a T-DCO fine-tuning bank.

DPLL loop bandwidth. The loop bandwidth affects mainly the settling time and noise rejection of the PLL, so a 5–25% variation would have a minimal effect on the system performance.

The measured DCO oscillation range is from 56.4 to 63.4 GHz with coarse-tuning of 400 MHz/bit, mid-coarse-tuning of 35 MHz/bit, and fine-tuning of 1.64 MHz/bit. The measured frequency resolution (K_{DCO}) of the CB for each thermometer code bit is shown in Figure 11.9 for five IC samples. The CB bank achieves an average K_{DCO} of 400 MHz/bit and an in-band mismatch of 15%, employing the digitally controlled TL. The measured mismatch of FB_{Mod} (see Figure 11.10) is within 5%.

The PN of the free-running 60 GHz DCO is measured from the divide-by-64 test output, shown in Figure 11.11. The measured PN from a 965 MHz carrier is −127.8 dBc/Hz at 1 MHz frequency offset and an extra 36 dB ($20 \log_{10} N$) should be added to account for the division ratio of N (64 in this case) to obtain the equivalent PN at 60 GHz output, i.e., −92 dBc/Hz at 1 MHz offset.

11.4 Time-to-Digital Converter

A TDC is used as the phase/frequency detector and charge-pump replacement in a digital PLL, as shown in Figure 11.1. A number of TDC architectures have been proposed for implementing DPLLs to obtain fine time resolution and good linearity [12,31–35]. The delay line-based TDC will be discussed in detail in this section as it is the simplest TDC

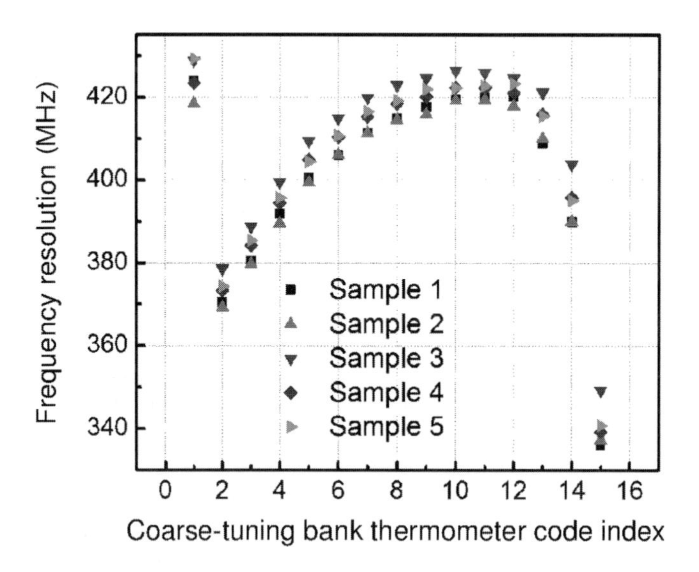

Figure 11.9 Measured K_{DCO} for each bit in coarse-tuning bank.

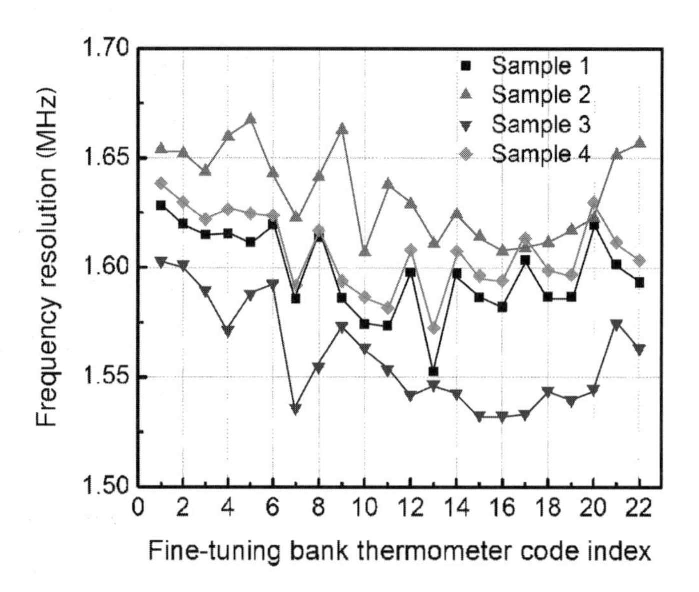

Figure 11.10 Measured K_{DCO} for FB_{Mod} over different samples (CB = 8).

implementation and serves as the basics of all other TDC circuits. The resolution of delay line TDCs [31] is limited by the achievable gate delay of the process technology at hand, making resolution below 10 ps difficult to achieve. To further improve the resolution, a Vernier TDC [32] can be employed, which takes use of the delay differences rather than the absolute delay. Since the resolution is set by the delay differences, mismatch between the delay lines gets amplified relative to the Vernier LSB. To ensure monotonicity, the delay lines are typically sized to achieve of less than 0.5 LSB. This sets the minimum power dissipation and area, which does not scale well with process technology. A 2D Vernier TDC [10] can reduce the number of delay elements. However, the TDC transfer function is still nonlinear due to the mismatch. Gated ring oscillators (GROs) [33,34] can serve as TDCs with noise-shaping characteristics, thus achieving time resolution of subpicoseconds at low frequencies. The quantization noise is shaped to higher frequencies and is filtered by the loop filter. Due to leakage, the GRO internal state could vary during the off-state and it translates to an elevated noise floor, which often limits the in-band phase noise of the DPLL.

In the 60 GHz DPLL prototype, the required TDC resolution is 12 ps for FMCW radar applications. Therefore, a pseudodifferential delay chain–based TDC architecture was employed for its simplicity. The created variable phase signal is a fixed-point digital word in which the fractional part is measured with a resolution of an inverter delay (~12 ps in 65 nm CMOS) by means of the TDC core, as shown in Figure 11.12 (bottom). The divided down DCO clock (CKV) gets delayed by a string of inverters whose outputs are sampled at the rising edge of FREF. A pseudodifferential delay chain is adopted to avoid mismatch between rising and falling edge transitions due to differing strengths of the NMOS and PMOS transistors. The delayed-clock replica vector is sampled by FREF using an array of 50 sense-amplifier-based flip-flops (SAFF) that are adapted from [36]. The 50-bit TDC output forms a pseudo thermometer code,

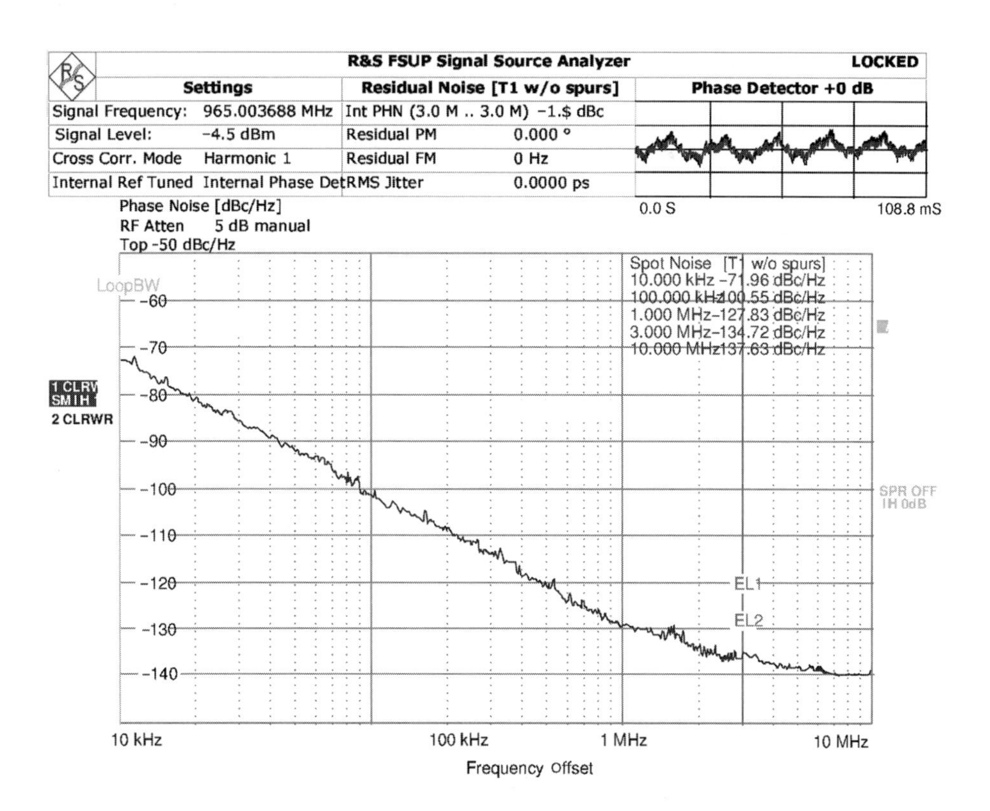

Figure 11.11 Measured 60 GHz DCO PN from divide-by-32 test output.

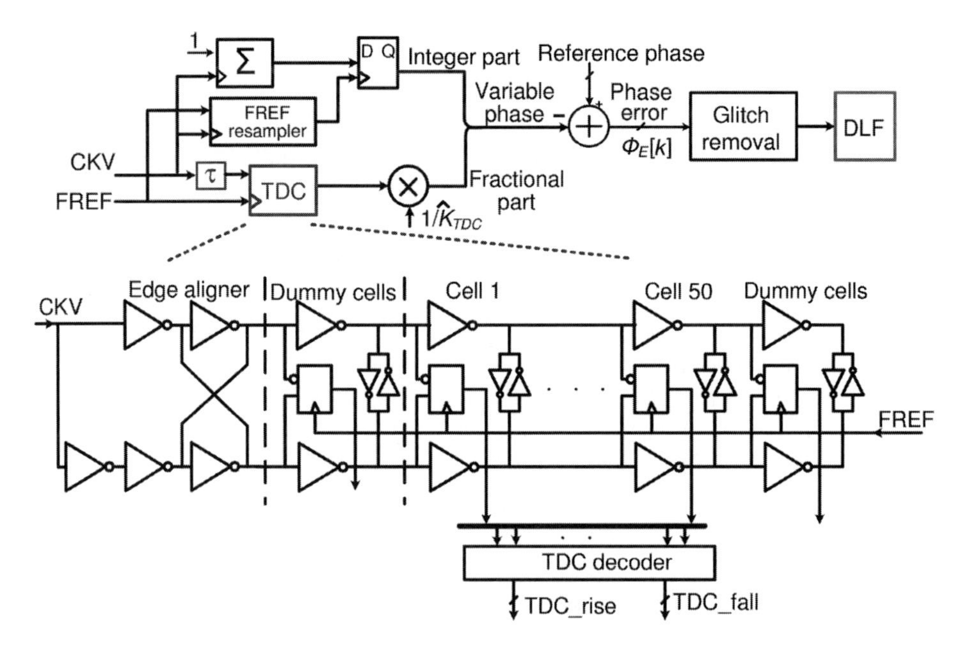

Figure 11.12 Simplified schematic of an inverter-chain-based TDC core.

which is then converted to binary (TDC_{rise} and TDC_{fall}) using a simple digital priority decoder [31]. The number of inverters is set to cover one T_v, which is the period of CKV. To increase the dynamic range arbitrarily, an edge counter (CKV) with sufficient word length is added, thus contributing the integer part of the variable phase, as shown in Figure 11.12 (top).

To compensate for any phase difference in the input high frequency clock (CKV) due to routing mismatches, they are first edge-aligned and then passed through a complementary string of 50 inverters. The edge aligner can tolerate up to 70 ps of skew, so the negating feed can simply be replaced with an inverter. Four dummy cells are placed at both the beginning and end of the TDC delay chain to improve the mismatch of unit delay cells. The simulated TDC time resolution t_{res} varies from 8.648 ps (the best case: fast process corner, 1.3 V supply, and 0°C) to 22.43 ps (the worst case: slow process corner, 1 V supply, and 100°C) over process, voltage, and temperature (PVT) variations. The mean value of t_{res} is 12.1 ps for the entire TDC delay chain with a standard deviation of 0.9 ps, derived from Monte Carlo simulations. The t_{res} for even and odd cells are 11.94 and 12.25 ps, respectively, due to the small asymmetric rising edge and falling edge of the unit inverter cell with layout parasitics. This even–odd mismatch can be calibrated and compensated in digital domain to reduce the fractional spurs.

In PLL applications, the absolute phase is more useful than the instantaneous frequency deviation. Also, the reference edge locations are quite predictable, so the power is significantly saved by gating off the TDC activity during 90% of the time between the reference edges, as shown in Figure 11.13. The TDC is also self-calibrating during the regular operation for the PVT inverter delay variations, as shown in Figure 11.14. The absolute difference between the measured rising-edge and falling-edge delays of CKV to FREF is the half-period of CKV in terms of number of inverters, i.e., half of $T_v/\Delta t_{inv}$. An accurate estimate of $T_v/\Delta t_{inv}$ is obtained through averaging over 2^{10} samples, with an error below 1%. Its inverse is used for the fixed-point period normalization multiplier with 19 fractional bits (W_F). This value divided by the CKV frequency, $1/T_v$, is the inverter delay in units of seconds. The normalized TDC output produces the fractional part of the variable phase in Figure 11.12, which will be used in the phase error detection.

The fixed-point TDC output timestamp (i.e., variable phase) consists of the sampled CKV edge count (integer part) and the normalized delay from CKV to FREF (fractional part). The FREF clock provides triggering moments that sample both the counter and TDC outputs. These different sampling instants could have a timing misalignment τ, indicated in Figure 11.12, and thus cause glitches in the phase error when the counter and TDC outputs are combined. Instead of correcting these glitches [35], they are removed by digital signal processing. Digital logic first detects a glitch by comparing the current phase error to that in the previous clock cycle. If the difference is larger than a threshold (e.g., 0.5), the input is assumed to contain the glitch. The phase error is frozen for this clock cycle by disregarding the current phase error to obtain a glitch-free output. In addition, the same logic can be reused as a lock indicator, or to generate a clock quality monitoring signal by setting a different comparator threshold.

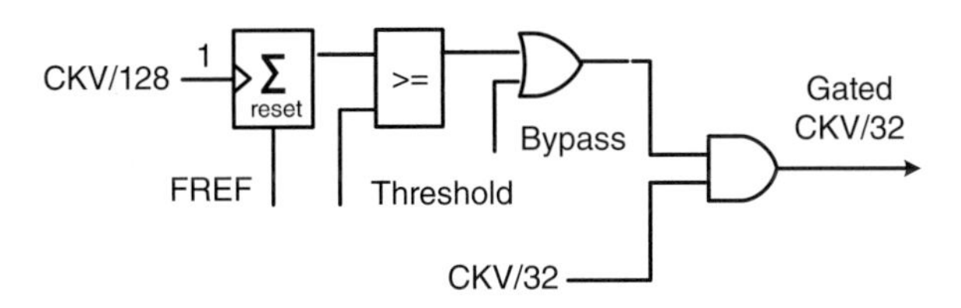

Figure 11.13 Clock gating logic in TDC to save power.

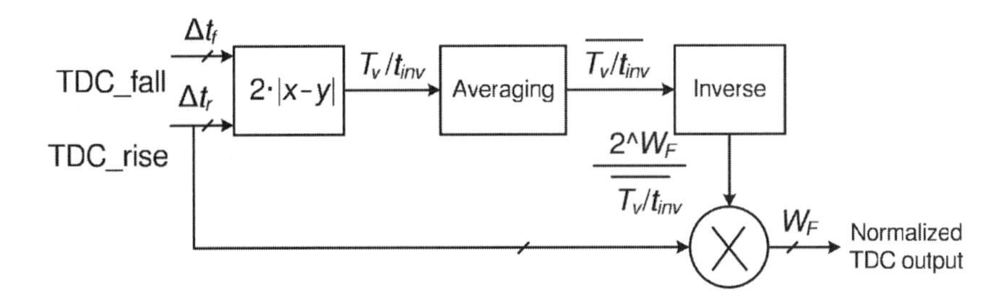

Figure 11.14 TDC gain calibration.

11.5 Digital Calibration Techniques for High RF Performance

The major calibration required in a high-performance DPLL consists of a DCO gain (i.e., DCO tuning step size) calibration, TDC gain calibration, and a calibration related to the two-point frequency modulation. As the DCO and TDC gains are calibrated, the closed-loop transfer function of the PLL can be controlled precisely to obtain the optimum integrated phase noise. The two-point FM-related calibration is the key to achieve a linear wideband modulation. A modulation range of several gigahertz is required for many mm-wave applications, such as FMCW radar and high-data-rate communications. Ideally, a single tuning bank with a constant K_{DCO} across the modulation range would be achieved. However, the DCO tuning must be segmented into CBs and FBs (i.e., each with different K_{DCO}) to practically realize both high resolution and a wide tuning range, as shown in the 60 GHz DCO design example. The measured tuning step mismatches for a 60 GHz transformer-coupled DCO prototype is ≈15 % in CB (see Figure 11.9), which is much larger than in FB (see Figure 11.10) since dummy cells are not employed there due to the limited LC budget when oscillating at 60 GHz. Moreover, the K_{DCO} in FB varies with the CB tuning word due to the wide coarse-tuning range (e.g., 7 GHz). When the capacitance increases by ΔC, the oscillation frequency (f_o) will decrease by Δf, or approximately $\Delta C \cdot f_0/(2C_0)$ (i.e., $\Delta f_0/\Delta C_0 \approx \partial f_0/\partial C_0 = -f_0/(2C_0)$). Therefore, Δf will vary with f_0 even for the same ΔC, which is the case for modulation frequency range up to a few gigahertz. As discussed in Section 11.2, the two-point modulation scheme relies on accurate DCO gain (K_{DCO}). Thus, not only does the DCO gain of the fine-tuning banks need to be calibrated, its tuning characteristics over multiple tuning banks need to be known and linearized to achieve linear frequency modulation across several gigahertz, as shown in Figure 11.15. The frequency step size mismatch within the fine-tuning bank is also critical when striving for an ultralinear FM. These techniques will be elaborated in this section.

11.5.1 DCO Gain Calibration and Linearization

The oscillator gain (K_{DCO}) dependence on PVT and frequency makes it necessary to estimate it on an as-needed basis within the actual operating environment. A digital normalization algorithm that measures the phase error present in the loop due to DCO control word change can be used to calibrate the K_{DCO} of a linear FB [5]. Alternatively, adaptive gain compensation by a sign-least mean square (LMS) loop [37,38] calibrates the K_{DCO} in the background without interrupting normal frequency modulation. However, they are difficult to apply to an mm-wave DCO for an FMCW application as both CB and FB are used for modulation. Each bit in CB has a different K_{DCO}, and the K_{DCO} in FB also varies with CB settings as explained earlier. It requires more than 10 K_{DCO} values to be calibrated in the background, which makes the adaption algorithm too complicated to implement and it may not converge to a stable solution. Correcting K_{DCO} via a look-up table for individual bits in each bank employing open-loop calibration algorithms [38] requires a long calibration time (up to hours) and an unacceptably large look-up table for a gigahertz range.

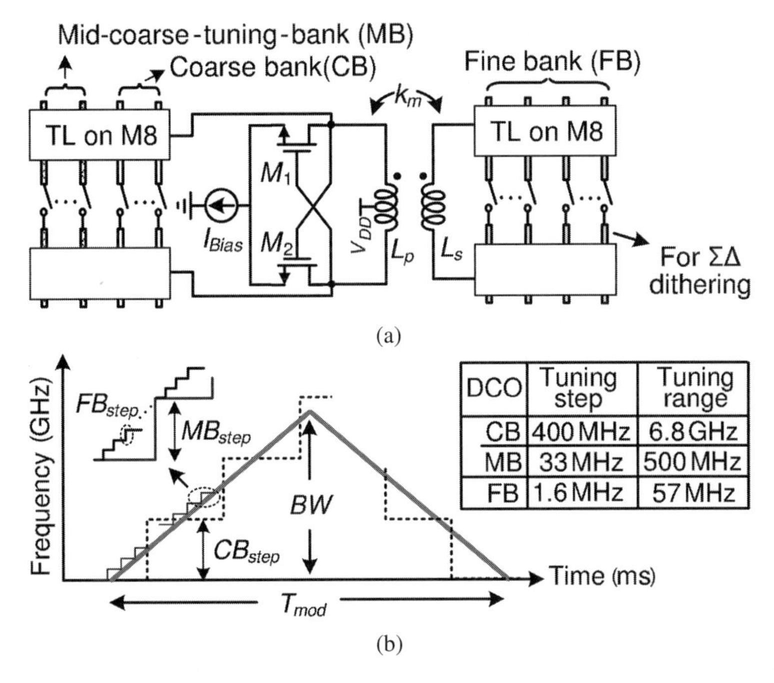

DCO	Tuning step	Tuning range
CB	400 MHz	6.8 GHz
MB	33 MHz	500 MHz
FB	1.6 MHz	57 MHz

Figure 11.15 (a) Schematic of the multibank 60 GHz DCO; and (b) wideband triangular modulation traversing three tuning banks.

As an alternative, a closed-loop DCO gain linearization technique for a linear FMCW generation was used in the 60 GHz DPLL prototype. For a triangular modulation of a slope $k_{mod} = 2BW/T_{mod}$ (BW is the modulation range and T_{mod} is the period of the triangular modulation, as shown in Figure 11.15), the output frequency change within each modulation clock (CKM) is k_{mod}/f_{CKM}, where f_{CKM} is the modulation sampling rate. Instead of finding and storing accurate DCO oscillator tuning words (OTWs) for each frequency along the triangular modulation trajectory, accurate OTWs are determined only in the vicinity of the bank-switchover points (see Figure 11.16). Thus, the size of the look-up table is determined by the number of bank-switchover points and independent of the FM rate and range. To ensure monotonic tuning against PVT, the total FB tuning range is set to 1.7 times the frequency step size in CB. The midpoint of the overlap region is a natural choice for a robust switchover. Between the two adjacent bank-switchover points, only FB is used for modulation, which is sufficiently linear, and one normalized K_{DCO} for each subrange is employed. When the upper and lower boundaries of the tuning word in FB are determined for $CB = c$ and $MB = m$ (i.e., $FB_{max}(c,m)$ and $FB_{min}(c,m)$), the normalized tuning step $FB_{step}(c,m)$ for each CKM is calculated by $FB_{step}(c,m) = \Delta FB(c,m)/\Delta n(c,m)$, in which $\Delta FB(c,m)$ is the frequency range for $CB = c$ and $MB = m$, and $\Delta n(c,m)$ is the number of CKM cycles needed to modulate across $\Delta FB(c,m)$ for a specific chirp slope, k_{mod}. Thus, only three variables – $FB_{max}(c,m)$, $FB_{min}(c,m)$, and $FB_{step}(c,m)$ – need to be saved in static random-access memory (SRAM) for each index (c,m). Note that two sets of the DCO tuning words are saved for each switchover point to implement hitless modulation, e.g., $FB_{max}(c,m)$

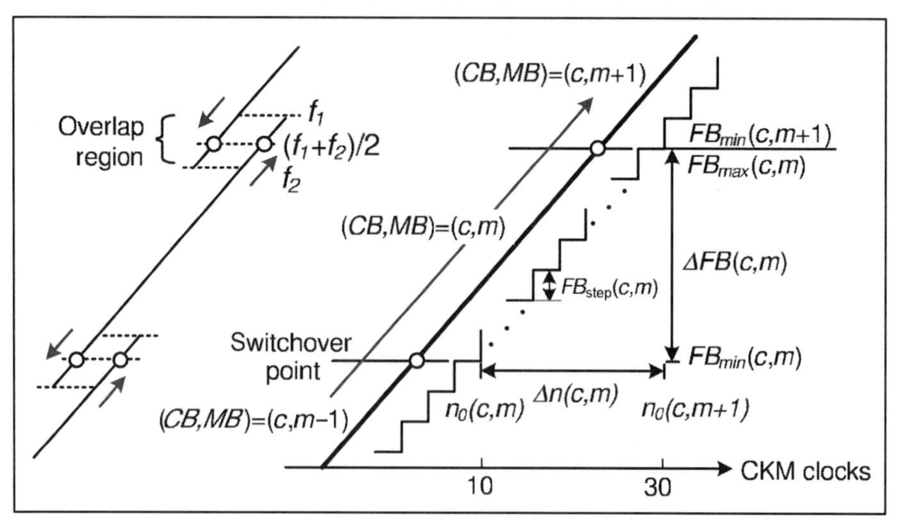

Figure 11.16 (a) Schematic of the multibank 60 GHz DCO; (b) wideband triangular modulation traversing three tuning banks.

and $FB_{min}(c, m + 1)$. The required calibration time at power-up is ensured to be no more than 4 seconds.

11.5.2 Mismatch Calibration of the Fine-Tuning Bank

The transformer-based fine-tuning bank achieves a raw frequency resolution of $\sim$2 MHz. A $\Delta\Sigma$-modulator with a higher dithering rate may be employed to obtain a resolution on the order of hundreds of hertz. The fractional bits (i.e., those undergoing dithering) are physically located at the end of the fine-tuning bank (see Figure 11.15), and thus there will be mismatches between the fractional and integer bits of the fine-tuning bank. This can introduce nonlinearity in the frequency modulation.

The mismatch of the fractional tuning bits with respect to the average K_{DCO} of the integer bits can be characterized in an open-loop manner by a forced on/off toggling of the fractional bit [39]. Since small capacitance fluctuations in the DCO tank result in proportional frequency fluctuations, changes in capacitance, resulting from on/off switching of the dithering bit, are evaluated by subtracting frequency measurements performed at each of the two states. This open-loop configuration is used since each toggling procedure addresses a specific fine-tuning bit, which could not be done through the normal modulation capability of the DPLL. The frequency measurements are based on a counter within the DPLL, and multiple readings of the counter (e.g., M readings) are averaged to reduce the quantization error in a single measurement of frequency deviation, especially in the presence of DCO phase noise, as shown in Figure 11.17.

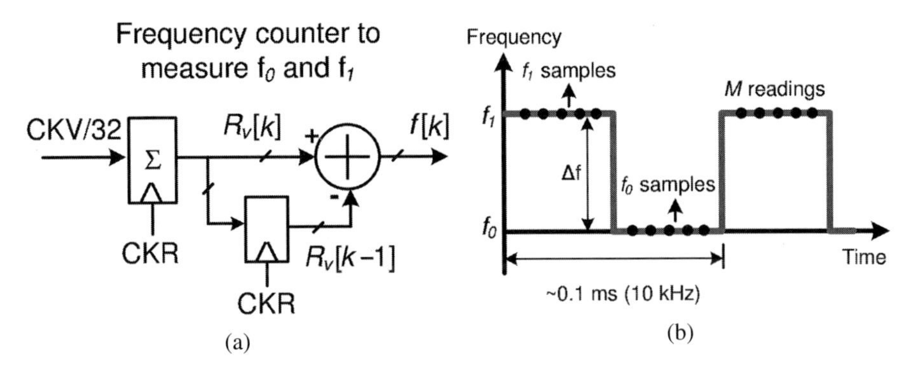

Figure 11.17 Open-loop DCO gain calibration based on toggling [40]. (©2014 IEEE. Reprinted, with permission, from *Proceedings of the IEEE 2014 Custom Integrated Circuits Conference*.)

The tuning step for a particular bit i is Δf_i, and can be calculated by $\Delta f_i = \frac{1}{M}[\sum_{k=1}^{M} f_{1_k} - \sum_{k=1}^{M} f_{0_k}]$. The number of measurements (N) used is typically on the order of 2^{15} to satisfy a 1% fine-tuning step accuracy. The resulting frequency tuning step after averaging is $\Delta f_{avg,i} = \frac{1}{N}\sum_{i=1}^{N} \Delta f_i$. Thus, the normalized tuning step mismatch between the given fractional bit and the average of the integer bits of the FB_{Mod} is $\varepsilon = (\Delta f_{avg,i} - \Delta f_{FB})/\Delta f_{FB}$, where Δf_{FB} is the estimated frequency tuning step of the integer bits of FB_{Mod}. The correction factor ε can be then applied to the dithering bit upon modulation.

11.5.3 Synchronization in a Multirate System

The block diagram of the 60 GHz DPLL-based FMCW transmitter is elaborated in Figure 11.18, putting emphasis on the frequency modulation path. As explained in Section 11.2, the direct modulation path operates at a high clock rate (f_{CKM}), which is a down-divided DCO clock to obtain high-modulation bandwidth. The CKM is configurable from CKV/128 ($\sim$450 MHz) to CKV/1024 ($\sim$56 MHz) to minimize power consumption according to the required modulation ramp slope ($k_{mod} = BW/T_{mod}$). The compensation path is applied to the frequency reference and operates at the retimed reference clock (CKR) rate, f_R. During FM, CKV varies linearly with time and so does CKM.

The two functional parts of the DPLL-based frequency modulator, which are the phase error calculator and the data modulator, have their own separate clock domains: FREF and CKV, respectively. Since their frequency relationship is a time-varying fractional number, their interfaces normally require sampling rate converters. However, this is not necessary in this architecture because system clock CKR is always synchronized with modulator clock CKM via resampling of FREF by CKV/128. The fine-tuning bank used for data modulation (FB_{Mod}) is physically separated from the FB_{Loop}, which is used for phase error correction. As for coarse-tuning banks *CB* and *MB*, they are controlled by the direct data path upon modulation by a multiplexer (see Figure 11.18). Therefore, no sampling rate conversion is required for the DCO tuning word.

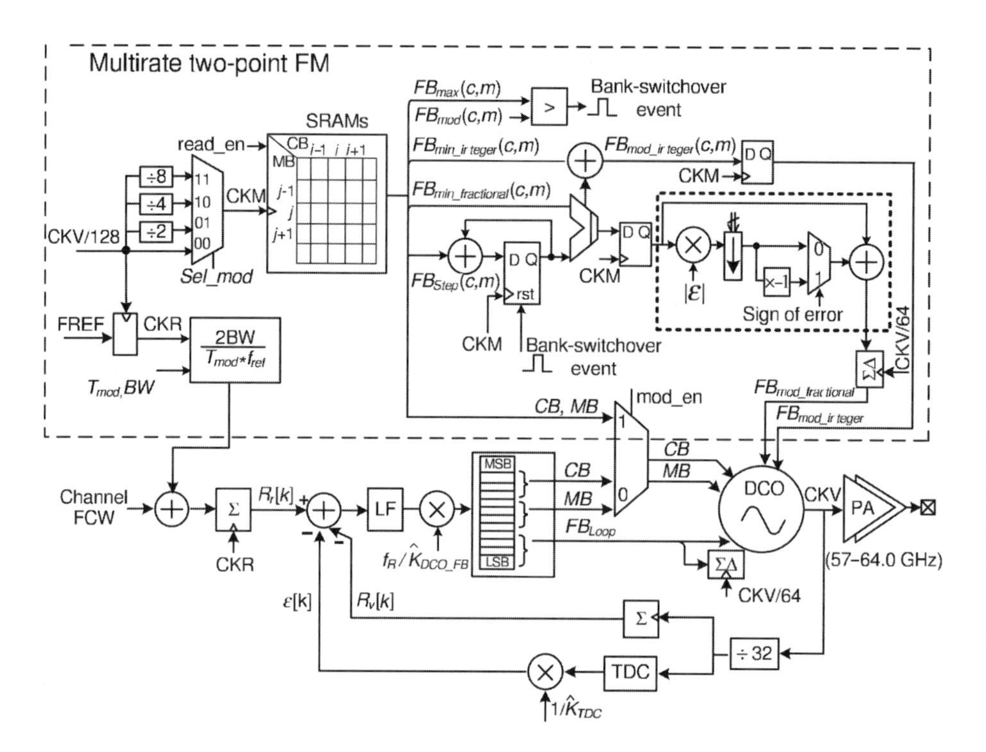

Figure 11.18 Multi-rate modulation for the 60 GHz DPLL-based FMCW transmitter.

During modulation, a state-machine controls the access to SRAMs and reads out the proper data before bank switchover. Operation at high speed is simplified to an accumulation of the FB_{step} and a comparator to generate the bank-switchover event. Meanwhile, a frequency step equal to $k_{mod}/f_R/32$ (32 is the division ratio in the feedback loop) is added as compensation to the frequency reference at every CKR to obtain the wideband FM output.

The mismatch of the dithering bit in FB is obtained via the open-loop calibration described earlier, and compensated in the direct path using the logic highlighted by the dotted line in Figure 11.18. The compensation mechanism is based on a digital gain correction factor that is applied to the 10-bit FB_{Mod} fractional tuning word before it is fed to the fractional tuning unit, where it is converted into the appropriate dithering signal. It is implemented using a reduced-size multiplier followed by an adder. The magnitude of the error ε correction is limited to 8 bits, allowing for a dynamic range of mismatch errors up to 25% and a theoretical resolution of 0.1%, which is more than sufficient. In addition, the fractional unit in FB_{Mod} (see Figure 11.15a) is sized in this design so that the compensating factor is always a fraction, thereby avoiding potential overflows.

11.5.4 Experimental Results

The 60 GHz DPLL-based transmitter employing the time-domain circuits and calibration techniques presented in this chapter was fabricated in TSMC 65 nm bulk

CMOS [41]. The die photo is shown in Figure 11.19. The DPLL core occupies 0.5 mm^2 of the 2.2 mm^2 total die area, including bondpads, output power amplifier, SRAMs (6×2^{13}-bit), and other digital circuitry for debugging. The SRAM is used to take a system snapshot for debugging purposes (which will be discussed in the next section) and is also used to store the K_{DCO} calibration data for the gigahertz range, linear FM. The DPLL chip consumes 40 mA: 11 mA by the DCO, 23 mA in the frequency prescaler (divide by 32), and 6 mA for the TDC and digital part, while the power amplifier dissipates 34 mA, all from a single 1.2 V supply. The TDC consumes 4.5 mA, which reduces to only 1.5 mA when edge prediction and power gating are enabled. The prescaler consists of an injection-locked frequency divider (ILFD) by-2 stage, current-mode logic (CML) to further divide by 4, and finally, a digital divider to perform a total divide by 32 to generate an RF signal of ~2 GHz driving the rest of the loop. The single-ended 60 GHz PA output is measured by on-die probing. In addition, a 2 GHz test output (after divide-by-32) is also accessed via the printed circuit board (PCB), providing a convenient way to characterize the 60 GHz DPLL without the on-die probing.

The fractional-N DPLL can generate arbitrary frequencies ranging from 56.4 to 63.4 GHz. The measured spectrum of the mm-wave output when locked at 60 GHz is plotted in Figure 11.20. A low reference spur level of −74 dBc is observed, with no other significant spurs detectable. The measured worst-case reference spur is −72.4 dBc across the 7 GHz locking range. The out-of-band fractional spurs are filtered out heavily by the type II, fourth-order infinite impulse response (IIR) loop filter. For some channels (e.g., near integer-N channel), the fractional spurs fall in-band but are always less than −60 dBc. A spectrum of the 60 GHz carrier close-in is shown in Figure 11.20b. It indicates that the PN at 1 MHz frequency offset is −88.5 dBc/Hz (i.e., $-41.58 - 10\log_{10}(50\,\text{kHz})$).

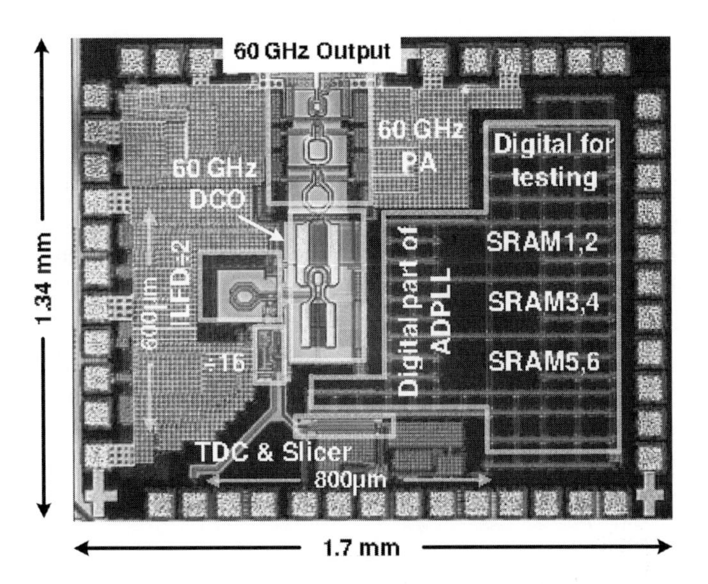

Figure 11.19 Die photo of the 60 GHz DPLL-based transmitter [41]. (©2014 IEEE. Reprinted, with permission, from *IEEE Journal of Solid-State Circuits*.)

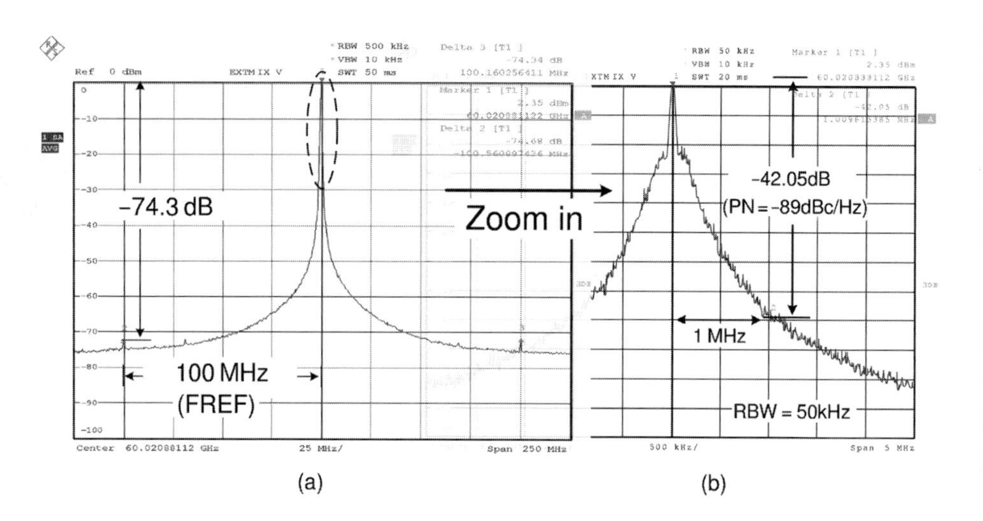

Figure 11.20 Measured 60 GHz DPLL output spectrum [41]. (©2014 IEEE. Reprinted, with permission, from *IEEE Journal of Solid-State Circuits*.)

The PN of the DPLL is measured from the divide-by-32 test output (CKV/32) and plotted in Figure 11.21 for various loop bandwidths. A 30.1 dB (i.e., $20\log_{10}(32)$) adjustment should be added to refer the PN to the mm-wave (i.e., 60 GHz) output. For a nominal loop bandwidth of 300 kHz, the measured PN is −118 dBc/Hz at 1 MHz offset, which agrees well with the PN obtained at the PA output shown in Figure 11.20a. The measured TDC resolution is 12.2 ps via the TDC self-calibrating algorithm, which corresponds to a theoretical in-band PN of −80 dBc/Hz for a 60 GHz carrier. From Figure 11.21, the measured in-band PN is −78 dBc/Hz at 60 GHz or a wide loop *BW* (∼1.5 MHz), and thus it is dominant by the TDC quantization noise.

The measured optimum integrated phase noise (IPN) is −45.9 dBc, integrated from 10 kHz to root mean square (rms) 10 MHz, which corresponds to rms jitter of 590.2 fs. This is sufficient for the targeted short-range FMCW radar applications [14–16]. To further reduce the IPN in order to meet more stringent PN requirements for other 60 GHz applications (e.g., IEEE 802.11ad with 16-QAM modulation), a TDC with finer resolution (e.g., 1 ps) can be used to lower the in-band PN. Consequently, the in-band PN will be reduced by $20\log_{10}(12.2\,\text{ps}/1\,\text{ps}) = 20.6\,\text{dB}$, and the optimal loop bandwidth for IPN is widened to a few megahertz to further suppress the PN of the DCO and improve the IPN at synthesizer output. According to the system analysis, up to 16 dB IPN reduction can be achieved (i.e., IPN of −31.8 dBc at 60 GHz band) by improving the TDC resolution from 12 to 1 ps, without any change in DCO PN. The timing resolution of 1 ps can be obtained in deep submicron CMOS by employing well-known high-resolution TDC techniques used in low-gigahertz DPLL, e.g., two-step TDC combing coarse and fine [12,35] and a gated ring oscillator with noise shaping [33], as discussed in Section 11.4.

The measured lock-in time is within 3 μs for a frequency step of up to ∼10% of the carrier frequency via the dynamic control on the loop parameters. During frequency

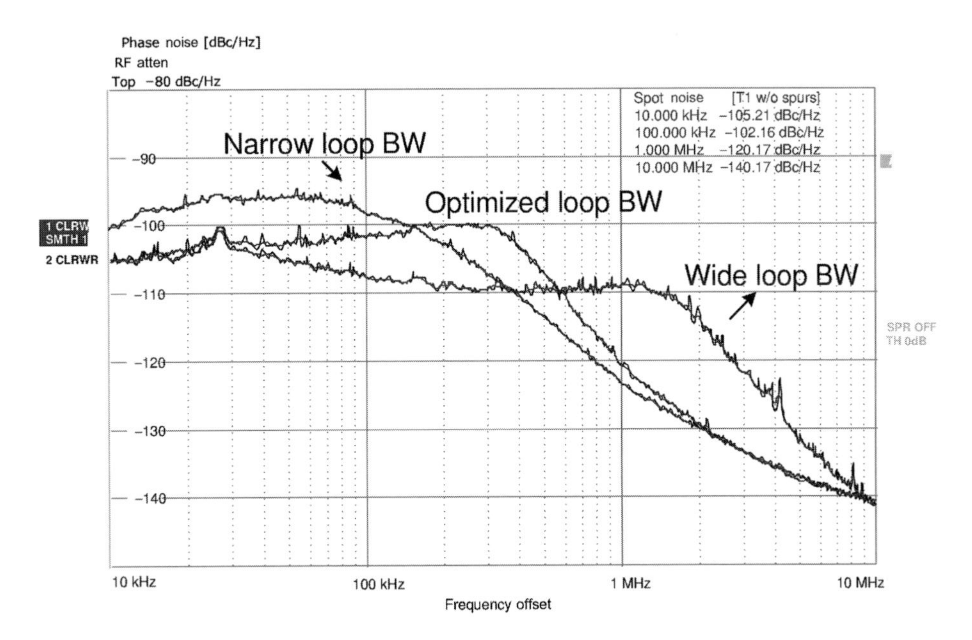

Figure 11.21 Measured DPLL PN at various loop bandwidths (at divide-by-32 output) [41]. (©2014 IEEE. Reprinted, with permission, from *IEEE Journal of Solid-State Circuits.*)

acquisition, the loop operates in type I with a wide bandwidth of 1.5 MHz. It is then switched hitlessly to type II, using a fourth-order IIR filter and 300 kHz bandwidth only in tracking mode.

Compared to the leading 60 GHz analog PLLs [42–44], the 60 GHz DPLL achieves fractional-N synthesis and exhibits excellent in-band and out-of-band PN performance, fast locking, and lower reference spur. Moreover, it is also capable of a wideband FM, which is demonstrated in Figure 11.22 for triangular FMCW generation.

Figure 11.22a plots the instantaneous output frequency of the DPLL-based FMCW transmitter when a triangular modulation across 1.22 GHz in range is applied (T_{mod} = 8.2 ms). The frequency error compared to an ideal triangular chirp is also shown in Figure 11.22a, with an rms value of only 117 kHz. Figure 11.22b shows the modulation results when the modulation speed is 16 times faster (i.e., 1 GHz change in 210 μs, and the measured frequency error is still smaller than 400 kHz_{rms}. The performance of the 60 GHz digitally intensive FMCW synthesizer is summarized in Table 11.1. Compared to state-of-the-art FMCW generators [14–16,45–48], the digitally intensive architecture achieves wider modulation range for varying modulation slopes, and better phase noise with lower power consumption.

11.6 Built-In Self-Test and Built-In Self-Characterization for DPLL

Frequency synthesizers and transmitters are tested for RF performance by measuring the carrier frequency, phase noise spectral density, integrated phase noise, spurious content, and modulated phase error trajectory at the RF output when modulation stimuli are applied [49,50]. In a debugging scenario, it is difficult to identify the root cause and provide a fix when the PLL fails to lock, or degraded performance is observed at the RF output (e.g., poor phase noise or high spur levels). Although functional testing of individual blocks can be conducted open loop, relating closed-loop PLL performance

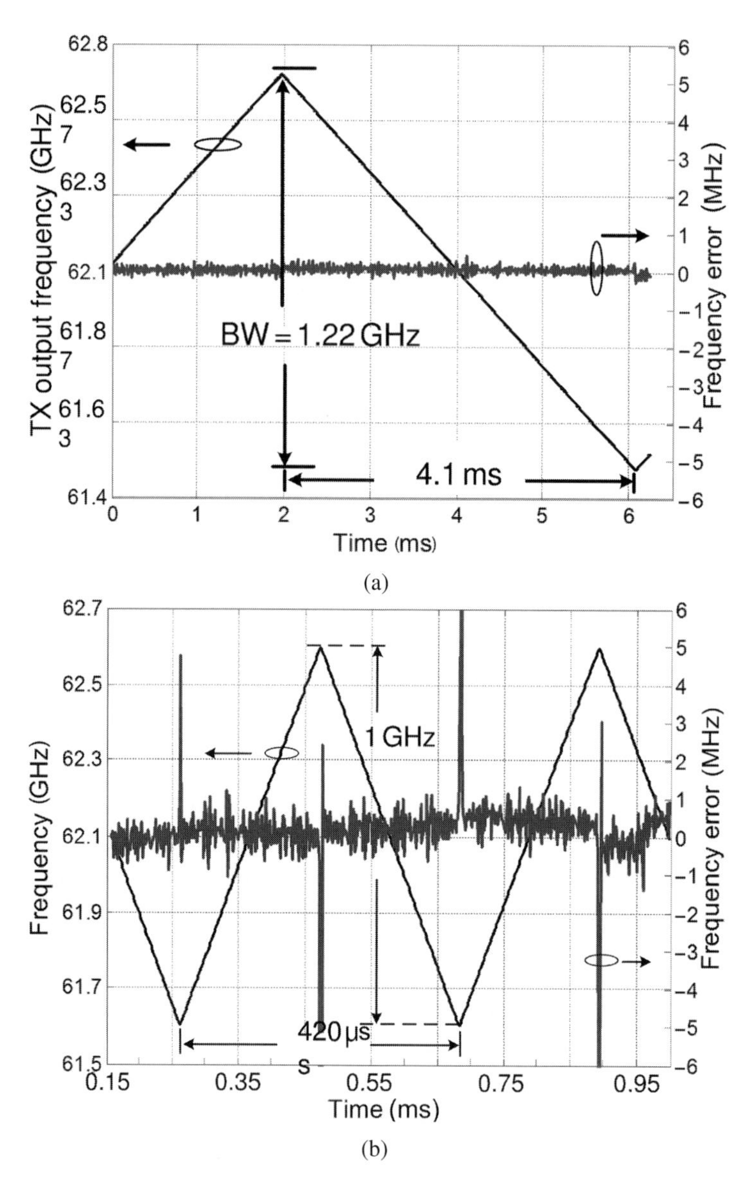

Figure 11.22 Measured time-domain frequency characteristics of the FMCW signal: (a) $T_{mod} = 8.2$ ms, $BW = 1.22$ GHz; (b) $T_{mod} = 0.42$ ms, $BW = 1$ GHz [41]. (©2014 IEEE. Reprinted, with permission, from *IEEE Journal of Solid-State Circuits.*)

to the performance of individual circuit blocks is nontrivial due to the tight feedback nature of a PLL [51].

Alternatively, built-in self-test (BIST) requires no external test equipment. It is widely used to reduce testing time and the cost of digital ICs while increasing the test coverage. Including BIST capabilities in mixed-signal radio frequency integrated circuits (RFICs) lessens the need for high-performance test equipment and provides data for debugging purposes [52]. These benefits are easier to realize on SoCs with digitally intensive PLLs, as the most of the PLL and its signal path can be accessed and evaluated by digital signal processing with little hardware overhead. By comparison, analog charge-pump PLLs are difficult to adapt for BIST because loading sensitive analog nodes for test purposes changes loop behavior and skews the measured data [49].

Aside from debugging, BIST applied to RF performance characterization of low-gigahertz DPLLs has been reported in [39,53–55], where digital signal processing of a lower-frequency internal signal is used to ascertain RF performance without external test equipment. A DPLL is always ultimately intended to be integrated in a digitally intensive SoC, consisting of a digital baseband processor, SRAM memories, and power management functions. Due to the reuse of SRAM and signal processing circuitry, very little hardware overhead is required to implement the design for test (DFT) and design for characterization (DFC) techniques into a DPLL. For example, the system snapshot can be triggered by a sequence of major internal events in a normal operation of the PLL to capture the transient behavior of the loop at a particular moment for observation, and to provide a means of analyzing loop operation analogous to simulation methodologies.

The 60 GHz DPLL prototype has implemented several BIST and built-in self-characterization (BISC) techniques, which are very useful for debugging, performance characterization, and production test of the DPLL and DPLL-based FM transmitter, especially operating at mm-wave frequencies.

11.6.1 Critical Signals in DPLL for BIST and BISC

A more detailed system diagram for the 60 GHz DPLL prototype is shown in Figure 11.23 in order to highlight the critical digital signals, which can be used for BIST and BISC. As shown in Figure 11.23, the phase error signal ϕ_E is the numerical difference between the reference and variable phases at the digital output of the phase detector. ϕ_E has a low-pass, unity-gain transfer characteristic to the variable phase at the DPLL RF output, that is flat up to the PLL bandwidth in type-I or type-II configurations with a large damping factor (e.g., 1) [55]. Therefore, the trajectory of ϕ_E correlates closely with the RF performance measured at the PLL output.

For a type-II PLL, ϕ_E, or its filtered version, has a zero mean once the loop is locked. Its variance represents the PN at the RF output with adequate accuracy. The trajectory of ϕ_E reveals the transient behavior of the loop, loop stability, and frequency response. For example, if there is an unwanted spurious tone in the RF spectrum, the tone frequency and its energy level can be sensed from spectral estimation of the ϕ_E

Table 11.1 Millimeter-wave FMCW synthesizer performance comparison.

	[22]	[16]	[14]	[15]	[45]	[47]	[48]	[46]
Architecture	ADPLL + multi rate 2-point mod.	Hybrid PLL	DDFS + PLL	Analog PLL	ADPLL + 2-point mod.	Analog PLL	Hybrid-PLL + 2-point mod	ADPLL
Frequency (GHz)	56.4–63.4	82.1–83.8	78.1–78.8	75.6–76.3	20.4–24.6	75–83	12.2–26.4	36.3–38.2
CMOS node (nm)	65	65	90	65	65	130 (SiGe)	32 (SOI)	40
Reference frequeny (MHz)	40	26	77	700	52	125	133	120
Chirp type	Triangular	Triangular	Triangular	Triangular	Triangular, Sawtooth	Sawtooth	Triangular	Triangular
Modulation slope (k_m)	Fast: 1 GHz/0.21 ms Slow: 1.22 GHz/4.1 ms	Fast: 1.5 GHz/1 ms Slow: 0.5 GHz/5 ms	614 MHz/0.5 ms	700 MHz/1.5 ms	173.3 MHz/1 μs	100 MHz/μs	10 MHz/μs	9.1 MHz/μs
Max. chirp BW (MHz)	1,220	1,505	614	700	208	8,000	8,000	500
Frequency error (rms)	Fast: 384 kHz Slow: 117 kHz	Fast: 179 kHz Slow: 170 kHz	1.05 MHz[a]	<300 kHz[1]	124 kHz	3.2 MHz	NA	820 kHz
PN (at 1 MHz)	−90 dBc/Hz	−84 dBc/Hz	−85 dBc/Hz	−85 dBc/Hz	−90 dBc/Hz	−97 dBc/Hz	−79.3 dBc/Hz	−73.6 dBc/Hz
P_{DC} (mW)	48 + 41 (PA)	152 with buffer	101	73 + 115 (PA)	19.7	590	30	68
P_{out} (into 50 Ω)	+5 dBm	NA	−13 dBm	+5.1 dBm	−13.3 dBm	NA	−6 dBm	−20.57 dBm
Area (mm^2)	2.2	1.5	NA	1.0	0.48	4.42	0.28	0.18[b]

[a] Includes turnaround point.
[b] Only active area.

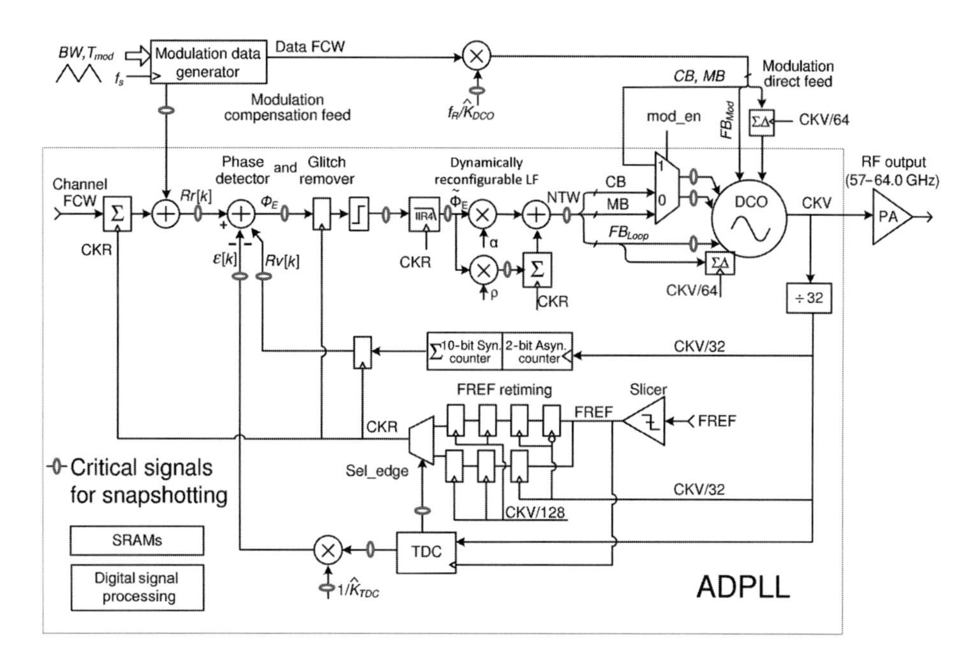

Figure 11.23 Block diagram of the 60 GHz DPLL synthesizer.

trajectory. For catastrophic errors during the DPLL operation (e.g., the PLL loses its lock), the filtered ϕ_E is no longer flat and its variance exceeds the normal operating bounds.

For two-point FM, the ϕ_E trajectory may be different from the CW mode because the modulation data affects the ϕ_E noise characteristics. The increased range for the ϕ_E variation indicates greater phase noise in the system, which could be due to the DCO gain calibration accuracy or due to the digital-to-frequency conversion nonlinearity in the DCO modulation bank.

Besides ϕ_E, both integer and fractional part of the variable phase (R_v and ϵ) and reference phase (R_r) are informative for debugging, as $\phi_{E[k]} = R_r[k] - (R_v[k] + \epsilon[k])$. When tracing fractional spurs, observing $\epsilon[k]$ is more effective. It reflects the periodic behavior in variable phase due to TDC nonlinearities. In addition, other signals along the path from the phase detector to the DCO could also be used for DFT/DFC, such as internal signals in the LF, the scaled-down and filtered version of ϕ_E at the LF output, and the DCO control word (OTW). A frequency deviation in the loop can be ascertained by observing an output of the integral path accumulator when in type-II operation. Alternatively, the output of the IIR filter, which is connected to the ϕ_E, could be observed. The OTW can be in a binary form, or an encoded number that matches the DCO interface.

The aforementioned critical internal signals (highlighted in Figure 11.23) are intrinsically present in any DPLL. Thanks to the digitally intensive nature of the DPLL, these signals can be monitored as well as processed on-chip for testing and characterization.

11.6.2 Snapshotting Internal Signals for Debugging

Although the spectrum measured at the RF output indicates the PLL performance, it is difficult to locate the source of a problem from an unlocked spectrum or when many spurious tones are observed at the RF output. In these cases, internal digital signals (raw phase error ϕ_E, filtered ϕ_E, oscillator tuning word, etc.), clocked at FREF rate, should be monitored.

Parallel outputs are normally used to monitor internal digital signals in real time. An on-chip mux selects the internal signals of interest for output. The number of test outputs is limited by the available bondpads (e.g., 8–16). As the digital signals internal to the DPLL are clocked at the FREF rate (e.g., 100 MHz), they should be output at the same rate or down-sampled synchronously to ensure signal integrity. Propagating multibit digital signals at 100 MHz rate requires attention to the PCB design and the use of properly shielded test cables to minimize crosstalk. Moreover, these parallel outputs toggle between 0 and 1, generating switching transients that can be coupled to the sensitive analog nodes on-chip via bondwires or the ESD/pad ring. Consequently, a higher noise floor and increased spurious tone levels are measured at the RF output when the digital test outputs are enabled.

To overcome the aforementioned limitations of the parallel test outputs, we can make use of on-chip SRAM to take a system snapshot that records one or more internal signals in the time frame of interest [40]. Subsequently, the saved data are read out from SRAM via a serial peripheral interface (SPI). Sharing on-chip SRAM that is used for other (nonconflicting) purposes in an SoC reduces hardware overhead. Consequently, the digital signals and the time frame recorded are selected carefully in order to utilize the limited word depth of SRAMs and to fulfill debugging needs.

The important digital signals to monitor were discussed earlier. To specify the proper time frame for recording, a series of events when the loop is prone to disruptions or even bugs are defined to trigger the snapshot. One or several trigger events are selected for monitoring by a control register. Some of these events indicate a loop status change and can be enabled intentionally during debugging, such as switching the loop from type-I to type-II loop, or increasing/decreasing loop bandwidth. Some indicate a different loop operation mode, e.g., enabling DCO/TDC gain calibration, enabling a phase error glitch remover, starting modulation, etc. Some events are flag signals generated internally (i.e., read only). For example, when no transition edges at the outputs of the TDC's inverter chain are detected, the TDC fail flag is asserted. When the DCO tuning word exceeds its range, an overflow flag turns on. Once the phase error variation exceeds a predefined window, the poor-clock-quality flag is activated. These flag signals report an abnormal status during the PLL operation. With the aid of system snapshots triggered by these flags, we are able to not only discover the loop abnormality, but also to infer the cause by examining associated digital signals.

Once the event of interest is triggered (externally/internally), the values of the associated digital signals (which can be one or several) are written into SRAM at a programmable clock rate (f_{clk_w}). Rate f_{clk_w} normally equals the clock rate of these digital signals (i.e., f_R) in order to capture their precise trajectory. To make the system snapshot more powerful, we can configure it into different operational modes. For example, we can save one digital signal into all SRAMs in sequence to maximize the snapshot depth, or save up to six different signals into six small SRAMs in parallel to observe multiple signals synchronously; we can also stop saving data when the SRAMs are full, capture the short moment when the trigger is enabled, or save data to SRAM cyclically until triggered by the specified event to freeze the moment just before the event happens.

11.6.3 DCO Tuning Step Analyzer

Two categories of tests are normally involved in an RFIC: structural-based and functional performance–based. While the former is used for block-level design verification and defect tests, the latter is used for PVT characterization.

For the 60 GHz FMCW transmitter example, frequency modulation traverses three tuning banks to obtain the desired GHz modulation range (see Figure 11.15). The tuning step mismatch within each unit-weighted bank and the K_{DCO} ratio between different banks affect the transmit spectrum and modulation distortion. It is important to establish the tolerable extent for these mismatches in the design phase and to verify that it is not exceeded by fabricated SoCs.

It should be emphasized that it is extremely difficult and time consuming to measure a free-running, mm-wave (e.g., 60 GHz) DCO's tuning step mismatch for the targeted accuracy, even with the aid of specialized test equipment. For a FB K_{DCO} of $\sim$1 MHz and 5% mismatch, the frequency difference is just 50 kHz at the 60 GHz carrier. Tremendous efforts are required to stabilize the free-running DCO and to reduce thermal noise in the test setup (e.g., noise of a harmonic mixer should be avoided). Fortunately, the BISC technique extracts the accurate tuning characteristic of the DCO without external resources. The K_{DCO} for each tuning bit is characterized by using the same technique as employed for mismatch calibration between $\Delta\Sigma$ dithering bit and the integer bits in the FB, as illustrated in Figure 11.17. The measured fine-tuning step size of the 60 GHz DCO is shown in Figure 11.10 via the built-in DCO tuning step analyzer.

BISC results can be compared on-chip to statistically chosen thresholds for defect detection. Furthermore, a "self-healing" capability can be implemented based on the BISC outcomes. For example, an on-chip look-up table can be built based on the BISC results and used to predistort the modulation data in order to compensate for the mismatch and retrieve adequate linearity.

11.7 Another Approach: DTC-Assisted DPLL Architecture

The digitally intensive PLLs discussed earlier in this chapter rely on a TDC to convert the phase error between the reference and variable clocks into a fixed-point digital word (ϕ_E). In a DPLL, the resolution of the TDC is critical for low phase noise. The TDC might need to provide fine time resolution of 1 ps to meet the stringent noise requirement of advanced wireless applications (e.g., WIFI). Meanwhile, the TDC has to provide a large dynamic range (e.g., ~500 ps) to cover more than one input clock period for the DPLL architecture in Figure 11.2 in order to operate in fractional-N mode (i.e., first-order $\Delta\Sigma$ architecture). For digital PLLs using second- or third-order $\Delta\Sigma$ for fractional-N synthesis, the required TDC dynamic range increases to two times and four times of the input clock period, respectively [33]. This implies that more than 10 bits are required for a TDC. Although many high-resolution TDC techniques are reported in literature, e.g., a two-step TDC combining coarse and fine [35], a gated ring oscillator [33], or an interpolation-based TDC [12], the high performance is achieved at the cost of circuit complexity as well as larger chip area and more power consumption.

Alternatively, the design challenge of the high-performance TDC can be mitigated by adding a DTC before the phase error comparison, as shown in Figure 11.24. The $\Delta\Sigma$ quantization is used to modulate the delay of a DTC on the reference clock path to cancel the quantization at the TDC input [11,56], such that the TDC sees only a small phase error after locking even in fractional-N mode. The DPLL scales the $\Delta\Sigma$ noise with the DTC gain and applies it to modulate the DTC delay. The DTC gain is background calibrated by correlating the quantization noise before/after the cancellation via an LMS algorithm. In such a DPLL architecture, a narrow TDC linear range (e.g., ~10 ps compared to ~500 ps previously) can thus be used, which greatly eases the TDC design.

One possible implementation of a narrow-range, high-resolution TDC is a sampling TDC, as shown in Figure 11.25. It consists of a slope generator and a successive approximation register (SAR) ADC clocked by the DTC output (CLKDTC). The feedback clock (CLKFB) triggers a voltage slope, whose dV/dt is defined by the resistor-capacitor (RC) time constant and designed to be about 6 GV/s around the zero crossing. The capacitor C is reused as part of the SAR ADC cap-bank. The ADC has 8-bit and 4 mV resolution, leading to a TDC resolution of 0.7 ps. For large phase errors, the TDC is naturally not linear due to the exponential RC response. However in the locked state, the TDC input only sees small variation around the zero crossing, where it is mostly linear, as the DTC prior to it cancels the majority of the quantization

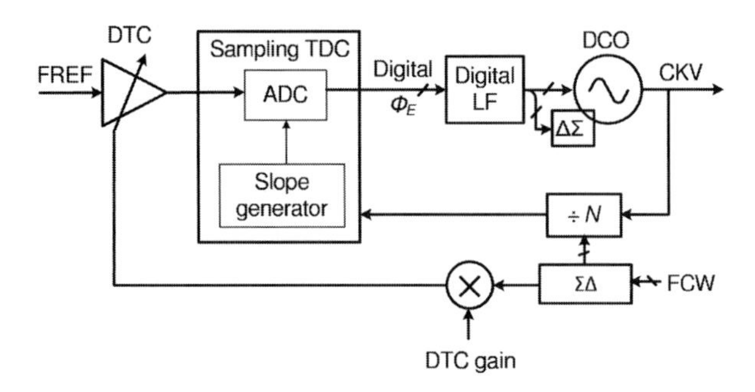

Figure 11.24 DTC-assisted DPLL architecture [11]. (©2016 IEEE. Reprinted, with permission, from *IEEE International Solid-State Circuits Conference (ISSCC) Digest of Technical Papers.*)

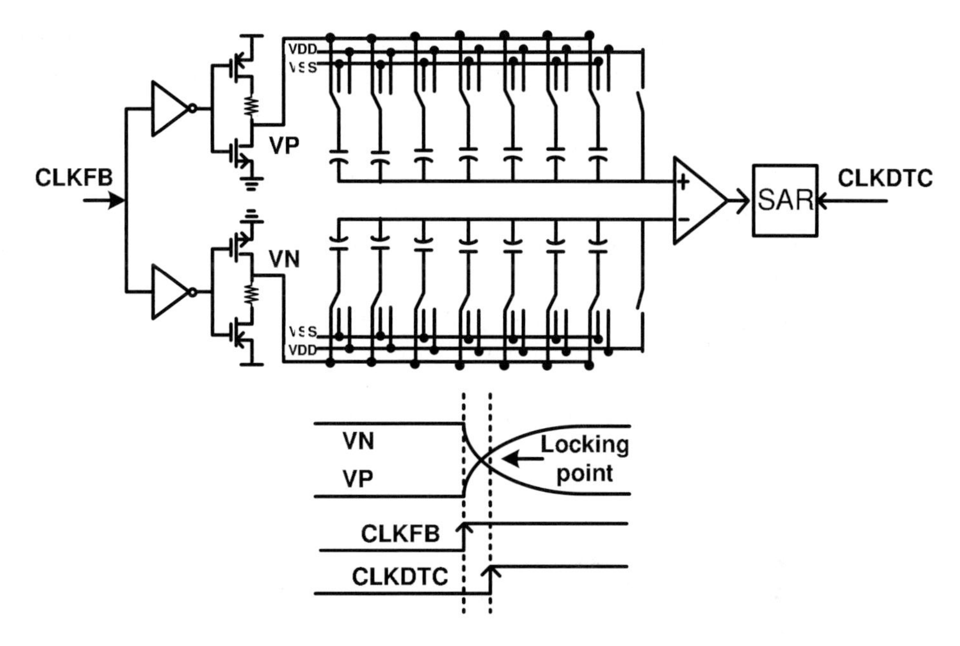

Figure 11.25 Schematic of the sampling TDC [11]. (©2016 IEEE. Reprinted, with permission, from *IEEE International Solid-State Circuits Conference (ISSCC) Digest of Technical Papers.*)

noise. During the initial phase locking, the phase error can be larger than the TDC's sampling detection range, resulting in a bang-bang behavior. To minimize this bang-bang behavior, an 8-bit counter running at DCO clock rate was used as a coarse TDC in [11] to assist locking.

In principle, the TDC could directly subsample the DCO without going through the multimodulus divider (MMDIV) in Figure 11.24 [56,57]. The fractional-N operation can be realized by directly modulating CLKREF through the DTC. However, in such a subsampling architecture the DTC gain needs to be very accurate in order to perform fractional-N operation at the exact target frequency. In the DTC-assisted architecture shown in Figure 11.24, limited DTC gain accuracy simply means some residual quantization noise is leaked to TDC. It is acceptable as long as it is still within the TDC linear range and low enough not to degrade the IPN. It is more robust at the cost of the extra MMDIV circuit.

Now let us take a close look at the DTC design requirement and possible circuit implementations. The DTC is used to cancel the $\Delta\Sigma$ quantization noise, but its own quantization noise also contributes to the PLL in-band PN. Therefore, 10-bit resolution is required for the DTC. Fortunately, it is much easier to implement a 10-bit DTC compared to a 10-bit TDC, as the DTC is operated at CLKREF rate instead of variable clock rate and the fine digitally controlled delay can be easily achieved by digitally controlled RC delay, as shown in Figure 11.26 [58]. A coarse calibration of the DTC delay is done by programming R, while the fine-tuning is accomplished via a switched metal capacitor bank. With an R of 500 Ω and switchable unit C_u of 2 fF, the DTC resolution is 700 fs, yielding sufficiently low quantization noise.

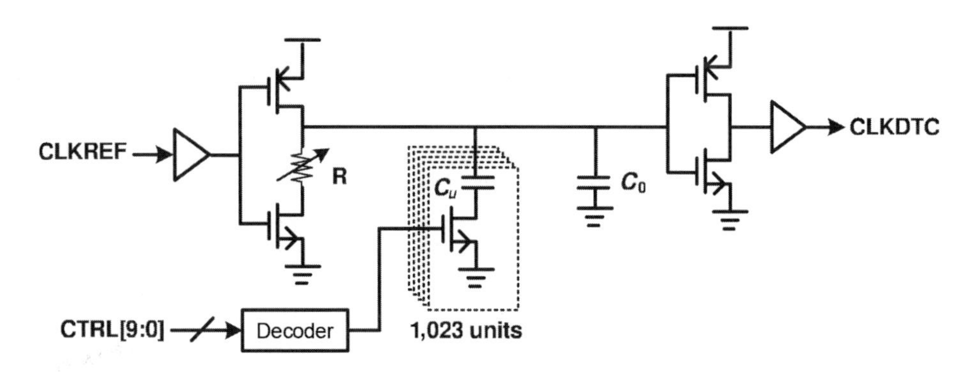

Figure 11.26 Schematic of the DTC.

With a 10-bit control, the total DTC range is about 700 ps, which is enough to cover two DCO periods in case a second-order $\Delta\Sigma$ is used. The DTC linearity is mainly limited by the output buffer, whose delay is dependent on the input signal slope. To lower the variation of input signal slope over the entire DTC range, a fixed capacitor can be added in the main path on top of the switched capacitor bank. With these techniques, the overall DTC integral nonlinearity (INL) is simulated to be about 1 LSB.

A DTC-assisted DPLL prototype was implemented in 28-nm CMOS targeted for IEEE 802.11ac applications. The measured DPLL achieved low rms jitter of 159 fs integrated from 10 kHz to 40 MHz. The reference spur at 40 MHz is -78 dBc and the worst fractional spur is −54 dBc at 100-kHz offset.

The DTC-assisted DPLL architecture, shown in Figure 11.24 can be applied to mm-wave band by two approaches. The first approach is to replace the low-gigahertz DCO by a mm-wave DCO together with a prescaler chain to divide down the DCO clock to a few gigahertz for MMDIV. The mm-wave DCO and prescaler used in the 60 GHz DPLL prototype can be employed for this purpose. The second approach is to cascade this low-gigahertz PLL by another integer-N PLL or a frequency multiplier to multiply up the output clock to a mm-wave band [59,60]. The architectural choice for a particular application depends on multiple design aspects: transceiver architecture and frequency planning, process technology, frequency tuning range, phase noise specifications, power consumption budget, and chip area.

For example, a PLL with a fundamental oscillator operating at mm-wave band is preferred for an FMCW radar application, in which a linear frequency modulation is normally obtained via a direct PLL modulation. Thus, the FMCW chirp generation and transmitting would be simpler and consume less power as demonstrated in the 60 GHz DPLL-based radar transmitter prototype. The frequency sweep linearity is directly controlled by the feedback loop and not subjected to any potential degradation due to the multiplier or harmonic generation. The local oscillator (LO) phase noise requirement in an FMCW radar is less demanding as compared to that in the communication systems since the phase noise of the transmitted and received signals is somewhat correlated, which reduces deleterious effects of synthesizer phase noise [61].

When used in a transmitter for mm-wave wireless communication (e.g., IEEE 802.11ad), the frequency multiplication–based approach is preferred as LO pulling is minimized since the LO is no longer at the same operating frequency as the power amplifier. The greater the multiplication factor, the smaller the chance for LO pulling. For phase noise performance, the final output phase noise is degraded by $20 \log_{10} N$, where N is the multiplication factor. Nevertheless, the phase noise of the Nth harmonic output may still be better than direct frequency generation by a PLL with DCO operating directly at mm-wave band, depending on the process employed. The tuning range may also be higher than in a fundamental PLL. However, the power consumption of the multiplication-based approach would be higher than a PLL with a fundamental oscillator (for the same output power) because it is less efficient to generate power at harmonic frequencies. Moreover, undesired harmonic/fundamental signal rejection is crucial in frequency multiplication–based architectures.

Several guidelines can be derived to help with the architectural choices for mm-wave clock generation based on literature study of recently published designs in nanoscale CMOS technology [59–65]. When implemented in, for example, 65 nm CMOS technology, a PLL with a fundamental oscillator is relatively simple and can provide moderate tuning range ($\sim$10%) and phase noise performance (-90 dBc/Hz at 1 MHz offset) with a smaller chip area in the 60 GHz band. Requirement of ultrawide tuning range ($>$15%) and superior phase noise (better than -100 dBc/Hz at 1 MHz offset) are extremely difficult (or perhaps even impossible) to satisfy with a fundamental oscillator at 60 GHz. Thus, a multiplication-based topology has better potential to satisfy the tougher specifications. This doesn't imply it is easier to design since the multiplier needs to have an ultrawide tuning range with an acceptable phase noise penalty. For a much higher RF carrier, such as 94 GHz or above, the harmonic generation has clear advantages over the fundamental PLL since the operating frequency is approaching the f_T and f_{MAX} of the transistor, which dramatically reduces the design margin. The loss and parasitics in the passive tank are also increasing rapidly with operating frequencies and limit the achievable phase noise and tuning range in a fundamental oscillator above 100 GHz. Another consideration is I/Q generation, which is needed in an in-phase/quadrature (I/Q) transmitter. To generate quadrature signals in the mm-wave regime, a multiplier-based approach is preferable to a direct quadrature generation at fundamental frequency due to less power consumption and better I/Q mismatch [62].

References

[1] Z. Luo, A. Steegen, R. M. M. Eller et al., "High performance and low power transistors integrated in 65nm bulk CMOS technology," in *IEDM Technical Digest IEEE International Electron Devices Meeting*, IEEE, 2004, pp. 661–664.

[2] A. Tomkins, R. Aroca, T. Yamamoto, S. Nicolson, Y. Doi, and S. Voinigescu, "A zero-IF 60GHz 65nm CMOS transceiver with direct BPSK modulation demonstrating up to 6Gb/s data rates over a 2m wireless link," *IEEE Journal of Solid-State Circuits*, vol. 44, no. 8, pp. 2085–2099, August 2009.

[3] K. Okada, N. Li, et al., "A 60-GHz 16QAM/8PSK/QPSK/BPSK direct-conversion transceiver for IEEE802.15.3c," *IEEE Journal of Solid-State Circuits*, vol. 46, no. 12, pp. 2988–3004, December 2011.
[4] S. Emami, R. Wiser, E. Ali, et al., "A 60 GHz CMOS phased-array transceiver pair for multi-Gb/s wireless communications," in *IEEE International Solid-State Circuits Conference Digest of Technical Papers*, pp. 164–165, February 2011.
[5] R. Staszewski and P. Balsara, *All-Digital Frequency Synthesizer in Deep Submicron CMOS*. Wiley-Interscience, 2006.
[6] B. Miller and B. Conley, "A multiple modulator fractional divider," in *44th Annual Symp. on Frequency Control*, IEEE, May 1990, pp. 559–568.
[7] T. Riley, M. Copeland, and T. Kwasniewski, "Delta-sigma modulation in fractional-N frequency synthesis," *IEEE Journal of Solid-State Circuits*, vol. 28, no. 5, pp. 553–559, May 1993.
[8] A. Swaminathan, K. J. Wang, and I. Galton, "Wide-bandwidth 2.4GHz ISM-band fractional-N PLL with adaptive phase-noise cancellation," in *IEEE International Solid-State Circuits Conference Digest of Technical Papers*, pp. 3–5, February 2007.
[9] R. B. Staszewski, K. Muhammad, D. Leipold, et al., "All-digital TX frequency synthesizer and discrete-time receiver for Bluetooth radio in 130-nm CMOS," *IEEE Journal of Solid-State Circuits*, vol. 39, no. 12, pp. 2278–2291, December 2004.
[10] L. Vercesi, L. Fanori, F. D. Bernardinis, et al., "A ditherless all digital PLL for cellular transmitters," *IEEE Journal of Solid-State Circuits*, vol. 47, no. 8, pp. 1908–1920, August 2012.
[11] X. Gao, O. Burg, H. Wang, et al., "A 2.7-4.3GHz 0.16psrms Jitter -246.8dB FOM digital fractional-N sampling PLL in 28nm CMOS," in *IEEE International Solid-State Circuits Conference (ISSCC) Digest of Technical Papers*, IEEE, February 2016, pp. 174–175.
[12] C. Yao, C. Lau, R. Ni, et al., "A 14-nm 0.14psrms fractional-N digital PLL with a 0.2 ps resolution ADC-assisted coarse/fine conversion chopping TDC and TDC nonlinearity calibration," *IEEE Journal of Solid-State Circuits* vol. 52, no. 12, pp. 3446–3457, December 2017.
[13] R. B. Staszewski, G. Shriki, and P. T. Balsara, "All-digital PLL with ultra fast acquisition," in *Proceedings of the IEEE Asian Solid-State Circuits Conference*, IEEE, November 2005, pp. 289–292.
[14] T. Mitomo, N. Ono, H. Hoshino, et al., "A 77 GHz 90 nm CMOS transceiver for FMCW radar applications," *IEEE Journal of Solid-State Circuits*, vol. 45, no. 4, pp. 928–937, April 2010.
[15] Y.-A. Li, M.-H. Hung, S.-J. Huang, and J. Lee, "A fully integrated 77GHz FMCW radar system in 65nm CMOS," in *IEEE International Solid-State Circuits Conference Digest of Technical Papers*, IEEE, February 2010, pp. 216–217.
[16] H. Sakurai, Y. Kobayashi, T. Mitomo, O. Watanabe, and S. Otaka, "A 1.5GHz-modulation-range 10ms-modulation-period 180kHzrms-frequency-error 26MHz-reference mixed-mode FMCW synthesizer for mm-wave radar application," in *IEEE International Solid-State Circuits Conference Digest of Technical Papers*, IEEE, February 2011, pp. 292–293.
[17] R. Winoto, A. Olyaei, M. Hajirostam et al., "A 2x2 WLAN and Bluetooth combo SoC in 28nm CMOS with on-chip WLAN digital power amplifier and integrated 2G/BT SP3T switch and BT pulling cancellation," in *IEEE International Solid State Circuits Conference (ISSCC) Digest of Technical Papers*, IEEE, February 2016, pp. 170–173.

[18] A. Elshazly, R. Inti, W. Yin, B. Young, and P. K. Hanumolu, "A 0.4-to-3 GHz digital PLL with PVT insensitive supply noise cancellation using deterministic background calibration," *IEEE Journal of Solid-State Circuits*, vol. 46, no. 11, pp. 2759–2771, December 2011.

[19] C. Ho and M. S. Chen, "A fractional-N DPLL with adaptive spur cancellation and calibration-free injection-locked TDC in 65nm CMOS," in *Proceedings of the IEEE Radio Frequency Integration Circuit Symposium*, IEEE, June 2014, pp. 97–100.

[20] W. Egan, *Phase Lock Basics*, Wiley, 1998.

[21] W. Wu, X. Bai, R. Staszewski, and J. Long, "A 56.4-63.4GHz spurious free all-digital fractional-N PLL in 65nm CMOS," in *IEEE International Solid-State Circuits Conference Digest of Technical Papers*, IEEE, February 2013, pp. 352–353.

[22] W. Wu, R. Staszewski, and J. Long, "A mm-wave FMCW radar transmitter based on a multirate ADPLL," in *Proceedings of the IEEE Radio Frequency Integrated Circuits Symposium*, IEEE, June 2013, pp. 107–110.

[23] R. Staszewski, D. Leipold, and P. Balsara, "A first multigigahertz digitally controlled oscillator for wireless applications," in *Proceedings of the IEEE Radio Frequency Integration Circuit Symposium*, vol. 51, no. 11, pp. 2154–2164, November 2003.

[24] J. Long, Y. Zhao, W. Wu, M. Spirito, L. Vera, and E. Gordon, "Passive circuit technologies for mm-wave wireless systems on silicon," *IEEE Transactions on Circuits and Systems I: Regular Papers*, vol. 59, no. 8, pp. 1680–1693, August 2012.

[25] W. Wu, J. Long, and R. Staszewski, "High-resolution millimeter-wave digitally-controlled oscillators with reconfigurable passive resonators," *IEEE Journal of Solid-State Circuits*, vol. 48, no. 11, pp. 2785–2794, November 2013.

[26] T. LaRocca, S.-W. Tam, D. Huang, et al., "Millimeter-wave CMOS digital controlled artificial dielectric differential mode transmission lines for reconfigurable ICs," in *IEEE International Microwave Symposium Digest*, IEEE, June 2008, pp. 181–184.

[27] H. Sjöland, "Improved switched tuning of differential CMOS VCOs," *IEEE Transactions on Circuits and Systems II: Analog and Digital Signal Processing*, vol. 49, no. 5, pp. 352–355, May 2002.

[28] R. Genesi, F. D. Paola, and D. Manstretta, "A 53 GHz DCO for mm-wave WPAN," in *Proceedings of the IEEE Custom Integrated Circuits Conference*, IEEE, September 2008, pp. 571–574.

[29] T. Lu, C. Yu, W. Chen, and C. Wu, "Wide tunning range 60 GHz VCO and 40 GHz DCO using single variable inductor," *IEEE Transactions on Circuits and Systems I: Regular Papers*, vol. 60, no. 2, pp. 257–267, February 2013.

[30] J. Long, "Monolithic transformers for silicon RF IC design," *IEEE Journal of Solid-State Circuits*, vol. 35, no. 9, pp. 1368–1382, September 2000.

[31] R. Staszewski, S. Vemulapalli, P. Vallur, J. Wallberg, and P. Balsara, "1.3 V 20 ps time-to-digital converter for frequency synthesis in 90-nm CMOS," *IEEE Transactions on Circuits and Systems II: Express Briefs*, vol. 53, no. 3, pp. 220–224, March 2006.

[32] P. Dudek, S. Szczepanski, and J. Hatfield, "A high-resolution CMOS time-to-digital converter utilizing a vernier delay line," *IEEE Journal of Solid-State Circuits*, vol. 35, no. 2, pp. 240–247, February 2000.

[33] C.-M. Hsu, M. Straayer, and M. Perrott, "A low-noise and wide-BW 3.6GHz digital $\Delta\Sigma$ fractional-N frequency synthesizer with a noise-shaping time-to-digital converter and quantization noise cancellation," in *IEEE International Solid-State Circuits Conference Digest of Technical Papers*, IEEE, February 2008, pp. 340–617.

[34] P. Lu, A. Liscidini, and P. Andreani, "A 3.6 mW and 90 nm CMOS gated-vernier time-to-digital converter with an equivalent resolution of 3.2 ps," *IEEE Journal of Solid-State Circuits*, vol. 47, no. 7, pp. 1626–1635, July 2012.

[35] M. Lee, M. Heidari, and A. Abidi, "A low-noise wideband digital phase-locked loop based on a coarse-fine time-to-digital converter with subpicosecond resolution," *IEEE Journal of Solid-State Circuits*, vol. 44, no. 10, pp. 2808–2816, October 2009.

[36] B. Nikolic, V. Stojanovic, V. G. Oklobdzija, W. Jia, J. Chiu, and M. Leung, "Sense amplifier-based flip-flop," in *IEEE International Solid-State Circuits Conference Digest of Technical Papers*, IEEE, 1999, pp. 282–283.

[37] R. Staszewski, J. Wallberg, G. Feygin, M. Entezari, and D. Leipold, "LMS-based calibration of an RF digitally controlled oscillator for mobile phones," *IEEE Transactions on Circuits and Systems II: Express Briefs*, vol. 53, no. 3, pp. 225–229, March 2006.

[38] G. Marzin, S. Levantino, C. Samori, and A. Lacaita, "A 20 Mb/s phase modulator based on a 3.6 GHz digital PLL with –36 dB EVM at 5 mW power," *IEEE Journal of Solid-State Circuits*, vol. 47, no. 12, pp. 2974–2988, December 2012.

[39] O. Eliezer, R. Staszewski, J. Mehta, F. Jabbar, and I. Bashir, "Accurate self-characterization of mismatches in a capacitor array of a digitally-controlled oscillator," in *Proceedings of the IEEE Dallas Circuits and Systems Workshop*, IEEE, October 2010, pp. 1–4.

[40] W. Wu, R. Staszewski, and J. Long, "Design for test of a mm-wave ADPLL-based transmitter," in *Proceedings of the IEEE Custom Integrated Circuits Conference (CICC)*, IEEE, September 2014, pp. 1–8.

[41] W. Wu, R. Staszewski, and J. Long, "A 56.4-to-63.4 GHz multi-rate all-digital fractional-N PLL for FMCW radar applications in 65-nm CMOS," *IEEE Journal of Solid-State Circuits*, vol. 49, no. 5, pp. 1081–1096, May 2014.

[42] K. Scheir, G. Vandersteen, Y. Rolain, and P. Wambacq, "A 57-to-66GHz quadrature PLL in 45nm digital CMOS," in *IEEE International Solid-State Circuits Conference Digest of Technical Papers*, IEEE, February 2009, pp. 494–495.

[43] A. Musa, R. Murakami, T. Sato, W. Chaivipas, K. Okada, and A. Matsuzawa, "A low phase noise quadrature injection locked frequency synthesizer for mm-wave applications," *IEEE Journal of Solid-State Circuits*, vol. 46, no. 11, pp. 2635–2649, November 2011.

[44] X. Yi, C. Boon, H. Liu, J. Lin, J. Ong, and W. Lim, "A 57.9-to-68.3GHz 24.6mW frequency synthesizer with in-phase injection-coupled QVCO in 65nm CMOS," in *IEEE International Solid-State Circuits Conference Digest of Technical Papers*, IEEE, February 2013, pp. 354–355.

[45] D. Cherniak, L. Grimaldi, L. Bertulessi, C. Samori, R. Nonis, and S. Levantino, "A 23GHz low-phase-noise digital bang-bang PLL for fast triangular and saw-tooth chirp modulation," in *IEEE International Solid-State Circuits Conference Digest of Technical Papers*, IEEE, 2018, pp. 248–250.

[46] D. Weyer, M. B. Dayanik, S. Jang, and M. P. Flynn, "A 36.3-to-38.2GHz -216dBc/Hz2 40nm CMOS fractional-N FMCW chirp synthesizer PLL with a continuous-time bandpass delta-sigma time-to-digital converter," in *IEEE International Solid-State Circuits Conference Digest of Technical Papers*, IEEE, 2018, pp. 250–252.

[47] J. Vovnoboy, R. Levinger, N. Mazor, and D. Elad, "A fully integrated 7583 GHz FMCW synthesizer for automotive radar applications with –97 dBc/Hz phase noise at 1 MHz offset and 100 GHz/mSec maximal chirp rate," in *2017 IEEE Radio Frequency Integrated Circuits Symposium (RFIC)*, IEEE, 2017, pp. 96–99.

[48] M. Ferriss, B. Sadhu, A. Rylyakov, H. Ainspan, and D. Friedman, "A 12-to-26GHz fractional-N PLL with dual continuous tuning LC-D/VCOs," in *IEEE International Solid-State Circuits Conference Digest of Technical Papers*, IEEE, 2016, pp. 196–198.

[49] S. Kim and M. Soma, "An all-digital built-in self-test for high-speed phase-locked loops," *IEEE Transactions on Circuits and Systems II*, vol. 48, no. 2, pp. 141–150, February 2001.

[50] M. Bums and G. Roberts, *An Introduction to Mixed Signal IC Test and Measurement*. Oxford University Press, 2001.

[51] P. Goteti, G. Devarayanadurg, and M. Soma, "DFT for embedded charge-pump PLL systems incorporating IEEE 1149.1," in *Proceedings of the IEEE Custom Integrated Circuits Conference*, IEEE, May 1997, pp. 210–213.

[52] M. Toner and G. Roberts, "On the practical implementation of mixed analog-digital BIST," in *Proceedings of the IEEE Custom Integrated Circuits Conference*, IEEE, May 1995, pp. 525–528.

[53] R. Staszewski, I. Bashir, and O. Eliezer, "RF built-in self test of a wireless transmitter," *IEEE Transactions on Circuits and Systems II*, vol. 54, no. 2, pp. 186–190, February 2007.

[54] O. Eliezer, O. Friedman, and R. Staszewski, "A built-in tester for modulation noise in a wireless transmitter," in *Proceedings of Fifth IEEE Dallas Circuits and Systems Workshop: Design and Application and Integration and Software (DCAS-06)*, IEEE, October 2006, pp. 59–62.

[55] O. Eliezer and R. Staszewski, "Built-in measurements in low-cost digital-RF transceivers," in *IEICE Transactions on Electronics*, IEEE, June 2011, pp. 930–937.

[56] K. Raczkowski, N. Markulic, B. Hershberg, and J. Craninckx, "A 9.2-12.7 GHz wideband fractional-N subsampling PLL in 28nm CMOS with 280fs RMS jitter," in *IEEE Radio Frequency Integrated Circuits Symposium (RFIC)*, IEEE, June 2014, pp. 89–92.

[57] W.-S. Chang, P. C. Huang, and T.-C. Lee, "A fractional-N divider-less phase-locked loop with a subsampling phase detector," *IEEE Journal of Solid-State Circuits*, vol. 49, no. 12, pp. 2964–2975, December 2014.

[58] N. Markulic, K. Raczkowski, P. Wambacq, and J. Craninckx, "A 10-bit and 550-fs step digital-to-time converter in 28nm CMOS," in *IEEE European Solid State Circuits Conference (ESSCIRC)*, IEEE, September 2014, pp. 79–82.

[59] S. Yoo, S. Choi, J. Kim, H. Yoon, Y. Lee, and J. Choi, "A PVT-robust -39dBc 1kHz-to-100MHz integrated-phase-noise 29GHz injection-locked frequency multiplier with a 600μW frequency-tracking loop using the averages of phase deviations for mm-band 5G transceivers," in *IEEE International Solid-State Circuits Conference Digest of Technical Papers*, IEEE, February 2014, pp. 324–325.

[60] W. El-Halwagy, A. Nag, P. Hisayasu, F. Aryanfar, P. Mousavi, and M. Hossain, "A 28GHz quadrature fractional-N synthesizer for 5G mobile communication with less than 100fs jitter in 65nm CMOS," in *2016 IEEE Radio Frequency Integrated Circuits Symposium (RFIC)*, IEEE, 2016, pp. 118–121.

[61] G. Brooker, "Understanding millimetre wave FMCW radars," *Proceedings of the First International Conference on Sensing Technology*, IEEE, November 2005, pp. 152–157.

[62] B. Catli and M. Hella, "Triple-push operation for combined oscillation/divison functionality in millimeter-wave frequency synthesizers," *IEEE Journal of Solid-State Circuits*, vol. 45, no. 8, pp. 1575–1589, August 2010.

[63] X. Zhang, X. Zhou, and A. Daryoush, "A theoretical and experimental study of the noise behavior of subharmonically injection locked local oscillators," *IEEE Transactions on Microwave Theory and Techniques*, vol. 40, no. 5, pp. 895–902, May 1992.

[64] W. Chan and J. Long, "A 56-65 GHz injection-locked frequency tripler with quadrature outputs in 90-nm CMOS," *IEEE Journal of Solid-State Circuits*, vol. 43, no. 12, pp. 2739–2746, December 2008.

[65] S. Shahramian, A. Hart, A. Tomkins, A. Carusone, P. Garcia, P. Chevalier, and S. Voinigescu, "Design of a dual W- and D-band PLL," *IEEE Journal of Solid-State Circuits*, vol. 46, no. 5, pp. 1011–1022, May 2011.

12 Practical VCO Design

Mohyee Mikhemar

Accurate and efficient clock generation is crucial for all forms of high-performance wireless communications. At least one voltage-controlled oscillator (VCO) is needed at the core of any clock generation system. In this chapter, the design of high-frequency CMOS VCO is discussed with emphasis on the practical aspects of the design that might be overlooked or deemphasized in textbooks and technical publications. The design approaches and techniques presented are valid over a wide range of frequency. This is done on purpose, because in a practical mm-wave transceiver the VCO might run at lower frequency to optimize the overall solution power consumption. Moreover, in 5G applications with backward compatibility, one VCO at an intermediate frequency can be used to cover low-frequency bands through frequency division and the mm-wave 5G bands using a frequency multiplier. The chapter starts by introducing a generic local oscillator (LO) chain and discussing the three most common LO architectures in use today, and shows how they affect the VCO specifications. Then in Section 12.2, the basics of VCO design are reviewed with emphasis on design parameters with practical significance. The effect of frequency scaling on VCO performance is discussed in Section 12.3. A step-by-step design procedure is presented in Section 12.4. The chapter concludes with Section 12.5, which discusses various practical considerations in modern VCO design, with special attention to parasitic effects that become more important for high-frequency designs.

12.1 LO Design

LO is a broad term that describes the clock generation circuits in a transceiver. The main function of the LO is to generate clocks for the up-conversion mixer in the transmitter and down-conversion mixer in the receiver. The final clocks driving the mixers should be available in quadrature phases to support complex modulation. A generic LO block diagram is shown in Figure 12.1; it consists of the VCO inside a phased-locked loop (PLL) to accurately set its average oscillation frequency to $f_{vco} = Nf_{ref}$, where N is the division ratio of the PLL, and is in general a fractional number. In the generic diagram of Figure 12.1, the PLL output frequency can be further scaled by a factor A/B, using a combination of frequency multipliers and dividers. The frequency scaling is done to decouple the VCO frequency from the final LO frequency. This can be beneficial in mitigating frequency pulling of the VCO by the

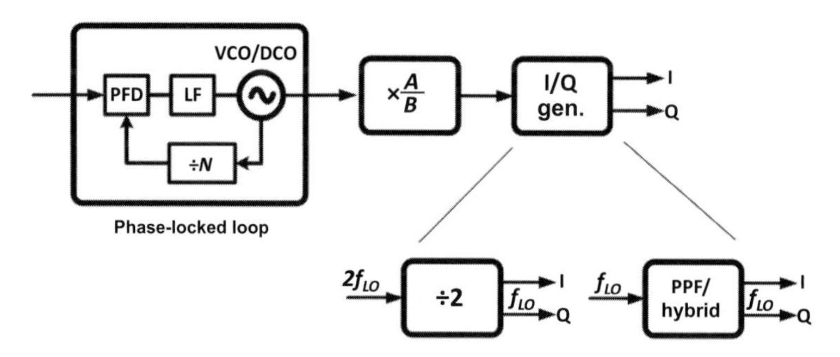

Figure 12.1 Generic block diagram of an LO chain.

transmitter (TX) or the power amplifier (PA) output fundamental or harmonics [1]. In high-frequency applications, it is also common to use a frequency multiplier to generate a high-frequency clock from a lower-frequency PLL to optimize the total power consumption. The final block of Figure 12.1 is the quadrature generation, which is commonly implemented as a divide-by-2 circuit when a $2f_{lo}$ frequency is available in the system. If the quadrature phases are required at higher frequency where $2f_{lo}$ is not available, then a resistor-capacitor (RC)-poly phase filter (PPF) can be used [2], or for even higher frequency a lumped or distributed hybrid could be the optimal choice. In some applications where more phases of the LO clock are required, a Johnson divider can be used if a higher-frequency input is available; otherwise, injection-locked ring oscillators [3] or delay-locked loops (DLL) can be used [4].

12.1.1 LO Architectures

An LO architecture is typically chosen to meet the system requirements while minimizing power and area. Three common LO architectures are shown in Figure 12.2, the direct conversion (DC) topology is the simplest with one LO path from the PLL to the mixers. It is always the first choice for transceivers below 6 GHz because of its power and area advantages [5]. At higher frequencies, quadrature generation is more difficult [6,7], and therefore a two-step frequency conversion scheme becomes more plausible. In a two-step topology, the quadrature generation occurs at an intermediate frequency (IF) to optimize for area and power and to achieve good quadrature accuracy. The second frequency conversion does not require quadrature clocks because the resulting image will be far enough, $2f_{IF}$ from the desired signal [8]. Two realizations of the two-step conversion are shown in Figure 12.2. If the IF is generated from the same PLL as the radio frequency (RF) clock, then it is a sliding-IF architecture [9,10], because the IF changes with the channel frequency. On the other hand, in a fixed-IF architecture a second PLL is used to generate the constant IF frequency [11], and the RF LO has to cover the entire tuning range, albeit at a lower center frequency, which results in a larger relative tuning range for the fixed-IF architecture. Despite the extra hardware cost of the fixed-IF topology, it remains an attractive option for the practically segmented architecture shown in Figure 12.3, where the RF or mm-wave antenna, or phased array

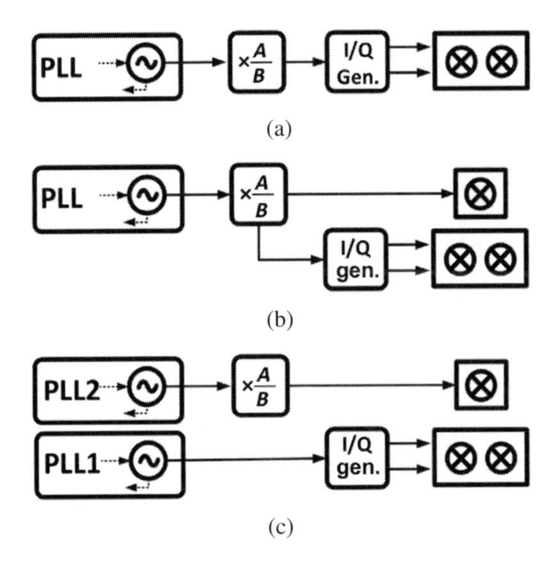

Figure 12.2 Common LO architectures: (a) direct conversion, (b) sliding IF, and (c) fixed IF.

(PA), has to be physically placed far from the application processor (AP). A single-chip solution would result in long routing either from AP to chip or from chip to antenna; both scenarios have a large power penalty and could result in unwanted interference in small form factor platforms, such as cell phones or tablets. Therefore, a typical practical segmentation is to have a modulator chip next to the AP that generates the modulated output at an IF or RF. The modulated output is then routed on a shielded thin coaxial cable to one front-end module (FEM) or multiple phased arrays. The cable typically carries other signals as well, for example the control signals exchanged between the modulator and the remote chips. It can also carry a low-frequency reference to remote PLLs, if needed. In a fixed-IF architecture, it is easier to combine and separate these signals at the two ends because the most critical signal on the cable is at a fixed frequency. Moreover, a fixed-IF frequency enables a more efficient amplification of the IF signal because a wider IF bandwidth would inevitably reduce the gain per stage for the same power, in the signal path. In general, the platform partitioning in Figure 12.3 can be used with a DC-LO or a sliding IF LO architectures in the modulator chip, given that the cable loss at the RF frequency is tolerable. This is definitely the case for lower-frequency signals. However, for mm-wave applications, a fixed-IF topology might offer a better overall solution [11].

12.1.2 Impact of LO Architecture on VCO Requirements

The VCO requirements for DC-LO architecture in Figure 12.2a is easily derived from the LO requirements. The tuning range and the phase noise (PN) requirements are the same as the LO when referred to the LO frequency. In a sliding-IF architecture, the VCO PN appears at both the IF and RF LO outputs and add coherently, resulting in a stringent

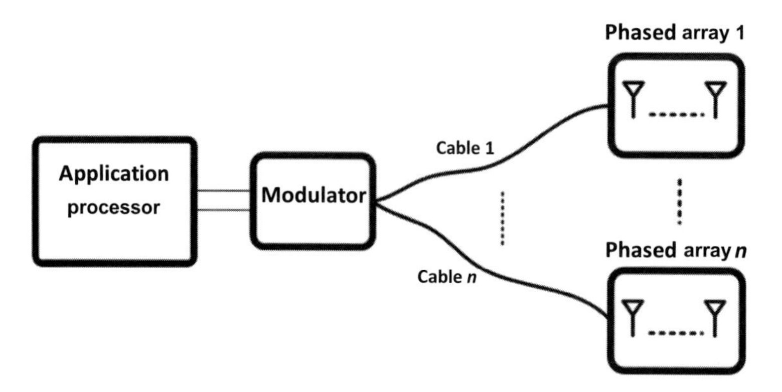

Figure 12.3 Platform segmentation into the modulator and front-end.

PN requirement compared to the fixed-IF case. On the other hand, the tuning range required from an RF-VCO, in a fixed-IF architecture, is wider than the LO tuning range.

$$TR_{RF\text{-}VCO,\ fixed\text{-}IF} = TR_{LO} \cdot \frac{f_{RF}}{(f_{RF} - f_{IF})} > TR_{LO} \tag{12.1}$$

A summary of the VCO requirements for the three LO architecture is given in Table 12.1. It is obvious that the VCO requirements vary depending on the topology. A fixed-IF topology has two VCOs, and one of them has relaxed noise and tuning range specifications. And the other VCO seems comparable to DC-VCO but with a wider tuning range and a relaxed PN. In Section 12.2, the trade-off between VCO tuning range and PN will be discussed and quantified. The design of a narrowband VCO for fixed-IF architecture is discussed in Section 12.5.4. Another critical decision in an LO topology is to decide whether to use a VCO at the required frequency or to use a VCO at half the frequency followed by a multiplier. The former arrangement always results in a better spectrum, with fewer spurs, because the number of frequencies in the system is lower. But the latter can be more power efficient. In fact, for a given tuning range, there is an optimal maximum frequency to implement a VCO for a given technology, as explained in Section 12.3.

12.2 Fundamentals of VCO Design

There are many types of oscillators that can be integrated on a silicon substrate. For RF applications, Colpitts and cross-coupled LC oscillators are the most-used topologies. In this chapter, we will focus on LC oscillators owing to their excellent performance for a wide range of requirements and applications. A good VCO design has to meet the following three requirements:

1. It has to start and sustain the oscillation over the frequency range and for all process, supply voltage, and temperature (PVT) corners.

Table 12.1 VCO requirements for different LO architectures.

Architecture	Tuning range	Phase noise	Quad. generation
DC VCO	LO-TR	LO-PN	Difficult
Sliding-IF VCO	LO-TR	LO-PN	Easier
Fixed-IF RF VCO	>LO-TR	LO-PN, relaxed	No need
Fixed-IF IF VCO	No need	Relaxed	Easier

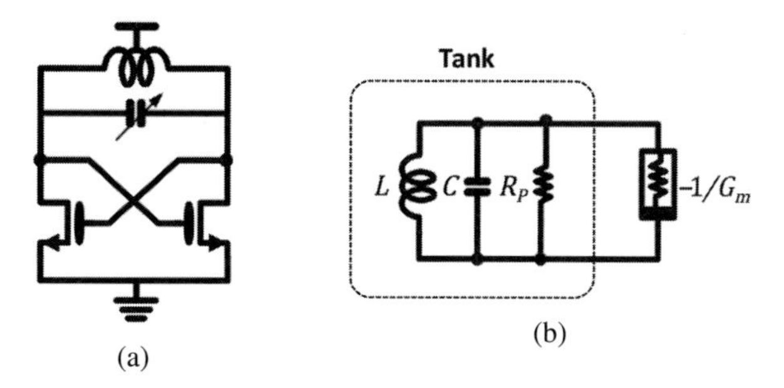

Figure 12.4 (a) Simplified NMOS VCO schematic and (b) its equivalent circuit.

2. A good VCO has to meet its noise specification, typically in the form of integrated PN over a bandwidth and spot PN for some frequency offsets.
3. Finally, the VCO has to meet the preceding requirements efficiently in terms of area and power, which are best expressed in terms of figure-of-merit (*FoM*).

A simplified schematic of an N-type metal-oxide-semiconductor (NMOS) VCO is shown in Figure 12.4a. The LC tank sets the oscillation frequency of the VCO, while the cross-coupled NMOS pair provides the negative resistance needed to compensate the tank loss and sustain the oscillation.

The equivalent circuit of the VCO is shown in Figure 12.4b, the resistor R_p represents the tank loss. The loop gain, *LG*, of the circuit at or around resonance is as follows:

$$LG = G_m \cdot R_p \tag{12.2}$$

The LG defined in (12.2) is a small-signal quantity that is valid only at the start of oscillation and when the VCO signals are around the zero-crossings. To satisfy the startup requirement, the condition $LG > 1$ has to be satisfied over all operating conditions. But to achieve good PN performance using common-mode resonance, the loop gain should be maximized [12].

The well-known Lesson's expression introduced in [13] expresses the PN at a frequency offset $\Delta\omega$ from the oscillation frequency of ω_o as follows:

$$L\{\Delta\omega\} = \frac{4KTFR_P}{A_c^2}\left(\frac{1}{2Q_t}\right)^2\left(\frac{\omega_o}{\Delta\omega}\right)^2 \tag{12.3}$$

where Q_t is the quality factor of the tank, A_c is the differential oscillation amplitude, and F is the noise factor of the circuit, which is the total noise of the VCO normalized to the tank noise. A quick look at (12.3) reveals that PN can be improved by maximizing the oscillation amplitude and tank quality factor, while minimizing the noise factor and the tank loss R_P. In many practical cases, the quality factor of the inductor is much smaller than the capacitor, thus dominating the tank quality factor: $Q_t \approx Q_{ind}$. In this case, (12.3) can be simplified to the following:

$$L\{\Delta\omega\}|_{Q_t \approx Q_{ind}} = \frac{\omega_o KT}{A_{c,max}^2} \left(\frac{\omega_o}{\Delta\omega}\right)^2 \left(\frac{FL}{Q_{ind}}\right) \tag{12.4}$$

where $A_{c,max}$ is the maximum oscillation amplitude and equals $2V_{DD}$ for the NMOS VCO in Figure 12.4a. The last bracket in (12.4) has all the remaining design parameters. And to reduce PN further, the inductance should be minimized, while maximizing its quality factor. In other words, the inductance should minimize the ratio L/Q_{ind}.

In summary, to improve PN, the designer should do the following:

1. Maximize the oscillation amplitude to the level allowed by the supply and reliability.
2. Maximize tank Q for a given area. Higher-Q inductors are typically larger.
3. Minimize tank inductance, at the expense of more power consumption.
4. Minimize noise factor, by preventing the switching pair from operating in triode as explained in Section 12.2.1.

The final requirement for a good VCO design is to achieve its PN target efficiently with the minimum power. Fortunately, the *FoM* defined in (12.5) is a fairly accurate and widely used metric for evaluating the power efficiency of a VCO design.

$$FoM = \frac{\left(\frac{\omega_o}{\Delta\omega}\right)^2}{L\{\Delta\omega\} P_{DC,mW}} \tag{12.5}$$

where $P_{DC,mW}$ is the DC power consumption of the oscillator in milliwatts. The *FoM* is a measure of how efficiently the PN performance was achieved. It is independent of oscillation frequency. The *FoM* should be frequency-independent if the source of VCO noise is thermal. In practice, the *FoM* is lower at low frequency because of flicker noise and becomes constant in the thermal noise region. In general, a better design has a higher *FoM*.

In [14] and [15], the *FoM* definition (12.5) was rewritten in terms of three main design parameters: efficiency η, noise factor F, and tank quality factor Q_t:

$$FoM = \frac{2\eta Q_t^2}{KTF} \cdot 10^{-3} \tag{12.6}$$

where η is the power efficiency of the VCO core and is defined as follows:

$$\eta = \frac{P_{Tank,mW}}{P_{DC,mW}} = \frac{\left(\frac{A_c^2}{2R_P}\right) \cdot 10^3}{P_{DC,mW}} \tag{12.7}$$

This efficiency resembles that of a power amplifier (PA). In fact, there are many similarities between power amplifiers and oscillators because both of them are large signal circuits. An efficiency comparison between Class-B and Class-C VCOs is listed in Table 12.2. Just like PAs, the class of a VCO is controlled via the bias of metal-oxide-semiconductor (MOS) devices. It should be noted also that for Class-B and Class-C modes, the peak efficiency of a complementary metal-oxide semiconductor (CMOS) VCO is the same as that of an NMOS or PMOS counterpart. This is in contrast to Class-A operation where the CMOS has higher efficiency than the NMOS because in a CMOS amplifier the bias current is shared between the NMOS and PMOS halves. Class-A operation should be avoided in oscillators because of its poor efficiency.

Now back to the *FoM* expression in (12.6). Note that the *FoM* is independent of inductance value, even though it is clear from (12.4) that PN improves with a lower inductance. This is because a lower inductance will require proportional increases in power consumption to maintain the output swing, resulting in no change to *FoM*.

The second factor affecting the *FoM* is the tank quality factor Q_t, which should be maximized by proper design of the inductor and capacitor bank. The quality factor of the inductor is a function of the available top metal layers and the area assigned for the design as explained in [16]. On the other hand, the switched capacitor quality factor depends on the required tuning range and the availability of a good switch. The performance metric of a switch is the product of its on resistance and off capacitance $\tau_{sw} = R_{on} \cdot C_{off}$. The unit switch quality factor can be written as follows:

$$Q_{cap} = \frac{1}{\omega R_{on} C_{on}} = \frac{1}{\omega \tau_{sw} \frac{C_{on}}{C_{off}}} \tag{12.8}$$

It is clear that cap quality factor is inversely proportional to the cap ratio C_{on}/C_{off}. A switched capacitor bank is essentially a digitally controlled capacitor. Like any good digital-to-analog converter (DAC), it should be segmented into thermometer and binary sections to optimize area and achieve target differential nonlinearity (DNL) [17]. The thermometer section typically dominates the area and quality factor. For an n-bit thermometer bank, the total capacitance varies from $C_{v,min} = (2^n - 1) \cdot C_{off}$ to a maximum value of $C_{v,max} = (2^n - 1) \cdot C_{on}$. The required ratio of $C_{v,max}$ to $C_{v,min}$ is a function of the target tuning range and fixed capacitance value:

$$\frac{C_{v,max}}{C_{v,min}} = \frac{(2+TR)^2}{(2-TR)^2} + \frac{8TR}{(2-TR)^2} \frac{C_{fixed}}{C_{v,min}} \tag{12.9}$$

Table 12.2 Comparison of Class-B and Class-C VCO efficiencies.

Parameter	NMOS VCO	CMOS VCO
Efficiency η Class-B	$\frac{1}{\pi}\frac{A_c}{V_{DD}}$	$\frac{2}{\pi}\frac{A_c}{V_{DD}}$
Efficiency η Class-C	$\frac{A_c}{2V_{DD}}$	$\frac{A_c}{V_{DD}}$
Peak swing $A_{c,max}$	$2V_{DD}$	V_{DD}
Peak efficiency η_{peak} Class-B	64%	64%
Peak efficiency η_{peak} Class-C	100%	100%

where *TR* is the target frequency tuning range defined as $TR = 2\frac{f_{max}-f_{min}}{f_{max}+f_{min}}$, and C_{fixed} is the total fixed capacitance, which is typically dominated by the switching-pair device capacitance. The required capacitance ratio from (12.9) is plotted in Figure 12.5. A higher fixed capacitance increases the tuning requirements on the variable cap, thus degrading its quality factor. For example, to achieve a target tuning range of 25%, with an extra margin of about (7 − 10) % to cover PVT variations, a capacitance ratio $\frac{C_{v,max}}{C_{v,min}}$ of about 3 is needed if $C_{fixed} = C_{v,min}$, and the required ratio exceeds 4 if $C_{fixed} = 2C_{v,min}$. In a 28 nm technology, a capacitor bank with a capacitance ratio of 3 has a quality factor of around 40 at 10 GHz, and the quality factor drops to about 30 for a tuning ratio of 4.

The third term in *FoM* is the noise factor, F, which is the toughest term to minimize, therefore it will be discussed in the following subsection.

12.2.1 Improving Noise Factor by Avoiding Triode Operation

The VCO noise factor is defined by the ratio of total VCO phase noise power PN_{VCO} divided by the tank phase noise power PN_{tank} as

$$F = \frac{PN_{VCO}}{PN_{tank}} \tag{12.10}$$

Therefore, an ideal VCO with noiseless transistors would have an $F_{noiseless} = 1$. If the tank is implemented using a transformer with voltage gain Av, as proposed by [18], then the optimal noise factor is $F_{optimal} = 1 + \gamma/Av^2$, where γ is the transistor channel thermal noise factor. This topology suffers from reliability limitations caused by the amplified voltage swing, and therefore is not suitable for nanometer CMOS.

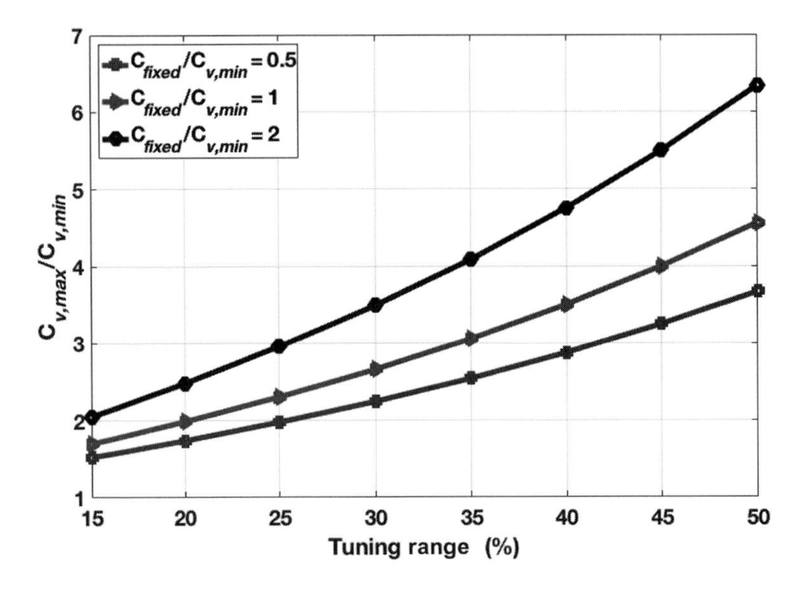

Figure 12.5 Required on/off capacitance ratio vs. tuning range.

For a practical VCO, like the one in Figure 12.6a, [19] have shown that the optimal noise factor is $F_{optimal} = 1 + \gamma$. The optimal F can be achieved only if the switching transistors did not enter triode region. The design of a good VCO, with high *FoM*, is all about avoiding triode operation one way or another. As shown in Figure 12.6b, when the gate-to-drain voltage of the NMOS device M1 exceeds the threshold voltage of the transistor, it enters triode region. The same happens to M2 in the second half of the cycle. The on resistance of the transistor in the triode region degrades the *FoM* in two ways: (1) it degrades tank Q by providing a low impedance path from the tank to ground, which increases tank loss and (2) the thermal noise from the triode resistance appears at the output and increases the noise factor.

The literature of VCO design is full of techniques to reduce noise factor by mitigating PN degradation by triode operation. In [20], a tail tank is introduced in the common-mode path to resonate at $2f_{VCO}$ and isolate the tank from ground, as shown in Figure 12.7a. The resonance has to be at twice the oscillation frequency because in one cycle both transistors enter triode region during one-half of the cycle (see Figure 12.6b), thus common-mode-wise, this happens twice every cycle, and can be prevented by the high impedance of the tail tank at $2f_{VCO}$.

The Class-C VCO [21] is another topology that is commonly used to mitigate triode operation (see Figure 12.7b). In a Class-C VCO, the switching pair bias is adjusted to reduce the conduction angle and avoid triode region. This is equivalent to increasing the threshold voltage in Figure 12.6b. The Class-C VCO improves both the noise factor and the efficiency. However, the bias resistors R_b appear in parallel with the tank and should be designed large enough to avoid degrading the tank Q, but not too large to dominate the thermal noise.

A more recent version of common-mode resonance was introduced by [12], and is shown in Figure 12.7c. In this topology, the tank capacitance is divided into differential and single-ended banks. A proper choice of the coupling factor k and the ratio of differential to common-mode capacitance C_{dm}/C_{cm} would result in a common-mode resonance at $2f_{VCO}$ in addition to the main differential-mode resonance at f_{VCO}.

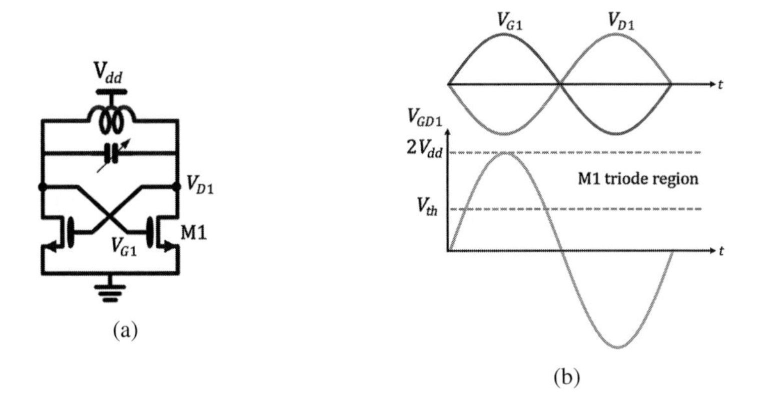

Figure 12.6 (a) NMOS VCO and (b) the associated voltage waveforms.

As clearly explained in [12], the three topologies in Figure 12.7 can achieve comparable performance. In fact, the best published *FoM* is for these three topologies and their variations. The same paper lists the conditions to maximize the efficacy of common-mode resonance: (1) a high loop gain to effectively suppress noise degradation from triode operation; (2) accurate modeling of the common-mode network of the VCO, which is discussed in more detail in Section 12.5.1; and (3) a smaller gap between the supply voltage and the device threshold voltage seems to minimize triode operation and improve noise factor, as is the case for a CMOS VCO where $V_{DD} \simeq V_{tn} + V_{tp}$, thus triode region degradation is minimal. The reader is encouraged to read [12] and [21] for a thorough discussion of common-mode resonance.

A well-designed production-worthy VCO that avoids triode operation should achieve a measured *FoM* within 2–3 dB of its optimal value. A summary of optimal *FoM* expressions is shown in Table 12.3. The *FoM* value has strong dependence on the tank quality factor, which is mostly dictated by the technology in two ways: (1) low, resistance, thick metal layers can be used to realize high-Q inductors and low loss routing and (2) a good nanometer process with excellent switches facilitates the design of a high-Q switched cap bank with the desired tuning range. Unfortunately, there is not much a designer can do to improve the tank quality factor beyond what the technology can offer. On-chip inductors have a quality factor the ranges from the low 10s at lower frequency to the high 20s for a process with a good metal stack and higher frequencies. The quality factor of switched caps varies greatly over process and tuning range and is inversely proportional to frequency. As a numerical example, some values of the optimal *FoM* for $Q_t = 12$ are listed in Table 12.3. As mentioned earlier, the measured *FoM* should be within 2–3 dB from the optimal value, so as a rule of thumb, a measured *FoM* of 190 dB or better indicates a well-designed VCO.

12.3 VCO Frequency Scaling

As discussed in Section 12.1, it is a common practice to design the VCO/PLL at a different frequency than the final LO. This is done in low-frequency applications to

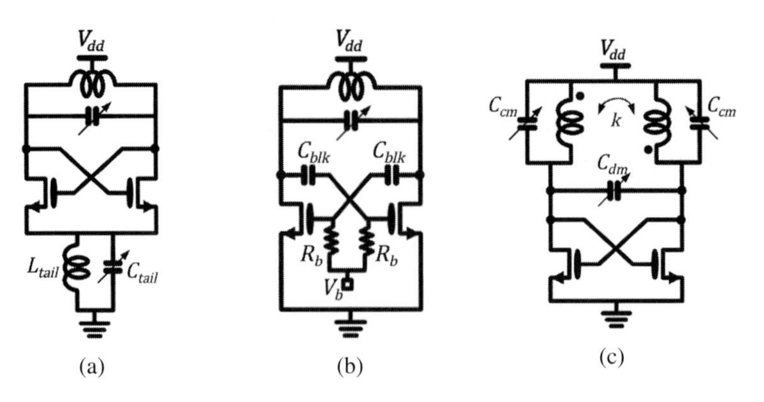

Figure 12.7 (a) VCO with tail tank tuned at $2f_{LO}$, (b) class-C VCO, and (c) implicit common-mode tank resonance.

Table 12.3 Optimal *FoM* for some VCO topologies.

Topology	Optimal *FoM*	$FoM_{opt}(Q_t = 12)$
Ideal noiseless VCO	$176.8 + 20 \log Q_t$	198.4 dB
Class-C VCO ($\gamma = 1$)	$173.8 + 20 \log Q_t$	195.4 dB
Class-B VCO ($\gamma = 1$)	$171.9 + 20 \log Q_t$	193.5 dB

mitigate VCO pulling or to avoid spurs, and in mm-wave designs to optimize the system power consumption. In mm-wave applications, a lower-frequency VCO/PLL output is multiplied to generate the mm-wave LO. This is done because conventional wisdom suggests that high-frequency VCOs are less power efficient than their lower-frequency counterparts. In Section 12.2, the *FoM* was introduced as a normalized measure of the oscillator power efficiency. The *FoM* frequency dependence follows its three main contributing factors: switching efficiency, noise factor, and tank quality factor, as shown in (12.6) and repeated here for convenience.

$$FoM = \frac{2\eta Q_t^2}{KTF} \cdot 10^{-3} \tag{12.11}$$

Ideally, the switching efficiency of the VCO should not change with frequency as long as the transistor resistive loss is not dominant, which is the case up to about $f_{max}/5$, where f_{max} is the maximum oscillation frequency of the transistor, which exceeds 150 GHz for 40 nm technology and below.

Similarly, one would expect that the noise factor should not degrade much with frequency up to $f_{max}/5$, but practically it does degrade much more than the efficiency, for three reasons. First, effective common-mode resonance, which is key to low F, requires large loop gain, which means large device size and more fixed capacitance added to the tank. As shown in Figure 12.5, a higher fixed capacitance requires a larger switching ratio and thus lower quality factor for a given tuning range. Moreover, as the oscillation frequency gets higher, the total capacitance budget is reduced and the loop gain is limited by the tuning range, unless the inductance is reduced to increase the total capacitance but at the expense of higher power consumption. The second reason for noise factor degradation over frequency is the increased difficulty of accurately modeling the common-mode path of the VCO at very high frequency, where second-order effects cannot be neglected. Therefore, for high-frequency VCOs, it might be better to design a low loop-gain VCO to minimize noise generation and do without the common-mode resonance. Finally, the third element affecting *FoM* is the tank quality factor. The two main reactances in the tank have different response with frequency. The quality factor of the capacitor bank is inversely proportional to frequency, $Q_{cap} \propto 1/f$. On the logarithmic scale in Figure 12.8, the capacitor quality factor versus frequency has a slope of -1. On the other hand, the quality factor of the inductor improves with frequency. The frequency response of the inductor quality factor depends on the inductor design and the dominant loss mechanism. If the ohmic metal loss of the inductor dominates, then $Q_{ind} \propto \sqrt{f}$, or a slope of $+0.5$ on a logarithmic scale. It is clear from Figure 12.8 there is an optimal frequency at which the tank quality factor

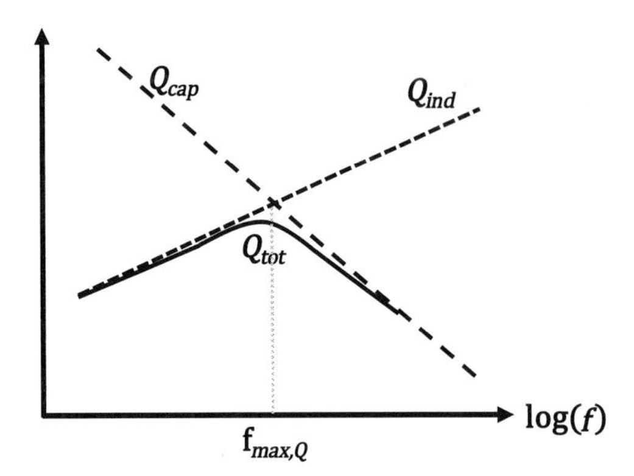

Figure 12.8 Asymptotes of the tank inductor and capacitors quality factors versus frequency and their intersection at the maximum quality factor frequency.

is maximized. It is desirable to operate around this frequency to maximize the *FoM*, assuming the noise factor and efficiency are holding up. As technology scales down and better switches are available, while inductors remain almost unchanged, the cap quality factor line moves to the right, resulting in higher optimal frequency and tank quality factor. It should be noted that the optimal frequency is also a function of tuning range. In a 28 nm technology, and with 35% tuning range, the optimal frequency is about 10 GHz.

There is another practical limitation on the value of the inductance, which depends on the PN specification: for very low noise requirements, the inductance value might be too low to implement or just comparable to the routing inductance, which has a much lower quality factor than the main inductor. For these cases, a lower-frequency VCO is a better option, or a multiple-core VCO where the small inductor is implemented as a multiple high-Q inductors in parallel to maintain the tank quality factor at the expense of larger area [25, 26].

12.4 Design Procedure

The following steps are given as a guideline to VCO designers to make the design process more structured:

1. Analyze PN requirements to find the two parameters defining the VCO PN: the spot PN at some offset, say 1 MHz; and the flicker noise corner f_{fnc}. The VCO PN can then be modeled by the following expression:

$$L\{\Delta\omega\} = 10\log_{10}\left[\frac{K_{PN}}{f^2}\left(1 + \frac{f_{fnc}}{f}\right)\right] \tag{12.12}$$

2. Using the target power consumption, calculate *FoM* from (12.5).
3. Calculate the tank quality factor needed to achieve the *FoM* with the aid of Table 12.3, and make sure it is achievable with the used technology.

(a)

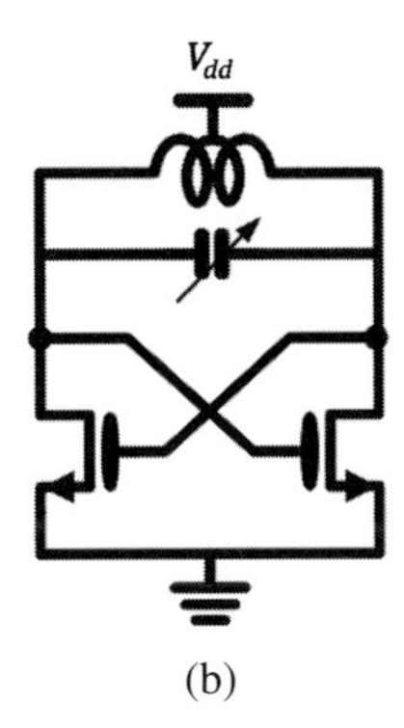

(b)

Figure 12.9 Topology choices: (a) CMOS architecture vs. (b) NMOS.

4. Decide on the VCO topology, NMOS versus CMOS, as shown in Figure 12.9. The first can make use of the large swing to achieve low phase noise, but the large swing comes with the cost; the transistors used in the switching pair and the switched-cap bank have to be thick-oxide to meet the reliability requirements. Thick-oxide transistors add more fixed capacitance to the tank and have a lower quality factor as switches. In a CMOS counterpart, core devices can be used and the benefit of down scaling is realized. Moreover, it is practically easier to get the common-mode resonance to work in a CMOS design, as explained in Section 12.5.1.

   Finally, the inductor value in a CMOS design is about one-fourth that of an NMOS design, so for very low phase noise requirements, the inductor value might be too low to implement or even comparable to the routing inductance, which has a poor quality factor.
5. Following the topology choice and the target frequency, decide whether a high LG is possible, or for high-frequency design, it is better to keep the LG low to minimize noise generation. Design the cross-coupled pair accordingly.
6. Plan the capacitance budget to achieve the target quality factor and tuning range, and add the varactor capacitance for analog PLLs. Design the best inductor you can in the given area.
7. Follow all practical considerations discussed in Section 12.5 to accurately simulate the performance and minimize the delta between simulations and measurements.

12.5 Practical Considerations in VCO Design

It seems from the previous sections that designing a VCO is a straightforward task. You understand the requirements and characterize the technology, then make the design decisions and execute. The first three steps are as easy as they sound, but the execution part is the tricky one where many designs fail to deliver the expected performance.

The measured *FoM* of a well-designed VCO is typically within 1 dB from simulations, in the thermal-noise region and up to 3 dB worse at the flicker noise corner frequency. In practice, however, an unexperienced designer might see much higher discrepancies. The main source of degradation is typically an incomplete simulation test bench that lacks proper modeling of all potential contributors. Here are a few recommendations:

1. Active circuits should be RC-extracted, while all internal routing, tank capacitors, and inductors should be modeled using electromagnetic (EM) tools.
2. Model the supply and ground network, including routing inductance, resistance, and bypass capacitance (a more detailed discussion follows in Section 12.5.1).
3. Include a full low-dropout regulator (LDO) circuit or, at least, source impedance and noise.
4. The VCO output should be probed after the first buffer and include the buffer supply network and noise.
5. Simulate the VCO single-ended PN to check for common-mode noise that might become a problem in the presence of mismatch.
6. Finally, if a large inductor is present close to the VCO, within 5× of the inductor larger dimension, the coupling should be investigated, but not necessarily added to the test bench. In extreme cases, an LO buffer might have excessive supply noise up-converted to the fundamental or the second harmonic, couple back to VCO, and inject noise into the VCO.

In the following subsections, some of the aforementioned effects will be discussed in some detail.

12.5.1 Tail Tuning and Bypass Capacitance

In a high-gain VCO, a good *FoM* can be only achieved with common-mode (CM) resonance, where the CM of *the entire VCO circuit* resonates at the second harmonic. The complete common-mode path of the NMOS VCO is shown in Figure 12.10. The CM current flows into the center tap of the load inductor and through the bypass can before it returns to the tail network. Ideally, the bypass capacitance should appear as a short circuit at the second harmonic, connecting the tail network directly to the center tap of the inductor. However, if the self resonance of the bypass capacitance is lower than the second harmonic, then the bypass capacitance will appear as a net inductor leading to a shift in the CM resonance and a significant degradation in PN. Moreover, the bypass capacitance and its routing is in parallel with the routing inductance to ground and to the LDO. Therefore, the designer should ensure that the parallel resonance of this circuit occurs at lower frequency, and that in the band of interest, the bypass capacitance shorts the routing, LDO, and ground paths. This issue is particularly challenging in NMOS VCOs because the NMOS devices are typically placed close to the inductor differential terminals to minimize the routing and hence the loss. This means that the center tap of the inductor is far from the NMOS devices and the tail network. The routing inductance from the inductor center tap to the tail network should be minimized by using wide

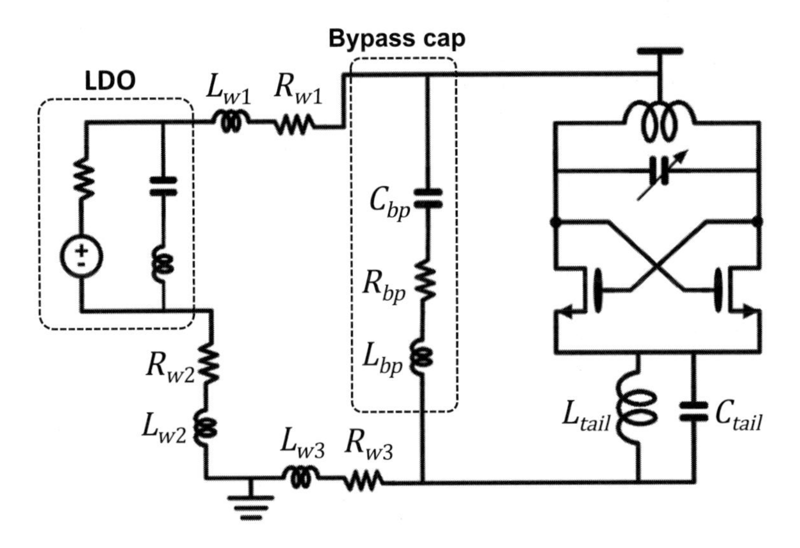

Figure 12.10 NMOS VCO with supply, ground, and bypass cap network.

metals [16] and placing them physically as close as possible to each other. The bypass capacitance should also be designed carefully to tap the low inductance line with minimum inductance. As a numerical example, a 10 pF bypass capacitance would resonate at 10 GHz if the routing inductance is 25 pH, so to operate well below resonance the routing inductance should be less than 10 pH, which is too low to realize. The common solution to this problem is to stagger banks of bypass capacitors, ranging from small caps with large self-resonance frequency (SRF) to large caps with smaller SRF. As a numerical example, it is possible to implement a 1 pF capacitor in 28 nm technology with SRF > 100 GHz and Q > 70 at 50 GHz. A good way to build the high SRF cap is to start with a small square unit cap of about 3–5 μm each side built from lower metal layers to achieve good quality factor and high density, then use low-inductance routing on higher metal layers to connect the units together as shown in Figure 12.11a. An accurate model of the bypass capacitance is crucial to ensure proper common-mode resonance. Ideally, the entire structure in Figure 12.11a should be modeled with an EM simulator; however, the computation time and resources could be excessive because the unit is built with minimum-width metal, while the routing uses wide upper metals. To alleviate this problem, the structure can be divided into two different cells, one for the routing and the other for the unit cap cell. The interaction between the two can be typically neglected. After each cell is extracted separately, in a much shorter simulation, the routing cell model is connected to the cap model we observing the correct port assignment. An example is shown in Figure 12.11; the ports marked on the routing diagrams, in (a), add up to 22 signal ports, therefore the routing model, in (b), has 22 ports that connect to the unit cap model. Now the bypass capacitance model consists of one 22-port model and 16 two-port model, but since the cap has only two terminals, a simple scattering parameter simulation of the network in Figure 12.11b would result in a compact, yet accurate, two-port representation of the bypass capacitor.

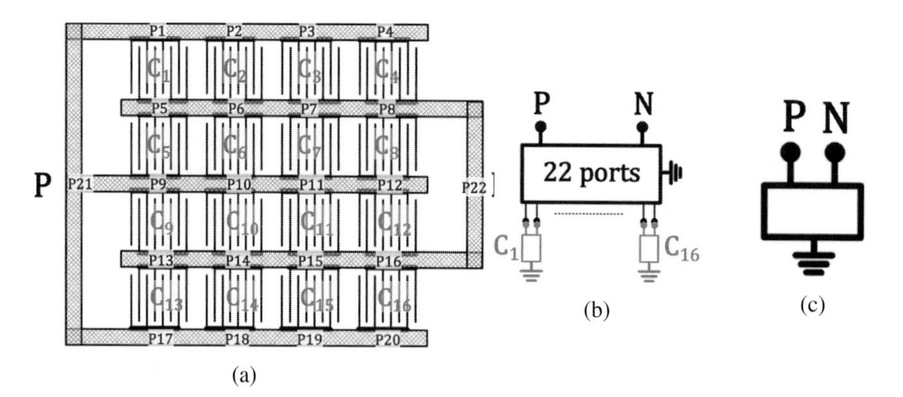

Figure 12.11 Example bypass capacitance layout and extraction, (a) typical cap-unit placement and routing, (b) combined schematic model, and (c) compact two-port model without loss of accuracy.

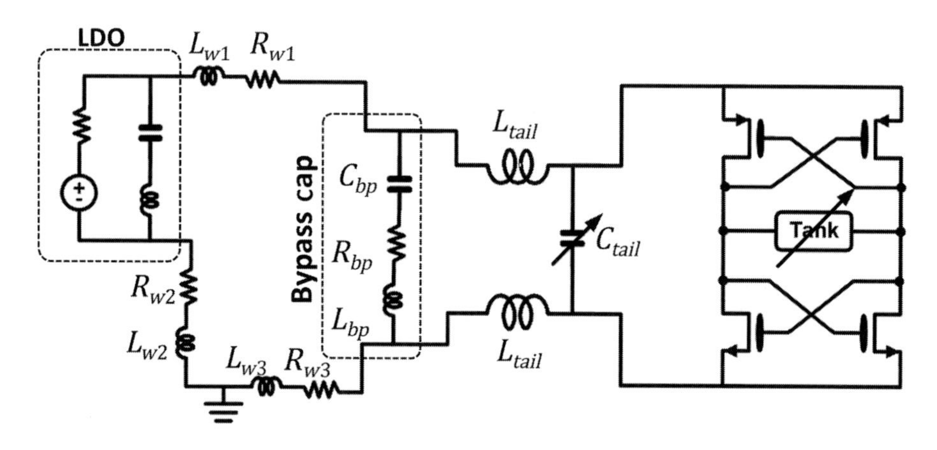

Figure 12.12 CMOS VCO with supply, ground, and bypass capacitance network.

As mentioned previously, controlling the return path of the common mode is particularly challenging in an NMOS VCO. The situation is better in a CMOS VCO, as shown in Figure 12.12; the NMOS and PMOS devices can be placed next to each other as they both connect to the same tank terminals, thus eliminating long routing in the common-mode path. Furthermore, the tail network is differential and can be implemented as coupled inductors [22], thus the bypass capacitance can be tightly connected to the tail network with minimal routing inductance. This is a major advantage for CMOS VCOs.

12.5.2 Kickback from the First VCO Buffer

The VCO tank is a high-Q RLC network. If the tank is loaded with a low-Q capacitance, the overall Q will be degraded, resulting in a lower *FoM*. Therefore, it is important to

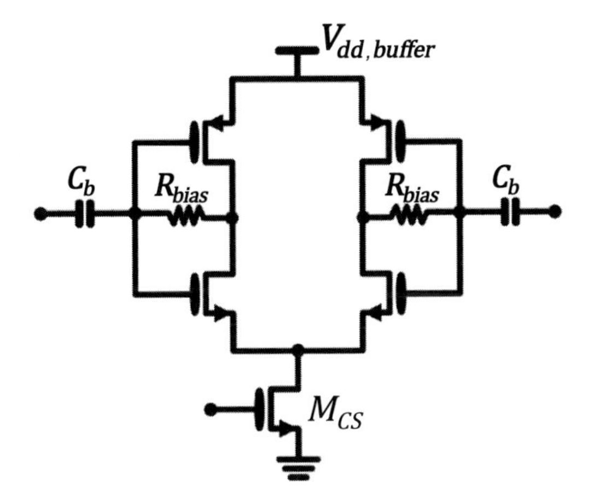

Figure 12.13 Example differential self-biased CMOS buffer with a tail current source that could be shorted.

add a buffer as close as possible to the tank to shield the VCO tank impedance and to protect the VCO core from unwanted coupling. An example VCO buffer circuit is shown in Figure 12.13. It is a self-biased inverter with an optional tail current source M_{CS}. The self-bias resistor R_{bias} ensures that the inverter is biased in the high gain region and therefore speeds up the transitions at the output of the buffer. It is expected that the buffer will have its own noise that modulates the zero crossing instance and appear as phase noise. The PN of a well-designed buffer is typically much lower than the VCO PN at lower and intermediate frequencies but starts to dominate only at large frequency offsets. This is due to the implicit phase integration in the VCO, which increases the close-in PN while attenuating far-out noise. The only case where the buffer might contribute to close-in PN is if it kicks back noise into the VCO. The buffer can kick back noise through the following mechanisms:

1. The tail current source noise, mainly flicker noise, modulates the operating point of the NMOS and PMOS transistors, including their gate capacitance. Since the gate capacitance of the inverter is part of the tank, the flicker noise modulates the VCO oscillation frequency and appears as PN at the VCO output. That is why the tail current source is not a good idea from a PN point of view.
2. The low frequency noise of the buffer might couple back to the VCO core and gets up-converted by nonlinear capacitors such as the varactor or the device capacitance. This is not issue in the topology of Figure 12.13 because the blocking cap C_b blocks all low-frequency noise.
3. Finally, the supply noise of the buffer $V_{dd,buffer}$ could also modulate the inverter gate capacitance and appear as PN. Therefore, it is advised to run the VCO buffer from the supply as the VCO core to ensure that the noise is correlated. Otherwise, a low-noise and clean LDO should be used to supply the buffer.

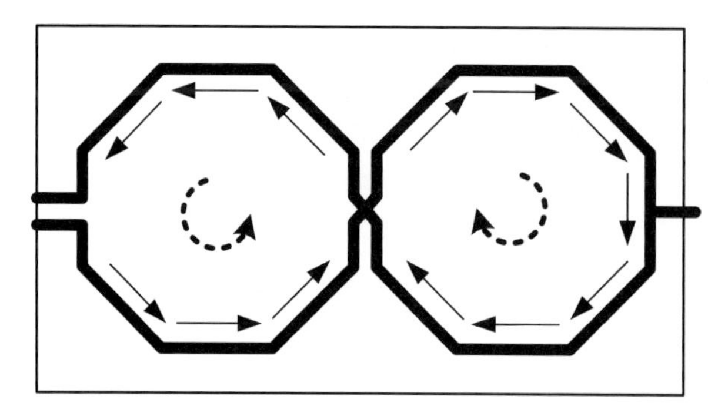

Figure 12.14 A bow-tie inductor with magnetic flux cancellation and shield loop.

12.5.3 Resilience to Pulling

The oscillation frequency of a VCO can be modulated, pulled, by a current injected into the tank [1]. There are three main coupling mechanisms: magnetic, capacitive, and through the substrate. Proper ground shielding of the VCO blocks diminishes capacitive coupling. Increasing the substrate resistivity by using the native layer and protecting sensitive circuits with grounded guard rings reduces the substrate coupling to a negligible level. The only mechanism that is really hard to eliminate is magnetic coupling. A common pulling problem in modern radios occurs when the power amplifier (PA) output transformer couples to a harmonically related VCO tank inductor. The modulated output of the PA changes the instantaneous frequency of the VCO and creates in-band distortion [23]. One way to mitigate magnetic coupling is to use a bow-tie inductor as shown in Figure 12.14. The two halves of the bow-ties inductor carry the signal current in opposite directions, which means that the resulting magnetic flux of each half will be canceled by the other. Therefore, an aggressor inductor far enough from the bow-tie will generate equal and opposite currents in the bow-tie that cancel each other and reduce the coupling, ideally, to zero. Practically, the bow-tie reduces the coupling by about 20 dB [24]. And if both the aggressor and the victim use bow-tie inductors, the coupling will be even lower. Typically a bow-tie inductor will occupy more area and achieve about 15–20% lower Q.

Another way of reducing magnetic coupling is the use of a shield loop around the inductor. The shield loop could be left floating or could be grounded. The idea simply is to use the shield loop to generate the aggressor current but in an opposite direction, which would cancel the aggressor flux inside the loop. More on the design of the shield can be found in [16]. Needless to say, the shield degrades the inductor Q by the same mechanism it uses to shield it.

12.5.4 AM/PM Conversion in Small Tuning-Range VCO Designs

In a fixed-IF LO architecture, a VCO with small tuning range is required to generate the fixed-IF frequency. An optimal design from a simplified *FoM* point of view would suggest that the switching pair of the VCO would dominate the tank capacitance to

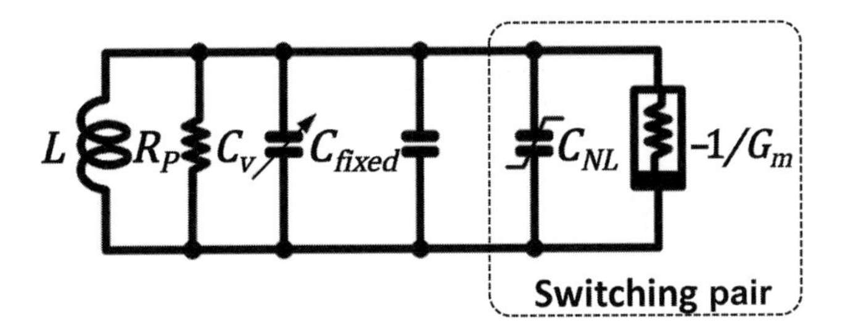

Figure 12.15 Equivalent circuit of a VCO with a fixed capacitor.

maximize the inductance and thus R_P as shown in Figure 12.15. In practice, however, such a design would perform poorly. The nonlinear device capacitance would increase the AM/PM conversion substantially, resulting in significant noise degradation from LDO, and bias circuits. The simplest solution to this problem is to add a large fixed metal capacitor to the tank to attenuate the AM-to-PM conversion transfer function and optimize the overall *FoM*. Another approach would be to use a lower-noise LDO and bias circuits, but this always comes at the cost of larger area and higher power consumption. Therefore, it is up to the designer to decide how to distribute the power consumption of the blocks to minimize the overall power consumption and maximize the *FoM*.

12.5.5 Bias Circuit Design for VCO

The switched-cap cell shown in Figure 12.16c requires a bias voltage V_{bias} to be applied through two bias resistors to set the DC voltage of the drain and source of the NMOS switch to a value sufficiently high to prevent the transistor from turning on in the off state, but not too high to cause overvoltage stress. The noise of the reference voltage is converted to PN through the source and drain nonlinear junction capacitance and the large voltage swing. Therefore, it is necessary to ensure that the noise of the reference voltage applied to the switch does not degrade the VCO PN. Two simple circuits for voltage generation are shown in Figure 12.16a and 12.16b. In the former, a calibrated reference current from the bandgap, is applied to a resistor, while in the latter, a resistive ladder from the supply to ground generates the programmable voltage. The current source approach can be realized with very low power consumption, while the voltage ladder approach can be achieved with lower noise. In both cases, the resulting voltage is applied to a passive RC filter, shown in Figure 12.16d, the filter corner frequency is typically set an order of magnitude lower than the lowest frequency of interest to guarantee proper rejection of in-band noise. Resistor values in $M\Omega$ is often used to reduce the capacitance and therefore the filter area. The output of the RC filter can be applied directly to the switch cells, except when the leakage currents of the source and drain junction diodes are large enough, for a worst-case corner of fast switch and high temperature, then a low-noise unity gain operational amplifier is used to avoid corner-dependent voltage drop on the filter resistor.

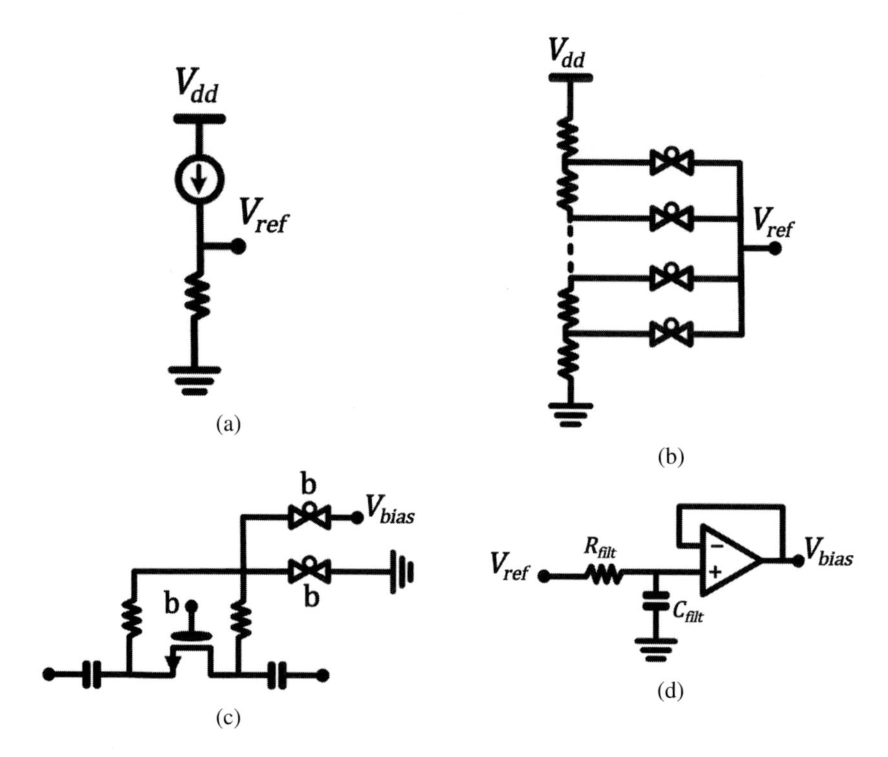

Figure 12.16 Example bias circuits: (a) voltage reference generated from band gap current, (b) reference generated from a voltage divider, (c) differential switched-cap circuit, and (d) noise filter followed by a unity-gain operational amplifier.

12.6 Conclusion

The theory of LC-VCO design has been developed and matured to the point where an optimal design can come within 1–2 dB of the best, physically possible, performance. Nevertheless, many designs still fall substantially short of the optimal performance due to subtle practical issues that are seldom discussed in the literature. In this chapter, a step-by-step design methodology for the LC-VCO has been presented. It takes into account the LO-architecture requirements as well as the technology limitations on both passive and active devices. Moreover, some of the most common practical issues in LC-VCO design have been discussed with suggestions on how to mitigate them. The VCO designer is encouraged to make a list of all the practical issues to check before a design is finalized and ready for tape-out. Expectedly, this will lead to better agreement between simulation and measurement results.

References

[1] B. Razavi, "A study of injection locking and pulling in oscillators," *IEEE Journal of Solid-State Circuits*, vol. 39, no. 9, pp. 1415–1424, September 2004.

[2] T. Zhang, A. Najafi, M. Taghivand, and J. C. Rudell, "A precision wideband quadrature generation technique with feedback control for millimeter-wave communication systems," *IEEE Transactions on Microwave Theory and Techniques*, vol. PP, no. 99, pp. 1–12, 2017.

[3] A. A. Hafez and C. K. K. Yang, "Design and optimization of multipath ring oscillators," *IEEE Trans. Circuits Syst. I Regul. Pap.*, vol. 58, no. 10, pp. 2332–2345, October 2011.

[4] X. Gao, E. A. M. Klumperink, and B. Nauta, "Advantages of shift registers over DLLs for flexible low jitter multiphase clock generation," *IEEE Transactions on Circuits and Systems II: Express Briefs*, vol. 55, no. 3, pp. 244–248, March 2008.

[5] A. A. Abidi, "Direct-conversion radio transceivers for digital communications," *IEEE Journal of Solid-State Circuits*, vol. 30, no. 12, pp. 1399–1410, December 1995.

[6] N. Saito, T. Tsukizawa, N. Shirakata, et al., "A fully integrated 60-GHz CMOS transceiver chipset based on WiGig/IEEE 802.11 ad with built-in self calibration for mobile usage," *IEEE Journal of Solid-State Circuits*, vol. 48, no. 12, pp. 3146–3159, 2013.

[7] A. Tomkins, A. Poon, E. Juntunen, et al. "A 60 GHz, 802.11ad/WiGig-compliant transceiver for infrastructure and mobile applications in 130 nm SiGe BiCMOS," *IEEE Journal of Solid-State Circuits*, vol. 50, no. 10, pp. 2239–2255, October 2015.

[8] B. Razavi, *RF Microelectronics*. Prentice-Hall, 2011.

[9] A. Natarajan, S. K. Reynolds, M. D. Tsai, et al., "A fully-integrated 16-element phased-array receiver in SiGe BiCMOS for 60-GHz communications," *IEEE Journal of Solid-State Circuits*, vol. 46, no. 5, pp. 1059–1075, May 2011.

[10] E. Cohen, M. Ruberto, M. Cohen, et al., "A CMOS bidirectional 32-element phased-array transceiver at 60 GHz with LTCC antenna," *IEEE Transactions on Microwave Theory and Techniques*, vol. 61, no. 3, pp. 1359–1375, March 2013.

[11] M. Boers, B. Afshar, I. Vassiliou, et al., "A 16TX/16RX 60 GHz 802.11 ad chipset with single coaxial interface and polarization diversity," *IEEE Journal of Solid-State Circuits*, vol. 49, no. 12, pp. 3031–3045, 2014.

[12] D. Murphy, H. Darabi, and H. Wu, "Implicit common-mode resonance in LC oscillators," *IEEE Journal of Solid-State Circuits*, vol. 52, no. 3, pp. 812–821, March 2017.

[13] D. B. Leeson, "A simple model of feedback oscillator noise spectrum," *Proceedings of the IEEE*, vol. 54, no. 2, pp. 329–330, Febrary 1966.

[14] M. Mikhemar, D. Murphy, A. Mirzaei, and H. Darabi, "A cancellation technique for reciprocal-mixing caused by phase noise and spurs," *IEEE Journal of Solid-State Circuits*, vol. 48, no. 12, pp. 3080–3089, December 2013.

[15] M. Garampazzi, S. D. Toso, A. Liscidini, et al., "An intuitive analysis of phase noise fundamental limits suitable for benchmarking LC oscillators," *IEEE Journal of Solid-State Circuits*, vol. 49, no. 3, pp. 635–645, March 2014.

[16] A. M. Niknejad, *Electromagnetics for High-Speed Analog and Digital Communication Circuits*. Cambridge University Press, 2007.

[17] C.-H. Lin and K. Bult, "A 10-b, 500-MSample/s CMOS DAC in 0.6 mm2," *IEEE Journal of Solid-State Circuits*, vol. 33, no. 12, pp. 1948–1958, December 1998.

[18] A. Mazzanti and A. Bevilacqua, "On the phase noise performance of transformer-based CMOS differential-pair harmonic oscillators," *IEEE Transactions on Circuits and Systems I: Regular Papers*, vol. 62, no. 9, pp. 2334–2341, September 2015.

[19] D. Murphy, J. J. Rael, and A. A. Abidi, "Phase noise in LC oscillators: A phasor-based analysis of a general result and of loaded Q," *IEEE Transactions on Circuits and Systems I: Regular Papers*, vol. 57, no. 6, pp. 1187–1203, June 2010.

[20] E. Hegazi, H. Sjoland, and A. A. Abidi, "A filtering technique to lower LC oscillator phase noise," *IEEE Journal of Solid-State Circuits*, vol. 36, no. 12, pp. 1921–1930, December 2001.

[21] A. Mazzanti and P. Andreani, "Class-C harmonic CMOS VCOs, with a general result on phase noise," *IEEE Journal of Solid-State Circuits*, vol. 43, no. 12, pp. 2716–2729, December 2008.

[22] M. Garampazzi, P. M. Mendes, N. Codega, et al., "Analysis and design of a 195.6 dBc/Hz peak FoM P-N Class-B oscillator with transformer-based tail filtering," *IEEE Journal of Solid-State Circuits*, vol. 50, no. 7, pp. 1657–1668, July 2015.

[23] A. Mirzaei and H. Darabi, "Pulling mitigation in wireless transmitters," *IEEE Journal of Solid-State Circuits*, vol. 49, no. 9, pp. 1958–1970, September 2014.

[24] L. Fanori, T. Mattsson, and P. Andreani, "21.6 A 2.4-to-5.3GHz dual-core CMOS VCO with concentric 8-shaped coils," in *2014 IEEE International Solid-State Circuits Conference Digest of Technical Papers (ISSCC)*, IEEE, February 2014, pp. 370–371.

[25] L. Iotti and A. Mazzanti and F. Svelto, "Insights Into Phase-Noise Scaling in Switch-Coupled Multi-Core LC VCOs for E-Band Adaptive Modulation Links," IEEE Journal of Solid-State Circuits. vol. no. 7, 52, pp. 1703–1718, July 2017. doi. 10.1109/JSSC.2017.2697442. ISSN. 0018-9200.

[26] D. Murphy and H. Darabi, "A 27-GHz Quad-Core CMOS Oscillator with No Mode Ambiguity," IEEE Journal of Solid-State Circuits, vol. 53, no. 11, pp. 1703–1718, November 2018. doi. 10.1109/JSSC.2018.2865460. ISSN 0018-9200

13 CMOS Power Amplifier Design for 5G Mobile Applications

Yang Zhang and Patrick Reynaert

13.1 Introduction

With 5G coming near, complementary metal-oxide semiconductor (CMOS) integrated radio frequency (RF) power amplifiers (PAs) have drawn increasing attention from both academic institutions and industries. On one hand, CMOS technology stands out for high yield, low cost, and low power while providing unparalleled integration level and digital processing capabilities. On the other hand, with the evolution of CMOS technology, which has followed Moore's law, the speed of CMOS transistors has greatly improved. Therefore, CMOS RF PAs enjoy great potential in the coming 5G radios for base stations and user equipment. However, significant challenges still exist in the PA designs from the system perspective.

To achieve 100-times higher data rate than 4G networks while ensuring better than 1 ms latency, 5G mobile networks will be implemented in both RF and millimeter-wave frequencies. A wide number of frequency bands in the 20–60 GHz spectrum have been proposed for 5G wireless communication systems around the world [1,2]. At these frequencies, the propagation loss dramatically increases; for this reason, high output power has to be delivered efficiently to the antenna or an antenna array from the PA. In the meanwhile, higher-order modulation schemes such as 64-quadratic amplitude modulation (QAM) are proposed to achieve the required data rate and spectral efficiency. From a PA design perspective, these pose clear design requirements: gain, output power, efficiency, linearity, etc. In the following sections, some key aspects in terms of design metrics, design challenges, and techniques will be clarified and discussed in more detail.

13.2 5G RF Front-End Requirement

RF PAs, as the most power-hungry components, need to be power efficient to maximize battery life for better user experience. 5G standards specify spectrally efficient linear modulation schemes that are sensitive to both amplitude and phase errors. On one hand, power amplifiers must be linear to preserve the signal quality. On the other hand, linearity is required to prevent spectral regrowth, which interferes with adjacent channels. A system-level study [3] shows that the unit PA has to deliver 7 dBm modulated power to cover around 50 m range.

To start with, we will first review key metrics from a system perspective.

13.2.1 Quantify the Signal Quality

To evaluate the modulated (QAM modulation) signal quality, the most commonly used metric is the error vector magnitude (EVM) [4]. However, there are multiple ways to calculate EVM, and these methods do not provide identical results. To have a valid comparison among designs, it is essential to apply the exact same metric. Figure 13.1 shows the normalized first quadrant constellation diagram for a 64-QAM signal. EVM measures how the actual constellation point is off from the ideal location. Unlike the bit error rate (BER) that merely focuses whether the received bits are right or wrong, EVM contains more information about the signal quality. For example, from the EVM evaluation it can be concluded if the signal quality degradation is due to either amplitude or phase distortion, which directly reflects the circuit limitation.

Two EVM definitions are commonly used in the literature [5]. The first one is a ratio of root mean square (rms) magnitudes:

$$EVM_{RMS} = \frac{\sqrt{\frac{1}{N}\sum_{i=1}^{N}|S_{ideal,i} - S_{meas,i}|^2}}{\sqrt{\frac{1}{N}\sum_{i=1}^{N}|S_{ideal,i}|^2}} = \frac{V_{error,RMS}}{C_{RMS}} \tag{13.1}$$

where C_{RMS} is the rms value of the constellation point magnitudes. The other one compares the rms magnitude of the error to the peak magnitude of the constellation

$$EVM_{max} = \frac{\sqrt{\frac{1}{N}\sum_{i=1}^{N}|S_{ideal,i} - S_{meas,i}|^2}}{|S_{max}|} = \frac{V_{error,RMS}}{C_{max}}. \tag{13.2}$$

From these two definitions in (13.1) and (13.2), respectively, it can be seen that EVM_{RMS} normalizes the rms value of the error vectors to the rms level of the M-ary signal constellation, while EVM_{max} adopts the maximum constellation magnitude as its normalization factor. The two definitions coincide for the constellations with constant magnitude (BPSK, QPSK, 8-PSK, etc.), while $EVM_{RMS} > EVM_{max}$ for constellations with multiple possible magnitudes (APSK, Star-QAM, 16-QAM, 64-QAM, etc.). A third less common definition of a metric for modulation quality is EVM_{peak}, which is the maximum value of the error vector magnitude that has occurred over sets of N symbols each.

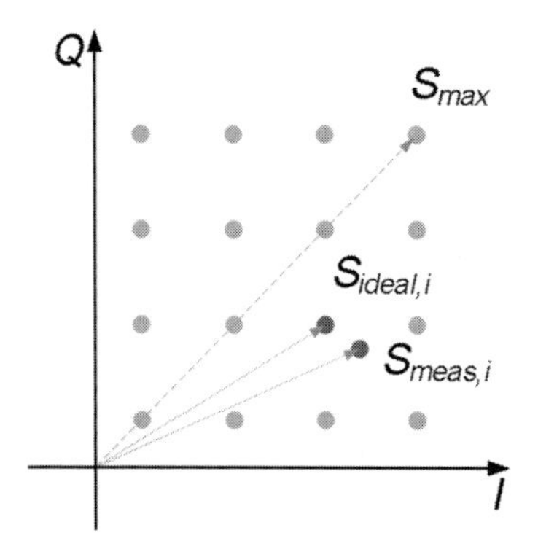

Figure 13.1 The first quadrant of the normalized 64-QAM constellation diagram.

The difference in EVM-metrics from (13.1) and (13.2) can be related to the peak-to-average power ratio (PAPR) of a signal, though the difference between EVM_{max} with EVM_{peak} is equal to the $PAPR$ of the ideal constellation diagram, i.e., before any Nyquist or channel filtering takes place. The average power of a constellation is calculated by evaluating the mean square of the constellation points with identical weights for each point [6]

$$d_{MS} = \frac{1}{N}\sum_{i=1}^{N} |P_i|^2 \tag{13.3}$$

where $|P_i|^2 = I_i^2 + Q_i^2$. When the max $\{|P_i|^2\}$ is normalized to 1, the PAPR is calculated as

$$PAPR = \frac{1}{d_{MS}} \tag{13.4}$$

Note that $PAPR$ increases by 4 dB for a root-raised cosine (RRC) filter (roll-off α equal to 0.35) applied. The $PAPR$ for analog signal is equal to the square of the peak instantaneous voltage divided by the square of the rms voltage value of the signal. But for RF designers, the $PAPR$ of a modulated carrier is defined differently. It is equal to the peak-envelope power (PEP) divided by the rms power of the signal. The PEP is the average power of a sine wave, having an amplitude equal to the peak instantaneous voltage of the modulated carrier.

Therefore, from an RF perspective, an unmodulated carrier has a $PAPR_{RF}$ of 0 dB, whereas that the same signal has a $PAPR_{BB}$ of 3 dB for an analog designer. This is expected, because a baseband amplifier needs excellent circuit linearity to properly amplify a sinusoidal signal (with constant envelope), while a band-pass RF PA does not require any circuit linearity to achieve the very same goal [7].

Based on the preceding discussion, $PAPR$, which depends on the modulation scheme, impacts EVM_{RMS}. E.g., EVM_{RMS} is higher versus EVM_{max} by 2.6, 3.7, and 4.2 dB for 16-, 64-, and 256-QAM signals [8]. Furthermore, EVM_{RMS} gives an improved comparison of the signal quality for different modulation schemes; also, in the specific cases of average white Gaussian noise (AWGN)-only signals, EVM_{RMS} is simplified to $EVM_{RMS} = -SNR$ [4].

It is worth noting that any power amplifier design entails a stringent trade-off between efficiency and linearity. The requirements on the output signal accuracy are often set by an EVM metric. To put things in perspective, −25 dB EVM allows 3 dB margin on the required SNR for a 64-QAM signal [3] when using EVM_{RMS}. This margin disappears if the EVM_{max} metric is used. To meet the specifications, a substantial power back-off from the maximum achievable output power is needed, compromising efficiency. Different manufacturers of measuring equipment who use different definitions for EVM and the normalization factor is not always clearly mentioned. Therefore, a 3 dB difference in the EVM definition immediately results in a very misleading comparison table, especially if the normalization used is omitted.

The $PAPR$ of the signal has crucial impact on the linearity requirements of the transmitter and the PA in particular. Moreover, the $PAPR$ of the signal sets the difference

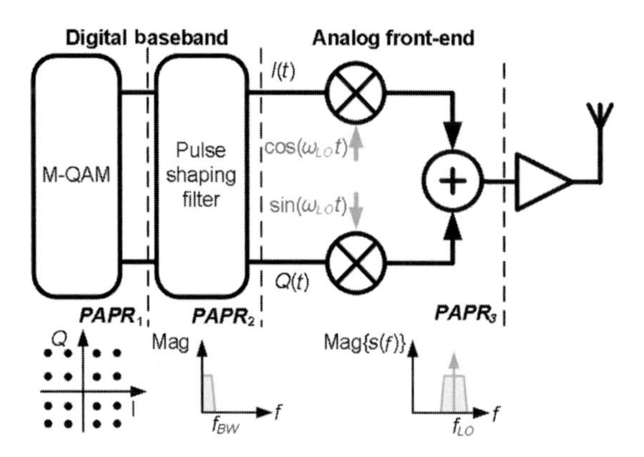

Figure 13.2 A simplified IQ direct conversion transmitter.

between the two discussed normalizations of EVM. However, it should be noted that the $PAPR$ in the aforementioned cases refers to two different signals and results in general in different values. To get more insight, Figure 13.2 shows the simplified block diagram of a direct-conversion transmitter for millimeter-wave applications, emphasizing different signals present at different sections. The signal $s(t)$ at the PA input can written according [6] as

$$s(t) = I(t)\cos(\omega_{LO}t) - Q(t)\sin(\omega_{LO}t) = \mathrm{Re}\left\{[I(t) + Q(t)]e^{j\omega_{LO}t}\right\} \quad (13.5)$$

$$= r(t)\cos(\omega_{LO}t + \theta(t)) \quad (13.6)$$

where $r(t) = \sqrt{I^2(t) + Q^2(t)}$ is the envelope of the baseband signal. It is possible to show that when the bandwidth of the baseband signal is $f_{BW} \ll f_{LO}$, the $PAPR$ of the RF signal at the PA input can be written according [9] as

$$PAPR(s) = 10\log_{10}\left(\frac{2P_{peak}(r)}{P_{RMS}(r)}\right) = PAPR(r) + 3\,\mathrm{dB}. \quad (13.7)$$

The difference between EVM_{RMS} and EVM_{max} is equal to the PAPR of the ideal constellation diagram, therefore the baseband signal with $PAPR_1$ depicted in Figure 13.2 should be considered. Before being up-converted to RF, this signal is low-pass filtered to limit its bandwidth [6,10]. Thus, the PAPR of the baseband signal envelop $PAPR_2$ is equal to $PAPR(r)$ in (13.7), and typically higher than $PAPR_1$ [6].

Finally, when the baseband definition of $PAPR$ is used and under the assumption of $f_{BW} \ll f_{LO}$, the up-converted signal shows a $PAPR_3 \approx 3$ dB higher than $PAPR_2$. This 3 dB difference actually comes from the calculation of the P_{RMS}, influenced by the local oscillator (LO), which is a sinusoidal wave with constant amplitude and has a 3 dB $PAPR$ by definition. However, recalling the definition of $PAPR$ in the RF domain, the 3 dB difference disappears. To an RF designer $PAPR_2 = PAPR_3$.[1]

[1] Later in this chapter, $PAPR_3$ will abbreviated as $PAPR$, because $PAPR = PAPR_{1,2}$ for baseband signals is defined differently.

The real challenge is to amplify a signal with a nonconstant envelope. When a PA is modeled as a hard limiter, to guarantee an ideally linear amplification, the back-off needed from the saturation point is indeed equal to the *PAPR* of the envelope of the baseband signal [11].

13.2.2 Signal Influenced by PA Nonlinearities

RF PAs can be modeled by amplitude-to-amplitude (AM–AM) and amplitude-to-phase (AM–PM) distortion curves at a certain frequency. Consider a modulated signal depicted in (13.6) that passes through a memoryless PA to produce an output. Due to AM–AM distortion, the constellation points in the complex plane will be displaced radially relative to the ideal positions, while these points will be skewed rotationally due to AM–PM distortion (see Figure 13.3). These two behaviors will degrade the EVM. For a phase error of 6 degrees, the normalized vector error magnitude is 0.1, which is as large as the normalized error due to 1 dB gain compression [12].

To further link the distortion characteristics to the EVM, Figure 13.4 provides a simulated EVM versus output power curve. Simulated AM–AM and AM–PM distortion curves of a 20 dB gain PA are extracted and applied to a memoryless transmitter model. The EVM performance is evaluated with different distortion combinations. To meet

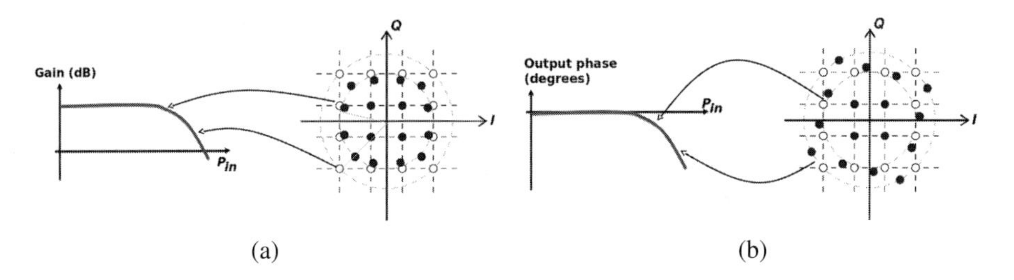

Figure 13.3 Modulated signals influenced by (a) amplitude and (b) phase distortions.

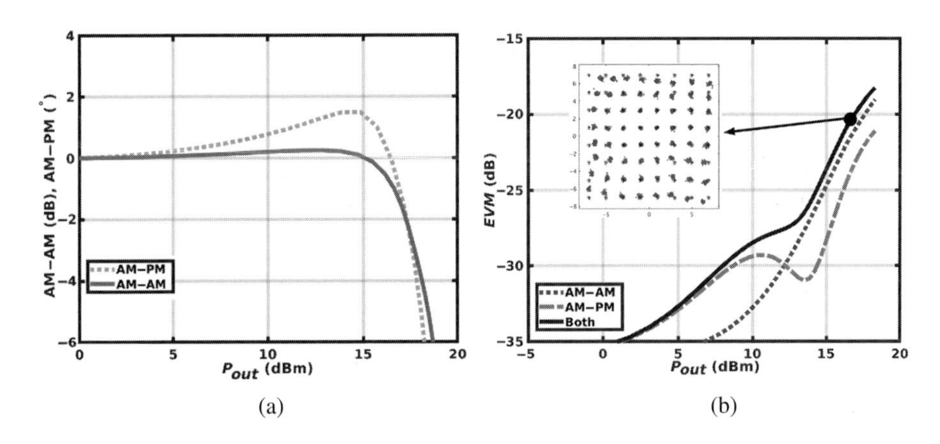

Figure 13.4 The linearity characteristics of (a) a 20 dB gain power amplifier and (b) corresponding *EVM* calculation from a memoryless model.

−25 dB EVM specification using 64-QAM signals, some amplitude and phase distortions of the PA can be tolerated.

13.3 Power Amplifier Basics

The PA can mainly be classified into two major categories depending on whether the transistor behaves as a current source or a switch. When acting as a current source, the PA is usually referred to as the conventional PA. Within this category, we have Class A/AB/B/C PAs. These four types of PAs are distinguished by the gate bias condition. For Class A type, the bias voltage is well above the threshold voltage of the transistor, thus the amplifier always conducts a direct conversion (DC) current, known as the quiescent current. This biasing provides the highest gain, but PA drains the battery of user's device. For a Class B PA, the bias voltage is exactly equal to the threshold voltage so that the conduction angle of the drain current is 180 degrees. Such a current that does not conduct all the time would result in higher efficiency. The Class AB realizes a compromise between Class A and B while the Class C PA is biased below the threshold voltage to further reduce the conduction angle. The drain efficiency η and relative output power P_{out} of Class A/AB/B/C PAs can be expressed according to [10] by

$$\eta = \frac{1}{4}\frac{\theta - \sin\theta}{\sin(\theta/2) - (\theta/2)\cos(\theta/2)}, \tag{13.8}$$

and

$$P_{out} \propto \frac{\theta - \sin\theta}{1 - \cos(\theta/2)}. \tag{13.9}$$

Both depend on the conduction angle θ and are illustrated in Figure 13.5.

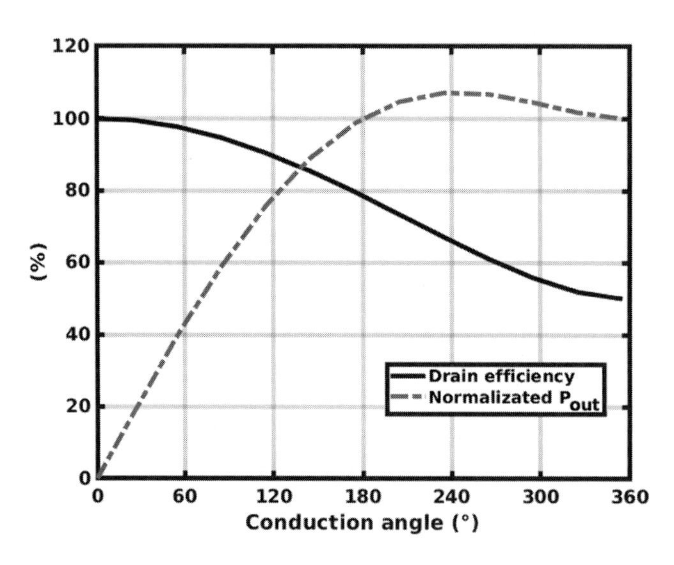

Figure 13.5 Normalized output power and drain efficiency as a function of conduction angle.

By turning off the transistor, higher harmonics are introduced at the drain current and voltage waveforms, causing linearity degradation. In Class AB operation, the largest component next to the fundamental is the second harmonic. While reducing the conduction angle even further in the Class C operation, all harmonics are present and the harmonic power cannot be neglected. This indicates that the Class C operation cannot be employed for linear amplification. Furthermore, the low conduction angle results in low gain and output power.

13.3.1 Transistor Optimization for PAs

Device and interconnect parasitics are the two dominant causes that limit the output power and efficiency of a millimeter wave PA. As the operating frequency increases, device parasitics present a large portion of the total impedance at each node, resulting in a reduced power gain. Layout optimization becomes essential to minimize the parasitics of the transistor cell, especially the gate resistance and gate-to-drain capacitance [13]. Moreover, the parasitic negative feedback path caused by gate–drain capacitance limits the power gain and reverse isolation, and potentially causes instability. For a differential amplifier, the neutralization technique is easy to implement by cross-connecting the interdigitated metal-oxide-metal (MOM) capacitors (C_N) between the drain and gate terminals of the differential stage [14,15], as shown in Figures 13.6 and 13.7. Note that this is also a broadband neutralization technique. The value of C_N is optimized for stability, and its dimension is also optimized to reduce the length of the interconnects. To deal with process, voltage, and temperature (PVT) variations, additional margin should be provided to ensure the stability. Series or shunt resistance can be added to the gate to further stabilize the transistor, but it certainly reduces the power gain and complicates the layout design. Considering the on-chip passives with limited quality factors, the stability will also be improved if the inherent loss resistance of the passive matching network is included in the simulation during the design phase of the single amplifier stage.

13.3.2 Passive Device in CMOS

In the following sections, an overview of the relevant passive devices. Their designs and key parameters are discussed.

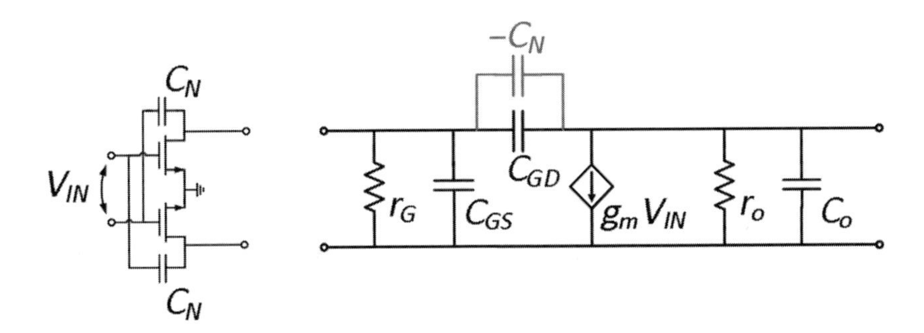

Figure 13.6 Neutralized differential amplifier and its small signal equivalent circuit.

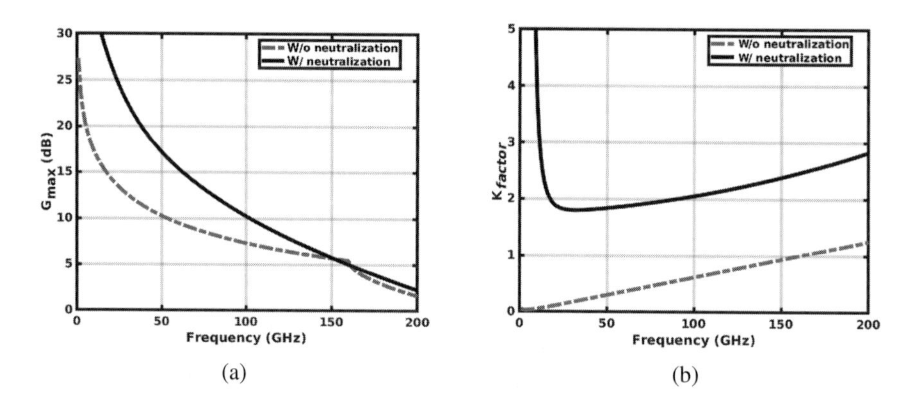

Figure 13.7 G_{max} and k_{factor} of a differential stage (a) with and (b) without the neutralization.

Figure 13.8 (a) Metal-oxide-metal capacitor and (b) inductor.

Inductors

Inductors are the most commonly used passive components in millimeter-wave circuit designs to tune out the capacitive parasitics or to improve the power matching. Compared to distributed transmission lines, inductors (Figure 13.8b) usually occupy less silicon area and ensure a compact floor plan. Skin effect of the metals, low resistance, and eddy current in the substrate are the potential causes to lower the Q factor at millimeter-wave frequencies. The skin depth of the metals can be calculated from the following expression:

$$\delta = \sqrt{\frac{2}{\omega \mu \rho_m}} \tag{13.10}$$

where μ and ρ_m are the permeability and conductivity of the metal. The performance decrease due to skin effect can be mitigated by increasing the metal width of a single wire or using multiple wires, but resulting in a lower self-resonance frequency. In advanced bulk CMOS technology, the silicon substrate usually has a resistance of about 10–15 Ω/cm, which leads to a relaxation frequency around 15 GHz. Therefore, the silicon substrate tends to be capacitive at frequencies above 20 GHz, and the conductive loss actually decreases with frequency for the same coupled energy in the substrate. Eddy currents can be minimized by placing the p-well blocking layer under the passive structure. Otherwise, the high doping p-well will result in substantial eddy current losses [16].

The performance metrics of an inductor can be extracted from the simulated Z-parameters using the following expressions:

$$L_S = \mathrm{Im}(Z_{1,1})/\omega \tag{13.11}$$

$$Q = \mathrm{Im}(Z_{1,1})/\mathrm{Re}(Z_{1,1}) \tag{13.12}$$

Capacitors

In an integrated system, the on-chip capacitor (Figure 13.8a) is often used in the signal path or the supply line. The former is for coupling the signal between stages or performing impedance transformation. The latter is to filter out supply noise and to damp the supply line for possible ringing and stability issues. To reduce the Q-factor of a decoupling capacitor within the frequency of interest, we can insert a small resistor in series with the capacitor, which is relatively easy to implement. The performance metrics of a capacitor can be extracted from the simulated Y-parameters using the following expressions:

$$C_S = \mathrm{Im}(Y_{1,1})/\omega \tag{13.13}$$

$$Q = \mathrm{Im}(Y_{1,1})/\mathrm{Re}(Y_{1,1}) \tag{13.14}$$

Transformers

Transformer-based passives are extensively used in both single-end and differential millimeter-wave amplifier designs, as they simplify signal and supply routing, enable a compact layout, and therefore reduce the losses of extra interconnects [14,17,18]. The performance metrics of a lumped transformer can be extracted from the Z-parameters according to the following expressions:

$$L_p = \mathrm{Im}(Z_{1,1})/\omega \tag{13.15}$$

$$L_s = \mathrm{Im}(Z_{2,2})/\omega \tag{13.16}$$

$$Q_p = \mathrm{Im}(Z_{1,1})/\mathrm{Re}(Z_{1,1}) \tag{13.17}$$

$$Q_s = \mathrm{Im}(Z_{2,2})/\mathrm{Re}(Z_{2,2}) \tag{13.18}$$

$$k_m = \sqrt{(\mathrm{Im}(Z_{1,2})\,\mathrm{Im}(Z_{2,1}))/(\mathrm{Im}(Z_{1,1})\,\mathrm{Im}(Z_{2,2}))} \tag{13.19}$$

where k_m is the coupling factor, L_p (L_s) the self-inductance of the primary (secondary) coil, and Q_p (Q_s) the corresponding quality factor. The maximum power transfer efficiency (η_{max}) of a transformer is given by [19]

$$\eta_{max} = \frac{1}{1 + 2\sqrt{\left(1 + \frac{1}{Q_p Q_s k_m^2}\right)\frac{1}{Q_p Q_s k_m^2}} + \frac{2}{Q_p Q_s k_m^2}} \tag{13.20}$$

Two typical transformer implementations are given in Figure 13.9. From Figure 13.10 it is clear that η_{max} can be improved by optimizing the k_{factor} and the Q-factors. As the width of metal windings is usually greater than its thickness, transformers with an stacked structure are preferred for a maximum magnetic coupling [20]. However, the

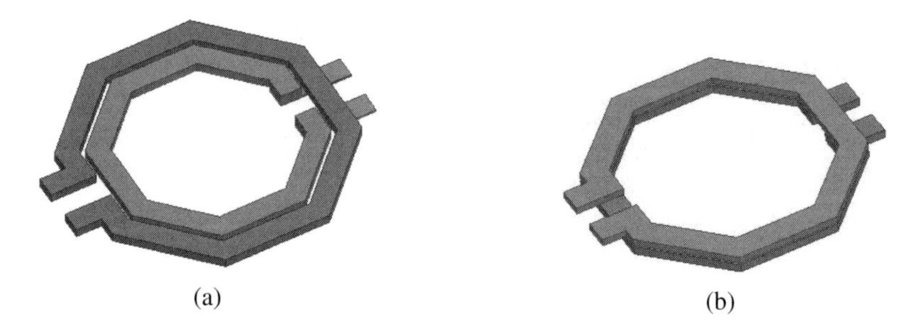

Figure 13.9 (a) Coplanar and (b) stacked transformer topology.

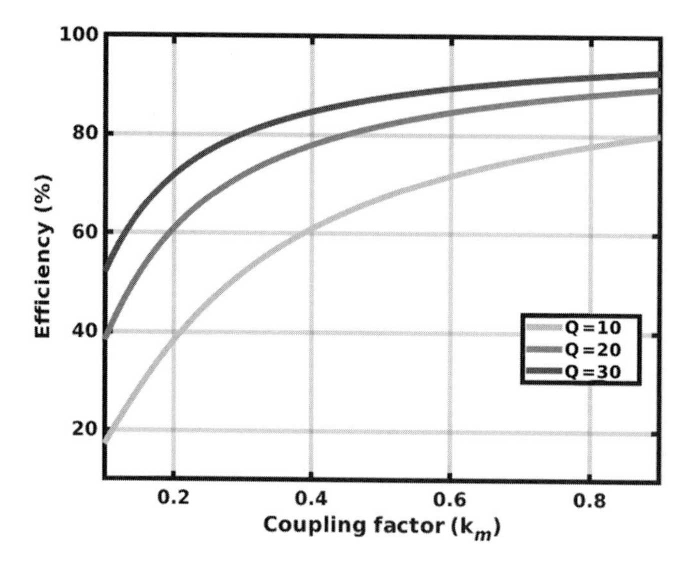

Figure 13.10 Loss of a transformer in function of k and Q-factor.

capacitive coupling is also higher, which could lead to a possible common-mode stability issue. The coplanar structure enjoys high Q-factors for both primary and secondary coils, as they can be implemented in the same ultra thick metal layer. Although the capacitive coupling is less, the coplanar structure generally has lower k_{factor} than the stacked one.

Transmission Lines

Transmission lines (T-lines) are important structures for mm-wave design. At millimeter-wave frequencies, T-line-based reactive components and matching circuits become physically small and are implementable on chip [21]. The advantage of using T-lines is that it is easy to build a scalable model so that the time-consuming EM simulation is not required. Although the lumped elements (e.g., transformers and inductors) are preferred for their small footprint, T-lines are still very often needed for distributing signals. In addition, all the long interconnects at mm-wave have to be modeled as T-line and the line impedance should be well controlled for optimum performance. To reduce

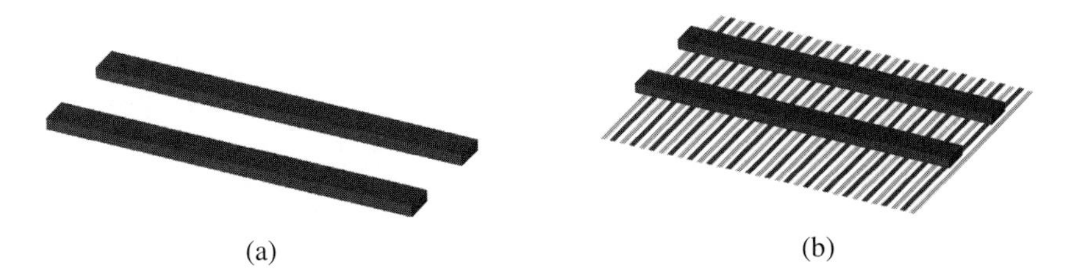

Figure 13.11 Differential T-lines (a) without and (b) with floating metal strips.

the influence of the lossy substrate, floating metal strips can be placed below the signal traces, as shown in Figure 13.11. It is also called slow-wave T-line, since the equivalent capacitance is increased.

The T-line can be characterized based on $ABCD$ parameters, which is converted from simulated S-parameters. The characteristic impedance and propagation constant of the T-line can be calculated from the matrix given by

$$[ABCD]_{T\text{-}line} = \begin{bmatrix} \cosh(\gamma l) & Z_o \sinh(\gamma l) \\ \sinh(\gamma l)/Z_o & \cosh(\gamma l) \end{bmatrix} \tag{13.21}$$

where Z_o is the characteristic line impedance, l is the length of the T-line, and $\gamma = \alpha + j\beta$ is the propagation constant.

13.4 Impedance Transformation and Power Combining

For millimeter-wave wireless communications, a target transmit power has to be specified in a link power budget to ensure a robust data link. However, the peak output power of a single-stage PA is limited in modern bulk CMOS technologies. Firstly, the low breakdown voltage restricts the maximum voltage swing at the drain of the transistor. Secondly, given a certain supply voltage, the RF transistor needs to be sized up for a large current. In this case, the required load impedance is very small. To provide a low load resistance to the PA output stage and thus increase the output power, a matching network with large impedance transformation ratio has to be used, which usually suffers from high loss. Finally, transistors with large size have to be employed in the PA while the associate long interconnects in the layout further degrade the performance.

There are generally two ways to enhance the PA output power in advanced CMOS technologies: stacking-field effect transistor (FET) and power-combining techniques. In bulk CMOS, the junction breakdown voltage limits the maximum number of transistors that can be stacked to two or three devices. More device stacking can be achieved using silicon-on-insulator (SOI) CMOS technology; however, along with its increased wafer cost, the output power gained by stacking more transistors is usually compromised by degraded linearity performance [22,23]. The power combining is a more commonly used technique for millimeter-wave circuits to enhance the output power [24,25]. The linearity of power-combining PAs will not be traded for output power as the power

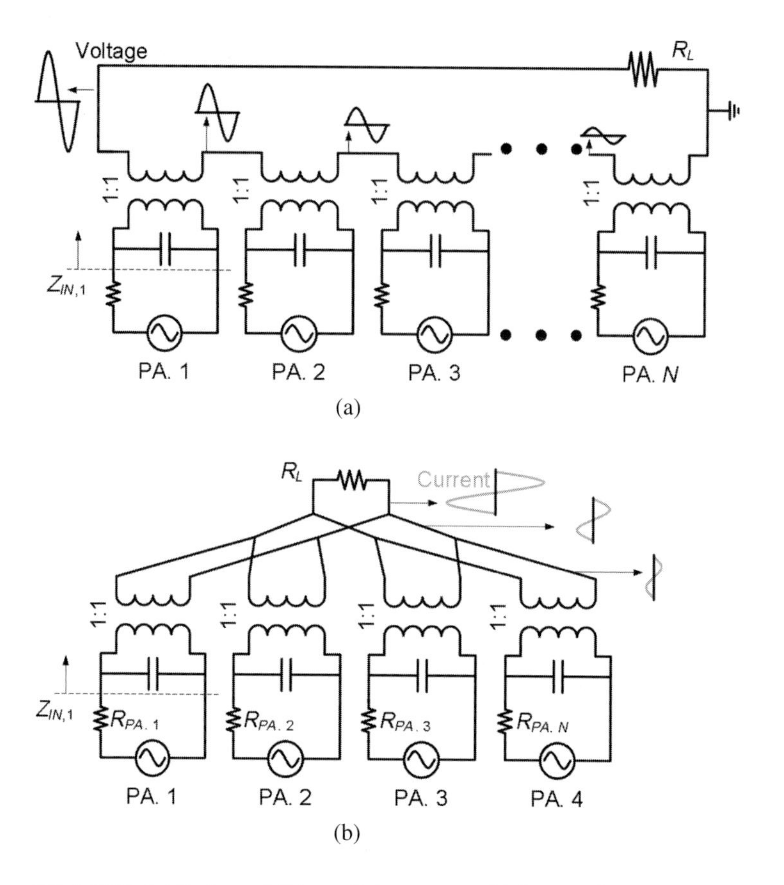

Figure 13.12 (a) Series and (b) parallel power combining topology.

is combined in a passive component. With an increased number of combining paths, the improvement in output power will diminish quickly, which is due to the increased complexity and insertion loss of the power-combining and distribution networks. The development of an efficient N-way power combiner becomes the key to further extend the power level of millimeter-wave PAs.

Figure 13.12 shows two power-combining topologies: series power combiner and parallel power combiner. These two techniques are extensively used CMOS RF PA designs. However, there is a significantly difference on the impedance transformation:

$$Z_{IN,ser} = \frac{R_L}{N} \tag{13.22}$$

$$Z_{IN,par} = N \cdot R_L \tag{13.23}$$

where N is the number of the combined paths. Based on the topology, the impedance can be multiplied or divided by N. Practically, a matching network with a low transformation ratio leads to low insertion loss and broad bandwidth. Therefore, it is more efficient to further boost the output power using a power-combining technique where a matching circuit with a moderate impedance transformation ratio can be applied.

To maximize the power extracted from the PA stage, ideally, each unit PA should see identical load impedance from the output power combiner. However, due to the practical layout limitation, the unit port could have an impedance variation of more than 40% of the four–way series power combiner in [24]. This issue will be exacerbated when more elements are combined. Additionally, it is difficult to evenly distribute the input signals among the unit PAs. The T-line-based parallel power combiner could easily distribute/combine the signal in a balanced way, leading to an identical load impedance seen by each unit PA. However, when implemented differentially, a large silicon area will be used to avoid the unwanted mutual coupling between each path.

13.4.1 PA Nonlinearity

It is well known that the AM–AM distortion is mainly due to gain compression for Class AB PAs, but the AM–PM distortion is more complex to analyze. The major causes for AM–PM distortion have been highlighted in several works [10,12,26–28]. In CMOS PAs, the key contributors to AM–PM are the device intrinsic nonlinear capacitors. Figure 13.13 shows the variation of gate-source capacitance in the function of gate voltage. The capacitance varies dramatically when the gate voltage exceeds the threshold and then tends to be saturated. When the transistor is biased in deep Class AB, the average capacitance increases as the instantaneous gate voltage increases, resulting in an amplitude-dependent phase shift. Moreover, due to the Miller effect, the linear feedback capacitance is present at the input node with gain dependency, also resulting

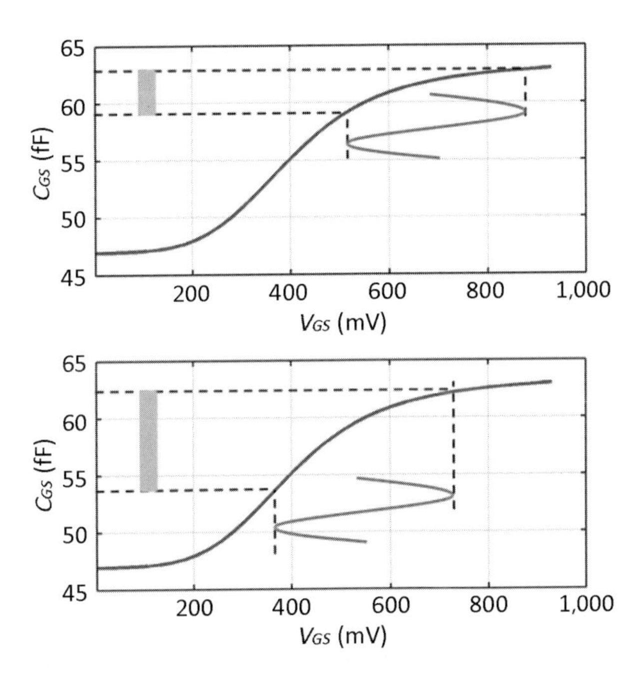

Figure 13.13 Gain-source capacitance in function of gate voltage. Under large signal operation, different biasing voltages could result in a large capacitance variation.

in an AM–PM conversion when the gain gets compressed at large signal operation. This phase shift due to nonlinear voltage-dependent capacitance is aggravated if a matching network with high Q-factor is present at the input or the interstage. Another major cause is the nonlinear drain current. At large signal operation, the transistor will partially work in the triode region and generate harmonic components [12]. First, the time delay caused by transistors will be different, causing AM–PM. Second, for Class AB PAs the first-order harmonic product is out of phase with the fundamental frequency, which leads to a variation of the drain current through a complex load. Moreover, even the load is purely resistive at the fundamental frequency, the second harmonic may still see a complex load (through output capacitance). Therefore, AM–PM distortion is generated [12].

From the design perspective, this implies that a de-Qing input and broadband interstage matching network will alleviate the influence from the nonlinear input capacitance, while a high-Q matching network is preferred at the output for improved AM–PM and high PA efficiency.

13.4.2 Linearity Enhancement Technology

With the major causes for AM–PM distortion discussed in the previous section, it is possible to devise circuit techniques to mitigate them. These techniques aim to linearize the devices, de-Qing the input resistor-capacitor (RC) product, or filter the harmonic components from the output, as shown in Figures 13.14 and 13.15. First, the nonlinear variation of the input capacitance of an N-type metal-oxide-semiconductor (PMOS) transistor can be compensated simply by adding a P-type metal-oxide-semiconductor (PMOS) device (transistor or varactor) [29–31]. Some tuning can be applied to compensate for PVT variations or inaccurate modeling. This technique will result in a relatively large but flat total input capacitance across the operating range. Although in Class AB biasing the input resistance is lower and closer to 50 Ω due to the low Q-factor of the PMOS device, the PA suffers a penalty in terms of gain.

Another distortion cancellation technique is the complementary N-PMOS amplifier [32]. The combination of NMOS and PMOS transistors results in both nonlinear

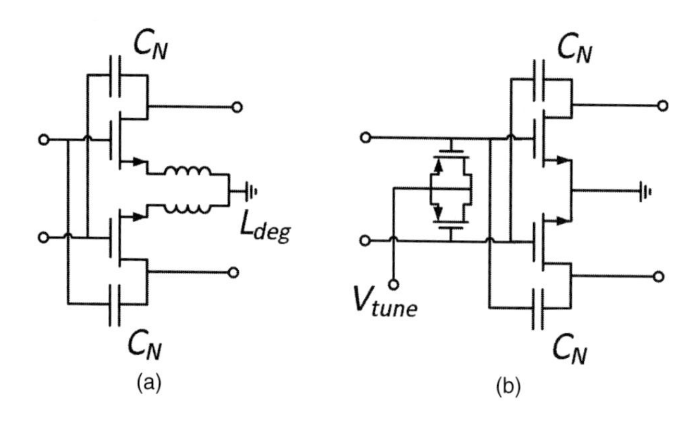

Figure 13.14 Linearity enhancement using (a) source degeneration inductor and (b) input PMOS varactor.

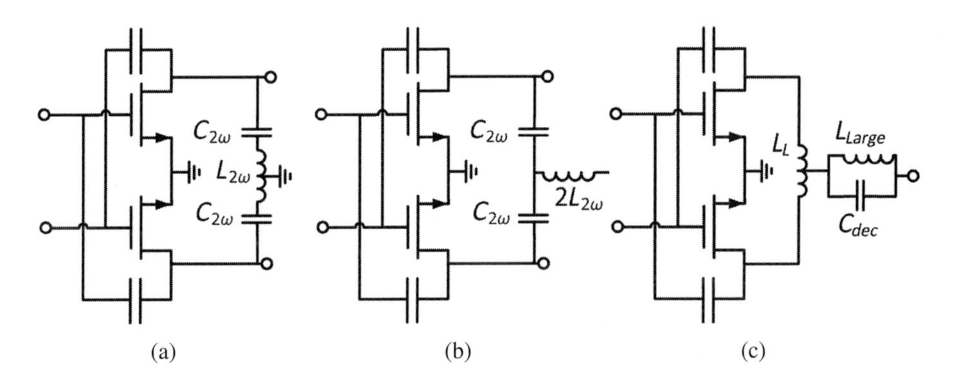

Figure 13.15 Possible implementation of second harmonic traps: (a) classical, (b) sharing inductor, and (c) solution utilizing circuit differential nature.

capacitance and current compensation. This technique is particularly powerful in deep-scaled CMOS, where PMOS devices benefit from more credits from scaling and thus have f_{max} closer to NMOS devices. However, some shortcomings include the following: (a) This technique is not favorable to supply scaling. (b) The PMOS transistors are here used as amplifiers, not just as nonlinear passive components for capacitive compensation. Therefore, the modeling at high frequencies needs to be accurate to correctly design the PMOS/NMOS size ratio. And (c) The supply V_x is sensitive to the PMOS biasing voltage, therefore a feedback loop may be added to this circuit to guarantee proper operation under PVTs, similarly to what is proposed in [33] for complementary N-PMOS Class-C oscillators.

A degeneration inductor can be added to the common-source (CS) amplifier for linearization [34,35]. This technique has been first proposed to improve the soft compression behavior of millimeter-wave PA designs in [36] and then to improve AM–PM distortion in [3]. As a series–series negative feedback, this technique compensates for the gain compression so that the $P_{1\text{-dB}}$ point can be boosted closer to P_{sat}, therefore $PAE_{1\text{-dB}}$ gets closer to PAE_{sat}. Additionally, this degeneration inductor value is reflected to the real part of the input impedance (i.e. $g_m L_{deg}/C_{gs}$), the Q-factor of the input port is reduced, and thus the PA is less sensitive to AM–PM distortion coming from the high-Q interstage matching. This technique comes with a penalty of gain as well.

In large-signal conditions, a Class-AB PA shows substantial AM–PM distortion, even when linearization techniques are applied. Moreover, it is well known that the second harmonic component of the output voltage is a key contributor [12, 37,38]. To further linearize the PA, harmonic traps can be introduced at the output, which shows a low-impedance return path for the second harmonic current. Three circuit solutiond for this condition are shown in Figure 13.15. The difference among these solutions is their influence on the fundamental operation. Since the second harmonic flows in common mode and seeks a low impedance return path and the fundamental operates in differential mode, the fundamental will see the whole trap or only the capacitors. Note that due to the limited quality factor of a practical on-chip implementation, the effect on the fundamental load impedance will not be negligible, further complicating the design and impairing the efficiency. A more elegant solution is to reuse the primary coil of the transformer, of which the trap only affects the second harmonic in common mode. The higher quality factor of the harmonic trap, the higher harmonic compression, and the narrower bandwidth will be obtained. To achieve a broadband operation, a harmonic-tuned network using high-order band-pass filter can be applied for the output matching network [39].

It has been shown both theoretically and with measurements that different distortion mechanisms in Class AB PAs may cancel each other, resulting in low AM–PM and third-order intermodulation distortion at specific power levels [40,41]. By properly choosing the bias point, it is possible to design PAs with high linearity close to the saturated power. However, it may depend heavily on PVT variations. A more elegant solution is to achieve a compensation using two or more power transistors with different biasings. This technique can be achieved between stages [42], cascode–cascade configuration within one stage [43] or between paths [44]. Note that the latter compensation between paths is normally referred to as the famous Doherty PAs, where the load modulation effect needs to be considered and optimized carefully.

13.5 Design Example of a 40 nm CMOS PA

13.5.1 Power Transistor with Source Degeneration Inductor

To minimize the transistor and interconnect parasitics, Figure 13.16 shows the proposed layout of unit transistor cell based on the work of [15]. It consists of 32 fingers with a finger width of 1.1 μm (32 × 1.1 μm/40 nm). The source node is connected on both side to metal-2 and connected to the bulk ring. Furthermore, the source connections are distributed on metal-1, 2, and 3 to minimize the impedance in the source network. Both gate and drain nodes are vertically connected to the transistor from metal-9. The overlap between the gate and drain is therefore minimized, resulting in a reduced extrinsic C_{gd}. A high output power is achieved by combining unit cells in a parallel structures with minimized interconnect parasitics. In the design of this PA, shown in Figure 13.17, a

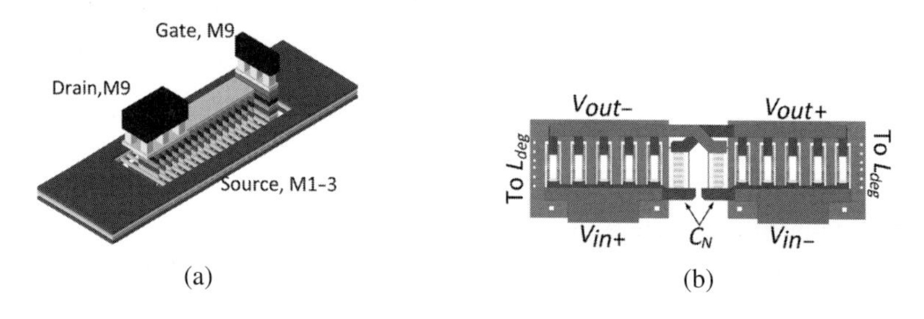

Figure 13.16 (a) Unit cell and (b) the differential power transistor layout with neutralization capacitors.

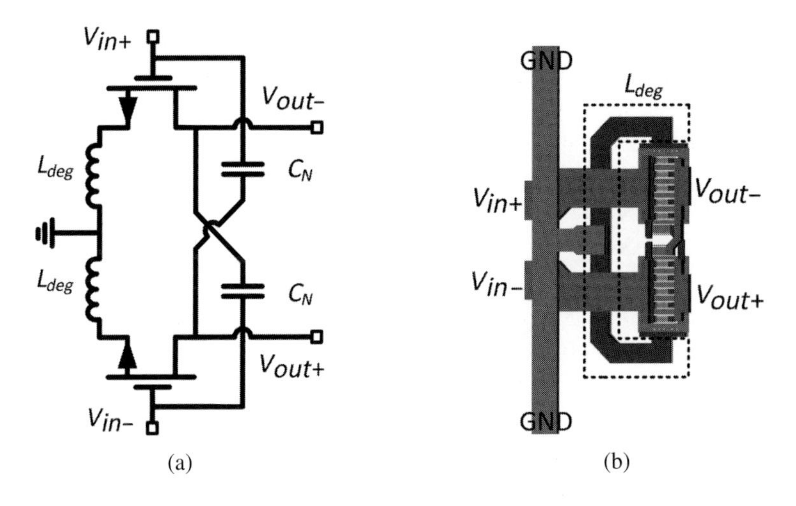

Figure 13.17 (a) Schematic of the unit PA with source degeneration inductor and (b) its layout [42]. (©2017 IEEE. Reprinted, with permission, from *2017 IEEE Radio Frequency Integrated Circuits Symposium.*)

combination of five units has been implemented as depicted in Figure 13.16. This PA delivers a total P_{out} of 18 dBm at a geometry of 176 μm/40 μm.

The parasitic negative feedback path caused by C_{gd} limits the power gain and reverse isolation, and may cause instability. Differential amplifier design allows implementation of an efficient neutralization technique in order to mitigate the impact of C_{gd}. The implementation is done by cross-connecting the interdigitated MOM capacitors between the drain and gate terminals of the differential stage. Note that neutralization is broadband, thus the benefit is gain, which can be used to trade off gain for linearity. The inductive source degeneration is applied to the differential stage to improve the linearity. The source inductor consists of metal strips from metal-2 to metal-8, which improves the Q-factor and meets the current density requirement. The simulated G_{max} and k_{factor} of the differential stage are shown in Figure 13.18, clearly showing that the use of source inductive degeneration results in a penalty of gain. However, the Q-factor of the input port is reduced which leads to a low impedance transformation ratio of the matching network (i.e., less loss and phase distortion). Loadpull simulation shows that the required optimal load impedance also increases, which enables low loss power combining network at the output. The simulated S-parameters and loadpull contours are given in Figure 13.19 using the 176 μm/40 nm transistors with 0.45 V gate biasing.

13.5.2 Design

The optimized architecture of the full PA is shown in Figure 13.20, where the driver amplifier (DA) and PA stage is given in Figure 13.17. In order to achieve flat AM–AM and AM–PM performance, the nonlinear driver is adopted in this design. Figure 13.21 shows the AM–AM and AM–PM distortion of a single-stage PA with degeneration inductor and gate biasing sweeping. Clearly, by combining unit PAs with different

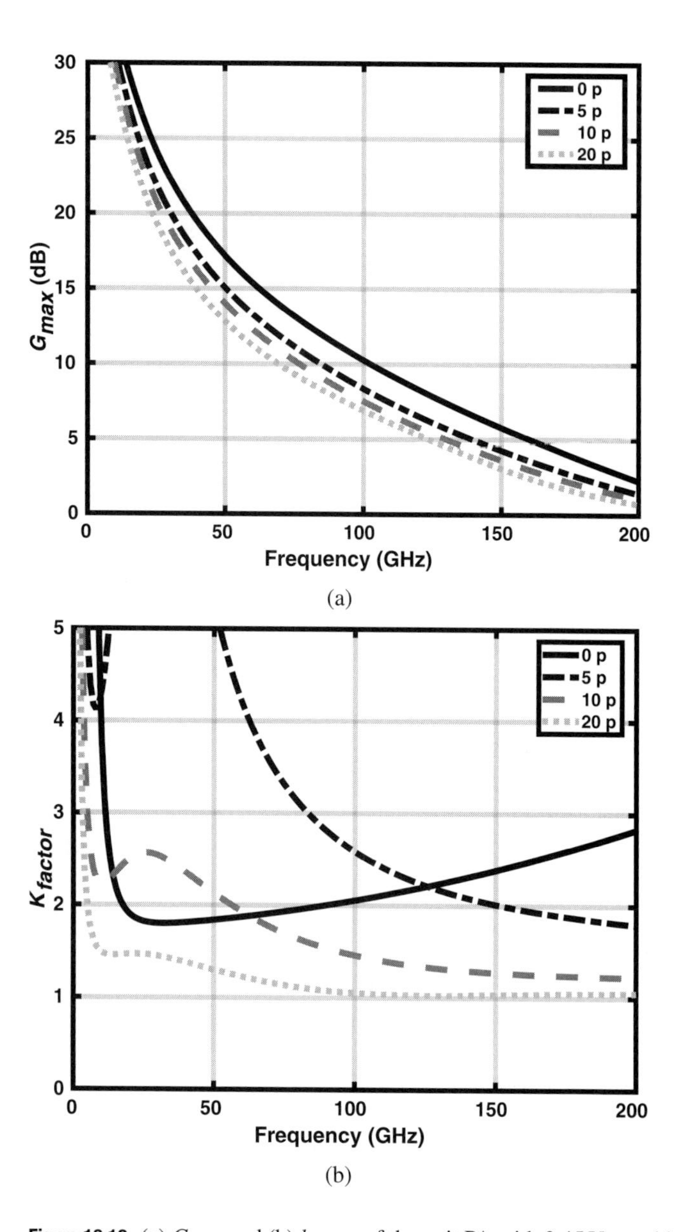

Figure 13.18 (a) G_{max} and (b) k_{factor} of the unit PA with 0.45 V gate biasing.

basing, equivalent to superposition of curves from Figure 13.21, an improved linear performance can be achieved. In this design, the power stage has a higher biasing voltage than the driver stage, which is biased in deep Class AB, in contrast to the condition where the drive has a much higher biasing voltage than the power stage to provide high gain. The reason of doing this is because the input impedance of the power stage is set to be much closer to the conjuration of the driver output, therefore the overall phase distortion is much less sensitive to the interstage matching network. This biasing topology also applies to the PA design with PMOS varactor compensation in [30].

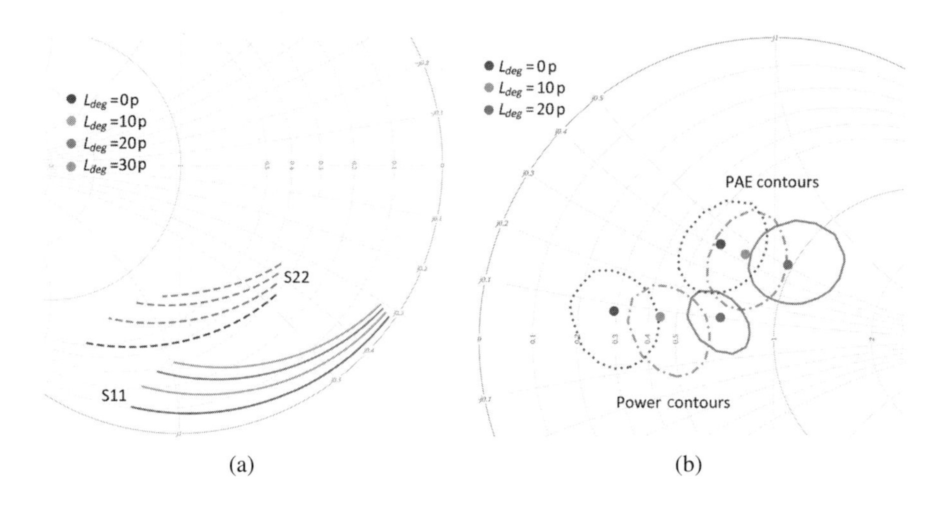

Figure 13.19 (a) Simulated S-parameters and (b) loadpull results of the unit PA with 0.45 V gate biasing.

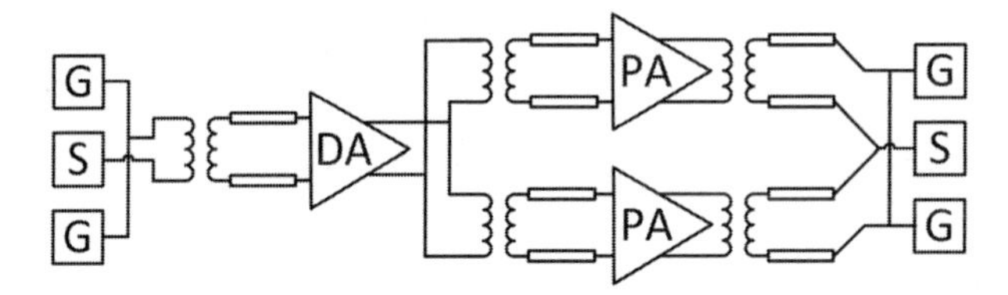

Figure 13.20 Simplified architecture of the full PA.

As the location of the matching network components gets closer to the output load, their losses have more impact on the efficiency. Therefore, special care is taken in the interstage and output network to minimize the losses. Figure 13.22 show the layout and equivalent circuit (one path) of the output combiner. To deliver the required linear output power, the 50 Ω load impedance needs to be scaled down for large transistors. In this design, such scaling is not extensive as the degeneration inductor is in series connection with the load. The output matching network consists of T-line-based power combiner and two 1:1 stacked transformers. Two MOM capacitors (C_o) are placed between signal and ground pads, which improves the matching with high-Q passives and improve the balance caused by the substrate capacitance. The 50 Ω load is transferred into two 58 Ω through the parallel power combiner and C_o, and then matched to the optimal transistor load impedance through high-k transformers. Since the layout is fully symmetrical, the two PA stages see the same differential load impedance. To further explore the balance property of the parallel combiner, we assume the combiner works in a single-ended way and simulate the port impedance of all four ports. Figure 13.23 gives the simulated total insertion loss and the four single-ended impedances of the combiner. Thanks to the low impedance transformation ratio and high Q-factor, the insertion loss at 27 GHz is less only 0.8 dB and better than 1 dB from 24 to 30 GHz. And output combiner almost presents identical load impedance for all four ports, which could maximize the output

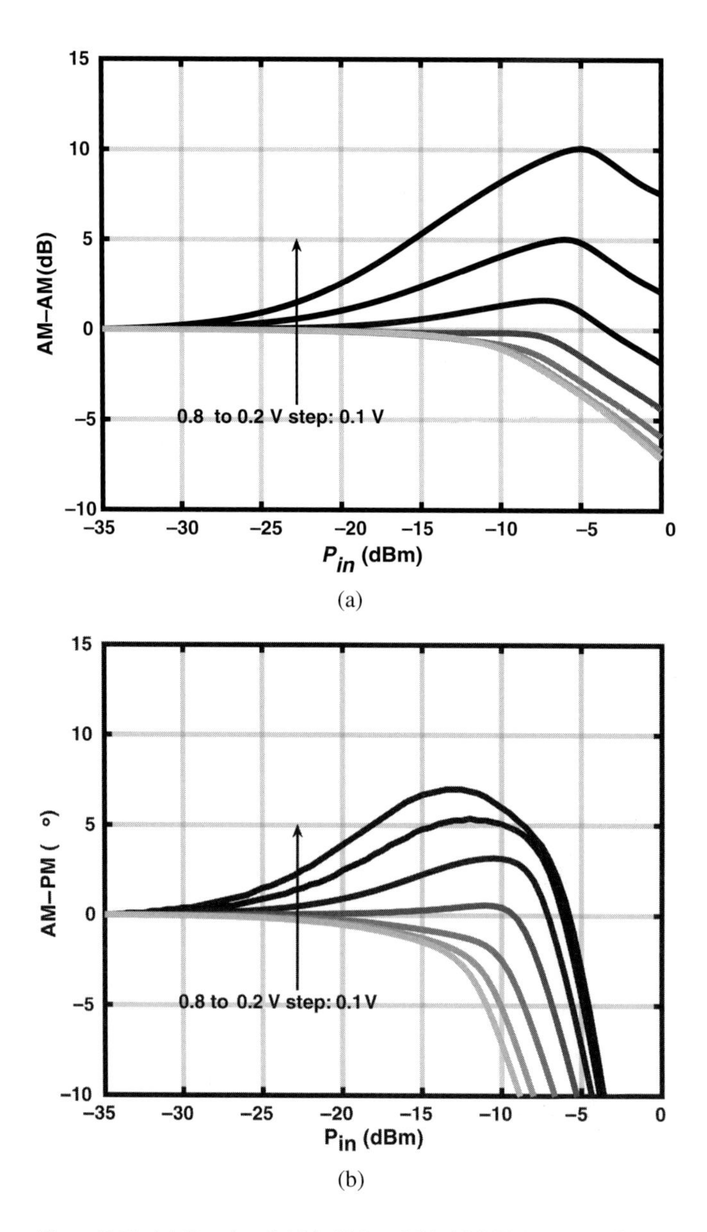

Figure 13.21 (a) Simulated AM–AM and (b) AM–PM curves with gate biasing sweep.

power extracted from the PA stages. The input matching and interstage power divider network are given in Figure 13.24. The interstage matching topology is similar to the output, while the matching transformers are separated from each other by the parallel divider to decrease the unwanted mutual coupling. The input matching is realized using a transformer and a series connected low-Q inductor. Although by doing this the loss will be higher, the phase distortion is greatly mitigated from the variation of the driver input capacitance. The final design layout and design values are shown in Figure 13.25.

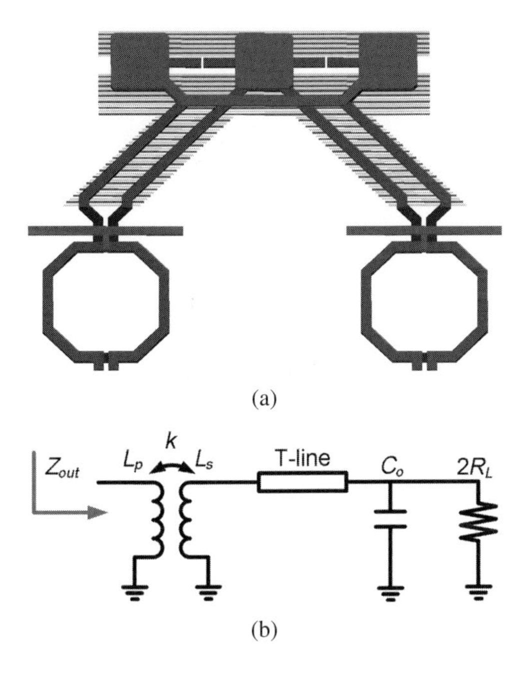

Figure 13.22 Parallel output power combining network: (a) layout and (b) schematic.

To evaluate the mutual coupling between stages, the entire layout is included in final EM simulations.

13.5.3 Measurement Results

The proposed design is fabricated in standard 40 nm CMOS technology and the die photo is shown in Figure 13.26. The ground–signal–ground (GSG) RF pads have a pitch of 150 μm, which is preferred for future packaging with antennas. During the measurement, the power supply was set to 1 V, while the driver stage and last stage were biased at 0.285 and 0.485 V, respectively. The supply and biasing pads are wire-bonded to an FR4 substrate. The vector network analyzer (VNA) and GSG probes are used to measure the signal input and output. The measured S-parameter and k_{factor} are shown in Figure 13.27. The measurement agrees well with simulation. The PA is unconditionally stable within the measured frequency range. The gain at 27 GHz is 20.5 dB with S12 lower than −50 dB. The dummy fillings in all layers slightly change the equivalent permittivity and generate parasitics, which are considered to be reflected on the frequency response.

Continuous-Wave Measurement

Figures 13.28 and 13.29 show the measured large-signal continuous-wave (CW) performance against input power at 27 GHz, using a Keysight PSG vector signal generator

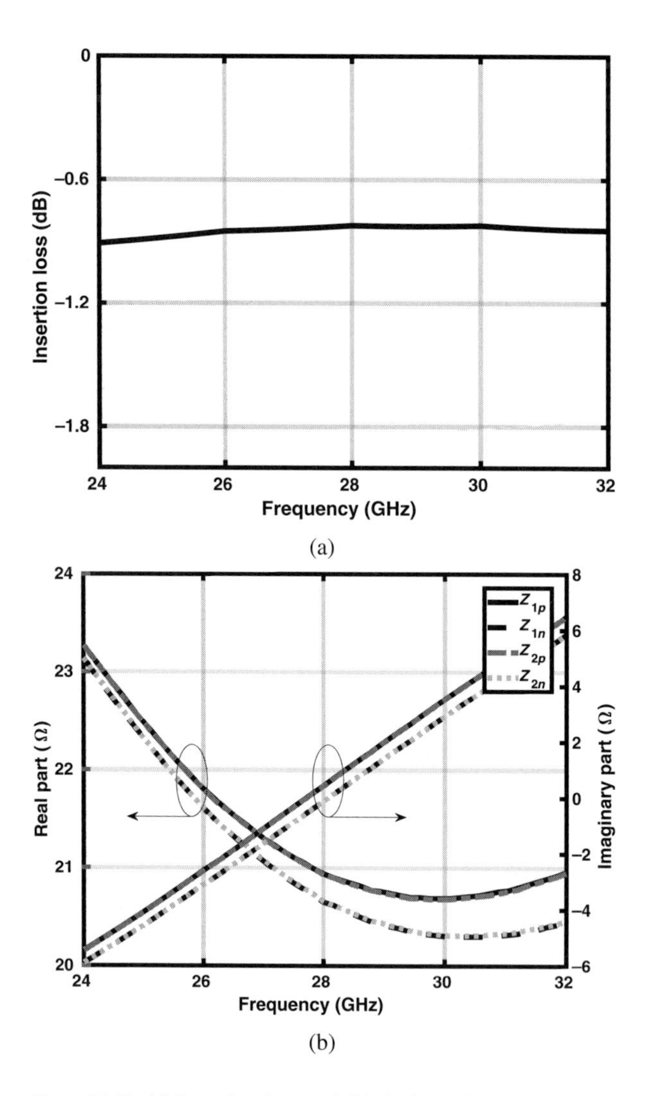

Figure 13.23 (a) Insertion loss and (b) the impedance seen by unit PA.

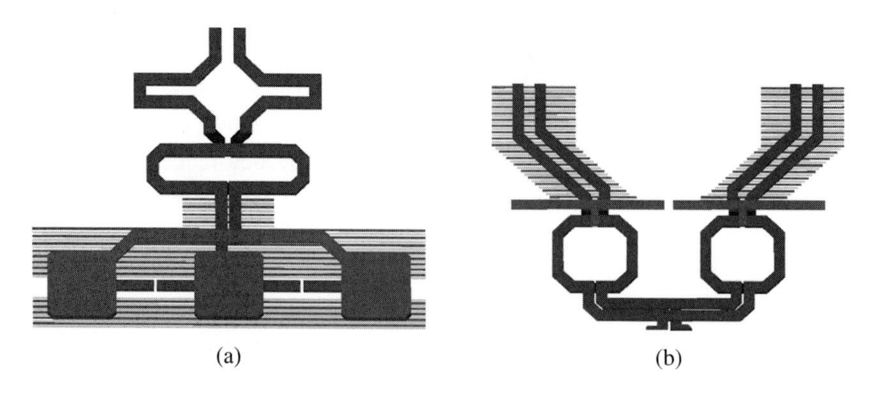

Figure 13.24 Layout of (a) the input matching and (b) the interstage power divider.

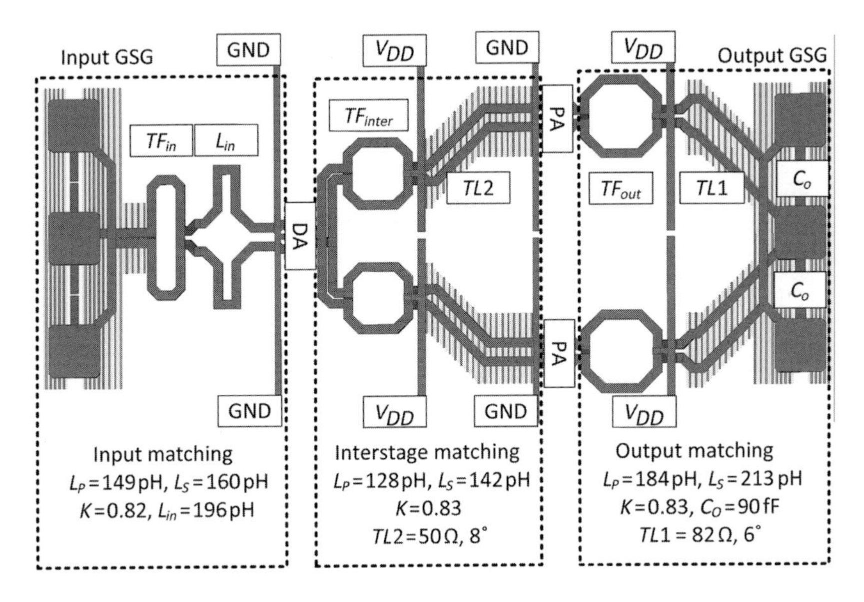

Figure 13.25 Full layout in EM simulation and design values.

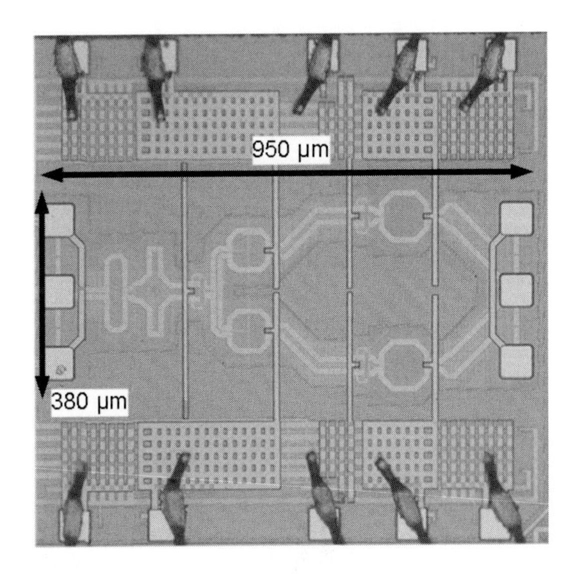

Figure 13.26 The photograph of the fabricated 40 nm CMOS PA [42]. (©2017 IEEE. Reprinted, with permission, from *2017 IEEE Radio Frequency Integrated Circuits Symposium.*)

and power meter. The implemented PA achieves 16.8 dBm output power at the $P_{1\text{-dB}}$ point with a *PAE* of 37.6%, the maximum output power $P_{out,max}$ is 18.1 dBm, and the measured peak *PAE* is 41.5%. The maximum AM–PM variation is less than 2 degrees for output power levels below $P_{1\text{-dB}}$, and the gain expansion in linear region is less

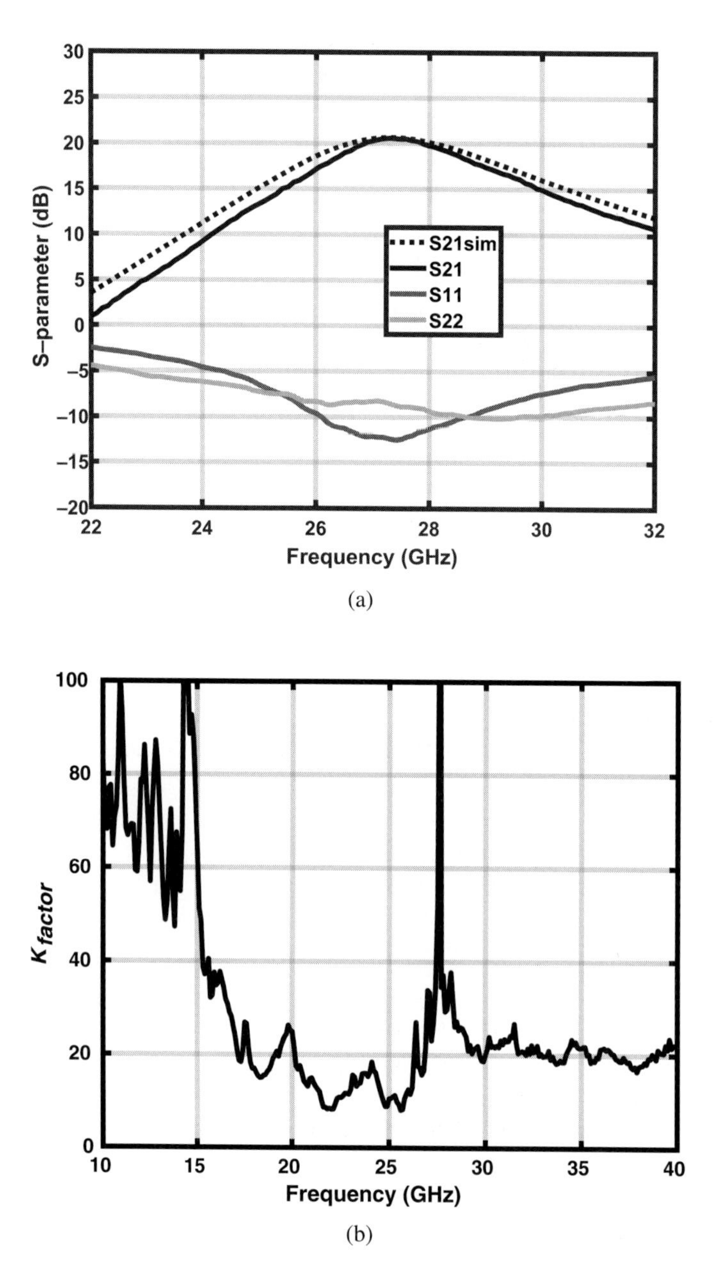

Figure 13.27 (a) Measured S-parameters and (b) k_{factor}.

than 0.35 dB. The results confirm that the PA linearity is greatly improved while the high efficiency and output power performance can still be maintained. Figure 13.29 implies that the modulated signal power back-off can be set closely to its *PAPR* while still maintaining an excellent EVM.

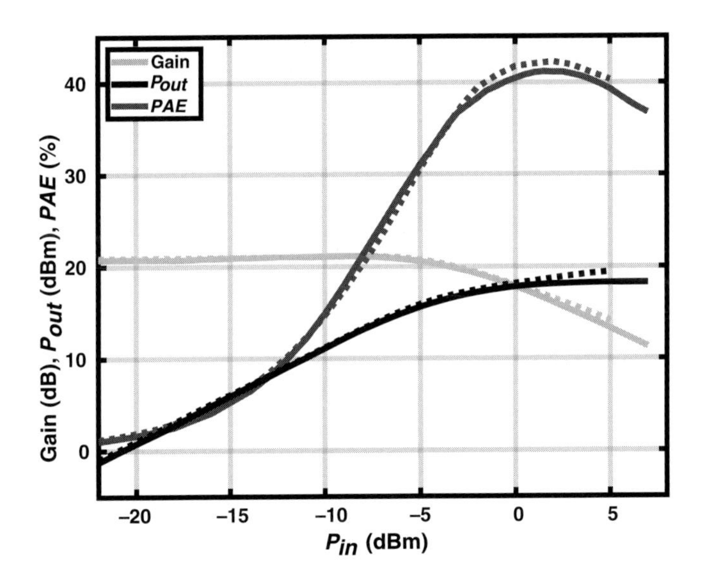

Figure 13.28 Continuous-wave measurement (solid) and simulation (dashed) results.

Modulated Signal Measurement

The PA was tested applying a 64-QAM modulated signal shaped with a raised-cosine filter (0.35 roll-off factor) and a *PAPR* of 8.3 dB. The modulated signals are up-converted and applied to the input of the PA by probes. The PA output was connected to an oscilloscope for evaluation. All the passive losses in the input and output paths are carefully handled. The setup is calibrated without the design under test (DUT) to eliminate the image effect and LO feed-through. Figure 13.30 shows the measured EVM, constellation diagram, and spectrum of 1.5, 4.5, and 6 Gbps data rates. The EVM is normalized to the reference rms power. The measured average output power at −25 dB EVM is 9.65 dBm with 11.8% *PAE*, 8.8 dBm with 9.6% *PAE*, and 8.4 dBm with 8.4% *PAE* for 1.5, 4.5, and 6 Gbps data rates, respectively. These measurement results verified the EVM and modulated power analysis in Section 13.2.1, and the results of low data rate (1.5 Gbps, 250 MHz modulation bandwidth) measurement correspond well with results from CW measurements. Compared with measured 16.8 dBm $P_{1\text{-dB}}$ output power, at low data rate the output power is 9.65 dBm, which is around 7 dB "back-off" from $P_{1\text{-dB}}$ and smaller than the 8.3 dB signal PAPR. Thanks to the low AM–AM and AM–PM distortions, the designed PA can be driven into a high-compression region by the modulated signals while still maintaining an excellent EVM. Additionally, similar to the analysis in [3], as the data rate increases the modulated output power has to decrease to maintain an EVM of −25 dB. The reason is that although the PA linearity allows a higher output power, the PA gain bandwidth starts to influence the signal quality when a wide modulation bandwidth is applied.

Table 13.1 summarizes the measurement results and provide a comparison with the latest published state-of-the-art PA designs for 5G applications. The proposed PA achieved the highest $P_{out} \cdot PAE$ at both 1.5 Gbps and the highest 6 Gbps data rate.

Table 13.1 Comparison with state-of-the-art 5G PAs.

Ref.	**This work**			ISSCC17 [45]	ISSCC17 [46]			JSSC16 [3]	TMTT16 [38]		RFIC17 [47]			RFIC17 [48]	TMTT17 [39]	JSSC17 [49]
Technology	**40 nm CMOS**			40 nm CMOS	130 nm SiGe			28nm CMOS	28 nm CMOS		28 nm CMOS			65 nm CMOS	130 nm SiGe	180 nm SiGe
V_{DD} (V)	**1**			1.1	1.5			1	1.1	2.2	0.9			1.1	3.6	3.8
Freq. (GHz)	**27**			27	28	37	39	30	28.5	28.5	34			28	28	28
Gain (dB)	**20.5**			22.4	18.2	17.1	16.6	15.7	10.1	13.6	20.8			10	15.5	28.6
P_{1dB} (dBm)	**16.8**			13.7	15.2	15.5	15.4	13.2	14	18.6	13.4			13.2	15.9	23.2
PAE_{1dB} (%)	**37.6**			31.1	19.5	21.6	20.7	34.3	35.2	41.4	12.6			NA	NA	32.7
P_{sat} (dBm)	**18.1**			15.1	16.8	17.1	17	14	14.8	19.8	16.6			14.75	18.8	23.7
PAE_{max} (%)	**41.5**			33.7	20.1	22.6	21.4	35.5	36.5	43.3	24.2			44.8[b]	35.3	32.7
AM–PM (°)	<2			NA	NA			<6[b]	NA		<1			NA	<2	>10
Modulation	**64-QAM**			64-QAM	64-QAM			64-QAM	64-QAM		64-QAM			NA	NA	NA
EVM (dB)[a]	**−25**			−25	−23.4	−26.7	−25.1	−25	−27.4	−27.5	−25			NA	NA	NA
Data rate (Gbps)	**1.5**	**4.5**	**6**	<4.8	3	3	3	1.5	<0.48	<0.48	1.5	3	6	NA	NA	NA
P_{out}@EVM (dBm)	**9.65**	**8.8**	**8.4**	6.7	9.2	9.5	9.3	4.2	6.77	10.97	10.1	8.9	5.9	NA	NA	NA
PAE@EVM (%)	**11.8**	**9.6**	**8.8**	11	8.5[b]	10[b]	8.5[b]	9	16.5	17.3	5.8	4.4	2.3	NA	NA	NA

[a] EVM is normalized to the reference rms power.
[b] Graphically estimated.

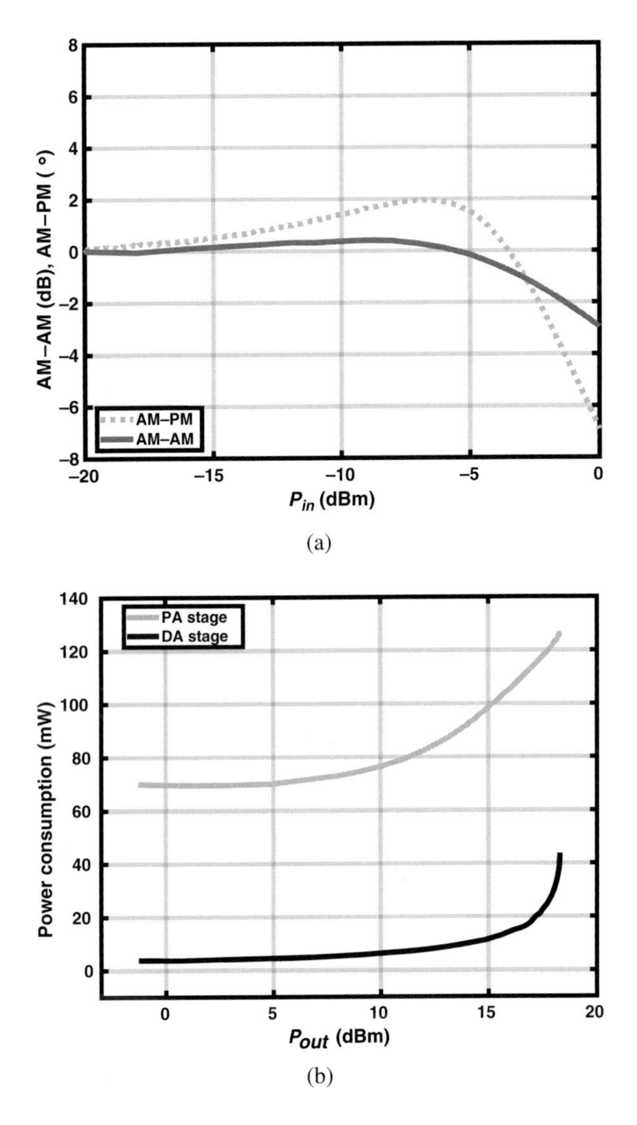

Figure 13.29 (a) Measured PA nonlinearities and (b) DC power consumption.

13.6 Conclusion

In this chapter, design techniques to enhance the linearity and output power of a power amplifier tailored for 5G mobile applications have been discussed. As part of this discussion, the design specifications from a system perspective have been investigated and the EVM and PAPR have been studied in detail. Moreover, some classical power amplifier basic theory and modern linearity enhancement design techniques are reviewed.

Finally, a design example of 40 nm CMOS linear power amplifier with power combining is presented in details. A differential stage with neutralized capacitors and source degeneration inductors is studied. A nonlinear driver is applied to the power amplifier,

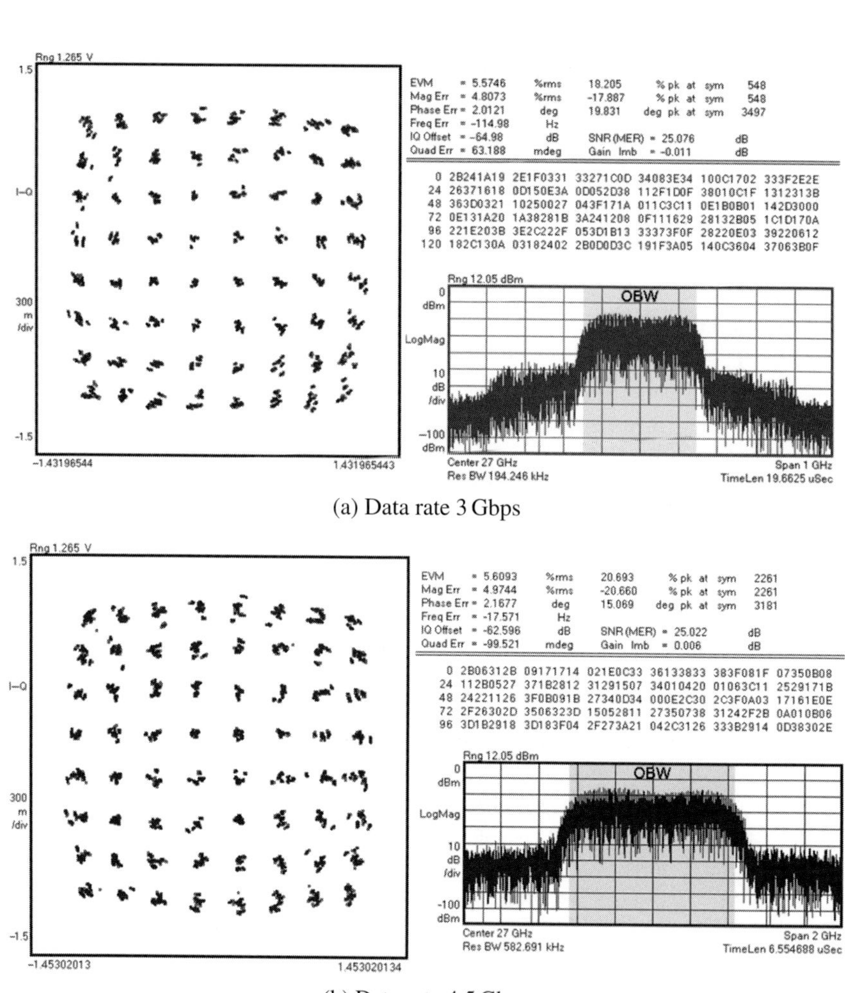

(a) Data rate 3 Gbps

(b) Data rate 4.5 Gbps

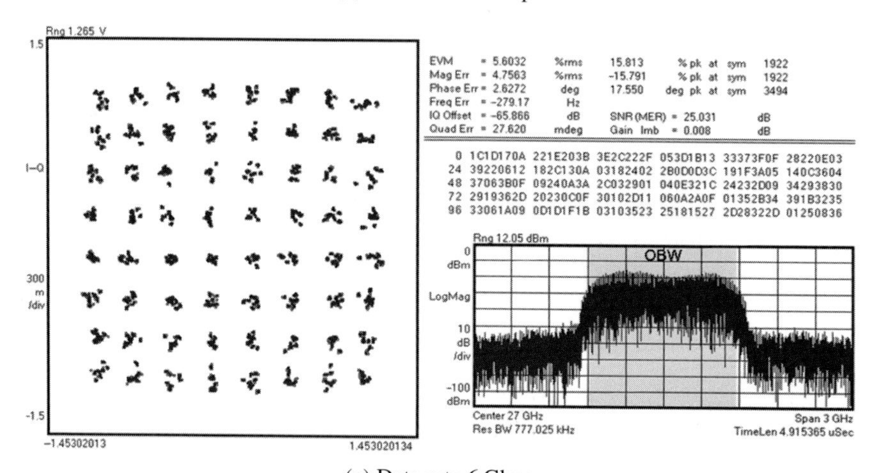

(c) Data rate 6 Gbps

Figure 13.30 64-QAM modulated signals measurement with (a) 3 Gbps, (b) 4.5 Gbps, and (c) 6 Gbps data rates [42]. (©2017 IEEE. Reprinted, with permission, from *2017 IEEE Radio Frequency Integrated Circuits Symposium.*)

which compensates the distortion and reduces the impedance transformation ratio of the interstage matching network, resulting in a flat AM–AM and AM–PM response. A parallel power combiner with 0.8 dB loss is designed to further enhance the linear output power. The measured power amplifier achieved the highest $P_{out} \cdot PAE$ at both 1.5 Gbps and the highest 6 Gbps data rate.

References

[1] F. C. Commission et al., "Use of spectrum bands above 24 GHz for mobile radio services," *Federal Register*, vol. 81, no. 164, pp. 58270–58308, 2016.

[2] A. Apostolidis, L. Campoy, K. Chatzikokolakis, et al., "Intermediate description of the spectrum needs and usage principles," *METIS Deliverable D*, vol. 5, p. 1, 2013.

[3] S. Shakib, H.-C. Park, J. Dunworth, V. Aparin, and K. Entesari, "A highly efficient and linear power amplifier for 28-GHz 5G phased array radios in 28-nm CMOS," *IEEE Journal of Solid-State Circuits*, vol. 51, no. 12, pp. 3020–3036, 2016.

[4] M. D. McKinley, K. A. Remley, M. Myslinski, J. S. Kenney, D. Schreurs, and B. Nauwelaers, "EVM calculation for broadband modulated signals," in *64th ARFTG Conference Digest*, ARFTG, 2004, pp. 45–52.

[5] Keysight Technologies. "Vector signal analysis basics. 2017. [Online]. Available: http://literature.cdn.keysight.com/litweb/pdf/5989-1121EN.pdf

[6] E. McCune, *Practical Digital Wireless Signals*. Cambridge University Press, 2010.

[7] E. McCune, "A technical foundation for RF CMOS power amplifiers: Part 5: Making a switch-mode power amplifier," *IEEE Solid-State Circuits Magazine*, vol. 8, no. 3, pp. 57–62, 2016.

[8] M. Vigilante, E. McCune, and P. Reynaert, "To EVM or two EVMs?: An answer to the question," *IEEE Solid-State Circuits Magazine*, vol. 9, no. 3, pp. 36–39, Summer 2017.

[9] T. J. Rouphael, *RF and Digital Signal Processing for Software-Defined Radio: A Multi-Standard Multi-Mode Approach*. Newnes, 2009.

[10] B. Razavi and R. Behzad, *RF Microelectronics*. Prentice Hall, 1998, vol. 1.

[11] P. Reynaert and M. Steyaert, *RF Power Amplifiers for Mobile Communications*. Springer Science & Business Media, 2006.

[12] S. Golara, S. Moloudi, and A. A. Abidi, "Processes of AM–PM distortion in large-signal single-FET amplifiers," *IEEE Transactions on Circuits and Systems I: Regular Papers*, vol. 64, no. 2, pp. 245–260, 2017.

[13] B. Heydari, M. Bohsali, E. Adabi, and A. M. Niknejad, "Millimeter-wave devices and circuit blocks up to 104 GHz in 90 nm CMOS," *IEEE Journal of Solid-State Circuits*, vol. 42, no. 12, pp. 2893–2903, 2007.

[14] W. L. Chan and J. R. Long, "A 58–65 GHz neutralized CMOS power amplifier with PAE above 10% at 1-V supply," *IEEE Journal of Solid-State Circuits*, vol. 45, no. 3, pp. 554–564, 2010.

[15] D. Zhao and P. Reynaert, "A 60-GHz dual-mode class AB power amplifier in 40-nm CMOS," *IEEE Journal of Solid-State Circuits*, vol. 48, no. 10, pp. 2323–2337, 2013.

[16] A. M. Niknejad, D. Chowdhury, and J. Chen, "Design of CMOS power amplifiers," *IEEE Transactions on Microwave Theory and Techniques*, vol. 60, no. 6, pp. 1784–1796, 2012.

[17] D. Chowdhury, P. Reynaert, and A. M. Niknejad, "A 60GHz 1V+ 12.3 dBm transformer-coupled wideband PA in 90nm CMOS," in *IEEE International Solid-State Circuits Conference, 2008 (ISSCC 2008), Digest of Technical Papers*. IEEE, 2008, pp. 560–635.

[18] T. LaRocca, J. Y.-C. Liu, and M.-C. F. Chang, "60 GHz CMOS amplifiers using transformer-coupling and artificial dielectric differential transmission lines for compact design," *IEEE Journal of Solid-State Circuits*, vol. 44, no. 5, pp. 1425–1435, 2009.

[19] I. Aoki, S. D. Kee, D. B. Rutledge, and A. Hajimiri, "Distributed active transformer – A new power-combining and impedance-transformation technique," *IEEE Transactions on Microwave Theory and Techniques*, vol. 50, no. 1, pp. 316–331, 2002.

[20] J. R. Long, "Monolithic transformers for silicon RF IC design," *IEEE Journal of Solid-State Circuits*, vol. 35, no. 9, pp. 1368–1382, 2000.

[21] C. H. Doan, S. Emami, A. M. Niknejad, and R. W. Brodersen, "Millimeter-wave CMOS design," *IEEE Journal of Solid-State Circuits*, vol. 40, no. 1, pp. 144–155, 2005.

[22] A. Agah, J. A. Jayamon, P. M. Asbeck, L. E. Larson, and J. F. Buckwalter, "Multi-drive stacked-FET power amplifiers at 90 GHz in 45 nm SOI CMOS," *IEEE Journal of Solid-State Circuits*, vol. 49, no. 5, pp. 1148–1157, 2014.

[23] J. Jayamon, A. Agah, B. Hanafi, H. Dabag, J. Buckwalter, and P. Asbeck, "A W-band stacked FET power amplifier with 17 dBm P sat in 45-nm SOI CMOS," in *2013 IEEE Topical Conference on Biomedical Wireless Technologies, Networks, and Sensing Systems (BioWireleSS)*, IEEE, 2013, pp. 79–81.

[24] U. R. Pfeiffer and D. Goren, "A 23-dBm 60-GHz distributed active transformer in a silicon process technology," *IEEE Transactions on Microwave Theory and Techniques*, vol. 55, no. 5, pp. 857–865, 2007.

[25] D. Zhao and P. Reynaert, "14.1 A 0.9 V 20.9 dBm 22.3%-PAE E-band power amplifier with broadband parallel-series power combiner in 40nm CMOS," in *2014 IEEE International Solid-State Circuits Conference, Digest of Technical Papers (ISSCC)*, IEEE, 2014, pp. 248–249.

[26] S. C. Cripps, *Advanced Techniques in RF Power Amplifier Design*. Artech House, 2002.

[27] J. P. Aikio and T. Rahkonen, "A comprehensive analysis of AM–AM and AM–PM conversion in an LDMOS RF power amplifier," *IEEE Transactions on Microwave Theory and Techniques*, vol. 57, no. 2, pp. 262–270, 2009.

[28] L. C. Nunes, P. M. Cabral, and J. C. Pedro, "AM/AM and AM/PM distortion generation mechanisms in Si LDMOS and GaN HEMT based RF power amplifiers," *IEEE Transactions on Microwave Theory and Techniques*, vol. 62, no. 4, pp. 799–809, 2014.

[29] C. Wang, M. Vaidyanathan, and L. E. Larson, "A capacitance-compensation technique for improved linearity in CMOS class-AB power amplifiers," *IEEE Journal of Solid-State Circuits*, vol. 39, no. 11, pp. 1927–1937, 2004.

[30] M. Vigilante and P. Reynaert, "A wideband Class-AB power amplifier with 29-57-GHz AM-PM compensation in 0.9-V 28-nm bulk CMOS," *IEEE Journal of Solid-State Circuits*, vol. PP, no. 99, pp. 1–14, 2017.

[31] W. Ye, K. Ma, and K. S. Yeo, "2.5 A 2-to-6GHz Class-AB power amplifier with 28.4% PAE in 65nm CMOS supporting 256QAM," in *2015 IEEE International Solid-State Circuits Conference-(ISSCC)*, IEEE, 2015, pp. 1–3.

[32] S. Kulkarni and P. Reynaert, "A 60-GHz power amplifier with AM–PM distortion cancellation in 40-nm CMOS," *IEEE Transactions on Microwave Theory and Techniques*, vol. 64, no. 7, pp. 2284–2291, 2016.

[33] L. Fanori and P. Andreani, "A high-swing complementary class-C VCO," in *2013 Proceedings of the ES-SCIRC (ESSCIRC)*, IEEE, 2013, pp. 407–410.

[34] K. L. Fong and R. G. Meyer, "High-frequency nonlinearity analysis of common-emitter and differential-pair transconductance stages," *IEEE Journal of Solid-State Circuits*, vol. 33, no. 4, pp. 548–555, 1998.

[35] P. R. Gray, P. Hurst, R. G. Meyer, and S. Lewis, *Analysis and Design of Analog Integrated Circuits*. Wiley, 2001.

[36] Y. He, L. Li, and P. Reynaert, "60GHz power amplifier with distributed active transformer and local feedback," in *ESSCIRC, 2010 Proceedings of the*. IEEE, 2010, pp. 314–317.

[37] B. François and P. Reynaert, "Highly linear fully integrated wideband RF PA for LTE-advanced in 180-nm SOI," *IEEE Transactions on Microwave Theory and Techniques*, vol. 63, no. 2, pp. 649–658, 2015.

[38] B. Park, S. Jin, D. Jeong, et al., "Highly linear mm-wave CMOS power amplifier," *IEEE Transactions on Microwave Theory and Techniques*, vol. 64, no. 12, pp. 4535–4544, 2016.

[39] A. Sarkar and B. A. Floyd, "A 28-GHz harmonic-tuned power amplifier in 130-nm SiGe BiCMOS," *IEEE Transactions on Microwave Theory and Techniques*, vol. 65, no. 2, pp. 522–535, 2017.

[40] D. Kang, B. Park, D. Kim, J. Kim, Y. Cho, and B. Kim, "Envelope-tracking CMOS power amplifier module for LTE applications," *IEEE Transactions on Microwave Theory and Techniques*, vol. 61, no. 10, pp. 3763–3773, 2013.

[41] C. Fager, J. C. Pedro, N. B. de Carvalho, H. Zirath, F. Fortes, and M. J. Rosário, "A comprehensive analysis of IMD behavior in RF CMOS power amplifiers," *IEEE Journal of Solid-State Circuits*, vol. 39, no. 1, pp. 24–34, 2004.

[42] Y. Zhang and P. Reynaert, "A high-efficiency linear power amplifier for 28GHz mobile communications in 40nm CMOS," in *2017 IEEE Radio Frequency Integrated Circuits Symposium (RFIC)*, IEEE, 2017, pp. 33–36.

[43] G. Jeong, T. Joo, and S. Hong, "A highly linear and efficient CMOS power amplifier with cascode–cascade configuration," *IEEE Microwave and Wireless Components Letters*, vol. 27 , no. 6, pp. 596–598, 2017.

[44] M. Elmala, J. Paramesh, and K. Soumyanath, "A 90-nm CMOS Doherty power amplifier with minimum AM-PM distortion," *IEEE Journal of Solid-State Circuits*, vol. 41, no. 6, pp. 1323–1332, June 2006.

[45] S. Shakib, M. Elkholy, J. Dunworth, V. Aparin, and K. Entesari, "2.7 A wideband 28GHz power amplifier supporting 8 × 100MHz carrier aggregation for 5G in 40nm CMOS," in *2017 IEEE International Solid-State Circuits Conference (ISSCC)*, IEEE, 2017, pp. 44–45.

[46] S. Hu, F. Wang, and H. Wang, "2.1 A 28GHz/37GHz/39GHz multiband linear Doherty power amplifier for 5G massive MIMO applications," in *2017 IEEE International Solid-State Circuits Conference (ISSCC)*, IEEE, 2017, pp. 32–33.

[47] M. Vigilante and P. Reynaert, "A 29-to-57GHz AM-PM compensated Class-AB power amplifier for 5G phased arrays in 0.9V 28nm bulk CMOS," in *2017 IEEE Radio Frequency Integrated Circuits Symposium (RFIC)*, IEEE, June 2017, pp. 116–119.

[48] S. N. Ali, P. Agarwal, S. Mirabbasi, and D. Heo, "A 42–46.4% PAE continuous Class-F power amplifier with Cgd neutralization at 26–34 GHz in 65 nm CMOS for 5G applications," in *2017 IEEE Radio Frequency Integrated Circuits Symposium (RFIC)*, IEEE, June 2017, pp. 212–215.

[49] A. Sarkar, F. Aryanfar, and B. A. Floyd, "A 28-GHz SiGe BiCMOS PA with 32% efficiency and 23-dBm output power," *IEEE Journal of Solid-State Circuits*, vol. 52, no. 6, pp. 1680–1686, June 2017.

14 FinFET Process Technology for RF and Millimeter-Wave Applications

Hyung-Jin Lee and Bernhard Sell

14.1 Overview of FinFET Technology

Fin field-effect transistor (FinFET) logic process technology has been widely adopted for SoC applications since its commercial introduction in 2012 [1]. The following section highlights the benefits of FinFET devices over planar transistors and describes how the structures can be optimized for each application. Mainstream logic technologies are optimized for high density, high logic performance, and low power [2,3], while specialized technologies have been introduced recently that are optimized for analog and radio frequency (RF) performance [4]. In FinFET technology, traditional two-dimensional transistors are replaced by three-dimensional geometries called FinFETs. While the structure and performance of these devices are different, the layout view is identical for planar and FinFET transistors, as shown in Figure 14.1. Figure 14.1a shows the standard layout view with the drawn dimensions for gate length and gate width.

Figure 14.1b shows a three-dimensional view of a standard planar transistor. The physical device width and device length are identical to the drawn dimensions and are typically tunable within a given range. Figure 14.1c shows the three-dimensional view of a FinFET transistor. The active device is no longer limited to the planar surface but instead is located along the side and the top of the fin. Defining the height of the silicon fin as h and the width of the fin as w_{si}, the total active device width per fin w_f is given by the following equation:

$$w_f = 2h + w_{si} \tag{14.1}$$

The fins are placed at a defined fin pitch p, and each device in the design has to be drawn with a width of an exact multiple of the fin pitch. This means that rather than dialing in the device width continuously as is done on planar devices, the physical device width of a FinFET device is an exact multiple of w_f. In modern FinFET technologies, the fin pitch is less than 50 nm, the height is 45–60 nm, and the fin width is around 10 nm. As a result, the physical width of a FinFET transistor is at least twice the drawn device width, leading to a much higher device density compared to the traditional planar device.

Different optimization of fin parameters is required based on the application. Logic technologies benefit most from tighter fin pitch as it improves the area scaling, while for applications where the vertical gate resistance is critical, such as the RF transistor, a slightly relaxed fin pitch is preferred. In addition to improving area scaling, FinFET devices exhibit excellent short channel effects and high drive current. Figure 14.2 shows

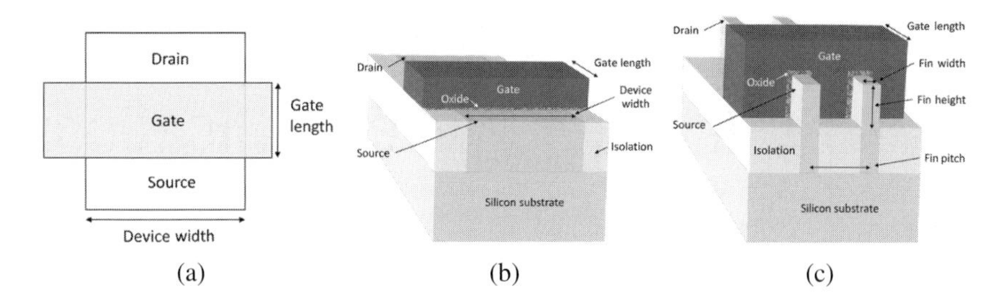

Figure 14.1 FinFET vs. planar structures.

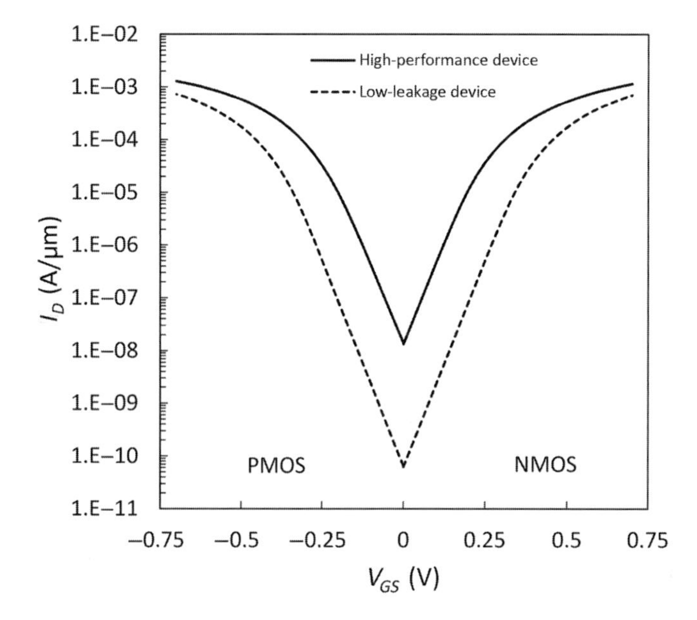

Figure 14.2 $I_D - V_{GS}$ characteristics of an RF-optimized FinFET technology.

the I_D–V_{GS} characteristics of a FinFET technology that has been optimized for RF applications. The subthreshold slope is near ideal and devices are targeted to have off currents for both N-type metal-oxide-semiconductor (NMOS) and P-type metal-oxide-semiconductor (PMOS) devices well below 100 pA/μm.

For best RF performance, devices with high drive current, low leakage, low gate capacitance, and low parasitic capacitance are paramount. To achieve this, the gate length is reduced to the point at which a sharp increase in the subthreshold slope is observed. The correlation between subthreshold slope and gate length is shown in Figure 14.3 for both planar and FinFET technologies. While at 30 nm gate length the planar device already indicates a sharp increase, the FinFET device is still very close to the ideal subthreshold slope of 60 mV/decade. As a result, RF devices on FinFET technology are showing best performance at less than 30 nm channel length.

Transistor drive currents of FinFET devices are significantly higher than on any commercial planar technology. I_D–V_{DS} characteristics of an RF-optimized technology

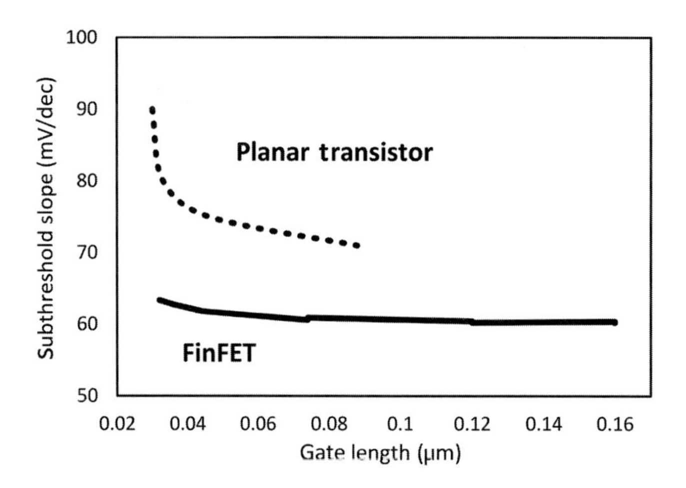

Figure 14.3 Subthreshold slope comparison between planar and FinFET transistors.

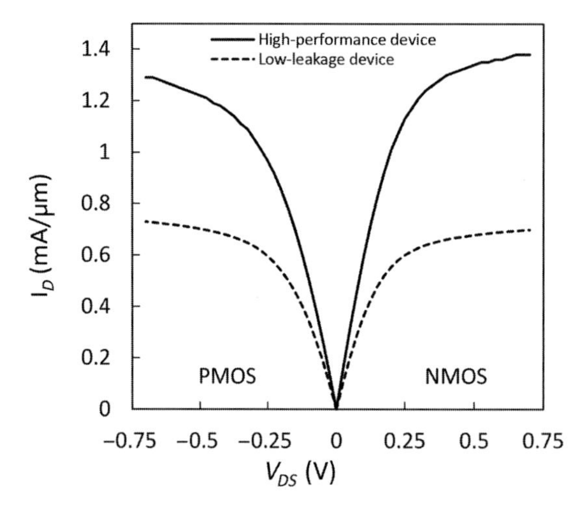

Figure 14.4 I_D–V_{DS} characteristics of an RF-optimized FinFET technology.

are shown in Figure 14.4, demonstrating a current density greater than 13 mA/μm at 700 mV supply voltage. On planar technologies, the supply voltage would need to be increased close to 1 V to achieve a similar drive current [5]. Even higher performance can be achieved when the technology is optimized for high-performance logic [3].

A well-targeted and optimized FinFET technology does not require any substrate doping to control short channel devices. Eliminating the substrate doping entirely and controlling the threshold voltage solely by the workfunction of the metal gate leads to devices with an extremely low random variation of the threshold voltage, ρV_T, as shown in Figure 14.5. For both NMOS and PMOS devices, ρV_T values of much less than 20 mV per fin can be achieved. The lack of substrate doping also improves flicker noise.

To understand the potential sources of variation, the simplified process flow of a FinFET technology is depicted in Figure 14.6. All modern FinFET technologies have

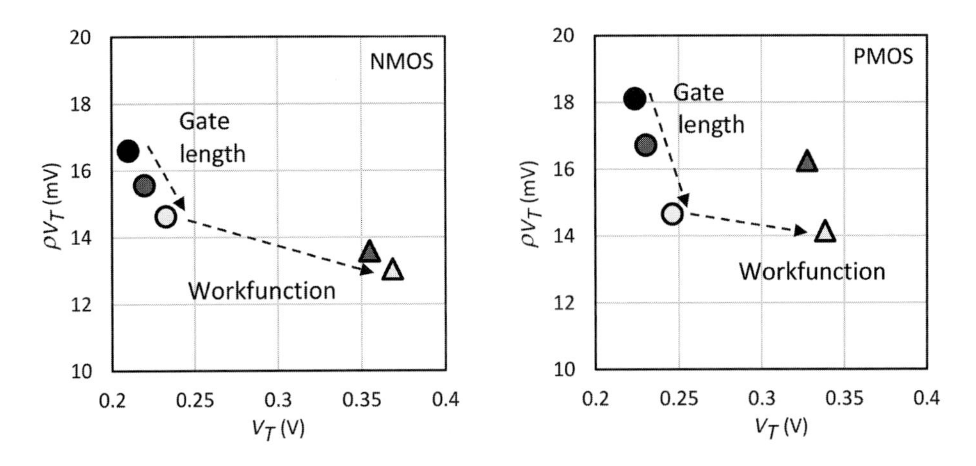

Figure 14.5 Random threshold voltage variation per n for an RF-optimized FinFET technology.

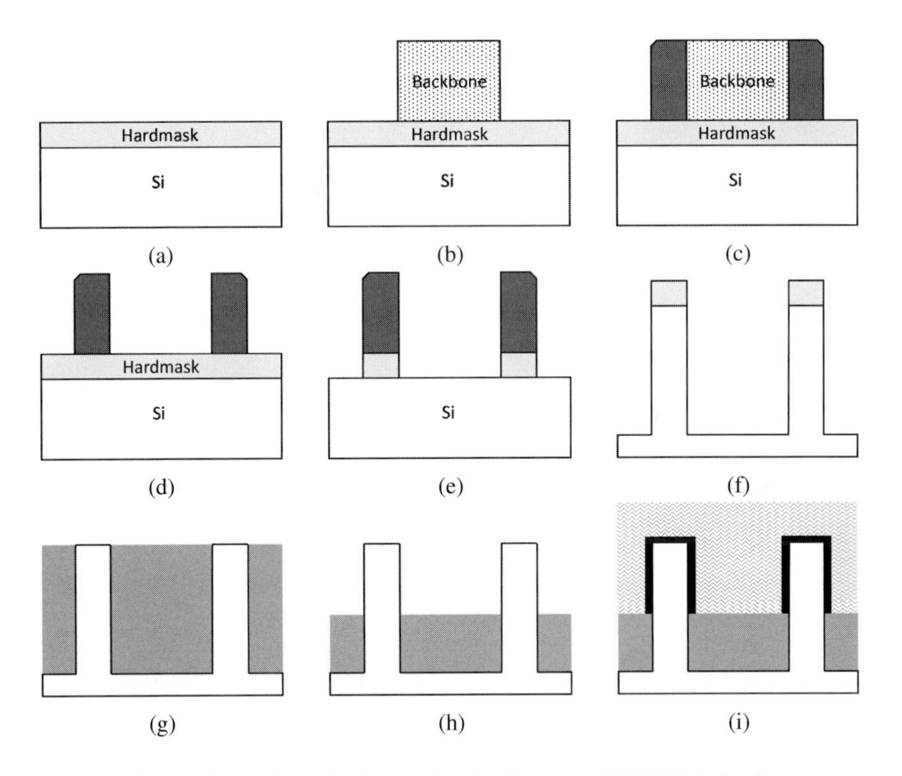

Figure 14.6 Process flow schematic for creating fins in a typical FinFET technology.

less than 50 nm fin pitch and therefore require the pitch division process described in the following discussion. Logic optimized technologies with fin pitches close to 30 nm necessitate pitch-quartering techniques that are not discussed here.

The process starts with depositing a hardmask on Si (Figure 14.6a) and patterning a backbone material on top of this hardmask (Figure 14.6b). This backbone is printed with twice the final fin pitch. For example, if the final fin pitch is 45 nm, this backbone is patterned at 90 nm pitch. This is followed by depositing and etching a spacer (Figure 14.6c). The thickness of this spacer is critical, as it determines the width of the final fin. Having the fin width defined by a spacer rather than by direct printing reduces the variation of this width. In modern FinFET technologies, the width of the final fin can be controlled to within a few Angstrom.

After spacer formation, the backbone layer is removed (Figure 14.6d) and the pattern is transferred into the hardmask (Figure 14.6e). The next step is to etch the hardmask pattern into the silicon (Figure 14.6f). This forms the final fin and determines the depth of the isolation. This step is typically followed by the cut of the fins in the other direction that is now shown here.

The isolation material is deposited and polished (Figure 14.6g) and recessed to determine the final height of the active part of the fin (Figure 14.6h). The recess step determines the final height of the fin, and the control of this step is therefore critical to minimize performance variation. The remainder of the patterning is similar to planar technologies with the exception that the gate dielectric and the gate have to be formed on the vertical channel region as shown in Figure 14.6i.

14.2 Unique Properties of FinFET Technology for RF/mm-Wave Design Consideration

The recent silicon technology has to inevitably evolve from planar to FinFET technology to ensure continuous logic scaling, though not all the new features are favorable to the RF/mm-wave circuit. Some will hurt RF performance, while some will bring unexpected benefit to RF performance. In this section, we will review the unique properties of FinFET technology, which should be under consideration for RF/mm-wave design optimization, and we will evaluate how the technology evolution will impact RF performance positively or negatively in the following section.

14.2.1 Transistor Scaling and Performance

For over the past 50 years, Moore's law has been sustained through the continual reduction of transistor channel length and hence increased transistor density and performance. The shorter transistor channel contributes to the higher transconductance g_m scaled by the square of channel length L^2 and the lower gate capacitance, thus higher unity gain frequency f_t. However, the transistor performance improvement trend has been significantly disturbed as the channel length is getting too short. The major two root causes are velocity saturation and drain induced barrier lowering (DIBL). The former effect simply slows down the transconductance improvement by the channel length scaling, as the g_m is scaled by $1/L$ and eventually no scaling for short channel rather than by $1/L^2$ as is found for long channel devices, where L is the channel length. Since no g_m improvement is expected for the extremely short channel length, the reduction of the gate capacitance C_{gg} by the channel length scaling is the only source of improvement. The latter effect plagues analog and RF circuit design more.

When gate length is reduced, the drain and source start to interact more, and the gate control of the device is reduced. This effect is termed as short channel effect degradation, and the metric to measure this degradation is DIBL. The potential barrier in the channel is supposed to be lowered only by the gate in an ideal device, which is true for long-channel devices. This assumption starts breaking down for short channel devices.

DIBL is measured as the reduction in threshold voltage when the drain voltage is increased $\partial V_t/\partial V_{DS}$. Planar technologies started having huge DIBL issues with gate lengths ~30 nm as shown in Figure 14.7, which implies less gate control to channel and more drain to source interaction resulting in higher leakage current. In order to dampen the DIBL degradation, planar technologies need to use high channel doping and halo channel doping that would reduce the drain fields from encroaching in the channel at the cost of mobility degradation, and hence weakening of the on-state drive current.

FinFET technologies, however, provide excellent gate control over the thin channel by surrounding fin with three-sided gate. FinFETs have been able to scale gate lengths down to 15–20 nm without short-channel effect degradation (Figure 14.7). The improvement of DIBL leads to lower threshold voltage and lower leakage current, which enables lower power design. The lower threshold voltage provides a higher overdrive gate voltage that increases drive currents at lower power supply, as shown in Figure 14.8. This has been a dominant benefit to digital circuits.

The lower DIBL also results in higher output resistance in the saturation mode R_{out}, as the drive current is insensitive to the drain voltage. This is also shown as a flatter $I_D \cdot V_{DS}$ curve. For short-channel devices, R_{out} is primarily driven by DIBL as opposed to channel-length modulation that dominates for long-channel devices. Higher output resistance leads to higher voltage gain such as $g_m R_{out}$, and to higher linearity for large signal amplification.

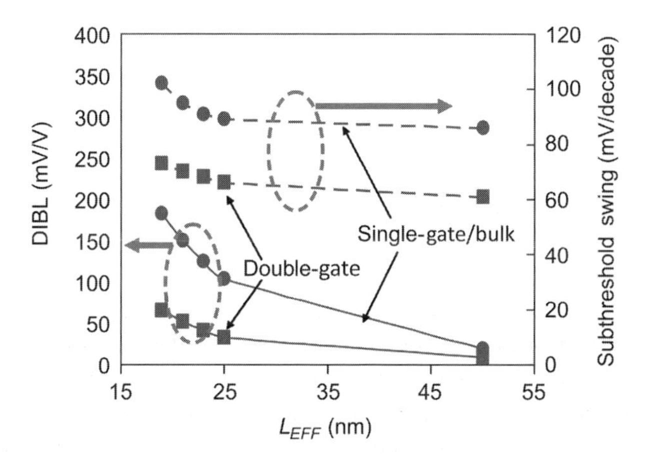

Figure 14.7 Medici-predicted DIBL and subthreshold swing versus effective channel length for double-gate (DG) and bulk-silicon nFETs [6]. (©2004 IEEE. Reprinted, with permission, from *IEEE Circuits and Devices Magazine*.)

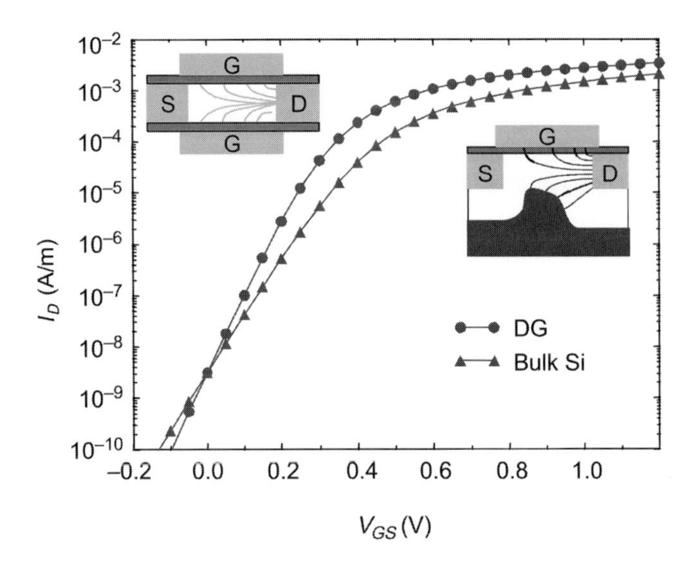

Figure 14.8 Medici-predicted DIBL and subthreshold swing versus effective channel length for DG and bulk-silicon nFETs [6]. (©2004 IEEE. Reprinted, with permission, from *IEEE Circuits and Devices Magazine*.)

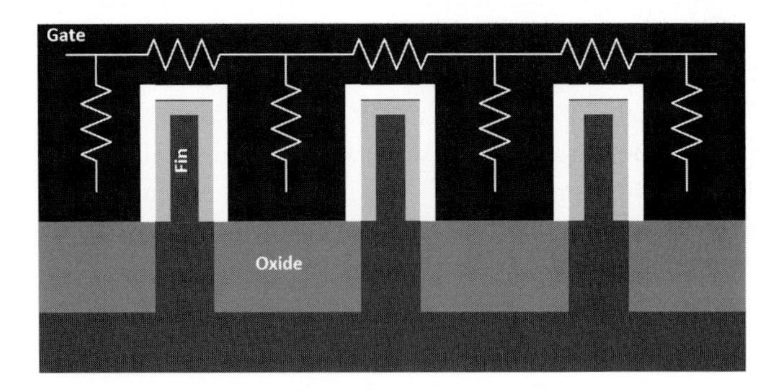

Figure 14.9 3D gate network.

14.2.2 Nonlinear Gate Resistance by 3D Structure

The FinFET transistor has a three-dimensional (3D) channel, called fin, wrapped around by gate material. By the nature of the 3D structure, the gate resistance has two major components, horizontal and vertical resistance, as shown in Figure 14.9.

To evaluate the behavior of gate resistance by device width equivalent to the number of fins for FinFET technology, Figure 14.11 suggests the simple decomposition of the resistance components. R_1 is the fixed amount of resistance representing the contact resistance at the edge of the gate material to the metal interface. The vertical resistance R_V is the combination of R_2, R_3, and the effective R_4 counting open-end resistance. The horizontal resistance R_h is the parallel connection of R_5 and R_6. Assumption of the power equivalent as shown in Figure 14.10 suggests the calculation of the total equivalent gate resistance, $R_{g,eq}$ or simply R_g.

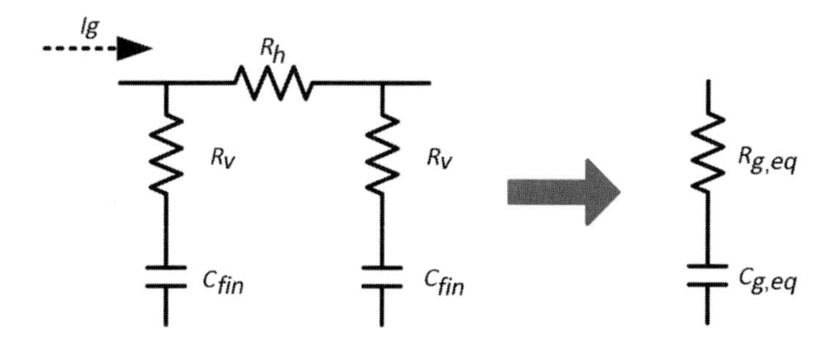

Figure 14.10 Power-equivalent RC network simplification.

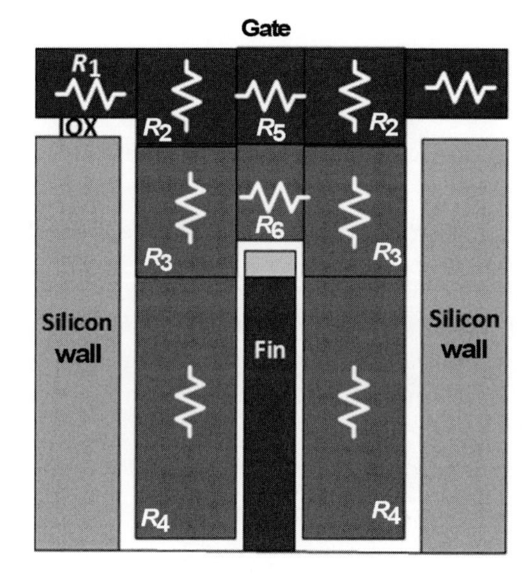

Figure 14.11 Decomposition of 3D resistance components.

For the single fin structure, one can calculate as follows:

$$R_{const} = R_1 \tag{14.2}$$

$$R_v = R_2 + R_3 + \frac{R_4}{3} \tag{14.3}$$

$$R_h = R_5 || R_6 \tag{14.4}$$

$$R_g = R_{const} + \frac{R_v}{2} + \frac{R_h}{4} \tag{14.5}$$

These equations can be extended to support multiple fins with the parameter n referring to the number of the fins, as follows:

$$R_g = R_{const} + \frac{n}{3}\left(1 - \frac{1}{4n^2}\right) R_h + \left(\frac{1}{n} - \frac{1}{2n^2}\right) R_v \tag{14.6}$$

If n is large enough, the equation for R_g can be further simplified as

$$R_g \approx R_{const} + \frac{nR_h}{3} + \frac{R_v}{n}. \tag{14.7}$$

Double gate contact will reduce only the horizontal resistance by a factor of 4, as follows:

$$R_g \approx R_{const} + \frac{nR_h}{12} + \frac{R_v}{n}. \tag{14.8}$$

Due to the nonlinear relation of R_v to the number of fins n, R_g starts decreasing as n increases until the horizontal resistance starts dominating. Similar observation has been reported, and the R_g behavior for FinFET is shown as Figure 14.12 [7].

14.2.3 Fin Self-Heating

Current flow through the device channel generates heat, and FinFETs are no exception. For FinFET transistors, heat degrades performance even more as the heat is trapped within the fin structures because the fin has only a narrow thermal conducting channel, called sub-fin, to the substrate.

In addition to excessive heat, the slow thermal network suggested as a low-pass filter, as shown in Figure 14.13 [8], causes significant modification from the R_{out} at direct conversion (DC). The deviation of the R_{out} increases as the frequency goes up until it meets the cutoff frequency of the thermal network response. Even though the intrinsic

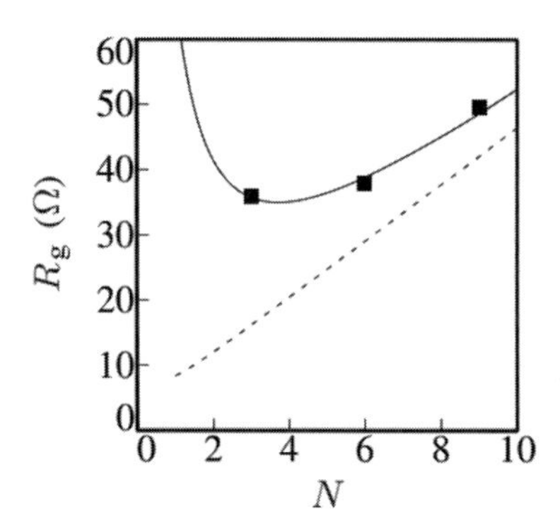

Figure 14.12 R_g by number of fins [7]. (©2010 IEEE. Reprinted, with permission, from *2010 International Electron Devices Meeting*.)

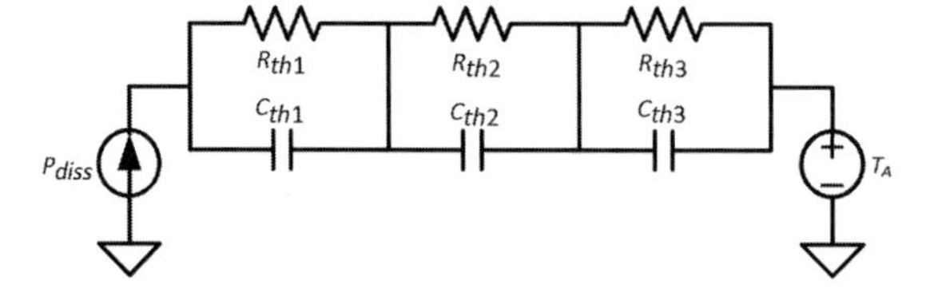

Figure 14.13 Thermal network model.

gain $g_m R_{out}$ is still superior to planar technologies, device models should account for the frequency-dependent R_{out} changes to maintain model accuracy.

$$T_{device} = T_{ambient} + Z_{th} P_{dis} \tag{14.9}$$

$$Z_{th} = R_{th} + \frac{1}{j\omega C_{th}} \tag{14.10}$$

As illustrated in Figure 14.13 and the preceding equations, the device temperature is not affected by the dynamic self-heating if the frequency is high enough beyond the cutoff frequency of the thermal network, as the deviation of R_{out} is being saturated as the frequency goes higher. The R_{out} also converges to DC value as the frequency goes lower, as shown in Figure 14.14.

$$R_{out,delta}(\%) = 100 \frac{R_{out,RF} - R_{out,DC}}{R_{out,DC}} \tag{14.11}$$

It is also worth noting that R_{out} deviation from DC is different for PMOS and NMOS. The deviation is determined by the amount of velocity saturation change from temperature and the amount of threshold voltage V_{th} shift by the device workfunction sensitivity to the temperature. There is an optimum bias condition where there is no frequency sensitivity. Circuit performance will be most stable to frequency shift at this bias condition. One can find zero $R_{out,delta}$ at certain bias conditions in Figure 14.15 and 14.16 for PMOS and NMOS, respectively.

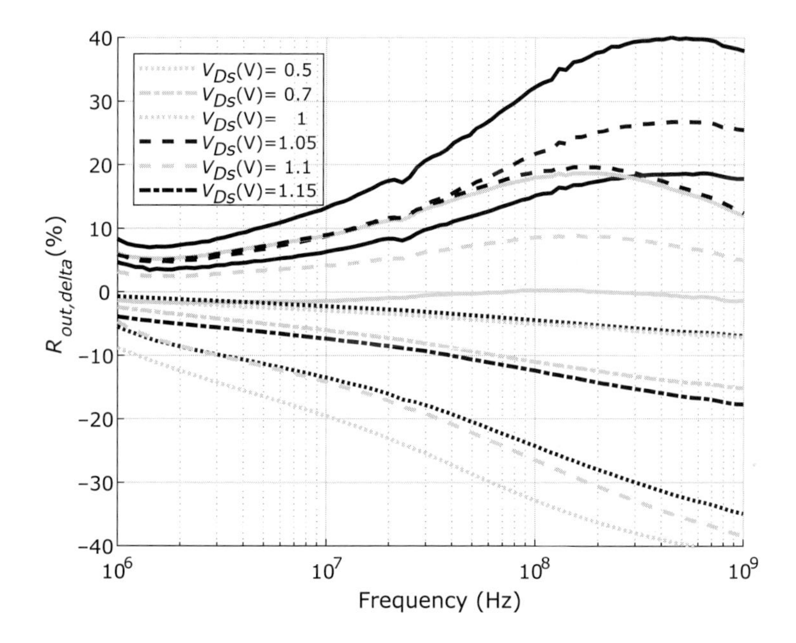

Figure 14.14 $R_{out,delta}$ by frequency.

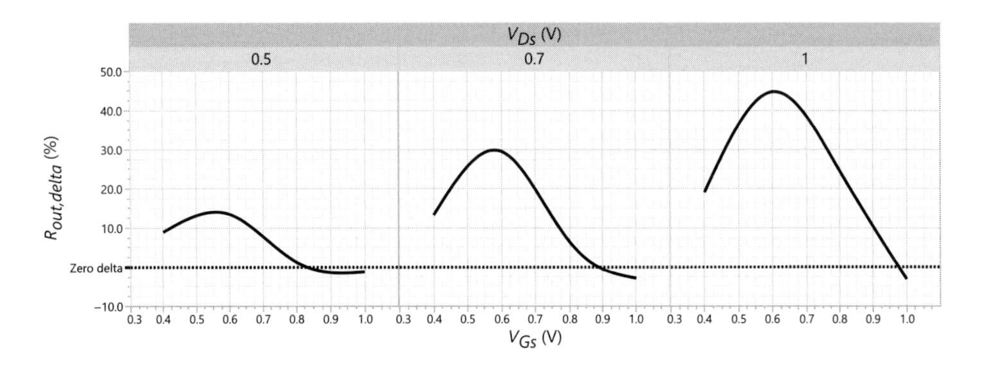

Figure 14.15 PMOS $R_{out,delta}$ at 100 GHz.

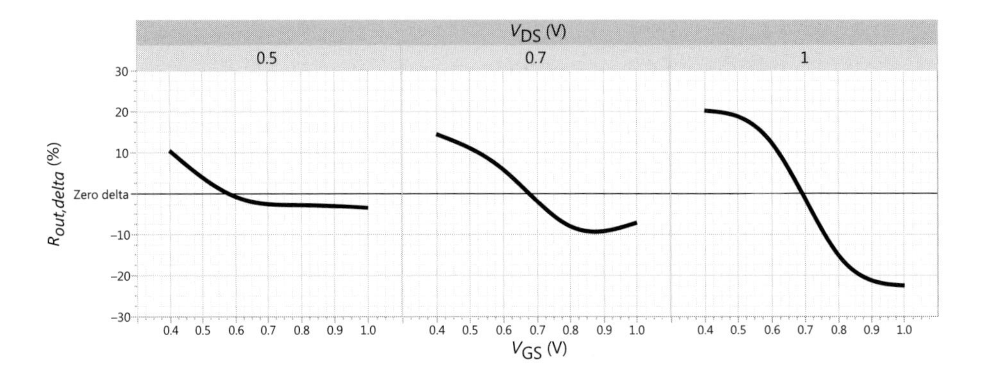

Figure 14.16 NMOS $R_{out,delta}$ at 10 GHz.

14.3 Assessment of FinFET Technology for RF/mm-Wave

FinFET transistors introduce higher parasitic capacitance by the nature of the three-dimensional device architecture. Therefore, the peak F_t of FinFET is shy to the planar. Even with all the superior DC characteristics achieved by FinFET technology and scaling benefit, one may hesitate to start exploring RF and mm-wave circuit design due to the relatively lower peak F_t number compared to planar technology. However, the question arises if the peak F_t and F_{max} indeed matter for RF and mm-wave circuit design. To answer this question, we need to consider how much gain we can drive from the transistor without placing too much stress on the transistor for reliable lifetime operation and how high of a frequency the device can support with meaningful power gain. Also, we will evaluate how the linearity of FinFET technology performs against planar technology, which is another critical property for RF and mm-wave system design. In next few sections, we will suggest different ways to compare device technologies for RF and mm-wave system design and provide the speculatively outstanding performance offered by FinFET technology that is favorable to RF and mm-wave circuit design.

14.3.1 Parasitics and RF Performance

In Section 14.2.2, we have calculated the total gate resistance for multiple numbers of fins. In the total gate material across multiple fins, some portion of gate material between fins is not functional, as it does not contribute to transconductance of the device, but still contributes to the parasitic load ($R_{out,DC}$). The nonfunctional section of the gate increases the total gate resistance and the parasitic capacitance between gate-to-source/drain interface ($C_{gd,par}$, $C_{gs,par}$) as well as the gate to the fin outside of the channel ($C_{g,finr}$). Therefore, the excessive gate material between fins degrades the peak F_t compared to the peak F_t of planar as shown in Figure 14.17. The recent silicon evidence suggests the peak F_t of FinFET is as high as 20% below that of planar.

However, the FinFET has improved its peak F_{max} performance over the planar thanks to the vertical portion of gate resistance, which reduces total gate resistance up to a certain number of fins as shown in Figure 14.18. The higher F_{max} is favorable for mm-wave circuit design.

According to the recent literature survey, the peak F_t and peak F_{max} start degrading in FinFET technology as poly pitch keeps tightening due to the excessive parasitic. In

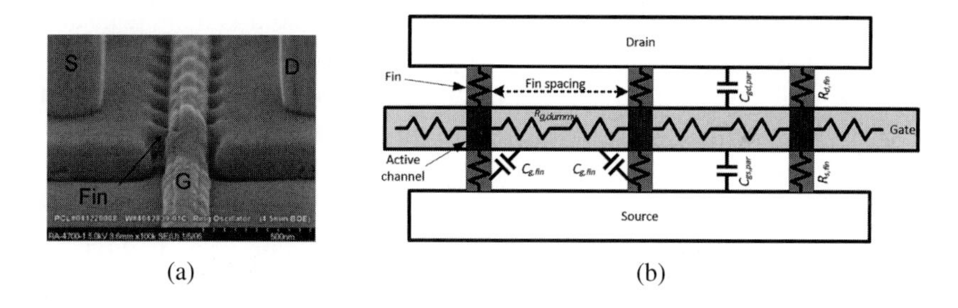

Figure 14.17 FinFET structure and parasitic.

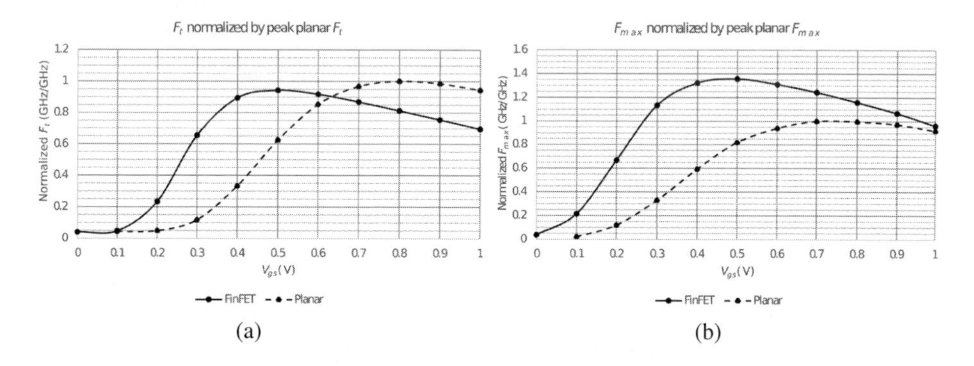

Figure 14.18 (a) F_t and (b) F_{max} of FinETs and planars.

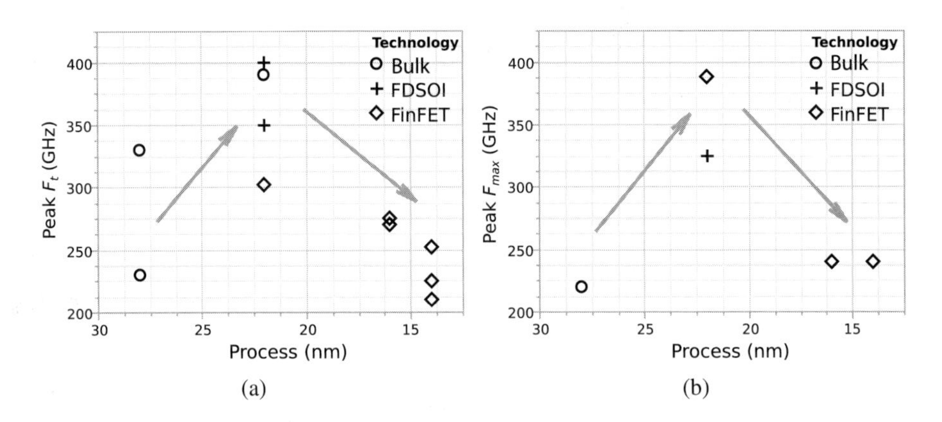

Figure 14.19 (a) F_t and (b) F_{max} by the channel length.

planar technology, peak F_t and peak F_{max} continue improving by scaling channel length down by enhancing the transconductance and reducing gate capacitance. However, in FinFET technology, the transconductance improvement has stalled with the channel length shrinking due to the velocity saturation, and the total gate capacitance does not improve since the increased gate to source/drain capacitance nulls the improvement of the gate to channel capacitance. The overall trend of the peak F_t and F_{max} has reached the peak number around 22 nm technology for both planar and FinFET technology as shown in Figure 14.19. The measurement data shown in Figure 14.19 include the most low two-level back-end interfaces down to 14 nm, so it is more realistic for the actual circuit design.

Despite the peak F_t reduction in FinFET technology due to the parasitics, FinFET technology still offers remarkable performance benefit for RF and mm-wave design over the planar. It is noticeable that the planar device reaches the peak F_t and peak F_{max} at around 60% higher V_{gs} than FinFET due to the short channel effect or DIBL. For the low power design, such as the bias condition kept, FinFET can reach over 50% higher F_t than planar, and the intrinsic gain $g_m/I_d > 10\,V^{-1}$ of FinFET is over four times higher than planar at $g_m/I_d > 10\,V^{-1}$ as shown in Figure 14.20. Improvement of the intrinsic device performance, such as short channel effect and full depletion mode, is more than enough to compensate for the parasitic degradation and still achieves better RF performance in the low power regime. We will discuss the details of power dissipation improvement for the given signal gain by FinFET in Section 14.4.

14.3.2 Noise Performance

To review the noise properties in FinFET technology, let us recall the two-port noise theory [9,10]. One can reconstruct the noise metal-oxide-semiconductor field-effect transistor (MOSFET) into the equivalent input-referred noise source and the noiseless MOSFET as shown in Figure 14.21.

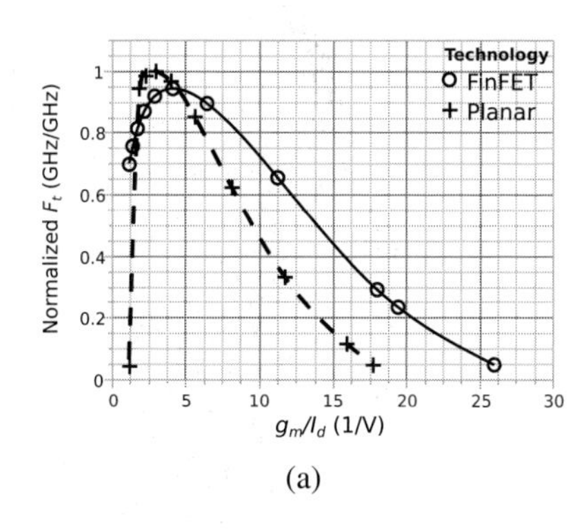

(a)

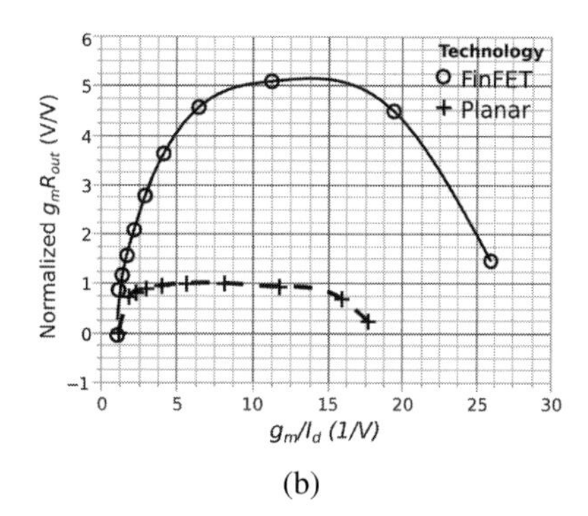

(b)

Figure 14.20 (a) F_t and (b) intrinsic gain for bias condition.

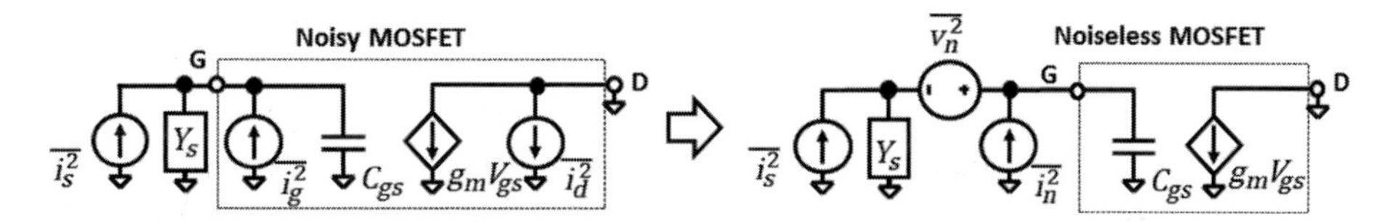

Figure 14.21 Two-port noise network.

The equivalent two-port noise network can derive noise factor, the minimum possible noise factor F_{min}, optimum noise impedance G_n, and equivalent noise conductance as

$$F = F_{min} + \frac{G_n}{R_s} \left| Z_s - Z_{opt} \right|^2 \tag{14.12}$$

$$F_{min} = 1 + 2\left(\frac{\omega}{\omega t}\right) b_3 \gamma \tag{14.13}$$

$$Z_{opt} = \frac{1}{\omega C_{gs}} \left(\frac{b_3}{b_2}\right) + j\frac{1}{\omega C_{gs}} \left(\frac{b_1}{b_2}\right) \tag{14.14}$$

$$G_N = \frac{\gamma(\omega C_{gs})^2}{g_m} \tag{14.15}$$

where

$$b_1 = 1 + \Delta|c| \tag{14.16}$$

$$b_2 = 1 + 2\Delta|c| + \Delta^2 \tag{14.17}$$

$$b_3 = \Delta\sqrt{1 - |c|^2} \tag{14.18}$$

$$\Delta = \sqrt{\frac{\delta}{5\gamma}} \tag{14.19}$$

and c is the noise correlation coefficient, γ and δ are the thermal noise excess factor (or noise gamma factor) and thermal noise parameter respectively. The noise correlation coefficient, c, is frequency independent and decreases as channel length decreases.

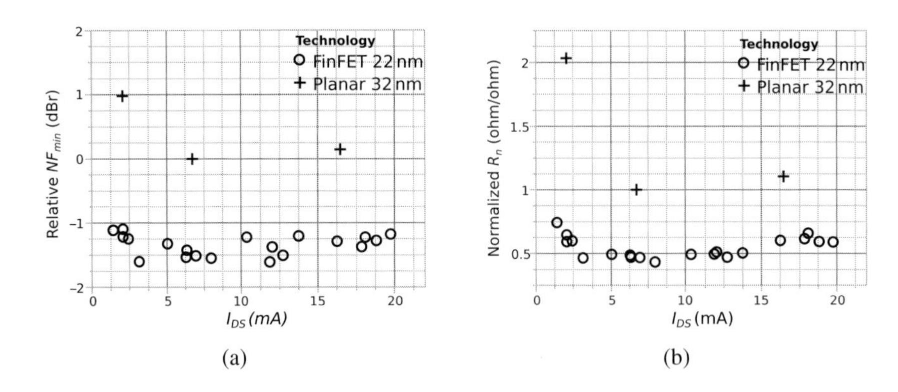

Figure 14.22 (a) NF_{min} and (b) R_n measurement of FinFET and planar.

From the preceding equation, it is worth of mentioning that overall NF_{min} reduces as the unity gain frequency $\omega t = 2\pi f_t$ increases, and G_n is even more sensitive to C_{gs} than w_t. One should note that G_n is the sensitivity of the device to noise mismatch by amplifying the amount of mismatch between source impedance and the optimum noise impedance.

To achieve overall better noise performance, lowering NF_{min} and G_n (or increasing R_n in the impedance form) will achieve the lower total noise factor. We already reviewed that unity gain frequency ωt of FinFET exceeds planar at the lower bias condition by a significant amount (>50%). Therefore, one can expect a significant improvement of the thermal noise for the gain at the lower current bias condition, and the measurement at 30 GHz with V_{cc} of 1 V confirms this, as shown in Figure 14.22.

On top of the ωt improvement at low current bias condition, the recent FinFET silicon also reports an R_n improvement over planar thanks to presumably γ improvement, and the γ improvement is partly due to the velocity saturation and channel length modulation improvement. The recent silicon evidence reports γ very close to 2/3, the same as classical long channel devices. The improvement of R_n desensitizes the noise-gain mismatch, hence there is less compromise between power matching at the cost of noise figure degradation.

14.3.3 Gain and Noise Matching at the mm-Wave Frequency

Figure 14.23 presents the classical low-noise amplifier (LNA) design problem of noise and input matching. Traditionally, there is a mismatch between Z_{in}^* and Z_{opt}, and Z_{in}^* is depicted as

$$Z_{in}^* = r_g + j\frac{1}{\omega C_{gs}} \tag{14.20}$$

where r_g is the gate resistance.

The mismatch between Z_{in}^* and Z_{opt} defines how much noise performance degradation there is when the input matching network is designed for minimum signal reflection and vice versa. The mismatch between the optimum noise impedance and the input

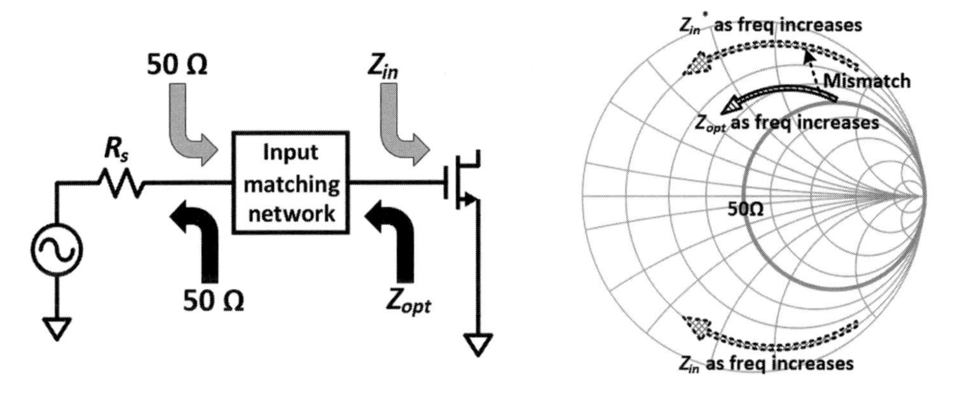

Figure 14.23 F_t and intrinsic gain for bias condition.

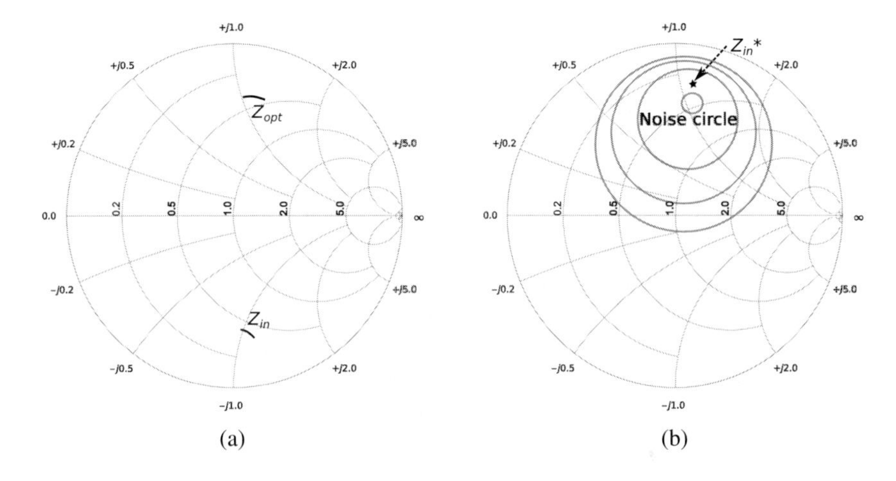

Figure 14.24 Z_{opt} and Z_{in} of FinFET at 82 GHz.

matching impedance is not negligible for planar. Therefore, it has been a major challenge in LNA design to manipulate the matching condition to bring the two points of Z_{in}^* and Z_{opt} closer by using various circuit techniques, such as inductive source degeneration, for example.

To make Z_{in}^* closer to Z_{opt}, the following relationship should be met:

$$r_g \approx \frac{1}{\omega C_{gs}} \frac{b_3}{b_2} \tag{14.21}$$

$$b_1 \approx b_2 \tag{14.22}$$

The preceding relationship requires a low-noise correlation coefficient c and a low-noise gamma factor γ. Low-noise gamma factors in FinFETs have already been reported, but further research is required to validate the low-noise correlation coefficient.

Figure 14.24 shows the optimum noise matching point Z_{opt}, and the input impedance Z_{in} of FinFET device at 82 GHz, and they are getting closer as the frequency goes higher. As one can see from the figure, Z_{opt} and Z_{in}^* are almost conjugate, and therefore no significant effort is required in designing an input matching network.

Furthermore, since gate resistance R_g is nonlinear with respect to the number of fins as described earlier, the real term of Z_{opt} and Z_{in} is finely adjustable by choosing a proper number of fins. This allows for modification of the resistive input impedance to compensate for inductive source degeneration or any update on the amount of feedback.

14.4 Design Methodology for RF/mm-Wave Performance Optimization with FinFET

14.4.1 Wireless Design Consideration in Cascade Chain

Wireless system design often aims for two primary performance targets: noise and linearity. Meeting the performance target is one thing, but market competitiveness of the product requires attention to power dissipation and area. In general, the silicon area of the wireless system is highly dominated by passive components, such as an embedded coil, and the coil technology is highly independent of process nodes. Also, transistor node scaling does not always bring positive impact to silicon area scaling. This is due to design rule complications and excessive parasitics in the device and metal interface within such a crowded accessible space. Therefore, the total radio frequency integrated circuit (RFIC) power dissipation is the premier differentiator to competitor products in the market.

During wireless system design, the signal link is optimized using the cascade chain shown in Figure 14.25. Noise factor (F) and the linearity ($IIP3$) in the cascade chain are well understood as expressed in (14.23). As the equations suggest, the very first stage predominates the noise performance (F_{total}), while the very last stage dictates the linearity performance ($IIP3_{total}$). The remaining stages should focus more on power efficiency to optimize the overall power dissipation.

$$F = F_1 + \frac{F_2}{G1} + \frac{F_3 - 1}{G_1 G_2} + \cdots \tag{14.23}$$

$$\frac{1}{IIP3} = \frac{1}{I_1} + \frac{G_1}{I_2} + \frac{G_1 G_2}{I_3} + \cdots \tag{14.24}$$

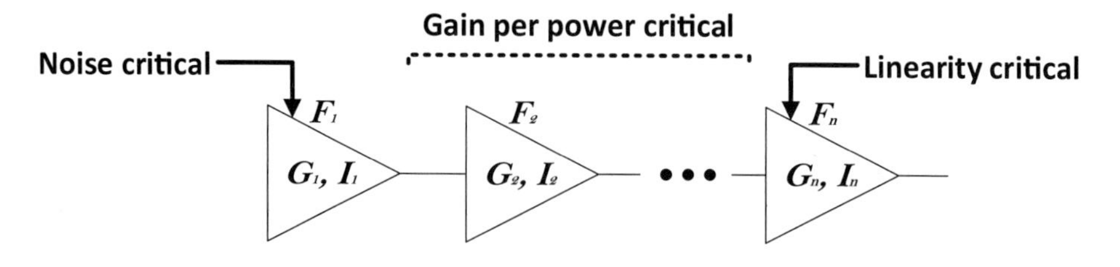

Figure 14.25 Cascaded chain.

In the next three sections, we will develop three bias conditions to optimize noise performance, gain per power efficiency, and linearity for gain and power performance respectively.

14.4.2 Optimizing NF with G_{max} for LNA within Self-Heat Limit

A simple figure of merit for the RF MOSFET in an LNA is suggested as [11]

$$FoM_{LNA} = \frac{G}{P(F-1)} \tag{14.25}$$

where G is the gain, P is the DC power, and F is the noise factor. The figure-of-merit (FoM) may include bandwidth as well, but this simple FoM is what we need to derive the bias condition for a mm-wave LNA design.

The FoM may include bandwidth as well, but this simple FoM is what we need to derive the bias condition for an mm-wave LNA design.

Mason gain U is the proposed reference device metric for mm-wave design methodology in case G_{max} extension is required to overcome the F_{max} limitation with the neutralization technique as explained in Section 14.4.5. As previously reviewed, gate resistance is a nonlinear function of the number of fins, so it is important to determine the number of fins for the design under consideration first based on the distance between Z_{in}^* and Z_{opt}. This implies that the peak G_{max} by the lowest reflection and the minimum of NF_{min} can be obtained in a relatively close bias condition.

It is also well known that lower noise figure requires higher g_m, therefore high current density is required for planar technology. In FinFET technology, the overall DC power in the device should be limited to avoid self-heating issues.

With the fin self-heating under consideration, NF_{min} and G_{max} under the constant power limit, the so-called FiSH limit, are swept for the current density J_D as shown in Figure 14.26. Since the parameter sweep assumes a constant DC power, V_{DS} decreases accordingly as increases.

In Figure 14.27, along with the G_{max} and NF_{max} plot by J_D under the constant power limit, Mason gain U is also shown for the constant power limit after being penalized by NF_{min}, listed as $U(\mathrm{dB}) - NF_{min}(\mathrm{dB})$. This is the modified FoM_{LNA}. Since the FiSH power limit is the consideration, the DC power term in the original FoM_{LNA} is constant, and hence removed. Without the DC power term in the equation, the NF_{min} adjusted Mason gain $U(\mathrm{dB}) - NF_{min}(\mathrm{dB})$ is equivalent to the original FoM_{LNA}. One can find a similar approach in the literature as well [12].

The modified FoM_{LNA} will reach highest value if the highest G_{max} biasing point and the lowest NF_{min} bias point are closest. According to Figure 14.27, the four-fin device reaches the highest FoM_{LNA} point as both G_{max} and NF_{min} reach their optimum performance at the closer bias condition ($J_D \approx 0.22$). S11 is still close to Z_{opt}, as we reviewed in Section 14.3.3. Impedance matching should be chosen based on where the G_{max} circle and the NF_{min} circle intersect at the selected current bias and the V_{DS} condition, which is around 0.6 V for the preceding example.

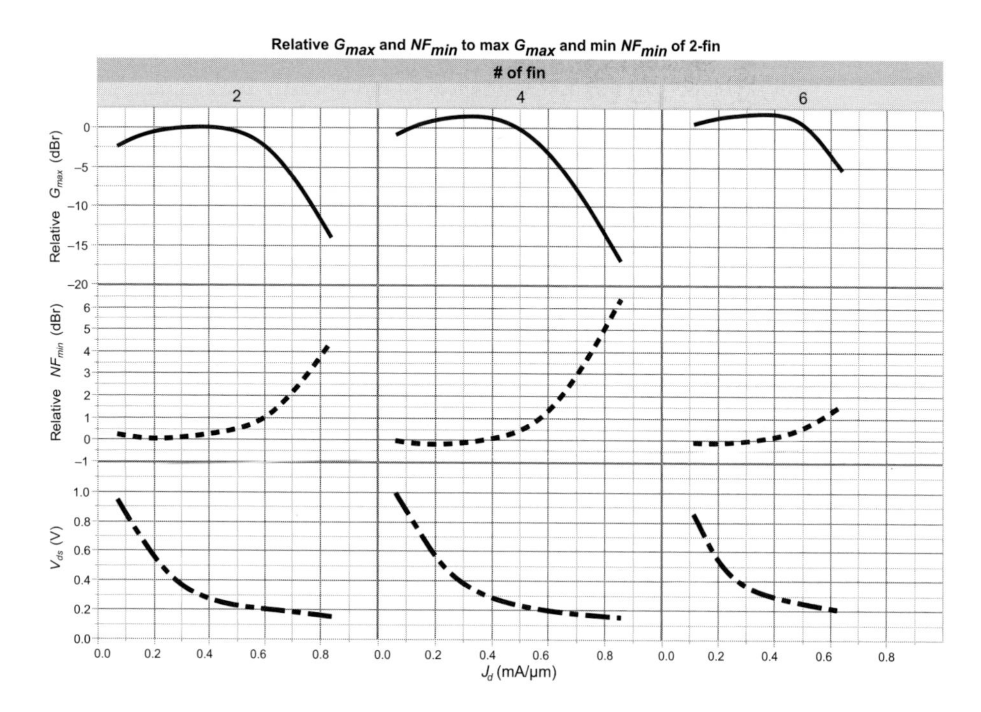

Figure 14.26 G_{max}, NF_{min}, and V_{DS} for the constant power limit.

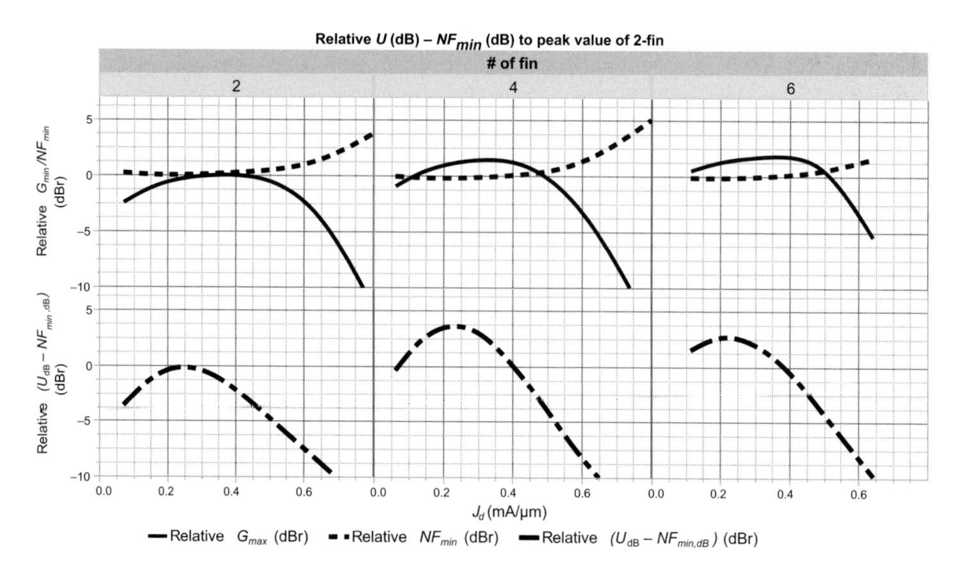

Figure 14.27 G_{max}, NF_{min}, and FoM_{LNA} for the FiSH limit.

As demonstrated in Figure 14.28, the G_{max} circle is much wider than the NF_{min} circle, which implies the noise figure is more sensitive to the matching condition. The crossing point of the G_{max} circle and NF_{min} circle is the matching point enabling the gain and the noise figure predicted by the bias conditioning, and the conjugate impedance matching

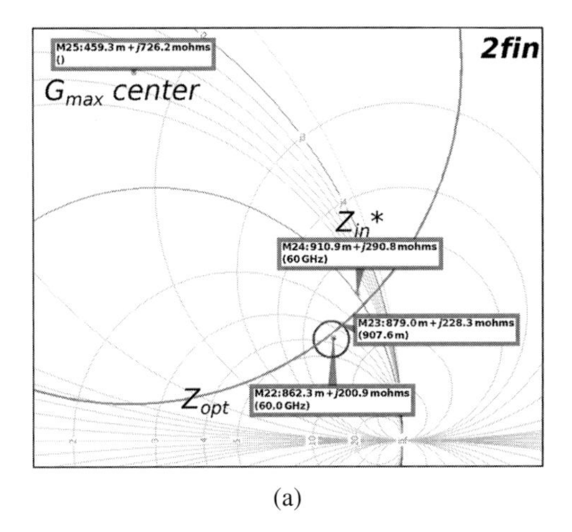

(a)

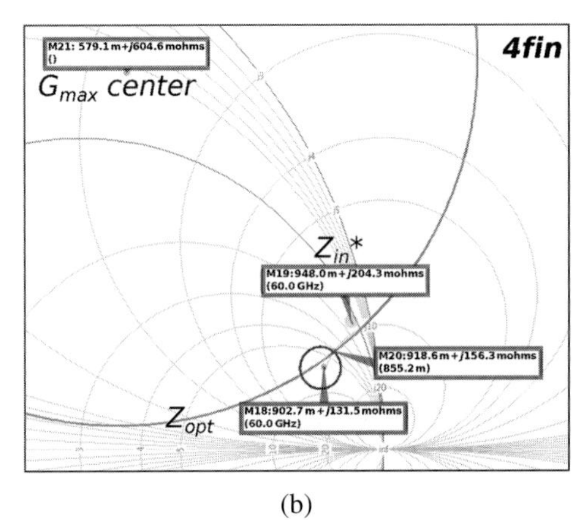

(b)

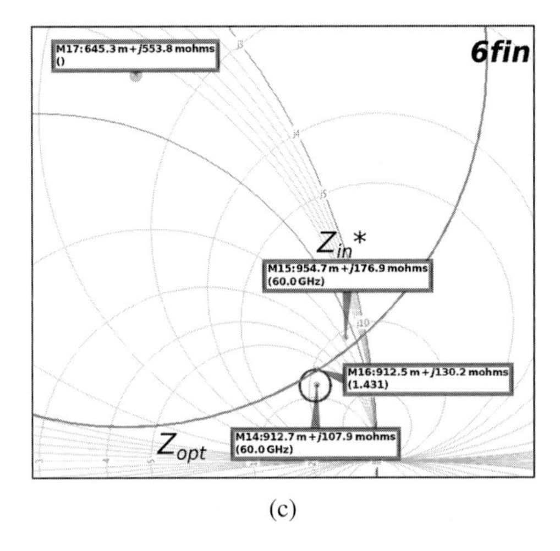

(c)

Figure 14.28 Gain and noise circle along with Z_{in}^* for (a) two, (b) four, and (c) six fin devices.

Table 14.1 Summary of NF optimization experiments.

Parameter	Two fins	Four fins	Six fins
Number of fingers	14	12	8
J_D [mA/μm]	0.2368	0.21.37	0.2002
V_{DS} [V]	0.55	0.6	0.65
Z_{opt} [Ω]	$0.86 + j0.22$	$0.90 + j0.13$	$0.91 + j0.11$
Z_{in}^* [Ω]	$0.91 + j0.29$	$0.95 + j0.20$	$0.95 + j0.18$
Z_{in}^* at crossing [V]	$0.88 + j0.23$	$0.92 + j0.16$	$0.91 + j0.13$
Z at G_{max} center [V]	$0.46 + j0.73$	$0.58 + j0.60$	$0.65 + j0.55$

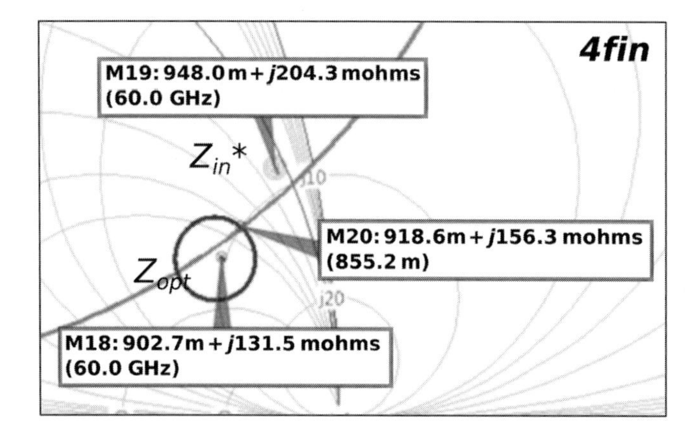

Figure 14.29 Close-up four-fin device matching conditions.

point Z_{in}^* is very close to the selected matching point. As a result, one can expect to meet the target gain and noise figure based on the G_{max} and NF_{min} number read from the current density J_D at the peak of FoM_{LNA}, and the input matching, S11, will be as low as -27 dB at 60 GHz. A summary of the experiments is shown in Table 14.1. A close-in of the circles, Z_{in}^*, and the selected matching point are shown in Figure 14.29.

14.4.3 Gain per Power Efficiency

Mason introduced the unilateral gain, also known as Mason gain, in 1954 [13]. This teaches us that one can obtain a maximum stable gain (MSG) if there is only forward gain in the lossless network. The Mason gain U can be also used to measure F_{max} by simply measuring the frequency where Mason gain is unity. The Mason gain is the true measure of the maximum frequency where the device can operate with positive power gain if there is a proper technique to compensate for any feedback that exists in the device. This compensation will be discussed in Section 14.4.5. One also should note that any active device inevitably produces less gain as frequency increases due to the increased passive loss in the device network as shown in Figure 14.30.

As modern wireless systems integrate more wireless standards supporting many frequency bands in a single chip, modern RFIC design emphasizes low-power design even

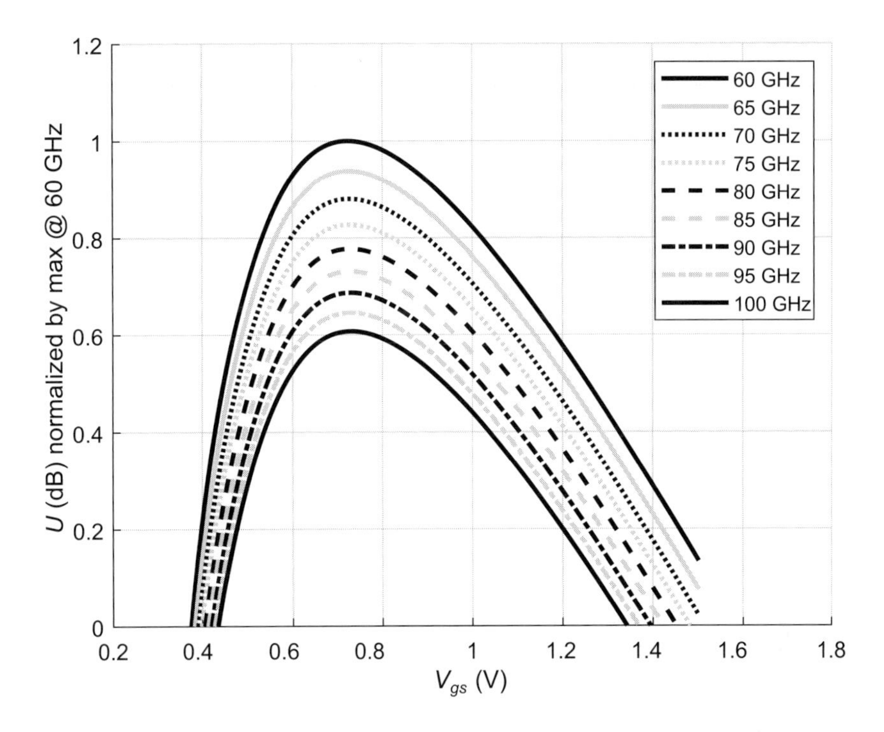

Figure 14.30 Mason gain at various frequencies.

more. In order to guide the low-power design at mm-wave frequency, one needs to observe the Mason gain by g_m/I_d, where g_m/I_d indicates the effectiveness of current usage to produce voltage-current gain in the device. Figure 14.31 shows the Mason gain at 60 and 100 GHz along with the g_m/I_d. Both Mason gain and g_m/I_d are normalized to the maximum value. Figure 14.31 hints that there is a certain bias condition maximizing the Mason gain for the given power dissipation target.

For low-power design, we will trade the number of amplifier stages for the total power dissipation by maximizing the gain per stage under the condition that avoids unnecessary power waste for the gain. One can estimate the total number of amplifier stages to achieve a total target gain by dividing the total target gain over the Mason gain of the device assuming the Mason gain is the maximum achievable gain per device at the specific frequency. The power dissipation per stage is analogue of I_d/g_m. Hence, the product of I_d/g_m and the number of total amplifier stages, which is the division of the total target gain over the Mason gain U. This indicates the total power dissipation required to achieve the target gain with a certain number of stages. Now we need to search for an optimum bias condition that minimizes the total power dissipation while achieving the target gain.

In other words, one can maximize the inverse of the product of I_d/g_m and the number of stages, which can be simplified as follows, and we define it as Gain-Power Figure-of-Merit (FoM_{GP}):

$$FoM_{GP} = U\frac{g_m}{I_d}. \tag{14.26}$$

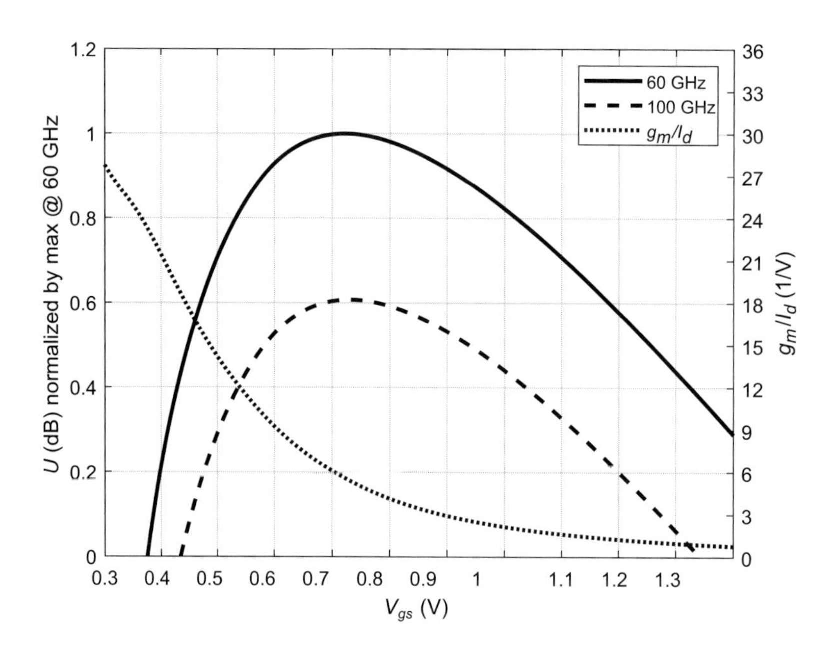

Figure 14.31 Mason gain at 60 and 100 GHz, and g_m/I_d.

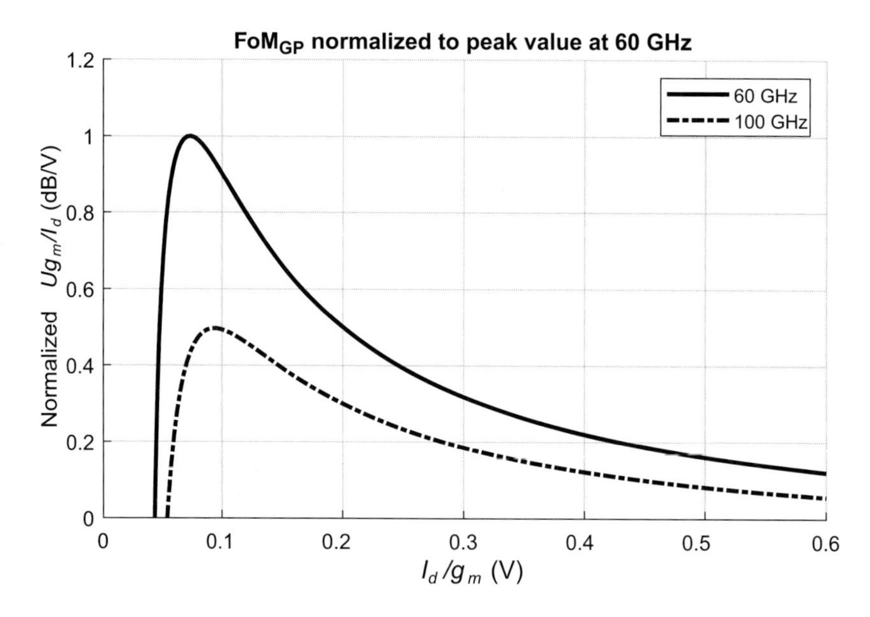

Figure 14.32 FoM_{GP} at 60 and 100 GHz.

Figure 14.32 plots the FoM_{GP} at 60 and 100 GHz, where both are normalized to the peak of FoM_{GP} at 60 GHz. One can notice the FoM_{GP} reaches at peak value at a certain point, which is the optimum bias condition to maximize device gain per power dissipation.

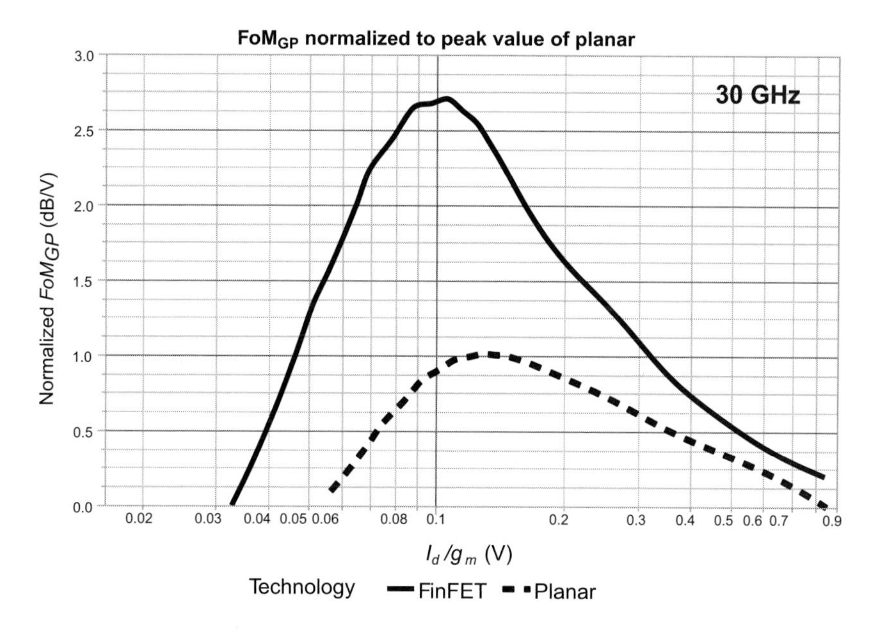

Figure 14.33 *FoM_{GP}* of FinFET and planar.

The recent FinFET technology shows about 70–160% improvement of FoM_{GP} over the latest generation of planar technology, and it implies 40–60% less power dissipation over the planar technology to achieve the same amount of gain. The FoM_{GP} of FinFET and planar technology is shown in Figure 14.33.

The significant FoM_{GP} improvement stems from the improved DIBL appearing as an intrinsic gain, $g_m R_{out}$. Also, one should notice that the planar devices have to drive higher I_d/g_m to reach the peak FoM_{GP} point. This implies worse power efficiency on the device for the gain, and the higher current density could cause device stress.

14.4.4 Linearity for Gain and Power Efficiency

Another critical property for the wireless system is the linearity represented by *IIP3* or P_{1-dB} as the linearity dictates the overall performance degradation by blockers and jammers. The simple expression of the third-order input interception point (IIP3) can be found as [14]

$$IIP3 = \frac{1}{6R_s} \frac{\left(1 + (\omega C_{gs} R_s)^2\right)}{\left|\frac{K_{3gm}}{g_m}\right|} \tag{14.27}$$

where

$$K_{3gm} = \frac{1}{6} \frac{\partial I_{DS}}{\partial V_{GS}^3} \tag{14.28}$$

$$g_m = \frac{\partial I_{DS}}{\partial V_{GS}} \tag{14.29}$$

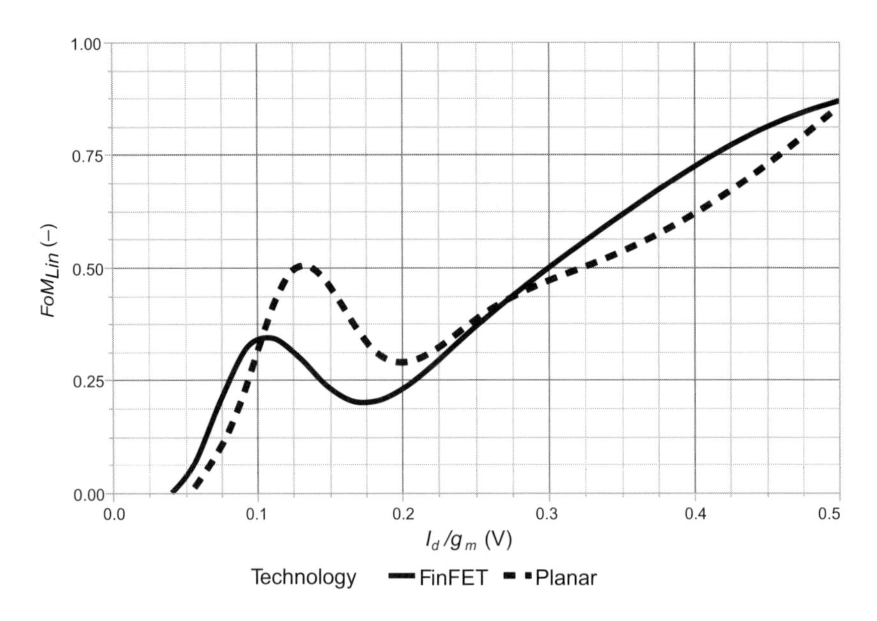

Figure 14.34 FoM_{Lin} for FinFET and planar.

In (14.27), the only bias dependent term is $\left|\frac{K_{3gm}}{g_m}\right|$, and *IIP3* improves by minimizing the ratio, $\left|\frac{K_{3gm}}{g_m}\right|$. In order to maximize the linearity, it should maximize the ratio defined as linearity figure-of-merit (FoM_{Lin}) depicted in (14.30), which approaches unity when the third-order term K_{3gm} is minimized by

$$FoM_{Lin} = \frac{g_m}{g_m + |K_{3gm}|}. \tag{14.30}$$

Figure 14.34 shows FoM_{Lin} for FinFET and planar for the comparison.

As suggested in Figure 14.34, FoM_{Lin} improves as I_d/g_m increases as expected, and it implies the improvement of the linearity requires extra power dissipation. However, there is one local sweet spot where I_d/g_m is around 0.1–0.15 A/S, and any higher level of I_d/g_m than 0.3 A/S would make the device unreliable because of extreme current density. The high current density region should be avoided for long-term device reliability.

According to Figure 14.34, planar may outperform FinFET at the local optimum point, I_d/g_m of ~0.14 A/S, which consumes about 20–30% more current than FinFET for FinFETs' local optimum point. If we consider both gain and linearity improvement per power dissipation, the product of FoM_{GP} and FoM_{Lin}, which derives another figure-of-merit defined gain–power–linearity figure-of-merit (FoM_{GPL}), guides bias condition for the optimum gain–linearity balanced performance per power dissipation as shown in Figure 14.35. This reports over 80% benefit by FinFET over planar even at lower current density requirements. It also suggests FinFET outperforms at $I_d/g_m >$ 0.25 A/S to planar, where FoM_{Lin} of FinFET starts exceeding planar's. Section 14.5

will demonstrate the effectiveness of the biasing method with simple design examples at 60 GHz.

14.4.5 Neutralization for mm-Wave Applications

In Section 14.4.3, we have analyzed and compared the FinFET technology against planar using unilateral gain, which assumes no feedback network. However, in reality, the FinFET has greater feedback capacitance, C_{gd}, than the planar, as described in Section 14.3.1, which is a key contributor for limiting the maximum oscillation frequency F_{max}. Therefore, it is advantageous to adapt a technique to reduce or eliminate

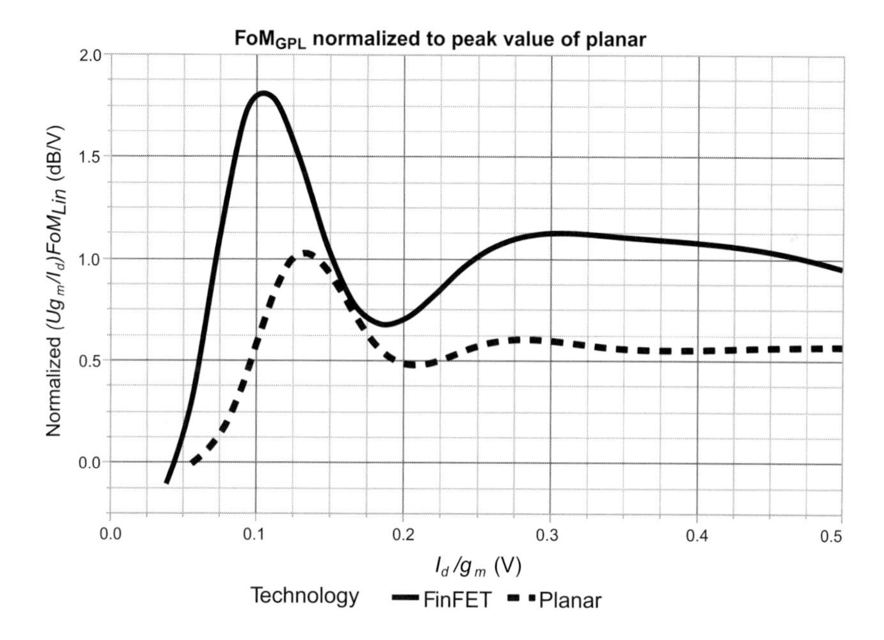

Figure 14.35 FoM_{GPL} for FinFET and planar.

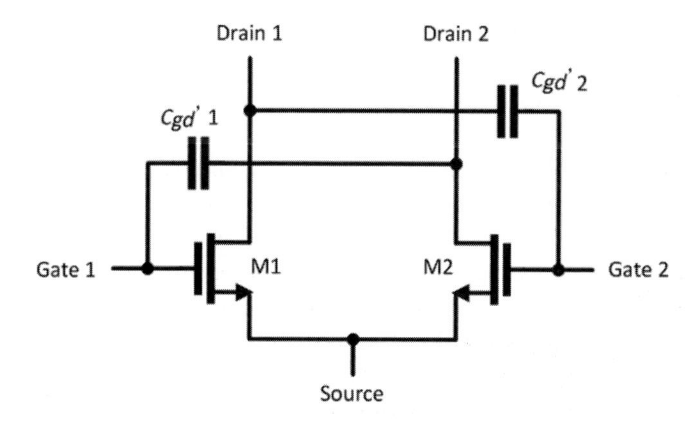

Figure 14.36 Differential neutralization technique.

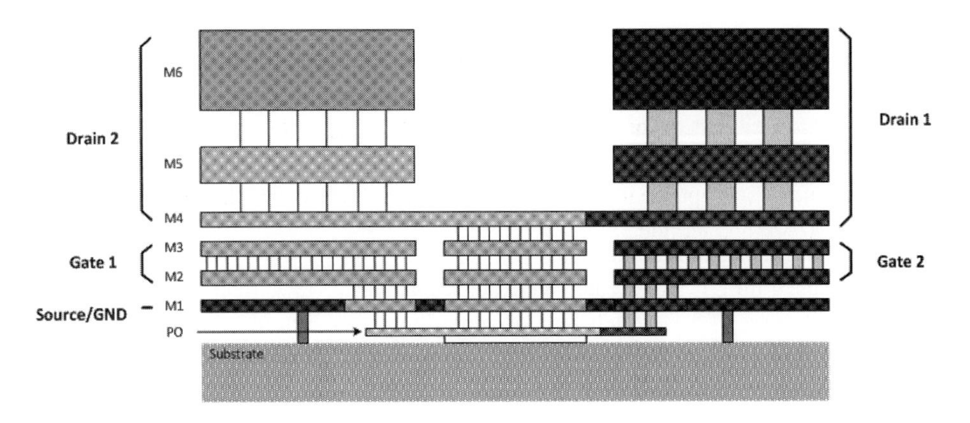

Figure 14.37 Neutralization capacitance using a metal interface.

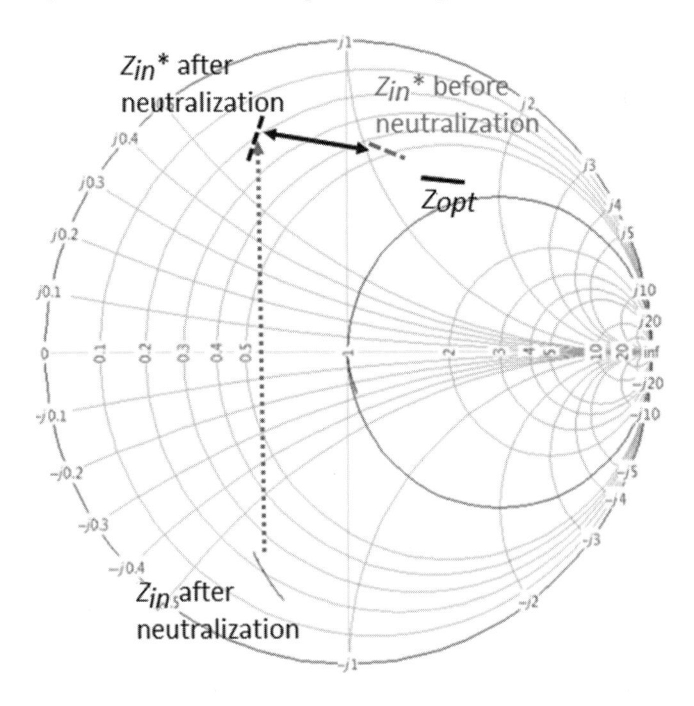

Figure 14.38 Impedance modification after neutralization.

the feedback capacitance to extend the F_{max}, which results in the higher gain, normally 4–5 dB boost in G_{max} with the neutralization [15].

The typical neutralization technique is shown in Figure 14.36. This technique relies on the cross-coupled capacitor to generate the negative C_{gd}, hence neutralize the device C_{gd}. This is only available in a differential configuration, but provides neutralization over a wide bandwidth. Several designs have been reported, most for mm-wave frequencies. One design utilized the metal back-end interface to create the negative feedback capacitance, and the metal interface configuration is shown in Figure 14.37 [16].

However, one should pay attention to the input impedance change after neutralization. After neutralization, Z_{in}^* and Z_{opt} are further apart, as shown in Figure 14.38, especially the real part of the input impedance. While Z_{opt} remains relatively constant before and after neutralization, Z_{in} yields a lower resistive term after neutralization, and hence is more susceptible to stability issues. This needs to be addressed when using the neutralization technique.

14.5 Design Example for an mm-Wave Amplifier with the Proposed Design Methodology

So far, we have reviewed three biasing conditions for noise figure optimization under the FiSH limit, gain-per-power optimization, and linearity-gain-per-power optimization. In this chapter, we will demonstrate the effectiveness of the biasing condition with simple amplifier design at 60 GHz.

Recent work has demonstrated an LNA design at the 71–76 GHz frequency range with the biasing condition for noise figure optimization under the FiSH limit using Intel 22FFL process technology [17]. The authors swept MAG and NF_{min} for I_{DS} and V_{DS}, and searched for the optimum supply voltage V_{DS} and the current I_{DS} for the stage. Since the authors were able to achieve the MAG and NF_{min} performance with lower V_{DS} of 0.5 V, two-stage stacking was suggested to reuse the DC current to save power. As one can see in the schematic of Figure 14.39, the differential neutralization technique is used to boost G_{max} as well.

As a result, the proposed design achieves 20 dB gain with 10.4 GHz 3 dB bandwidth (BW) and 4 dB noise figure at 73.5 GHz with 10.8 mW DC power [17].

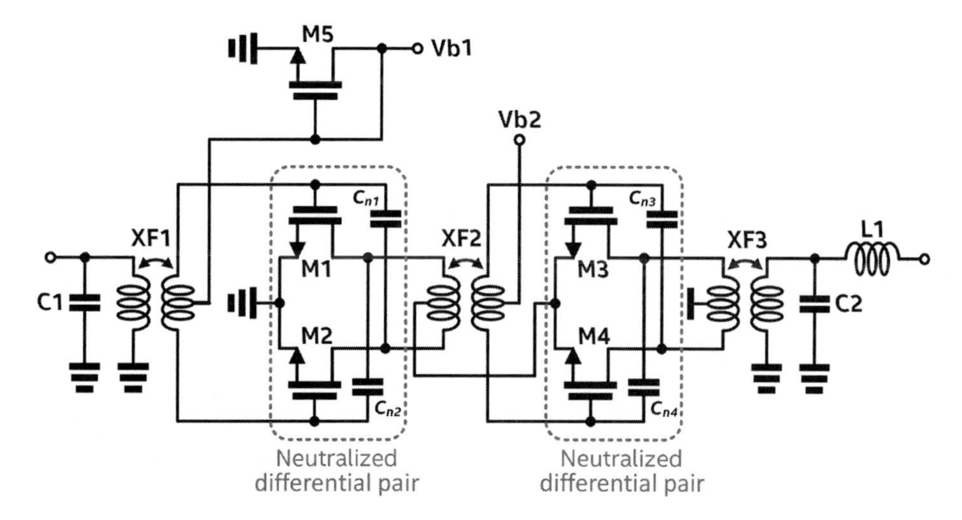

Figure 14.39 71–76 GHz LNA design [17]. (©2018 IEEE. Reprinted, with permission, from *2018 IEEE Radio Frequency Integrated Circuits Symposium*.)

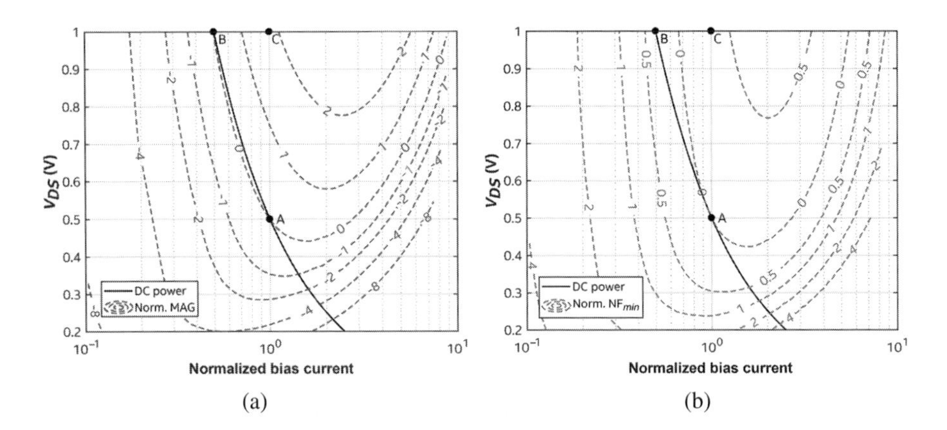

Figure 14.40 (a) G_{max} and (b) NF_{min} sweep for current under FiSH limit [17]. (©2018 IEEE. Reprinted, with permission, from *2018 IEEE Radio Frequency Integrated Circuits Symposium.*)

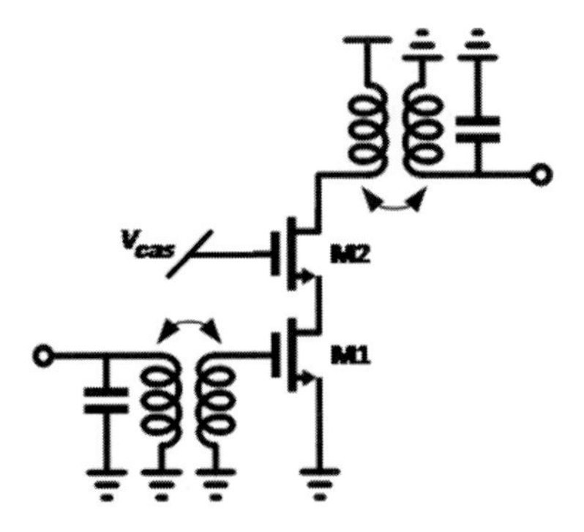

Figure 14.41 Example design – single-ended cascode amplifier.

Figure 14.41 shows an example design of a simple single-ended cascode amplifier. For the demonstration purposes, the design adopts simple lossless transformer-based input and output matching without BW optimization. For the fair comparison, the same topology was used for three biasing conditions, and the number of fins was kept constant, while the number of fingers was adjusted to achieve the same level of gain (14 dB) across the three biasing conditions. Since the neutralization technique would alter the matching condition, we avoid using the neutralization technique for this demonstration, and the decent gain is still achievable at 60 GHz without utilizing the neutralization technique.

The cascode bias voltage V_{cas} is set to adjust V_{DS} of M1 transistor properly to meet the FiSH limit for the noise figure optimized design, but it is tied to the power supply for two other designs as the power supply was assumed to be 1 V for the gain–power optimization and the linearity–gain–power optimization.

The simulation includes a metal interface as additional parasitic to the device, and Intel 22FFL low-leakage RF transistors were used for the simulation. The simulation

Table 14.2 Summary of design example, relative performance to gain-power optimization design with the performance data at 60 GHz.

Parameter	Gain-power-linearity optimization	Noise-figure optimization
NF (dBr)	−0.89	−1.44
$IIP3$ (dBr)	13.17	9.2
$IP_{1\text{-dB}}$ (dBr)	11.47	8.54
P_{DC} (dBr)	1.29	5.50

results of gain–power–linearity optimum design and noise–figure optimum design are compared to gain–power optimum design, which is supposed to consume the lowest power level for the given gain level. The results are normalized to the gain–power optimum design for the comparison purpose; thus, the units are [dBr] as shown in Table 14.2.

As listed in Table 14.2, the gain–power–linearity optimum bias condition achieves over 13 dB higher at the cost of 1.3 dB DC power. Meanwhile, the noise–figure optimum design manages 1.44 dB lower NF by burning over 5 dB higher DC power, even though the design is constrained by the FiSH limit. One should note that the noise performance optimization is very power costly.

14.6 Conclusion

As the semiconductor industry keeps pushing Moore's law scaling by adopting FinFET technology primarily targeting logic design, the adaptation of FinFET technology for RF and mm-wave applications is slowly catching up, though it is slow due to the lack of device understanding in RF circuit operation perspectives. The excellent gate-channel controllability by improving DIBL with the FinFET structure is the significant improvement over the planar technology, helping to overcome the scaling limitation and keep Moore's law alive. Besides improvement of the DC behavior, a significant improvement and optimization of the FinFET process technology targeting RF and mm-wave applications were engineered over the last several FinFET technology generations. The identification of major causes of degradation of RF critical metrics such as fin self-heating and excessive parasitic by complicated structural interaction is continuously driving the process technology improvement for RF and mm-wave applications.

This chapter advocates the superiority of RF critical performance of FinFET technology to planar transistors in three aspects; noise, power efficiency, and linearity. The suggested three figure-of-merits (FoM_{LNA}, FoM_{GP}, and FoM_{GPL}) self-describe the improvement of the noise performance (Sections 14.3.2, 14.3.3, and 14.4.2), the power efficiency to deliver the signal gain (Section 14.4.3), and the higher linearity support at the lower power dissipation (Section 14.4.4) than planar technology. The given simple design examples demonstrate noise figure and linearity improvement relative to the most power-efficient design option by following the suggested RF/mm-wave amplifier design methodologies explicitly developed for the FinFET transistors.

Acknowledgments

Special thanks to Wireless Integration and Circuit Technology team members, especially Surej Ravikumar, Daniel Yeh, Vijaya Neeli, and Saurabh Morarka for data preparation and helpful conversation.

References

[1] C. Auth et al., "A 22nm high performance and low-power CMOS technology featuring fully-depleted tri-gate transistors, self-aligned contacts and high density MIM capacitors," in *2012 Symposium on VLSI Technology (VLSIT)*, IEEE, June 2012, pp. 131–132.

[2] S. Natarajan et al., "A 14nm logic technology featuring 2nd-generation FinFET transistors, air-gapped interconnects, self aligned double patterning and a 0.0588m^2 SRAM cell size," in *2014 IEEE International Electron Devices Meeting (IEDM 2014)*, IEEE, December 2012, pp. 71–73.

[3] C. Auth et al., "A 10nm high performance and low-power CMOS technology featuring 3rd-generation FinFET transistors, self-aligned quad patterning, contact over active gate and cobalt local interconnects," in *2017 IEEE International Electron Devices Meeting (IEDM 2017)*, IEEE, December 2017, pp. 29.1.1–29.1.4.

[4] B. Sell et al., "A high performance and ultra low power FinFET technology for mobile and RF applications," in *2017 IEEE International Electron Devices Meeting (IEDM 2017)*, IEEE, December 2017, pp. 29.4.1–29.4.4.

[5] P. VanDerVoorn et al., "A 32nm low power RF CMOS SOC technology featuring high-k/metal gate," in *2010 Symposium on VLSI Technology*, IEEE, June 2010, pp. 137–138.

[6] E. J. Nowak, I. Aller, T. Ludwig, et al., "Turning silicon on its edge [double gate CMOS/FinFET technology]," *IEEE Circuits and Devices Magazine*, vol. 20, no. 1, pp. 20–31, January 2004.

[7] A. J. Scholten et al., "FinFET compact modelling for analogue and RF applications," in *2010 IEEE International Electron Devices Meeting (IEDM 2010)*, December 2010.

[8] *How to Improve PA Performance and Reliability Using Electro-Thermal Analysis*, Keysight.

[9] S. Voinigescu, *High-Frequency Integrated Circuits*, 1st ed. Cambridge University Press, 2013.

[10] T. Lee, *The Design of CMOS Radio-Frequency Integrated Circuits*, 2nd ed. Cambridge University Press, 2004.

[11] I. Song, J. Jeon, H. S. Jhon, et al., "A simple figure of merit of RF MOSFET for low-noise amplifier design," *IEEE Electron Device Letters*, vol. 29, no. 12, pp. 1380–1382, December 2008.

[12] F. Fadhuile, T. Taris, Y. Deval, C. Enz, and D. Belot, "Design methodology for low power RF LNA based on the figure of merit and the inversion coefficient," in *2014 21st IEEE International Conference on Electronics, Circuits and Systems (ICECS)*, IEEE, December 2014, pp. 478–481.

[13] S. Mason, "Power gain in feedback amplifier," *Transactions of the IRE Professional Group on Circuit Theory*, vol. CT-1, no. 2, pp. 20–25, June 1954.

[14] X. Wei, G. Niu, Y. Li, M. T. Yang, and S. S. Taylor, "Modeling and characterization of intermodulation linearity on a 90-nm RF CMOS technology," *IEEE Transactions on Microwave Theory and Techniques*, vol. 57, no. 4, pp. 965–971, April 2009.

[15] Z. Deng and A. M. Niknejad, "A layout-based optimal neutralization technique for mm-wave differential amplifiers," in *2010 IEEE Radio Frequency Integrated Circuits Symposium*, IEEE, May 2010, pp. 355–358.

[16] C. Hull, "Leading the way in 5G: Challenges for millimeter wave MIMO in CMOS technology," in *The Brooklyn 5G Summit 2015*, IEEE, April 2015.

[17] W. Shin, S. Callender, S. Pellerano, and C. Hull, "A compact 75 GHz LNA with 20 dB gain and 4 dB noise figure in 22nm FinFET CMOS technology," in *2018 IEEE Radio Frequency Integrated Circuits Symposium*, IEEE, June 2018, pp. 284–287.

Author Index

Alon, Elad, 55, 112

Bourdoux, André, 162

Calderin, Lucas Albert, 112
Craninckx, Jan, 162

Dinc, Tolga, 84

Guermandi, Davide, 162

Heydari, Payam, 273
Hueber, Gernot, 1

Krishnaswamy, Harish, 84

LaCaille, Greg, 55
Lee, Hyung-Jin, 400

Mikhemar, Mohyee, 347

Natarajan, Arun, 146
Niknejad, Ali M., 1, 55, 112
Nikolić, Borivoje, 55, 112

Okada, Kenichi, 193

Pärssinen, Aarno, 18
Puglielli, Antonio, 55

Ramakrishnan, Sameet, 112
Rexberg, Leonard, 243
Reynaert, Patrick, 369

Sadhu, Bodhisatwa, 243
Sell, Bernhard, 400
Staszewski, R. Bogdan, 305

Wambacq, Piet, 162
Wu, Rui, 193
Wu, Wanghua, 305

Zhang, Yang, 369

Subject Index

ADC, 89
  ENOB, 89
AM–AM distortion, 198
AM–PM distortion, 198
Angle-of-incidence, 148
Antenna
  Omnidirectional, 249
Antenna array, 244
  1D, 247
  2D, 250
Antenna cointegration, 153
Antenna suppression, 92
Antenna-in-package, 152
Array factor, 248
Augmented reality, 88

Backhaul, 86
Baseband amplifier, 211
  Open-loop, 211
Beam control, 252
Beam shaping, 244, 249
Beam steering, 249
Beam-tracking, 243
Beamformer, 37
Beamforming, 36, 65, 243
  Hybrid, 36
  Linear, 65
Built-in self-test, 333
  DCO step analyzer, 337
  DPLL, 333
  Snapshotting, 336

Calibration, 210
  g_m biasing, 210
  Cutoff frequency, 222
  Gain, 221
  Gain calibration loop, 210
  I/Q mismatch, 195
  Phase, 221
  Voltage regulator, 210
  Wideband, 195
Carrier recovery, 200
Carrier tracking, 201
Carrier-to-noise ratio (CNR), 200
CDMA, 56
Channel
  Estimation, 64
  Matrix, 64
Channel bonding, 198
Channel model, 38
Channel Rank, 62
Chirp, 164
Clock generation, 347
Clock jitter, 198
CML, *see* Current-mode logic
CMRR, *see* Common-mode rejection ratio
CNR, *see* carrier-to-noise ratio
Colpitts oscillator, 350
Common-mode rejection ratio, 215
Continuous Source of Radiation, 246
Conventional MIMO, 64
Cross-polarization, 158
Current-mode logic (CML), 203

DAC, 116
  Linearity, 126
  Power consumption, 117
  Thermal noise cancellation, 127
DCO
  Capacitor-based tuning, 315
  Controlled transmission line, 314
DIBL, *see* Drain-induced barrier lowering
Digital PLL, 305
Direct-conversion transceiver architecture, 195
Doppler
  Mesh, 186
  Profile, 186
  Range, 186
  Sidelobe, 188
Doppler shift, 165
DPLL
  Digital calibration, 324
  DTC-assisted DPLL, 338
  Gain calibration, 325
  Linearization, 325
  Mismatch calibration, 326
Drain-induced barrier lowering, 404
Dual-polarization, 157

Electrostatic discharge, 215
Element factor, 248
Equalizer, 197
  Recursive least square, 197
Error vector magnitude, 370
  Max, 370
  Peak, 370
  Rms, 370
ESD, *see* Electrostatic discharge
EVM, *see* Error vector magnitude

FD, *see* Full-duplex
FDD, *see* Frequency division multiplex
FDE, *see* Frequency domain equalization
FDMA, 56
FET stacking, 379
FI, *see* Frequency-interleaved transceiver architecture
FinFET, 400
  Noise performance, 412
  Nonlinear gate resistance, 406
  Parasitics, 411
  Self-heating, 408, 417
Fixed intermediate frequency, 349
Flipped voltage follower (FVF), 201, 211
FMCW, 164, 168
Fractional-N PLL, 306
Fraunhofer distance, 188
Frequency detector (FD), 306
Frequency division multiplex, 84
Frequency domain equalization, 98
Frequency modulated continuous wave, 164, 168
Frequency modulation (FM), 311
Frequency-interleaved transceiver architecture, 223
Full-duplex, 84
Full-duplex antenna interface, 85

Hadamard, 173
Harmonic-based frequency tripler, 287
Hartley architecture, 258
HBFT, *see* Harmonic-based frequency tripler

I/Q mismatch, 218
  Frequency dependent, 218
I/Q phase mismatcch, 220
ILFT, *see* Injection-locking frequency tripler
ILO, 175
Image rejection ratio, 198
IMMR, *see* Image rejection ratio
Impedance transformation, 379
Impedance translation, 149
Injection architecture, 258
Injection-locked oscillator, 175
  I/Q mismatch, 203
  Subharmonic, 175
Injection-locked oscillator (QILO), 201
Injection-locking freqency tripler, 280
Injection-locking technique, 195

LC oscillator, 350
LCP, *see* Liquid crystal polymer
Lesson's expression, 351
Line of sight, 40, 57
Linear phase shifter, 259
Linearity enhancement, 382
Link budget, 170
Liquid crystal polymer, 155
LNA, *see* Low-noise amplifier
LO, 214, *see* Local oscillator
  Current-bleeding, 214
  Phase noise, 175
  Pulling, 175
  Tuning range, 175
LO leakage, 195
LO phase noise, 172, 195
Local oscillator, 347
  Architectures, 348
  Direct conversion, 348
  Intermediate frequency, 348
  Quadrature, 348
Lorentzian phase noise spectrum, 141
LOS, 40, 57
Loss
  propagation, 40
Loss invariant linear phase shifter, 259
Low-density parity check (LPCD), 199
Low-noise amplifier (LNA), 215
  Common-source common-gate, 215
  Common-source common-source, 215
  CS-CG, 215
  CS-CS, 215
LPCD, *see* Low-density parity check

Mason gain, 417
Massive MIMO, 55
Millimeter-wave backhaul, 86
Millimeter-wave relaying, 86
MIMO, 55, 64
  Capacity, 40
  Channel, 39
  Conventional, 64
  Successive interference cancellation, 67
MIMO spatial filtering, 148
Mixer, 214
  Capacitive cross-coupling, 214
  Double-balanced, 206, 214
  Mixer-first, 206
Mixer-first topology, 201
Modulation scheme, 199
  Single-carrier, 199
MU-MIMO, *see* Multiuser MIMO
Multipath, 60

Multiplexing
  Spatial, 58
Multirate DPLL, 309
Multiuser MIMO, 55, 63
MUSIC, 189

N-path filtering, 96, 149
  Spatiospectral, 149
Neutralization, 425
Noise matching, 414
Nonlinear gate resistance, 406
Notch filter, 151

Oscillator
  LC, 203

PA
  AM–AM, 373
  AM–PM, 373
  Broadband neutralization, 375
  FET stacking, 379
  Linearity enhancement, 382
  Neutralization, 375
  Nonlinearity, 373, 381
  Power combining, 379
Packaging, 152
Path loss, 40
Phase and frequency detector (PFD), 306
Phase detector (PD), 306
Phase modulated continuous wave, 166, 168
Phase noise, 198
Phased array, 243
  Beam steering, 257
  Calibration, 257
  Gain control, 257
  Phase control, 257
  Sidelobe suppression, 257
PMCW, 166, 168
Polarization, 157, 254
  Cross, 158
  Dual, 157, 254
  Orthogonal, 254
Power amplifier
  AM–AM, 373
  AM–PM, 373
  Nonlinearity, 373, 381
Power combining, 379
Propagation
  Multipath, 60

QILO, *see* Injection-locked oscillator

Radar, 86, 162
  Chirp, 164
  Continuous wave, 166
  Frequency modulated continuous wave (FMCW), 164, 168
  MIMO, 172
  Modulation, 164
  Phase modulated continuous wave (PMCW), 86, 166, 168
  Range resolution, 165
  Sawtooth wave, 164
  Self-interference, 166
  SoC, 162
  Spillover, 86, 166
Radiation pattern, 250
Radiation source, 247
  Continuous, 247
  Sampled, 247
Reflection type phase shifter, 259

Sawtooth wave, 164
SCPA, *see* Switched-capacitor power amplifier
Self-heating, 408, 417
Self-interference, 84
Self-interference cancellation, 84, 113
  Active cancellation, 114
  Antenna, 92
  Dual-path, 99
  FDE, 98
  Hybrids, 113
  Vector-modulator based, 98
Self-interference suppression, 90
Self-interference suppression techniques, 92
SH-ILO, 175
Shared antenna interface, 94
Shunt filter, 149
Sidelobe suppression, 252
Sidelobe-level control, 244
Singular value decomposition, 43
Sliding intermediate frequency, 349
Small-cell base stations, 88
Spatial filtering, 147
Spatial Fourier transform, 244
Spatial multiplexing, 58
Spatiospectral filtering, 149
Spillover, 166
Subharmonic injection-locked oscillator, 175
Subharmonic quadrature ILO, 175
SVD, 44
Switched-capacitor power amplifier, 124

Tapering control, 260
TDD, *see* Time division multiplex
TDMA, 56
Time division multiplex, 84
Time-to-digital converter (TDC), 308, 321
Transformer
  Complanar, 377
  Stacked, 377
Transformers, 377
Transistor neutralization, 425
Transmission lines, 378

True-time delay, 259
Two-point frequency modulation, 311
TX orthogonality, 173
  Orthogonal sequences, 173
  Outer codes, 173
TX-to-RX spillover, 166, 167, 169

VCO
  AP/PM conversion, 364
  Bias circuit, 365
  Bypass capacitance, 360
  Class-B, 353
  Class-C, 353
  Common-mode resonance, 360
  Frequency scaling, 356
  Kick-back, 362
  Noise factor, 354
  Phase noise, 354
  Pulling, 175, 364
  Tail tuning, 360
Virtual reality, 88

Walsh function sequence, 149
Walsh–Hadamard, 173